TK 7871.58 .O6 F55 2000
Fiore, James M.
Op amps and linear
 integrated circuits

MHCC WITHDRAWN

D1614030

Op Amps and Linear Integrated Circuits

Theory and Application

Op Amps and Linear Integrated Circuits

Theory and Application

James M. Fiore

Mohawk Valley Community College

DELMAR
THOMSON LEARNING

Africa • Australia • Canada • Denmark • Japan • Mexico
New Zealand • Philippines • Puerto Rico • Singapore
Spain • United Kingdom • United States

Op Amps and Linear Integrated Circuits:
Theory and Application
by James M. Fiore

Business Unit Director:	**Developmental Editor:**	**Channel Manager:**	**Production Manager:**
Alar Elken	Michelle Ruelos Cannistraci	Mona Caron	Larry Main
Executive Editor:	**Editorial Assistant:**	**Marketing Coordinator:**	**Senior Project Editor:**
Sandy Clark	Jennifer A. Thompson	Paula Collins	Christopher Chien
Senior Acquisitions Editor:	**Executive Marketing Manager:**	**Executive Production Manager:**	**Art & Design Coordinator:**
Gregory L. Clayton	Maura Theriault	Mary Ellen Black	David Arsenault

COPYRIGHT © 2001
Delmar is a division of Thomson Learning. The Thomson Learning logo is a registered trademark used herein under license.

Printed in the United States of America
1 2 3 4 5 6 7 8 9 10 XXX 05 04 03 02 01 00

For more information, contact Delmar at
3 Columbia Circle, PO Box 15015, Albany, New York 12212-5015; or find us on the World Wide Web at http://www.delmar.com

ALL RIGHTS RESERVED. No part of this work covered by the copyright hereon may be reproduced or used in any form or by any means—graphic, electronic, or mechanical, including photocopying, recording, taping, or information storage and retrieval systems—without the written permission of the publisher.

For permission to use material from this text or product, contact us by
Tel (800) 730-2214
Fax (800) 730-2215
www.thomsonrights.com

Library of Congress Cataloging-in-Publication Data
Fiore, James M.
Op Amps and Linear Integrated Circuits: Theory and Application/James M. Fiore.
p. cm.
Includes bibliographical references and index.
ISBN 0-7668-1793-8
1. Operational amplifiers.
2. Linear integrated circuits.
I. Title.
TK7871.58.O6 F55 2000
621.39′5—dc21   00-060210

Asia
Thomson Learning
60 Albert Street, #15-01
Albert Complex
Singapore 189969

Australia/New Zealand
Nelson/Thomson Learning
102 Dodds Street
South Melbourne, Victoria 3205
Australia

Canada
Nelson/Thomson Learning
1120 Birchmont Road
Scarborough, Ontario
Canada M1K 5G4

International Headquarters
Thomson Learning
International Division
290 Harbor Drive, 2nd Floor
Stamford, CT 06902-7477
USA

Japan
Thomson Learning
Palaceside Building 5F
1-1-1 Hitotsubashi, Chiyoda-ku
Tokyo 100 0003
Japan

Latin America
Thomson Learning
Seneca, 53
Colonia Polanco
11560 Mexico D. F. Mexico

Spain
Thomson Learning
Calle Magallanes, 25
28015-Madrid
Espana

UK/Europe/Middle East
Thomson Learning
Berkshire House
168-173 High Holborn
London
WC1V 7AA United Kingdom

Thomas Nelson & Sons Ltd.
Nelson House
Mayfield Road
Walton-on-Thames
KT 12 5PL United Kingdom

NOTICE TO THE READER

Publisher does not warrant or guarantee any of the products described herein or perform any independent analysis in connection with any of the product information contained herein. Publisher does not assume, and expressly disclaim, any obligation to obtain and include information other than that provided to it by the manufacturer.

The reader is expressly warned to consider and adopt all safety precautions that might be indicated by the activities herein and to avoid all potential hazards. By following the instructions contained herein, the reader willingly assumes all risks in connection with such instructions.

The publisher makes no representation or warranties of any kind, including but not limited to, the warranties of fitness for particular purpose or merchantability, nor are any such representations implied with respect to the material set forth herein, and the publisher takes no responsibility with respect to such material. The publisher shall not be liable for any special, consequential, or exemplary damages resulting, in whole or part, from the readers' use of, or reliance upon, this material.

For *Karen*

"Information is not knowledge, knowledge is not wisdom."
—Frank Zappa

CONTENTS

PREFACE		xv
CHAPTER 1	INTRODUCTORY CONCEPTS AND FUNDAMENTALS	1
	Chapter Objectives	1
1.1	INTRODUCTION	1
	Variable Naming Convention	2
1.2	THE DECIBEL	3
	Decibel Representation of Power and Voltage Gains	3
	Signal Representation in dBW and dBV	8
	Items of Interest in the Laboratory	11
1.3	BODE PLOTS	12
	Lead Network Gain Response	12
	Lead Network Phase Response	15
	Lag Network Response	17
	Risetime Versus Bandwidth	20
1.4	COMBINING THE ELEMENTS—MULTISTAGE EFFECTS	22
1.5	CIRCUIT SIMULATIONS USING COMPUTERS	24
1.6	THE DIFFERENTIAL AMPLIFIER	26
	DC Analysis	27
	Input Offset Current and Voltage	29
	AC Analysis	30
	Common Mode Rejection	36
	Current Mirror	38
	Summary	40
	Review Questions	41
	Problems	41
CHAPTER 2	OPERATIONAL AMPLIFIER INTERNALS	44
	Chapter Objectives	44

	2.1	INTRODUCTION	44
	2.2	WHAT IS AN OP AMP?	45
		Block Diagram of an Op Amp	46
		A Simple Op Amp Simulation Model	50
		Op Amp Data Sheet and Interpretation	52
		Simple Op Amp Comparator	55
	2.3	OP AMP MANUFACTURE	60
		Monolithic Construction	61
		Hybrid Construction	63
		Summary	63
		Review Questions	63
		Problems	64
CHAPTER 3		NEGATIVE FEEDBACK	66
		Chapter Objectives	66
	3.1	INTRODUCTION	66
	3.2	WHAT NEGATIVE FEEDBACK IS AND WHY WE USE IT	67
	3.3	BASIC CONCEPTS	67
		The Effects of Negative Feedback	68
	3.4	THE FOUR VARIANTS OF NEGATIVE FEEDBACK	70
		Series-Parallel (SP)	72
		SP Impedance Effects	79
		Distortion Effects	84
		Noise	88
		Parallel-Series (PS)	89
		PS Impedance Effects	93
		Parallel-Parallel (PP) and Series-Series (SS)	95
	3.5	LIMITATIONS ON THE USE OF NEGATIVE FEEDBACK	96
		Summary	97
		Review Questions	97
		Problems	97
CHAPTER 4		BASIC OP AMP CIRCUITS	100
		Chapter Objectives	100
	4.1	INTRODUCTION	100
	4.2	INVERTING AND NONINVERTING AMPLIFIERS	101
		The Noninverting Voltage Amplifier	101
		Inverting Voltage Amplifier	105
		Inverting Current-to-Voltage Transducer	110
		Noninverting Voltage-to-Current Transducer	111

		Inverting Current Amplifier	114
		Summing Amplifiers	117
		Noninverting Summing Amplifier	120
		Differential Amplifier	123
		Adder/Subtractor	125
		Adjustable Inverter/Noninverter	126
	4.3	SINGLE-SUPPLY BIASING	128
	4.4	CURRENT BOOSTING	131
		Summary	132
		Review Questions	133
		Problems	134
CHAPTER 5		PRACTICAL LIMITATIONS OF OP AMP CIRCUITS	138
		Chapter Objectives	138
	5.1	INTRODUCTION	138
	5.2	FREQUENCY RESPONSE	139
	5.3	GAIN-BANDWIDTH PRODUCT	139
		Multistage Considerations	146
		Low-Frequency Limitations	150
	5.4	SLEW RATE AND POWER BANDWIDTH	152
		The Effect of Slew Rate on Pulse Signals	153
		The Effect of Slew Rate on Sinusoidal Signals and Power Bandwidth	153
		Design Hint	157
		Slew Rate and Multiple Stages	158
		Noncompensated Devices	161
		Feedforward Compensation	162
		Decompensated Devices	163
	5.5	OFFSETS	164
		Offset Sources and Compensation	164
	5.6	DRIFT	170
	5.7	CMRR AND PSRR	173
	5.8	NOISE	175
		Summary	180
		Review Questions	181
		Problems	181
CHAPTER 6		SPECIALIZED OP AMPS	183
		Chapter Objectives	183
	6.1	INTRODUCTION	183

	6.2	INSTRUMENTATION AMPLIFIERS	184
	6.3	PROGRAMMABLE OP AMPS	193
	6.4	OP AMPS FOR HIGH CURRENT, POWER, AND VOLTAGE APPLICATIONS	197
		High-Current Devices	197
		High-Voltage Devices	201
	6.5	HIGH-SPEED AMPLIFIERS	201
	6.6	VOLTAGE FOLLOWERS AND BUFFERS	202
	6.7	OPERATIONAL TRANSCONDUCTANCE AMPLIFIER	204
	6.8	NORTON AMPLIFIER	207
	6.9	CURRENT FEEDBACK AMPLIFIERS	214
	6.10	OTHER SPECIALIZED DEVICES	217
		Summary	220
		Review Questions	220
		Problems	221
CHAPTER 7		NONLINEAR CIRCUITS	223
		Chapter Objectives	223
	7.1	INTRODUCTION	223
	7.2	PRECISION RECTIFIERS	224
		Peak Detector	227
		Precision Full-Wave Rectifier	232
	7.3	WAVE SHAPING	235
		Active Clampers	235
		Active Limiters	240
	7.4	FUNCTION GENERATION	244
	7.5	COMPARATORS	254
	7.6	LOG AND ANTI-LOG AMPLIFIERS	266
		Four-Quadrant Multiplier	270
	7.7	EXTENDED TOPIC: A PRECISION LOG AMP	273
		Summary	276
		Review Questions	277
		Problems	277
CHAPTER 8		VOLTAGE REGULATION	280
		Chapter Objectives	280
	8.1	INTRODUCTION	280
	8.2	THE NEED FOR REGULATION	281

	8.3	LINEAR REGULATORS	283
		Three Terminal Devices	287
		Current Boosting	292
		Low Dropout Regulators	292
		Programmable and Tracking Regulators	293
	8.4	SWITCHING REGULATORS	307
	8.5	HEAT SINK USAGE	317
		Physical Requirements	317
		Thermal Resistance	317
	8.6	EXTENDED TOPIC: PRIMARY SWITCHER	322
		Summary	323
		Review Questions	324
		Problems	324
CHAPTER 9		**OSCILLATORS AND FREQUENCY GENERATORS**	326
		Chapter Objectives	326
	9.1	INTRODUCTION	326
	9.2	OP AMP OSCILLATORS	327
		Positive Feedback and the Barkhausen Criterion	327
		A Basic Oscillator	328
		Wien Bridge Oscillator	329
		Phase Shift Oscillator	335
		Square/Triangle Function Generator	343
	9.3	SINGLE-CHIP OSCILLATORS AND FREQUENCY GENERATORS	354
		Voltage-Controlled Oscillator	354
		Phase-Locked Loop	359
		555 Timer	364
		555 Monostable Operation	365
		555 Astable Operation	367
		Waveform Generator	370
		Summary	376
		Review Questions	376
		Problems	377
CHAPTER 10		**INTEGRATORS AND DIFFERENTIATORS**	380
		Chapter Objectives	380
	10.1	INTRODUCTION	380
	10.2	INTEGRATORS	381

		Accuracy and Usefulness of Integration	383
		Optimizing the Integrator	383
		Analyzing Integrators with the Time-Continuous Method	387
		Analyzing Integrators with the Time-Discrete Method	388
	10.3	DIFFERENTIATORS	393
		Accuracy and Usefulness of Differentiation	394
		Optimizing the Differentiator	395
		Analyzing Differentiators with the Time-Continuous Method	397
		Analyzing Differentiators with the Time-Discrete Method	398
	10.4	ANALOG COMPUTER	407
	10.5	ALTERNATIVES TO INTEGRATORS AND DIFFERENTIATORS	409
	10.6	EXTENDED TOPIC: OTHER INTEGRATOR AND DIFFERENTIATOR CIRCUITS	410
		Summary	412
		Review Questions	412
		Problems	413
CHAPTER 11		ACTIVE FILTERS	416
		Chapter Objectives	416
	11.1	INTRODUCTION	416
	11.2	FILTER TYPES	417
	11.3	THE USE AND ADVANTAGES OF ACTIVE FILTERS	419
	11.4	FILTER ORDER AND POLES	420
	11.5	FILTER CLASS OR ALIGNMENT	422
		Butterworth	423
		Bessel	423
		Chebyshev	424
		Elliptic	425
		Other Possibilities	426
	11.6	REALIZING PRACTICAL FILTERS	426
		Sallen and Key VCVS Filters	426
		Sallen and Key Low-Pass Filters	428
		Sallen and Key High-Pass Filters	439
		Filters of Higher Order	442
	11.7	BAND-PASS FILTER REALIZATIONS	452
		Multiple-Feedback Filters	454
		State-Variable Filter	459
	11.8	NOTCH FILTER (BAND-REJECT) REALIZATIONS	465
		A Note on Component Selection	467
		Filter Design Tools	468

	11.9	Audio Equalizers	469
	11.10	Switched-Capacitor Filters	474
	11.11	Extended Topic: Voltage-Controlled Filters	479
		Summary	481
		Review Questions	482
		Problems	482
CHAPTER 12		Analog-to-Digital-to-Analog Conversion	484
		Chapter Objectives	484
	12.1	Introduction	484
		The Advantages and Disadvantages of Working in the Digital Domain	485
	12.2	The Sampling Theorem	486
	12.3	Resolution and Sampling Rate	488
	12.4	Digital-to-Analog Conversion Techniques	492
		Practical Digital-to-Analog Converter Limits	496
		Digital-to-Analog Converter Integrated Circuits	501
		Applications of Digital-to-Analog Converter Integrated Circuits	506
	12.5	Analog-to-Digital Conversion	512
		Analog-to-Digital Conversion Techniques	515
		Analog-to-Digital Converter Integrated Circuits	520
		Applications of Analog-to-Digital Converter Integrated Circuits	526
	12.6	Extended Topic: Digital Signal Processing	529
		Summary	532
		Review Questions	533
		Problems	533
APPENDIX A		Data Sheets	535
APPENDIX B		Answers to Odd-Numbered Review Questions	536
APPENDIX C		Answers to Odd-Numbered Problem Sets	542
		Index	589

PREFACE

Intended Audience

The goal of this text, as its name implies, is to allow the reader to become proficient in the analysis and design of circuits utilizing modern linear ICs. It progresses from the fundamental circuit building blocks through to analog/digital conversion systems. The text is intended for use in a second-year Operational Amplifiers course at the associate level or for a junior-level course at the baccalaureate level. In order to make effective use of this text, students should have already taken a course in basic discrete transistor circuits and have a solid background in algebra and trigonometry, along with exposure to phasors. Calculus is used in certain sections of the text, but for the most part, its use is kept to a minimum. For students without a calculus background these sections may be skipped without a loss of continuity. The sole exception to this is Chapter 10, Integrators and Differentiators, which hinges on knowledge of calculus.

Approach

In writing this text, I have tried to make it ideal for both the teacher and the student. Instead of inundating the student with page after page of isolated formulas and collections of unconnected facts and figures, this text relies on building a sound foundation first. Although it may take just a little bit longer to "get into" the operational amplifier than a more traditional approach, the initial outlay of time is rewarded with a deeper understanding and better retention of the later material. I tried to avoid creating formulas out of thin air, as is often done for the sake of expediency in technical texts. Instead, I strove to provide sufficient background material and proofs so that the student is never left wondering where particular formulas came from, or worse, coming to the conclusion that they are either too difficult to understand completely, or are somehow "magic."

Organization

The text can be broken into two major sections. The first section, comprised of Chapters 1 through 6, can be seen as the foundation of the operational amplifier. Here, a methodical, step-by-step presentation is used to introduce the basic idealized operational amplifier and eventually to examine its practical limitations with great detail. These chapters should be presented in order. The remaining six chapters comprise a selection of popular applications, including voltage regulation, oscillators, and active filters, to name a few. Although it is not imperative that these chapters be presented in the order given (or for that matter, that they all be covered), the present arrangement will probably result in the most natural progression. Treat these chapters as application reference material and shape the presentation to your needs.

There are a few points worth noting about certain chapters. First, Chapter 1 presents material on decibels, Bode plots, and the differential amplifier, which is the heart of most operational amplifiers. This chapter may be skipped if your curriculum covers these topics in a discrete semiconductor or circuit analysis course. In this case, it is recommended that students scan the chapter to familiarize themselves with the nomenclature used later in the text. As part of the focus on fundamentals, a separate chapter on negative feedback (Chapter 3) is presented. This chapter examines the four basic negative feedback connections that might be used with an operational amplifier and details the action of negative feedback in general. It stresses the popular series-parallel (VCVS or noninverting voltage amplifier) form to show how bandwidth, distortion, input impedance, and so forth are affected. Chapter 6 presents a variety of modern special-purpose devices for specific tasks, such as high load current, programmable operation, and very high speed. An extensive treatment of active filters is given in Chapter 11. Unlike many texts that use a "memorize this" approach to this topic, Chapter 11 strives to explain the underlying operation of the circuits, yet it does so without the more advanced math requirement of the classical engineering treatment. The final chapter, Chapter 12, serves as a bridge between the analog and digital worlds, covering analog-to-digital and digital-to-analog conversion schemes. Please bear in mind that entire books have been written on the topics of active filters and analog/digital systems. Chapters 11 and 12 are designed as solid introductions to these topics, not as the final, exhaustive word.

Chapter Pedagogy

At the beginning of each chapter is a set of chapter objectives. These point out the major items of importance that will be discussed. Following the chapter objectives is an introduction. The introduction sets the stage for the upcoming discussion and puts the chapter into perspective with the text as a whole. At the end of each chapter is a summary and a set of review questions. The review questions are of a general nature and are designed to test for retention of circuit and system concepts. Finally, each chapter has a problem set. For most chapters, the problems are broken into four categories: analysis problems, design problems, challenge problems, and computer simulation problems. An important point here was to include problems of both sufficient variety and number. An optional coverage "extended topic" may also be found at the end of many chapters. An extended topic is designed to present extra discussion and detail on a certain portion of a chapter. This allows you to customize the chapter presentation to a certain degree. Not everyone will want to cover every extended topic. Even if they aren't used as part of the normal run of the course, they do allow for interesting side reading, possible outside assignments, and launching points for more involved discussions.

Computer Simulation

Integrated with the text are examples that can be used with popular SPICE-based circuit simulation packages and derivatives such as Electronics Workbench MultiSIM, Orcad PSpice, IntuSoft ICAP, Beige Bag SPICE A/D 2000, and others. If you are not using computer simulations in your courses, these items may be skipped over with little loss of continuity; however, if you are using simulations, I think that you will find that this integrated approach can be very worthwhile. Generally, simulations should not be used in place of a traditional analysis, and with this in mind, many of the examples are used to verify the results of manual calculations or to investigate second-order effects that may be too time consuming otherwise. Computer simulations are also very valuable in the classroom and laboratory as a means of posing "what if" questions. Some of the items explored include the effects of different op amp models, sensitivity to parameter sweeps, and the usage of Monte Carlo analysis to investigate typical production spreads. The circuits are presented using the schematic capture window of Electronics Workbench MultiSIM™. Although MultiSIM's schematic editor and output graphing tools are shown, the figures would be very similar in the other popular packages. At this level, the differences between the input and output figures of the various packages are largely cosmetic. In general, any quality SPICE-based simulator will be sufficient for the circuits presented in this book.

Applications and Clear Presentation

Finally, there are two elements included in this text that I always longed for when I was a student—real-world applications and a friendly writing style. First, besides the traditional circuit examples used to explain circuit operation, a number of schematics have been included to show how a given type of circuit might be used in the real world. It is one thing to read how a precision rectifier works; it is quite another thing to see how one is being used in an audio power amplifier as part of an overload protection scheme.

The second element deals with the writing style of this text. I have tried to be direct and conversational without being overly cute or "chatty." The body of this text is not written in a passive formal voice. Instead, it is meant to sound as if someone were explaining the topics to you over your shoulder. It is intended to draw you into the topic and to hold your attention. Over the years, I have noticed that some people feel that in order to be taken seriously, a topic must be addressed in a detached, almost antiseptic, manner. Although I will agree that this mind-set is crucial in order to perform a good experiment, it does not translate particularly well to textbooks, especially at the undergraduate level. The result, unfortunately, is a thorough but thoroughly unreadable book. If teachers find a text uninteresting, it shouldn't be surprising if the students feel the same. I hope that you will find this text to be serious, complete, and engaging.

Supplements

Instructor's Guide—This ancillary contains answers and worked-out solutions to the textbook questions and problems (ISBN: 0-7668-1914-0).

Lab Manual—This lab manual contains 21 exercises. All major topics are represented and cross-referenced to the text. Blank tables and graphs are included for experimental results, along with a complete parts listing. (ISBN: 0-7668-1794-6)

Online Companion—This textbook has a companion Internet Web site intended for use by both educators and students. It provides ongoing assistance in the form of additional circuit files, problems, and text updates. Please visit our Web site at *www.electronictech.com* for more details.

Acknowledgments

In a project of this size, there are many people who have done their part to shape the result. Some have done so directly, others indirectly, and still others without knowing that they did anything at all. First, I'd like to thank my family and friends for being such "good eggs" about the project, even if it did mean that I disappeared for hours on end in order to work on it. For their encouragement and assistance, I thank my fellow faculty members and students. For turning my ideas into reality, I thank the folks at Delmar Thomson Learning: Greg Clayton, Michelle Ruelos Cannistraci, Larry Main, David Arsenault, Chris Chien, and Jennifer Thompson. Also, I'd like to thank the various IC and equipment manufacturers for providing pertinent data sheets, schematics, and photographs. Finally, I'd like to thank the following reviewers for their constructive comments on the early manuscript:

> Leah Akins, Dutchess Community College, Poughkeepsie, NY
> Yolanda Guran, Oregon Institute of Technology, Portland, OR
> Mark Hughes, Cleveland Community College, Shelby, NC
> Seyed Jalali, DeVry Institute of Technology, Long Beach, CA
> A. Kent Johnson, Brigham Young University, Provo, UT
> Solomon Oldak, DeVry Institute of Technology, Pomona, CA
> Predrag Pesikan, DeVry Institute of Technology, Mississauga, Ontario
> Ronald Rockland, New Jersey Institute of Technology, Newark, NJ
> Lloyd Stallkamp, Montana State University, Havre, MT

For authors and those interested in the mechanics, the original manuscript was created using a variety of tools including numerous text editors and word processors, TeX, and a few utilities I created myself. The filter graphs found in Chapter 11 were created using Tony Richardson's public domain port of PLPLOT.

Finally, it is still no lie to say that I couldn't have completed this project without my personal computers and a stack of Frank Zappa, Kate Bush, and King Crimson CDs. Also, I received considerable fuel in the form of Rintrona's tomato pie (a local delicacy) and the endless stream of healthy and tasty goodies flowing from Karen's kitchen.

JIM FIORE, SUMMER 2000

CHAPTER 1

Introductory Concepts and Fundamentals

CHAPTER OBJECTIVES

After completing this chapter, you should be able to:

- Convert between ordinary and decibel-based power and voltage gains.
- Utilize decibel-based voltage and power measurements during circuit analysis.
- Define and graph a general Bode plot.
- Detail the differences between lead and lag networks and graph Bode plots for each.
- Combine the effects of several lead and lag networks together in order to determine a system Bode plot.
- Describe the use of digital computers in the area of circuit simulation.
- Analyze differential amplifiers for a variety of AC characteristics, including single-ended and differential voltage gains.
- Define *common-mode gain* and *common-mode rejection*.
- Describe a current mirror and note typical uses for it.

1.1 INTRODUCTION

Before we can begin our study of the operational amplifier, it is important that certain background elements be in place. The purpose of this chapter is to present the useful analysis concepts and tools associated with the decibel measurement scheme and the frequency domain. We will also examine the differential amplifier, which serves as the heart of most operational amplifiers. With a thorough working knowledge of these items, you will find that circuit design and analysis will proceed more quickly and efficiently. Consider this chapter as an investment in time, and treat it appropriately.

The decibel measurement scheme is in wide use, particularly in the field of communications. We will examine its advantages over the ordinary system of measurement and how to convert values of one form into the other. One of the more important parameters of a circuit is its frequency response. To this end, we will look at the general frequency domain representations of a circuit's gain and phase. This will include both manual and computer-generated analysis and graphing techniques. Although our main emphasis will eventually concentrate on application with operational amplifiers, the techniques explored can be applied equally to discrete circuits. Indeed, our initial examples will use simple discrete or black box circuits exclusively.

Finally, we will examine the DC and AC operation of the differential amplifier. This is an amplifier that utilizes two active devices and offers dual inputs. It offers certain features that make it suitable as the first section of most operational amplifiers.

Variable Naming Convention

One item that often confuses students of almost any subject is nomenclature. It is important, then, that we decide on a consistent naming convention at the outset. Throughout this text, we will examine numerous circuits containing several passive and active components. We are interested in a variety of parameters and signals. Although we utilize the standard conventions, such as f_c for critical frequency and X_c for capacitive reactance, a great number of other possibilities exist. To keep confusion to a minimum, we use the following conventions in our equations for naming devices and signals that haven't been standardized.

R Resistor (DC, or actual circuit component)
r Resistor (AC equivalent, where phase is 0 or ignored)
C Capacitor
L Inductor
Q Transistor (Bipolar or FET)
D Diode
V Voltage (DC)
v Voltage (AC)
I Current (DC)
i Current (AC)

Resistors, capacitors, and inductors are differentiated via a subscript, which usually refers to the active device it is connected to. For example, R_E is a DC bias resistor connected to the emitter of a transistor, whereas r_C refers to the AC equivalent resistance seen at a transistor's collector. C_E refers to a capacitor connected to a transistor's emitter lead (most likely a bypass or coupling capacitor). Note that the device-related subscripts are always shown in upper case, with one exception: If the resistance or capacitance is part of the device model, the subscript will be shown in lowercase to distinguish it from the external circuit components. For example, the AC dynamic resistance of a diode would be called r_d. If no active devices are present, or if several items exist in the circuit, a simple numbering scheme is used, such as R_1. In very complex circuits, a specific name will be given to particularly important components, as in R_{source}.

Voltages are normally given a two-letter subscript indicating the nodes at which it is measured. V_{CE} is the DC potential from the collector to the emitter of a transistor, and v_{BE} indicates the AC signal appearing across a transistor's base-emitter junction. A single-letter subscript, as in V_B, indicates a potential relative to ground (in this case, base to ground potential). The exceptions to this rule are power supplies, which are given a double-letter subscript indicating the connection point (V_{CC} is the collector power supply), and particularly important potentials that are directly named, as in v_{in} (AC input voltage) and V_{R_1} (DC voltage appearing across R_1). If an equation for a specific potential is valid for both the AC and DC equivalent circuits, the uppercase form is preferred (this makes things much more consistent with the vast majority of op amp circuits, which are directly coupled, and thus can amplify both AC and DC signals). Currents are named in a similar way, but generally use a single subscript referring to the measurement node (I_C is the DC collector current). All other items are directly named. By using this scheme, you will always be able to determine whether the item expressed in an equation is a DC or AC equivalent, its approximate circuit location, and other factors about it.

1.2 THE DECIBEL

Most people are familiar with the term *decibel* in reference to sound pressure. It's not uncommon to hear someone say something such as, "It was 110 decibels at the concert last night, and my ears are still ringing." This popular use is somewhat inaccurate, but does show that decibels indicate some sort of quantity—in this case, sound pressure level.

Decibel Representation of Power and Voltage Gains

In its simplest form, the decibel is used to measure some sort of gain, such as power or voltage gain. Unlike the ordinary gain measurements that you may be familiar with, the decibel form is logarithmic. Because of this, it can be very useful for showing ratios of change, as well as absolute change. The base unit is the Bel. To convert an ordinary gain to its Bel counterpart, just take the common log (base 10) of the gain. In equation form, it appears as:

$$\text{Bel gain} = \log_{10}(\text{ordinary gain})$$

Note that on most hand calculators, common log is denoted as `log`, but the natural log is given as `ln`. Unfortunately, many programming languages use `log` to indicate natural log and `log10` for common log. More than one student has been bitten by this bug, so be forewarned! As an example, if a circuit produces an output power of 200 milliwatts for an input of 10 milliwatts, we would normally say that it has a power gain of:

$$G = \frac{P_{out}}{P_{in}}$$

$$G = \frac{200 \text{ mW}}{10 \text{ mW}}$$
$$G = 20$$

For the Bel version, just take the log of this result.

$$G' = \log_{10} G$$
$$G' = \log_{10} 20$$
$$G' = 1.301$$

The Bel gain is 1.3 Bels. The term *Bels* is not a unit in the strict sense of the word (as in *watts*), but is simply used to indicate that this is not an ordinary gain. In contrast, ordinary power and voltage gains are sometimes given units of W/W and V/V to distinguish them from Bel gains. Also, note that the symbol for Bel power gain is G' and not G. All Bel gains are denoted with the following prime (') notation to avoid confusion. Bels tend to be rather large, so we typically use one-tenth of a Bel as the norm. The result is the decibel (one-tenth Bel). To convert to decibels, simply multiply the number of Bels by 10. Our gain of 1.3 Bels is equivalent to 13 decibels. The units are commonly shortened to dB. Consequently, we may say:

$$G' = 10 \log_{10} G, \text{ where the result is in dB} \qquad (1.1)$$

At this point, you may be wondering what the big advantage of the decibel system is. To answer this, recall a few log identities. Normal multiplication becomes addition in the log system, and division becomes subtraction. Likewise, powers and roots become multiplication and division. Because of this, two important things show up. First, ratios of change become constant offsets in the decibel system, and second, the entire range of values diminishes in size. The result is that a very wide range of gains may be represented within a fairly small scope of values, and the corresponding calculations can become quicker. There are a couple of dB values that are useful to remember. With the aid of your hand calculator, it is very easy to show the following:

Factor	dB Value (using $G' = 10 \log_{10} G$)
1	0 dB
2	3.01 dB
4	6.02 dB
8	9.03 dB
10	10 dB

We can also look at fractional factors (i.e., losses instead of gains.)

Factor	dB Value
.5	−3.01 dB
.25	−6.02 dB
.125	−9.03 dB
.1	−10 dB

If you look carefully, you will notice that a doubling is represented by an increase of approximately 3 dB. A factor of 4 is, in essence, two doublings. Therefore, it is equivalent to 3 dB + 3 dB, or 6 dB. Remember, because we are using logs, multiplication turns into simple addition. In a similar manner, a halving is represented by approximately −3 dB. The negative sign indicates a reduction. To simplify things a bit, think of factors of 2 as ±3 dB, the sign indicating whether you are increasing (multiplying), or decreasing (dividing). As you can see, factors of 10 work out to a very convenient 10 dB. By remembering these two factors, you can often estimate a dB conversion without the use of your calculator. For example, we could rework our initial conversion problem.

> The amplifier has a gain of 20.
> 20 can be written as 2 × 10.
> The factor of 2 is 3 dB, the factor of 10 is 10 dB.
> The answer must be 3 dB + 10 dB, or 13 dB.
> This verifies our earlier result.

Example 1-1

An amplifier has a power gain of 800. What is the decibel power gain?

$$G' = 10 \log_{10} G$$
$$G' = 10 \log_{10} 800$$
$$G' = 10 \times 2.903$$
$$G' = 29.03 \text{ dB}$$

We could also use our estimation technique.

> $G = 800 = 8 \times 10^2$.
> 8 is equivalent to 3 factors of 2, or 2 × 2 × 2, which can be expressed as 3 dB + 3 dB + 3 dB, which is, of course, 9 dB.
> 10^2 is equivalent to 2 factors of 10, or 10 dB + 10 dB = 20 dB.
> The result is 9 dB + 20 dB, or 29 dB.

Note that if the leading digit is not a power of 2, the estimation will not be as precise. For example, if the gain is 850, you know that the decibel gain is just a bit over 29 dB. You also know that it must be less than 30 dB (1000 = 10^3, which is 3 factors of 10, or 30 dB.) As you can see, by using the dB form, you tend to concentrate on the magnitude of gain, and not so much on trailing digits.

Example 1-2

An attenuator reduces signal power by a factor of 10,000. What is this loss expressed in dB?

$$G' = 10 \log_{10} \frac{1}{10,000}$$
$$G' = 10 \times (-4)$$
$$G' = -40 \text{ dB}$$

By using the approximation, we can say,

$$\frac{1}{10,000} = 10^{-4}$$

The negative exponent tells us we have a loss (negative dB value), and 4 factors of 10.

$$G' = -10 \text{ dB} - 10 \text{ dB} - 10 \text{ dB} - 10 \text{ dB} = -40 \text{ dB}$$

Remember, if an increase in signal is produced, the result will be a positive dB value. A decrease in signal will always result in a negative dB value. A signal that is unchanged indicates a gain of unity, or 0 dB.

To convert from dB to ordinary form, just invert the steps.

$$G = \log_{10}^{-1} \frac{G'}{10}$$

On most hand calculators, base 10 antilog is denoted as 10^x. In most computer languages, you just raise 10 to the appropriate power, as in `G = 10.0^(Gprime / 10.0)` (BASIC), or use an exponent function, as in `pow(10.0, Gprime / 10.0)` (C).

EXAMPLE 1-3

An amplifier has a power gain of 23 dB. If the input is 1 mW, what is the output?

In order find the output power, we need to find the ordinary power gain, G.

$$G = \log_{10}^{-1} \frac{G'}{10}$$

$$G = \log_{10}^{-1} \frac{23}{10}$$

$$G = 199.5$$

Therefore, $P_{out} = 199.5 \times 1$ mW, or 199.5 mW.

You could also use the approximation technique in reverse. To do this, break up the dB gain in 10 dB and 3 dB chunks:

$$23 \text{ dB} = 3 \text{ dB} + 10 \text{ dB} + 10 \text{ dB}$$

Now replace each chunk with the appropriate factor, and multiply them together (remember, when going from log to ordinary form, addition turns into multiplication).

$$3 \text{ dB} = 2X, \ 10 \text{ dB} = 10X, \text{ so}$$

$$G = 2 \times 10 \times 10$$

$$G = 200$$

Although the approximation technique appears to be slower than the calculator, practice will show otherwise. Being able to quickly estimate dB values can prove to be a very handy skill in the electronics field. This is particularly true in larger, multistage designs.

EXAMPLE 1-4

A three-stage amplifier has gains of 10 dB, 16 dB, and 14 dB per section. What is the total dB gain?

Because dB gains are a log form, just add the individual stage gains to arrive at the system gain.

$$G'_{total} = G'_1 + G'_2 + G'_3$$
$$G'_{total} = 10 \text{ dB} + 16 \text{ dB} + 14 \text{ dB}$$
$$G'_{total} = 40 \text{ dB}$$

As you may have noticed, all of the examples up to this point have used power gain and not voltage gain. You may be tempted to use the same equations for voltage gain. In a word, DON'T. If you think back for a moment, you will recall that power varies as the square of voltage. In other words, a doubling of voltage will produce a quadrupling of power. If you were to use the same dB conversions, a doubling of voltage would be 3 dB, yet, because the power has quadrupled, this would indicate a 6 dB rise. Consequently, voltage gain (and current gain as well) are treated in a slightly different fashion. We would rather have our doubling of voltage work out to 6 dB, so that it matches the power calculation. The correction factor is very simple. Power varies as the second power of voltage, thus the dB form should be twice as large for voltage (remember, exponentiation turns into multiplication when using logs). Applying this factor to Equation 1.1 yields:

$$A'_v = 20 \log_{10} A_v \qquad (1.2)$$

Be careful though, the Bel voltage gain only equals the Bel power gain if the input and output impedances of the system are matched. (You may recall from your earlier work that it is quite possible to design a circuit with vastly different voltage and power gains. A voltage follower, for example, exhibits moderate power gain with a voltage gain of unity. It is quite likely that the follower will not exhibit matched impedances.) If we were to recalculate our earlier table of common factors, we would find that a doubling of voltage gain is equivalent to a 6 dB rise, and a tenfold increase is equivalent to a 20 dB rise, twice the size of their power gain counterparts.

Note that current gain may be treated in the same manner as voltage gain (although this is less commonly done in practice).

EXAMPLE 1-5

An amplifier has an output signal of 2 V for an input of 50 mV. What is A'_v?

First find the ordinary gain.

$$A_v = \frac{2}{.05} = 40$$

Now convert to dB form.

$$A'_v = 20 \log_{10} 40$$
$$A'_v = 20 \times 1.602$$
$$A'_v = 32.04 \text{ dB}$$

The approximation technique yields $40 = 2 \times 2 \times 10$, or $6 \text{ dB} + 6 \text{ dB} + 20 \text{ dB} = 32 \text{ dB}$.

To convert A'_v to A_v, reverse the process.

$$A_v = \log_{10}^{-1} \frac{A'_v}{20}$$

EXAMPLE 1-6

An amplifier has a gain of 26 dB. If the input signal is 10 mV, what is the output?

$$A_v = \log_{10}^{-1} \frac{A'_v}{20}$$
$$A_v = \log_{10}^{-1} \frac{26}{20}$$
$$A_v = 19.95$$

$$V_{out} = A_v V_{in}$$
$$V_{out} = 19.95 \times 10 \text{ mV}$$
$$V_{out} = .1995 \text{ V}$$

The final point to note in this section is that, as in the case of power gain, a negative dB value indicates a loss. Therefore, a 2:1 voltage divider would have a gain of −6 dB.

Signal Representation in dBW and dBV

As you can see from the preceding section, it is possible to spend considerable time converting between decibel gains and ordinary voltages and powers. The decibel form does offer advantages for gain measurement, so it would make sense to use a decibel form for power and voltage levels as well. This is a relatively straightforward process. There is no reason why we can't express a power or voltage in a logarithmic form. As a dB value just indicates a ratio, all we need to do is decide on a reference (i.e., a comparative base for the ratio). For power measurements, a likely choice would be 1 watt. In other words, we can describe a power as being X dB above or below 1 watt. Positive values will indicate powers greater than 1 watt, whereas negative values will indicate powers less than 1 watt. In general equation form:

$$P' = 10 \log_{10} \frac{P}{\text{reference}} \quad (1.3)$$

The answer will have units of dBW, that is, decibels relative to 1 watt.

EXAMPLE 1-7

A power amplifier has a maximum output of 120 W. What is this power in dBW?

$$P' = 10 \log_{10} \frac{P}{1 \text{ Watt}}$$
$$P' = 10 \log_{10} \frac{120 \text{ W}}{1 \text{ W}}$$
$$P' = 20.8 \text{ dBW}$$

There is nothing sacred about the 1 watt reference, short of its convenience. We could just as easily choose a different reference. Other common reference points are 1 milliwatt (dBm) and 1 femtowatt (dBf). Obviously, dBf is used for very low signal levels, such as those coming from an antenna. dBm is in very wide use in the communications industry. To use these other references, just divide the given power by the new reference.

EXAMPLE 1-8

A small personal audiotape player delivers 200 mW to its headphones. What is this output power in dBW and in dBm?

For an answer in units of dBW, use the 1 watt reference

$$P' = 10 \log_{10} \frac{P}{1 \text{ Watt}}$$
$$P' = 10 \log_{10} \frac{200 \text{ mW}}{1 \text{ W}}$$
$$P' = -7 \text{ dBW}$$

For units of dBm, use a 1 milliwatt reference

$$P' = 10 \log_{10} \frac{P}{1 \text{ mW}}$$
$$P' = 10 \log_{10} \frac{200 \text{ mW}}{1 \text{ mW}}$$
$$P' = 23 \text{ dBm}$$

200 mW, −7 dBW, and 23 dBm are three ways of saying the same thing. Note that the dBW and dBm values are 30 dB apart. This will always be true, because the references are a factor of 1000 (30 dB) apart.

In order to transfer a dBW or similar value into watts, reverse the process.

$$P = \log_{10}^{-1} \frac{P'}{10} \times \text{reference}$$

Example 1-9

A studio microphone produces a 12 dBm signal while recording normal speech. What is the output power in watts?

$$P = \log_{10}^{-1} \frac{P'}{10} \times \text{reference}$$

$$P = \log_{10}^{-1} \frac{12 \text{ dBm}}{10} \times 1 \text{ mW}$$

$$P = 15.8 \text{ mW, or } .0158 \text{ W}$$

For voltages, we can use a similar system. A logical reference is 1 V, with the resulting units being dBV. As before, these voltage measurements will use a multiplier of 20 instead of 10.

$$V' = 20 \log_{10} \frac{V}{\text{reference}} \tag{1.4}$$

Example 1-10

A test oscillator produces a 2 V signal. What is this value in dBV?

$$V' = 20 \log_{10} \frac{V}{\text{reference}}$$

$$V' = 20 \log_{10} \frac{2 \text{ V}}{1 \text{ V}}$$

$$V' = 6.02 \text{ dBV}$$

When both circuit gains and signal levels are specified in dB form, analysis can be very quick. Given an input level, simply add the gain to it in order to find the output level. Given input and output levels, subtract them in order to find the gain.

Example 1-11

A floppy disk read/write amplifier exhibits a gain of 35 dB. If the input signal is −42 dBV, what is the output signal?

$$V'_{out} = V'_{in} + A'_v$$

$$V'_{out} = -42 \text{ dBV} + 35 \text{ dB}$$

$$V'_{out} = -7 \text{ dBV}$$

Note that the final units are dBV and not dB, indicating a voltage and not merely a gain.

Figure 1.1

Multistage

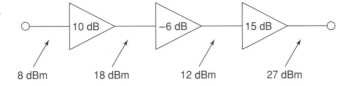

Example 1-12

A guitar power amp needs an input of 20 dBm to achieve an output of 25 dBW. What is the gain of the amplifier in dB?

First, it is necessary to convert the power readings so that they share the same reference unit. As dBm represents a reference 30 dB smaller than the dBW reference, just subtract 30 dB to compensate.

$$20 \text{ dBm} = -10 \text{ dBW}$$
$$G' = P'_{out} - P'_{in}$$
$$G' = 25 \text{ dBW} - (-10 \text{ dBW})$$
$$G' = 35 \text{ dB}$$

Note that the units are dB and not dBW. This is very important! Saying that the gain is "so many" dBW is the same as saying the gain is "so many" watts. Obviously, gains are "pure" numbers and do not carry units such as watts or volts.

The usage of a dB-based system is shown graphically in Figure 1.1. Note how the stage gains are added to the input signal to form the output. Even large circuits can be quickly analyzed in this form. To make life in the lab even easier, it is possible to take measurements directly in dB form. By doing this, you need never convert while troubleshooting a design. For general-purpose work, voltage measurements are the norm, and therefore a dBV scale is often used.

Items of Interest in the Laboratory

When using a digital meter on a dBV scale it is possible to "underflow" the meter if the signal is too weak. This will happen if you try to measure around zero volts, for example. If you attempt to calculate the corresponding dBV value, your calculator will probably show "error." The effective value is negative infinite dBV. The meter will certainly have a hard time showing this value! Another item of interest revolves around the use of dBm measurements. It is common to use a voltmeter to make dBm measurements, in lieu of a wattmeter. Although the connections are considerably simpler, a voltmeter cannot measure power. How is this accomplished then? As long as the circuit impedance is known, power can be derived from a voltage measurement. A common impedance in communication systems (such as recording studios) is 600 Ω, so a meter can be calibrated to give correct dBm readings by using Power Law. If this meter is used on a non–600 Ω circuit, the readings will no longer reflect accurate dBm values (but will still properly reflect relative changes in dB).

FIGURE 1.2

Gain plot

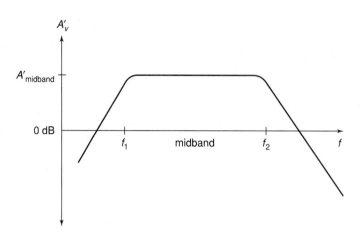

1.3 Bode Plots

The Bode plot is a graphical response prediction technique that is useful for both circuit design and analysis. A Bode plot is in actuality a pair of plots: one graphs the gain of a system versus frequency, and the other details the circuit phase versus frequency. Both of these items are very important in the design of well-behaved, optimal operational amplifier circuits.

Generally, Bode plots are drawn with logarithmic frequency axes, a decibel gain axis, and a phase axis in degrees. First, let's take a look at the gain plot. A typical gain plot is shown Figure 1.2.

Note how the plot is relatively flat in the middle, or midband, region. The gain value in this region is known as the midband gain. At either extreme of the midband region, the gain begins to decrease. The gain plot shows two important frequencies, f_1 and f_2. The lower break frequency is f_1, whereas f_2 is the upper break frequency. The gain at the break frequencies is 3 dB less than the midband gain. These frequencies are also known as the half-power points, or corner frequencies. Normally, amplifiers are only used for signals between f_1 and f_2. The exact shape of the rolloff regions will depend on the design of the circuit. It is possible to design amplifiers with no lower-break frequency (i.e., a DC amplifier); however, all amplifiers will exhibit an upper break. The break points are caused by the presence of circuit reactances, typically coupling and stray capacitances. The gain plot is a summation of the midband response with the upper and lower frequency limiting networks. Let's take a look at the lower break, f_1.

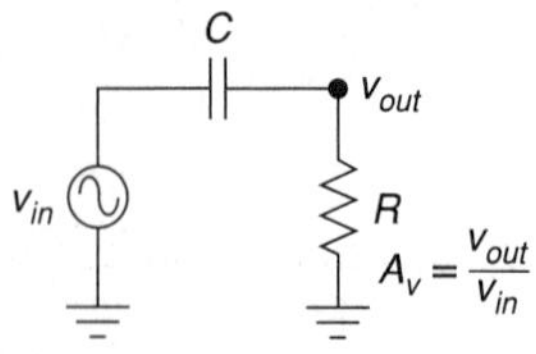

Lead Network Gain Response

Reduction in low-frequency gain is caused by lead networks. A generic lead network is shown in Figure 1.3. It gets its name from the fact that the output voltage developed across R leads the input. At very high frequencies the circuit will be essentially resistive. Conceptually, think of this as a simple voltage divider. The divider ratio depends on the reactance of C. As the input frequency drops, X_c increases. This makes V_{out} decrease.

FIGURE 1.3

Lead network

FIGURE 1.4

Lead gain plot (exact)

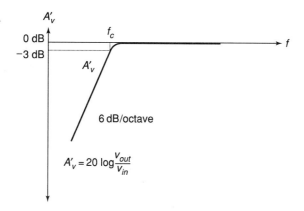

At very high frequencies, where $X_c \ll R$, V_{out} is approximately equal to V_{in}. This can be seen graphically in Figure 1.4. The break frequency (i.e., the frequency at which the signal has decreased by 3 dB) is found via the standard equation,

$$f_c = \frac{1}{2\pi RC}$$

The response below f_c will be a straight line if a decibel gain axis and a logarithmic frequency axis are used. This makes for very quick and convenient sketching of circuit response. The slope of this line is 6 dB per octave. (An octave is a doubling or halving of frequency, e.g., 800 Hz is 3 octaves above 100 Hz.)[1] This slope may also be expressed as 20 dB per decade, where a decade is a factor of 10 in frequency. With reasonable accuracy, this curve may be approximated as two line segments, called *asymptotes*, as shown in Figure 1.5. The shape of this curve is the same for any lead network. Because of this, it is very easy to find the approximate gain at any given frequency as

FIGURE 1.5

Lead gain plot (approximate)

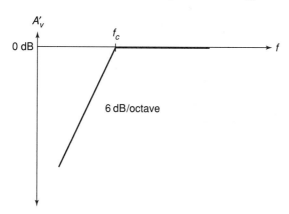

---

[1] The term *octave* is borrowed from the field of music. It gets its name from the fact that there are eight notes in the standard western scale, do-re-me-fa-so-la-ti-do. This range covers a factor of two in frequency.

long as f_c is known. It is not necessary to go through reactance and phasor calculations. To create a general response equation, start with the voltage divider rule to find the gain:

$$A_v = \frac{V_{out}}{V_{in}} = \frac{R}{R - jX_c}$$

$$A_v = \frac{R \angle 0}{\sqrt{R^2 + X_c^2} \angle -\arctan\frac{X_c}{R}}$$

The magnitude of this is

$$|A_v| = \frac{R}{\sqrt{R^2 + X_c^2}}$$

$$|A_v| = \frac{1}{\sqrt{1 + \frac{X_c^2}{R^2}}} \tag{1.5}$$

Recalling that

$$f_c = \frac{1}{2\pi RC}$$

we may say

$$R = \frac{1}{2\pi f_c C}$$

For any frequency of interest, f,

$$X_c = \frac{1}{2\pi f C}$$

Equating the two preceding equations yields

$$\frac{f_c}{f} = \frac{X_c}{R} \tag{1.6}$$

Substituting Equation 1.6 in Equation 1.5 gives

$$A_v = \frac{1}{\sqrt{1 + \frac{f_c^2}{f^2}}} \tag{1.7}$$

To express A_v in dB, substitute Equation 1.7 into Equation 1.2

$$A_v' = 20 \log_{10} \frac{1}{\sqrt{1 + \frac{f_c^2}{f^2}}}$$

After simplification, the final result is

$$A'_v = -10\log_{10}\left(1 + \frac{f_c^2}{f^2}\right) \qquad (1.8)$$

Where

f_c is the critical frequency,
f is the frequency of interest, and
A'_v is the decibel gain at the frequency of interest.

EXAMPLE 1-13

An amplifier has a lower break frequency of 40 Hz. How much gain is lost at 10 Hz?

$$A'_v = -10\log_{10}\left(1 + \frac{f_c^2}{f^2}\right)$$

$$A'_v = -10\log_{10}\left(1 + \frac{40^2}{10^2}\right)$$

$$A'_v = -12.3 \text{ dB}$$

In other words, the gain is 12.03 dB lower than it is in the midband. Note that 10 Hz is 2 octaves below the break frequency. Because the cutoff slope is 6 dB per octave, each octave loses 6 dB. Therefore, the approximate result is -12 dB, which double-checks the exact result. Without the lead network, the gain would stay at 0 dB all the way down to DC (0 Hz.)

Lead Network Phase Response

At very low frequencies, the circuit of Figure 1.3 is largely capacitive. Because of this, the output voltage developed across R leads by 90 degrees. At very high frequencies the circuit will be largely resistive. At this point V_{out} will be in phase with V_{in}. At the critical frequency, V_{out} will lead by 45 degrees. A general phase graph is shown in Figure 1.6. As with the gain plot, the phase plot shape is the same for any lead network. The general phase equation may be obtained from the voltage divider:

$$\frac{V_{out}}{V_{in}} = \frac{R}{R - jX_c}$$

$$\frac{V_{out}}{V_{in}} = \frac{R\angle 0}{\sqrt{R^2 + X_c^2}\angle - \arctan\frac{X_c}{R}}$$

The phase portion of this is

$$\theta = \arctan\frac{X_c}{R}$$

FIGURE 1.6
Lead phase (exact)

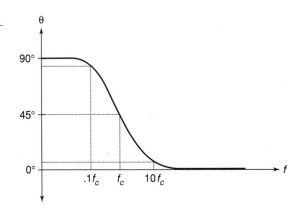

By using Equation 1.6, this simplifies to

$$\theta = \arctan \frac{f_c}{f} \tag{1.9}$$

Where

f_c is the critical frequency,
f is the frequency of interest, and
θ is the phase angle at the frequency of interest.

Often, an approximation such as Figure 1.7 is used in place of Figure 1.6. By using Equation 1.9, you can show that the approximation is off by no more than 6 degrees at the corners.

EXAMPLE 1-14

A telephone amplifier has a lower break frequency of 120 Hz. What is the phase response one decade below and one decade above?

FIGURE 1.7
Lead phase (approximate)

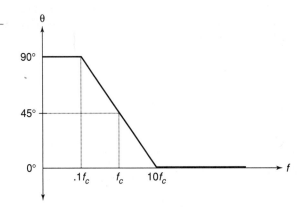

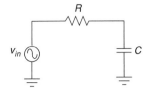

FIGURE 1.8

Lag network

One decade below 120 Hz is 12 Hz, and one decade above is 1.2 kHz.

$$\theta = \arctan \frac{f_c}{f}$$

$$\theta = \arctan \frac{120 \text{ Hz}}{12 \text{ Hz}}$$

$\theta = 84.3$ degrees one decade below f_c (i.e., approaching 90 degrees)

$$\theta = \arctan \frac{120 \text{ Hz}}{1.2 \text{ kHz}}$$

$\theta = 5.71$ degrees one decade above f_c (i.e., approaching 0 degrees)

Remember, if an amplifier is direct-coupled and has no lead networks, the phase will remain at 0 degrees right back to 0 Hz (DC).

Lag Network Response

Unlike its lead network counterpart, all amplifiers will contain lag networks. In essence, it's little more than an inverted lead network. As you can see from Figure 1.8, it simply transposes the R and C locations. Because of this, the response tends to be inverted as well. In terms of gain, X_c is very large at low frequencies, and thus V_{out} equals V_{in}. At high frequencies, X_c decreases, and V_{out} falls. The break point occurs when X_c equals R. The general gain plot is shown in Figure 1.9. Like the lead network response, the slope of this curve is −6 dB per octave (or −20 dB per decade.) Note that the slope is negative instead of positive. A straight-line approximation is shown in Figure 1.10. We can derive a general gain equation for this circuit in virtually the same manner as we did for the lead network. The derivation is left as an exercise.

$$A'_v = -10 \log_{10}\left(1 + \frac{f^2}{f_c^2}\right) \quad (1.10)$$

FIGURE 1.9

Lag gain (exact)

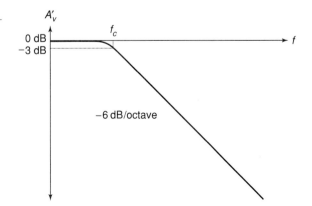

FIGURE 1.10

Lag gain (approximate)

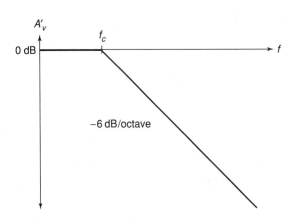

Where

f_c is the critical frequency,
f is the frequency of interest, and
A'_v is the decibel gain at the frequency of interest.

Note that this equation is almost the same as Equation 1.8. The only difference is that f and f_c have been transposed.

In a similar vein, we may examine the phase response. At very low frequencies, the circuit is basically capacitive. The output is taken across C, so V_{out} will be in phase with V_{in}. At very high frequencies, the circuit is essentially resistive. Consequently, the output voltage across C will lag by 90 degrees. At the break frequency the phase will be -45 degrees. A general phase plot is shown in Figure 1.11, with the approximate response detailed in Figure 1.12. As with the lead network, we may derive a phase equation. Again, the exact steps are very similar and left as an exercise.

$$\theta = -90 + \arctan \frac{f_c}{f} \tag{1.11}$$

FIGURE 1.11

Lag phase (exact)

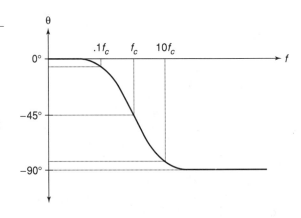

FIGURE 1.12

Lag phase (approximate)

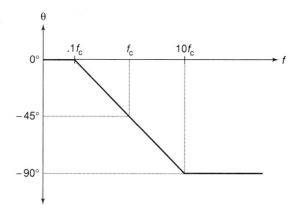

Where

f_c is the critical frequency,
f is the frequency of interest, and
θ is the phase angle at the frequency of interest.

EXAMPLE 1-15

A medical ultrasound transducer feeds a lag network with an upper break frequency of 150 kHz. What are the gain and phase values at 1.6 MHz?

This represents a little more than a 1 decade increase, so the approximate values are −20 dB and −90 degrees, from Figures 1.10 and 1.12, respectively. The exact values are

$$A'_v = -10 \log_{10}\left(1 + \frac{f^2}{f_c^2}\right)$$

$$A'_v = -10 \log_{10}\left(1 + \frac{1.6 \text{ MHz}^2}{150 \text{ kHz}^2}\right)$$

$$A'_v = -20.6 \text{ dB}$$

$$\theta = -90 + \arctan\frac{f_c}{f}$$

$$\theta = -90 + \arctan\frac{150 \text{ kHz}}{1.6 \text{ MHz}}$$

$$\theta = -84.6 \text{ degrees}$$

The complete Bode plot for this network is shown in Figure 1.13. It is very useful to examine both plots simultaneously. In this manner you can find the exact phase change for a given gain quite easily. This information is very important when the application of negative feedback is considered (Chapter 3). For example, if you look carefully at the plots of Figure 1.13, you will note that at the critical frequency of 150 kHz, the total phase change is −45 degrees. This circuit involved the use of a single lag network, so this is exactly what you would expect.

FIGURE 1.13

Bode plot for 150 kHz lag

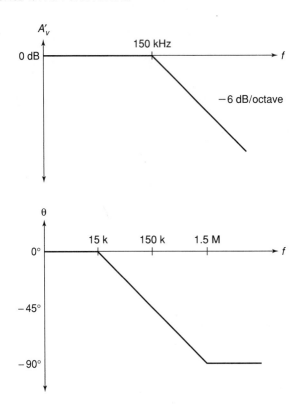

Risetime versus Bandwidth

For pulse-type signals, the speed of an amplifier is often expressed in terms of its *risetime*. If a square pulse such as Figure 1.14a is passed into a simple lag network, the capacitor charging effect will produce a rounded variation, as seen in Figure 1.14b. This effect places an upper limit on the duration of pulses that a given amplifier can handle without producing excessive distortion. By definition, risetime is the amount of time it takes for the signal to traverse from 10% to 90% of the peak value of the pulse. The shape of this pulse is defined by the standard capacitor charge equation examined in earlier course work and is valid for any system with a single clearly dominant lag network.

$$V_{out} = V_{peak}\left(1 - \epsilon^{\frac{-t}{RC}}\right) \qquad (1.12)$$

In order to find the time interval from the initial starting point to the 10% point, set V_{out} to $.1V_{peak}$ in Equation 1.12 and solve for t_1.

$$.1V_{peak} = V_{peak}\left(1 - \epsilon^{\frac{-t_1}{RC}}\right)$$

$$.1V_{peak} = V_{peak} - V_{peak}\epsilon^{\frac{-t_1}{RC}}$$

FIGURE 1.14

Pulse-risetime effect

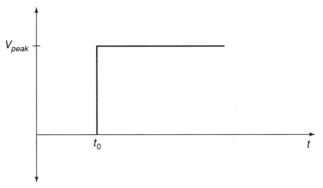

a. Input to lag network

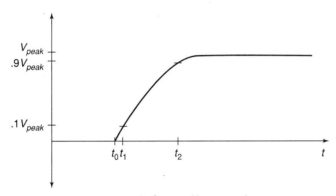

b. Output of lag network

$$.9V_{peak} = V_{peak}\epsilon^{\frac{-t_1}{RC}}$$
$$.9 = \epsilon^{\frac{-t_1}{RC}}$$
$$\log .9 = \frac{-t_1}{RC}$$
$$t_1 = .105\,RC \tag{1.13}$$

To find the interval up to the 90% point, follow the same technique using $.9V_{peak}$. Doing so yields

$$t_2 = 2.303\,RC \tag{1.14}$$

The risetime, T_r, is the difference between t_1 and t_2

$$T_r = t_2 - t_1$$
$$T_r = 2.303\,RC - .105\,RC$$
$$T_r \approx 2.2\,RC \tag{1.15}$$

Equation 1.15 ties the risetime to the lag network's R and C values. These same values also set the critical frequency f_2. By combining Equation 1.15 with the basic critical frequency relationship, we can derive an equation relating f_2 to T_r.

$$f_2 = \frac{1}{2\pi RC}$$

Solving Equation 1.15 in terms of RC and substituting yields

$$f_2 = \frac{2.2}{2\pi T_r}$$

$$f_2 = \frac{.35}{T_r} \tag{1.16}$$

Where

f_2 is the upper critical frequency, and
T_r is the risetime of the output pulse.

EXAMPLE 1-16

Determine the risetime for a lag network critical at 100 kHz.

$$f_2 = \frac{.35}{T_r}$$

$$T_r = \frac{.35}{f_2}$$

$$T_r = \frac{.35}{100 \text{ kHz}}$$

$$T_r = 3.5 \text{ } \mu S$$

1.4 COMBINING THE ELEMENTS—MULTISTAGE EFFECTS

A complete gain or phase plot combines three elements: (1) the midband response, (2) the lead response, and (3) the lag response. Normally, a particular design will contain multiple lead and lag networks. The complete response is the summation of the individual responses. For this reason, it is useful to find the dominant lead and lag networks. These are the networks that affect the midband response first. For lead networks, the dominant one will be the one with the highest f_c. Conversely, the dominant lag network will be the one with the lowest f_c. It is very common to approximate the complete system response by drawing straight-line segments such as those given in Figures 1.5 and 1.10. The process goes something like this:

1. Locate all f_cs on the frequency axis.
2. Draw a straight line between the dominant lag and lead f_cs at the midband gain. If the system does not contain any lead networks, continue the midband gain line down to DC.

3. Draw a 6 dB per octave slope between the dominant lead and the next lower lead network.
4. Because the effects of the networks are cumulative, draw a 12 dB per octave slope between the second lead f_c and the third f_c. After the third f_c, the slope should be 18 dB per octave, after the fourth it is 24 dB per octave, and so on.
5. Draw a −6 dB per octave slope between the dominant lag f_c and the next highest f_c. Again, the effects are cumulative, so increase the slope by −6 dB at every new f_c.

EXAMPLE 1-17

Draw the Bode gain plot for the following amplifier: A'_v midband = 26 dB, one lead network critical at 200 Hz, one lag network critical at 10 kHz, and another lag network critical at 30 kHz.

The dominant lag network is 10 kHz. There is only one lead network, so it's dominant by default.

1. Draw a straight line between 200 Hz and 10 kHz at an amplitude of 26 dB.
2. Draw a 6 dB per octave slope below 200 Hz. To do this, drop down one octave (100 Hz) and subtract 6 dB from the present gain (26 dB − 6 dB = 20 dB.) The line will start at the point 200 Hz/26 dB and pass through the point 100 Hz/20 dB. Because there are no other lead networks, this line may be extended to the left edge of the graph.
3. Draw a −6 dB per octave slope between 10 kHz and 30 kHz. The construction point will be 20 kHz/20 dB. Continue this line to 30 kHz. The gain at the 30 kHz intersection should be around 16 dB. The slope above this second f_c will be −12 dB per octave. Therefore, the second construction point should be at 60 kHz/4 dB (one octave above 30 kHz and 12 dB down from the 30 kHz gain). As this is the final lag network, this line may be extended to the right edge of the graph.

A completed graph is shown in Figure 1.15. There is one item that should be noted before we leave this section, and that is the concept of narrowing. Narrowing occurs when

FIGURE 1.15

Gain plot for complete amplifier

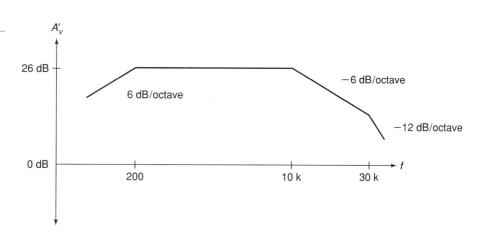

two or more networks share similar critical frequencies, and one of them is a dominant network. The result is that the true −3 dB breakpoints may be altered. Here is an extreme example. Assume that a circuit has two lag networks, both critical at 1 MHz. A Bode plot would indicate that the breakpoint is 1 MHz. This is not really true. Remember, the effects of lead and lag networks are cumulative. Because each network produces a 3 dB loss at 1 MHz, the net loss at this frequency is actually 6 dB. The true −3 dB point will have been shifted. The Bode plot only gives you the approximate shape of the response.

1.5 Circuit Simulations Using Computers

With the advent of low-cost personal computers, there are many alternatives to hand sketching of plots. One method involves the use of commercial or public domain software packages designed for circuit analysis. One common package is the public domain program SPICE. SPICE is an acronym that stands for Simulation Program with Integrated Circuit Emphasis and was originally written in the mid-1970s by Dr. Laurence Nagel of the University of California. This program is available for many different computing platforms for minimal cost. SPICE also serves as the core for a number of commercial packages. The commercial versions generally add features such as schematic capture (the ability to "draw" circuits using the computer's mouse), graphical input and output of data, interactive analysis, analog-digital mixed signal analysis, and large device libraries. Examples of popular simulation packages include Orcad PSpice and Electronics Workbench MultiSIM. The choice of a specific simulation tool depends on the needs of the user, available facilities, costs, and so forth. Of course, everyone has their own working style, so personal preference also plays a role. Generally, any quality SPICE-based simulator will be sufficient for the circuits presented in this book.

In order to use a simulator, a circuit is "described" with a special data file. For nongraphical simulators, the data file can be created using an ordinary text editor. The data file is then used by the simulation program to estimate the circuit response. Simulation results can include items such as Bode gain and phase plots.

With graphical input in packages such as MultiSIM and PSpice, components are usually dragged onto the work area and interconnected through the use of a mouse or other pointing device. Consequently, text-based input files are not needed, although many programs can import them and create the circuit from there. Similarly, simulation results are generally shown in graphical form using a plotting window or virtual oscilloscope instead of using a text-based output file. Graphical input and output using Electronics Workbench MultiSIM is shown in Figure 1.16 (see pp. 25–26). Here a simple lag network is drawn on the worksheet. A Bode plot is then generated by selecting the AC Analysis option of Simulate.

Note that accurate models for specific op amps or other devices that are not included in the simulator's libraries may be obtained from their manufacturer. This is often as simple as downloading them from the manufacturer's Web site.

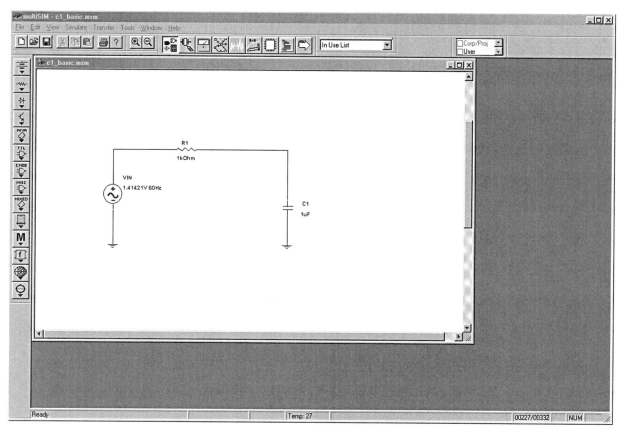

FIGURE 1.16a

MultiSIM schematic for a simple lag network

Simulators are by no means small or trivial programs. They have many features and options, not to mention the variations produced by the many commercial versions. This text does not attempt to teach all of the intricacies of SPICE-based simulators. For that, you should consult your simulator user's manual or one of the books available on the subject. The examples in this book assume that you already have some familiarity with computer circuit simulators.

We will be using simulations in the following chapters for various purposes. One thing to bear in mind is that simulation tools should not be used in place of a normal "human" analysis. Doing so can cause no end of grief. Simulations are only as good as the models used with them. It is easy to see that if the description of the circuit or the components within it are not accurate, the simulation will not be accurate. Simulation tools are best used as a form of double-checking a design, not as a substitute for proper analysis.

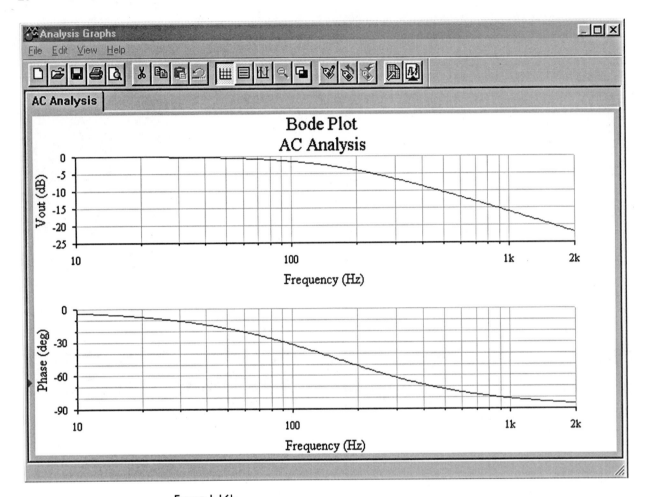

FIGURE 1.16b

MultiSIM gain and phase output graphs

1.6 THE DIFFERENTIAL AMPLIFIER

Most modern operational amplifiers utilize a differential amplifier front end. In other words, the first stage of the operational amplifier is a differential amplifier. This circuit is commonly referred to as a diff amp or as a long-tailed pair. A diff amp utilizes a minimum of two active devices, although four or more may be used in more complex designs. Our purpose here is to examine the basics of the diff amp so that we can understand how it relates to the larger operational amplifier. Therefore, we shall not be investigating the more esoteric designs. To approach this in an orderly fashion, we will examine the DC analysis first, and then follow with the AC small-signal analysis.

FIGURE 1.17

Simplified diff amp

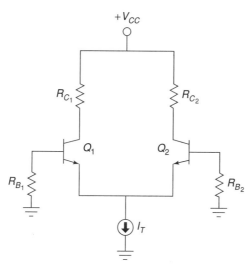

DC Analysis

A simplified diff amp is shown in Figure 1.17. This circuit utilizes a pair of NPN bipolar transistors, although the circuit could just as easily be built with PNPs or FETs. Note the inherent symmetry of the circuit. If you were to slice the circuit in half vertically, all of the components on the left half would have a corresponding component on the right half. Indeed, for optimal performance, we will see that these component pairs should have identical values. For critical applications, a matched pair of transistors would be used. In this case, the transistor parameters, such as β, would be very closely matched for the two devices.

In Figure 1.18, the circuit currents are noted, and the generalized current source has been replaced with a resistor/negative power supply combination. This is in essence

FIGURE 1.18

Diff amp analysis of Figure 1.17

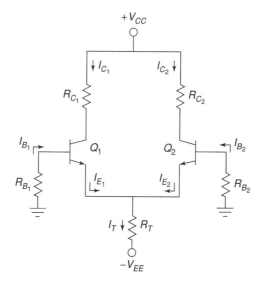

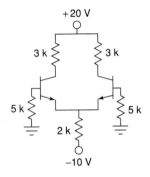

Figure 1.19

Diff amp for Example 1-16

an emitter bias technique. Assuming that the base voltages are negligible and that V_{BE} is equal to .7 V, we can see that the emitter of each device is at approximately −.7 V. Kirchhoff's Voltage Law indicates that the bulk of the negative supply potential must drop across R_T.

$$V_{R_T} = |V_{EE}| - .7 \text{ V}$$

Knowing this, we may find the current through R_T, which is known as the tail current, I_T.

$$I_T = \frac{|V_{EE}| - .7 \text{ V}}{R_T}$$

If the two halves of the circuit are well matched, the tail current will split equally into two portions, I_{E1} and I_{E2}. Given identical emitter currents, it follows that the remaining currents and voltages in the two halves must be identical as well. These potentials and currents are found through the application of Kirchhoff's Voltage and Current Laws, just as in any other transistor bias analysis.

Example 1-18

Find the tail current, the two emitter currents, and the two collector to ground voltages in the circuit of Figure 1.19. You may assume that the two transistors are very closely matched.

The first step is to find the tail current:

$$I_T = \frac{|V_{EE}| - .7 \text{ V}}{R_T}$$

$$I_T = \frac{10 \text{ V} - .7 \text{ V}}{2 \text{ k}\Omega}$$

$$I_T = \frac{9.3 \text{ V}}{2 \text{ k}\Omega}$$

$$I_T = 4.65 \text{ mA}$$

The tail current is the combination of the two equal emitter currents, so

$$I_{E1} = I_{E2} = \frac{I_T}{2}$$

$$I_{E1} = I_{E2} = \frac{4.65 \text{ mA}}{2}$$

$$I_{E1} = I_{E2} = 2.325 \text{ mA}$$

If we make the approximation that collector and emitter currents are equal, we may find the collector voltage by calculating the voltage drop across the collector resistor and subtracting the result from the positive power supply.

$$V_C = V_{CC} - I_C R_C$$

$$V_C = 20 \text{ V} - 2.325 \text{ mA} \times 3 \text{ k}\Omega$$

$$V_C = 20\text{ V} - 6.975\text{ V}$$
$$V_C = 13.025\text{ V}$$

Again, because we have identical values for both halves of the circuit, $V_{C1} = V_{C2}$. If we continue with this and assume a typical β of 100, we find that the two base currents are identical as well.

$$I_B = \frac{I_C}{\beta}$$
$$I_B = \frac{2.325\text{ mA}}{100}$$
$$I_B = 23.25\ \mu\text{A}$$

Noting that the base currents flow through the 5 kΩ base resistors, we may find the base voltages. Note that this is a negative potential because the base current is flowing from ground into the transistor's base.

$$V_B = -I_B R_B$$
$$V_B = -23.25\ \mu\text{A} \times 5\text{ k}\Omega$$
$$V_B = -116.25\text{ mV}$$

(This result indicates that the actual emitter voltage is closer to $-.8$ V than $-.7$ V, and thus the tail current is actually a little less than our approximation of 4.65 mA. This error is probably within the error we can expect by using the .7 V junction potential approximation.)

Input Offset Current and Voltage

As you have no doubt guessed, it is impossible to make both halves of the circuit identical, and thus, the currents and voltages will never be exactly the same. Even a small resistor tolerance variation will cause an upset. If the base resistors are mismatched, this will cause a direct change in the two base potentials. A variation in collector resistance will cause a mismatch in the collector potentials. A simple β or V_{BE} mismatch can cause variations in the base currents and base voltages, as well as smaller changes in emitter currents and collector potentials. It is desirable, then, to quantify the circuit's performance so that we can see just how well balanced it is. We can judge a diff amp's DC performance by measuring its input offset current and its input and output offset voltages. In simple terms, the difference between the two base currents is the input offset current. The difference between the two collector voltages is the output offset voltage. The DC potential required at one of the bases to counteract the output offset voltage is called the input offset voltage (this is little more than the output offset voltage divided by the DC gain of the amplifier). In an ideal diff amp all three of these factors are equal to 0. We will take a much closer look at these parameters and how they relate to operational amplifiers in later chapters. For now, it is only important to understand that these inaccuracies exist and what can cause them.

FIGURE 1.20

A typical diff amp with input and output connections

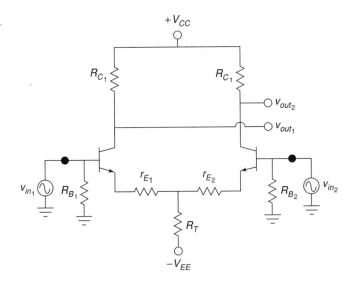

AC Analysis

Figure 1.20 shows a typical circuit with input and output connections. In order to minimize confusion with the DC circuit, AC equivalent values will be shown in lowercase. Small emitter degeneration resistors, r_{E1} and r_{E2}, have been added to this diff amp. This circuit has two signal inputs and two signal outputs. It is possible to configure a diff amp so that only a single input and/or output is used. This means that there are four variations on the theme:

1. Differential (also called dual- or double-ended) input, differential output
2. Differential input, single-ended output
3. Single-ended input, differential output
4. Single-ended input, single-ended output

These variations are shown in Figure 1.21. For use in operational amplifiers, the differential input/single-ended output variation is the most common. We will examine the most general case, the differential input/differential output version.

Because the diff amp is a linear circuit, we can use the principle of Superposition to independently determine the output contribution from each of the inputs. Utilizing the circuit of Figure 1.20, we will first determine the gain equation from V_{in1} to either output. To do this, we replace V_{in2} with a short circuit. The AC equivalent circuit is shown in Figure 1.22. For the output on collector 1, transistor 1 forms the basis of a common emitter amplifier. The voltage across r_{C1} is found via Ohm's Law.

$$v_{r_{C1}} = -i_{C1} r_{C1}$$

The negative sign comes from the fact that AC ground is used as our reference (i.e., for a positive input, current flows from AC ground down through r_{C1}, and into the

The Differential Amplifier

Figure 1.21

The four different diff amp input/output configurations

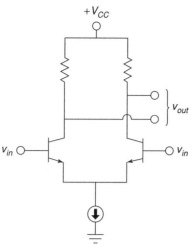

a. Differential input and output

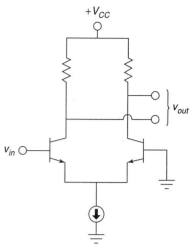

b. Single-ended input and differential output

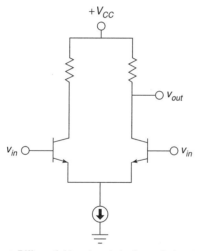

c. Differential input and single-ended output

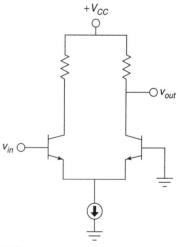

d. Single-ended input and output

Figure 1.22

The circuit of Figure 1.20 redrawn for AC analysis

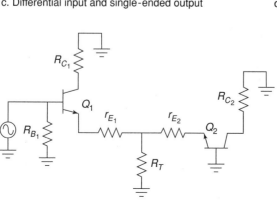

FIGURE 1.23

AC analysis

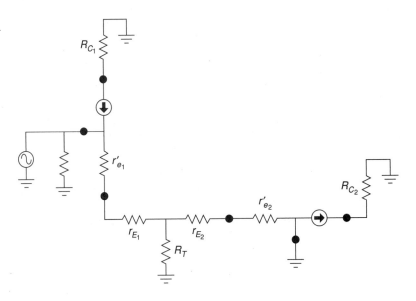

collector). To a reasonable approximation, we can say that the collector and emitter currents are identical.

$$v_{r_{C1}} = -i_{E1} r_{C1}$$

We must now determine the AC emitter current in relation to V_{in1}. In order to better visualize the process, the circuit of Figure 1.22 is altered to include simplified transistor models, as shown in Figure 1.23. r'_e is the dynamic resistance of the base emitter junctions and is inversely proportional to the DC emitter current. You may recall the following equation from your prior course work:

$$r'_e = \frac{26 \text{ mV}}{I_E}$$

Where

r'_e is the dynamic base-emitter junction resistance and
I_E is the DC emitter current.

For typical circuits, the values of r'_e and r_E are much smaller than the tail current biasing resistor, R_T. Because of its large size, we can ignore the parallel effect of R_T. By definition, the AC emitter current must equal the AC emitter potential divided by the AC resistance in the emitter section. If you trace the signal flow from the base of transistor 1 to ground, you find that it passes through r'_{e1}, r_{E1}, r'_{e2}, and r_{E2}. You will also notice that the magnitude of i_{E1} is the same as i_{E2}, although they are out of phase.

$$i_E = \frac{V_{in1}}{r'_{e1} + r_{E1} + r'_{e2} + r_{E2}}$$

Because the circuit values should be symmetrical for best performance, this equation may be simplified to

$$i_E = \frac{v_{in1}}{2(r'_e + r_E)}$$

If we now solve for voltage gain,

$$A_v = \frac{v_{out}}{v_{in}}$$

$$A_v = \frac{-i_E r_C}{v_{in}}$$

$$A_v = -\frac{\frac{v_{in}}{2(r'_e + r_E)} r_C}{v_{in}}$$

$$A_v = -\frac{r_C}{2(r'_e + r_E)}$$

Where

A_v is the voltage gain,
r_C is the AC equivalent collector resistance,
r_E is the AC equivalent emitter resistance, and
r'_e is the dynamic base-emitter junction resistance.

The final negative sign indicates that the collector voltage at transistor number 1 is 180 degrees out of phase with the input signal. Earlier, we noted that i_{E2} is the same magnitude as i_{E1}, the only difference being that it is out of phase. Because of this, the magnitude of the collector voltage at transistor number 2 will be the same as that on the first transistor. Because the second current is out of phase with the first, it follows that the second collector voltage must be out of phase with the first. This means that the voltage at the second collector is in phase with the first input signal. Its gain equation is

$$A_v = \frac{r_C}{2(r'_e + r_E)}$$

The various waveforms are depicted in Figure 1.24. The equation above is often referred to as the single-ended input/single-ended output gain equation because it describes the single change from one input to one output. The output signal will be in phase if we are examining the opposite transistor and out of phase if we are looking at the input transistor. Because the circuit is symmetrical, we will get similar results when we examine the second input. The voltage between the two collectors is 180 degrees apart. If we were to use a differential output, that is, derive the output from collector to collector rather than from one collector to ground, we would see an effective doubling of the output signal. If the reason for this is not clear to you, consider the following. Assume that each collector has a 1 V peak sine wave riding on it. When collector 1 is at +1 V, collector 2 is at −1 V, making +2 V total. Likewise, when collector 1 is

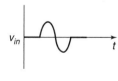
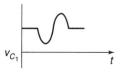
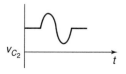

FIGURE 1.24

Waveforms for a single input

at its negative peak, collector 2 is at its positive peak, producing a total of -2 V. The single-ended input/differential output gain therefore is

$$A_v = \frac{r_C}{r'_e + r_E}$$

EXAMPLE 1-19

Using the circuit of Figure 1.20, determine the single-ended input/differential output and single-ended input/single-ended output voltage gains. Use the following component values: $V_{CC} = 15$ V, $V_{EE} = -8$ V, $R_T = 10$ kΩ, $R_C = 8$ kΩ, $r_E = 30$ Ω.

In order to find r'_e, we must find the DC current.

$$I_T = \frac{|V_{EE}| - .7 \text{ V}}{R_T}$$

$$I_T = \frac{7.3 \text{ V}}{10 \text{ k}\Omega}$$

$$I_T = .73 \text{ mA}$$

$$I_E = \frac{I_T}{2}$$

$$I_E = \frac{.73 \text{ mA}}{2}$$

$$I_E = .365 \text{ mA}$$

$$r'_e = \frac{26 \text{ mV}}{I_E}$$

$$r'_e = \frac{26 \text{ mV}}{.365 \text{ mA}}$$

$$r'_e = 71.2$$

For the single-ended output gain,

$$A_v = \frac{r_C}{2(r'_e + r_E)}$$

$$A_v = \frac{8 \text{ k}\Omega}{2(71.2 \text{ } \Omega + 30 \text{ } \Omega)}$$

$$A_v = \frac{8 \text{ k}\Omega}{202.4 \text{ } \Omega}$$

$$A_v = 39.5$$

The differential output gain is twice this value, or 79.

As it is possible to drive a diff amp with two distinct inputs, a wide variety of outputs may be obtained. It is useful to investigate two specific cases:

1. Two identical inputs in both phase and magnitude
2. Two inputs with identical magnitude, but 180 degrees out of phase

Let's consider the collector potentials for the first case. Assume that a diff amp has a single-ended input/single-ended output gain of 100, and a 10 mV signal is applied to both bases. Using Superposition, we find that the outputs due to each input are 100 × 10 mV, or 1 V in magnitude. For the first input, the voltages are sketched in Figure 1.25a. For the second input, the voltages are sketched in Figure 1.25b. Note that each collector sees both a sine wave and an inverted sine wave, both of equal amplitude. When these two signals are added, the result is zero, as seen in Figure 1.25c. In equation form,

$$v_{C1} = v_{in1}(-A_v) + v_{in2} A_v$$
$$v_{C1} = A_v(v_{in2} - v_{in1})$$

FIGURE 1.25

Input-output waveforms for common mode

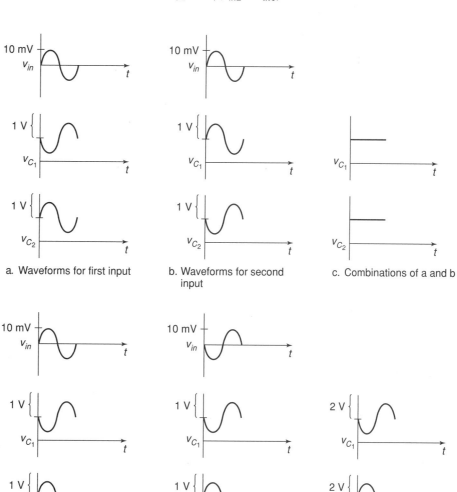

Because v_{in1} and v_{in2} are identical, the output is ideally zero given a perfectly matched and biased diff amp. The exact same effect is seen on the opposite collector. This last equation is very important. It says that the output voltage is equal to the gain times the difference between the two inputs. This is how the differential amplifier got its name. In this case, the two inputs are identical, and thus their difference is zero. On the other hand, if we were to invert one of the input signals (case 2), we find a completely different result.

$$v_{in1} = -v_{in2}$$
$$v_{C1} = A_v(v_{in2} - v_{in1})$$
$$v_{C1} = A_v(v_{in2} - (-v_{in2}))$$
$$v_{C1} = 2A_v v_{in2}$$

Thus, if one input is inverted, the net result is a doubling of gain. This effect is shown graphically in Figures 1.25d through 1.25f. In short, a differential amplifier suppresses in phase signals while simultaneously boosting out of phase signals. This can be a very useful attribute, particularly in the area of noise reduction.

Common Mode Rejection

By convention, in phase signals are known as common-mode signals. An ideal differential amplifier will perfectly suppress these common-mode signals, and thus, its common-mode gain is said to be zero. In the real world, a diff amp will never exhibit perfect common-mode rejection. The common-mode gain may be made very small, but it is never zero. For a common-mode gain of zero, the two halves of the circuit have to be perfectly matched, and all circuit elements must be ideal. This is impossible to achieve, as errors may arise from several sources. The most obvious error sources are resistor tolerance variations and transistor parameter spreads. The basic design of the circuit will also affect the common-mode gain. With some circuit rearrangements, it is possible to determine a common-mode gain for the circuits we have been using. The circuit of Figure 1.20 has been redrawn in Figure 1.26 in order to emphasize its parallel symmetry. Because the DC potentials are identical in both halves, and identical signals drive both inputs, we can combine resistors in parallel in order to arrive at the circuit of Figure 1.27. Although it is not shown explicitly on the diagram, the internal dynamic resistances (r'_e) may also be combined ($\frac{r'_e}{2}$). This circuit has been effectively reduced to a simple common emitter stage. Based on our earlier work, the gain for this circuit is

$$A_{v_{cm}} = \frac{\frac{r_C}{2}}{R_T + \frac{r'_e}{2} + \frac{r_E}{2}}$$

This is the common-mode voltage gain. If R_T is considerably larger than r_C, then this circuit will exhibit good common mode rejection (assuming that the other parts are matched, naturally). R_T is the effective resistance of the tail current source. A very high

FIGURE 1.26

The circuit of Figure 1.20 redrawn for common mode rejection ratio (CMRR) analysis

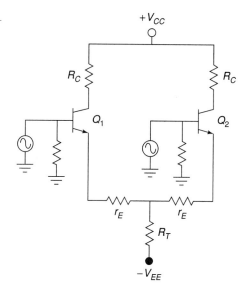

internal resistance (i.e., an ideal current source) is desirable. There are many ways of creating a more ideal current source. One way is to use a third bipolar transistor, as shown in Figure 1.28. The tail current is found by determining the potential across R_2 and subtracting the .7 V V_{BE} drop. The remaining potential appears across R_3. Given the voltage and resistance, Ohm's Law will let you find the tail current. In this circuit, R_2 is sometimes replaced with a Zener diode. This can help to reduce temperature

FIGURE 1.27

Common mode gain analysis

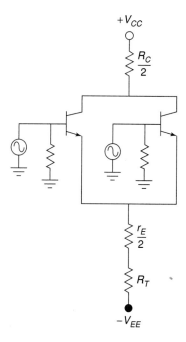

FIGURE 1.28

Improved current source

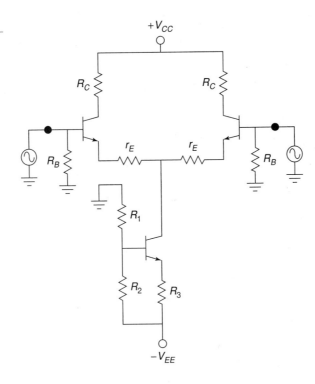

induced current fluctuations. In any case, the effective resistance of this current source is considerably larger than the simple tail resistor variation. It is largely dependent on the characteristics of the tail current transistor and can easily be in the megohm region.

Current Mirror

A very popular biasing technique in integrated circuits involves the current mirror. Current mirrors are also employed as active loads in order to optimize a circuit's gain. A simple current mirror is shown in Figure 1.29. This circuit requires that the transconductance curves of the diode and the transistor be very closely matched. One way to guarantee this is to use two transistors and form one of them into a diode by shorting its collector to its base. If we use an approximate forward bias potential of .7 V and ignore the small base current, the current through the diode is

$$I_D = \frac{V_{CC} - .7\text{ V}}{R}$$

In reality, the diode potential will probably not be exactly .7 V. This will have little effect on I_D, however. The diode is in parallel with the transistor's base-emitter junction, so we know that $V_d = V_{BE}$. If the two devices have identical transconductance curves, the transistor's emitter current will equal the diode current. You can think of the

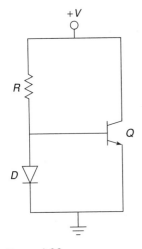

FIGURE 1.29

Current mirror

FIGURE 1.30

Transfer curve mismatch

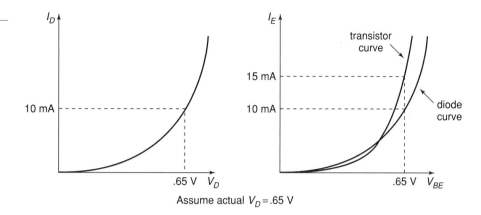

Assume actual $V_D = .65$ V

transistor as mirroring the diode's current, hence the circuit's name. If the two device curves are slightly askew, then the two currents will not be identical. This is shown graphically in Figure 1.30.

A current mirror could be used in the circuit of Figure 1.28. The result is shown in Figure 1.31. If the positive power supply is 15 V, the negative supply is -10 V, and R is 10 kΩ, the tail current will be

$$I_D = \frac{V_{CC} - V_{EE} - V_D}{R}$$

$$I_D = \frac{15 \text{ V} - (-10 \text{ V}) - .7 \text{ V}}{10 \text{ k}\Omega}$$

$$I_D = 2.43 \text{ mA}$$

FIGURE 1.31

Current mirror bias

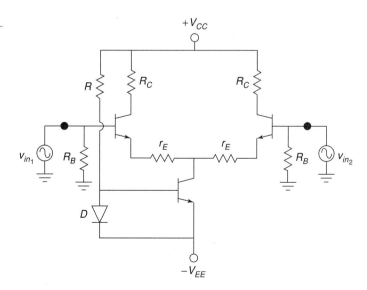

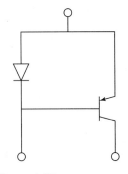

FIGURE 1.32

Active load current mirror

Because the tail current is the mirror current,

$$I_T = I_D$$
$$I_T = 2.43 \text{ mA}$$

Biasing of this type is very popular in operational amplifiers. Another use for current mirrors is in the application of active loads. Instead of using simple resistors for the collector loads, a current mirror may be used instead. A PNP-based current mirror suitable for use as an active load in our previous circuits is shown in Figure 1.32. To use this, we simply remove the two collector resistors from a circuit such as Figure 1.31 and drop in the current mirror. The result of this operation is shown in Figure 1.33. The current mirror active load produces a very high internal impedance, thus contributing to a very high differential gain. In effect, by using a constant current source in the collectors, all AC current is forced into the following stage. You may also note that the number of resistors used in the circuit has decreased considerably.

FIGURE 1.33

Current mirrors for bias and active load

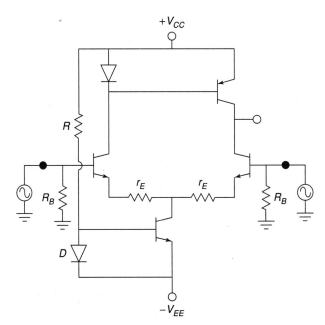

Summary

We have seen how to convert gains and signals into a decibel form for both powers and voltages. This is convenient because what would require multiplication and division under the ordinary scheme only requires simple addition and subtraction in the dB scheme. Along with this, dB measurement is used almost exclusively for Bode gain plots. A Bode plot details a system's gain magnitude and phase response. For gain, the amplitude is measured in dB, whereas the frequency is normally presented in log form. For a phase plot, phase is measured in degrees, and again, the frequency axis is logarithmic. The changes in gain and phase at the frequency extremes are caused by lead and lag networks. Lead networks cause the low frequency gain to roll off. The rolloff rate is 6 dB per octave per network. The phase will change from +90 degrees to 0 degrees per network. Lag networks cause the high-frequency gain to roll off at a rate of −6 dB per octave per network. The phase change per lag network is from 0 degrees to −90 degrees. It was noted that computers may be used to quickly tabulate the response of complex networks such as these. Many computer circuit simulators are based on the original SPICE program. Among the popular simulation programs are Orcad PSpice and Electronic Workbench MultiSIM. Packages such as these offer graphical schematic capture tools and large component libraries.

Differential amplifiers are symmetrical circuits, employing a minimum of two active devices. They may be configured with single or dual inputs and single or dual outputs. Diff amps are commonly used as the first stage of an operational amplifier. They tend to amplify the difference in the input signals while simultaneously suppressing in-phase, or common-mode, signals. Current mirrors are widely used for biasing purposes and as active loads. Active loads offer the advantage of producing higher gains than ordinary resistive loads.

REVIEW QUESTIONS

1. What are the advantages of using the decibel scheme over the ordinary scheme?
2. How do decibel power and voltage gain calculations differ?
3. What does the third letter in a decibel-based signal measurement indicate (as in dBV or dBm)?
4. What is a Bode plot?
5. What is a lead network? What general response does it yield?
6. What is a lag network? What general response does it yield?
7. What do the terms f_1 and f_2 indicate about a system's response?
8. What are the rolloff slopes for lead and lag networks?
9. What are the phase changes produced by individual lead and lag networks?
10. How is risetime related to upper-break frequency?
11. How do multiple lead or lag networks interact to form an overall system response?
12. What is SPICE?
13. What is common-mode rejection?
14. What is a current mirror?
15. What is the advantage of using an active load?

PROBLEMS

Analysis Problems—dB Emphasis

1. Convert the following power gains into dB form: (a) 10, (b) 80, (c) 500, (d) 1, (e) .2, (f) .03.
2. Convert the following dB power gains into ordinary form: (a) 0 dB, (b) 12 dB, (c) 33.1 dB, (d) .2 dB, (e) −5.4 dB, (f) −20 dB.
3. An amplifier has an input signal of 1 mW and produces a 2 W output. What is the power gain in dB?
4. A hi-fi power amplifier has a maximum output of 50 W and a power gain of 19 dB. What is the maximum input signal power?
5. An amplifier with a power gain of 27 dB is driven by a 25 mW source. Assuming the amplifier doesn't clip, what is the output signal in Watts?
6. Convert the following voltage gains into dB form: (a) 10, (b) 40, (c) 250, (d) 1, (e) .5, (f) .004.
7. Convert the following dB voltage gains into ordinary form: (a) .5 dB, (b) 0 dB, (c) 46 dB, (d) 10.7 dB, (e) −8 dB, (f) −14.5 dB.
8. A guitar preamp has a gain of 44 dB. If the input signal is 12 mV, what is the output signal?
9. A video amplifier has a 140 mV input and a 1.2 V output. What is the voltage gain in dB?
10. The preamp in a particular tape deck can output a maximum signal of 4 V. If this amplifier has a gain of 18 dB, what is the maximum input signal?
11. Convert the following powers into dBW: (a) 1 W, (b) 23 W, (c) 6.5 W, (d) .2 W, (e) 2.3 mW, (f) 1.2 kW, (g) .045 mW, (h) .3 μW, (i) 5.6 × 10^{-18} W.
12. Repeat Problem 11 for units of dBm.
13. Repeat Problem 11 for units of dBf.

14. Convert the following voltages into dBV: (a) 12.4 V, (b) 1 V, (c) .25 V, (d) 1.414 V, (e) .1 V, (f) 10.6 kV, (g) 13 mV, (h) 2.78 μV.
15. A two-stage power amplifier has power gains of 12 dB and 16 dB. What is the total gain in dB and in ordinary form?
16. If the amplifier of Problem 15 has an input of −18 dBW, what is the final output in dBW? in dBm? in Watts?
17. Referring to Figure 1.1, what are the various stages' outputs if the input is changed to −4 dBm? to −34 dBW?
18. Which amplifier has the greatest power output? (a) 50 Watts, (b) 18 dBW, (c) 50 dBm.
19. Which amplifier has the greatest power output? (a) 200 mW, (b) −10 dBW, (c) 22 dBm.
20. A three-stage amplifier has voltage gains of 20 dB, 5 dB, and 12 dB respectively. What is the total voltage gain in dB and in ordinary form?
21. If the circuit of Problem 20 has input of voltage of −16 dBV, what are the outputs of the various stages in dBV? In volts?
22. Repeat Problem 21 for an input of 12 mV.
23. Which amplifier produces the largest output voltage? (a) 15 V, (b) 16 dBV.

Analysis Problems—Bode Plot Emphasis

24. Given a lead network critical at 3 kHz, what are the gain and phase values at 100 Hz, 3 kHz, and 40 kHz?
25. Given a lag network tuned to 700 kHz, what are the gain and phase values at 50 kHz, 700 kHz, and 10 MHz? What is the risetime?
26. A noninverting amplifier has a midband voltage gain of 18 dB and a single-lag network at 200 kHz. What are the gain and phase values at 30 kHz, 200 kHz, and 1 MHz? What is the risetime?
27. Repeat Problem 26 for an inverting (−180 degrees) amplifier.
28. Draw the Bode plot for the circuit of Problem 26.
29. Draw the Bode plot for the circuit of Problem 27.
30. An inverting (−180 degrees) amplifier has a midband gain of 32 dB and a single-lead network critical at 20 Hz (assume the lag network f_c is high enough to ignore for low-frequency calculations). What are the gain and phase values at 4 Hz, 20 Hz, and 100 Hz?
31. Repeat Problem 29 with a noninverting amplifier.
32. Draw the Bode plot for the circuit of Problem 30.
33. Draw the Bode plot for the circuit of Problem 31.
34. A noninverting amplifier used for ultrasonic applications has a midband gain of 41 dB, a lag network critical at 250 kHz, and a lead network critical at 30 kHz. Draw its gain Bode plot.
35. Find the gain and phase at 20 kHz, 100 kHz, and 800 kHz for the circuit of Problem 34.
36. If the circuit of Problem 34 has a second lag network added at 300 kHz, what are the new gain and phase values at 20 kHz, 100 kHz, and 800 kHz?
37. Draw the gain Bode plot for the circuit of Problem 36.
38. What are the maximum and minimum phase shifts across the entire frequency spectrum for the circuit of Problem 36?
39. A noninverting DC amplifier has a midband gain of 36 dB and lag networks at 100 kHz, 750 kHz, and 1.2 MHz. Draw its gain Bode plot.
40. What are the maximum and minimum phase shifts across the entire frequency spectrum for the circuit of Problem 39?
41. What is the maximum rate of high-frequency attenuation for the circuit of Problem 39 in dB/Decade?
42. If an amplifier has two lead networks, what is the maximum rate of low frequency attenuation in dB/Octave?

Analysis Problems—Differential Amplifier Emphasis

43. Given the circuit of Figure 1.20, determine the single-ended input/single-ended output gain for the following values: $R_B = 5$ kΩ, $R_T = 7.5$ kΩ, $R_C = 12$ kΩ, $V_{CC} = 25$ V, $V_{EE} = -9$ V, $r_E = 50$ Ω.
44. Determine the differential voltage gain in the circuit of Figure 1.28 if $R_B = 15$ kΩ, $R_1 = 5$ kΩ, $R_2 = 7$ kΩ, $R_3 = 10$ kΩ, $R_c = 20$ kΩ, $V_{CC} = 22$ V, $V_{EE} = -12$ V, $r_E = 75$ Ω.
45. For the circuit of Problem 44, determine the output at collector 2 if $V_{in1}(t) = .001 \sin 2\pi 1000t$ and $V_{in2}(t) = -.001 \sin 2\pi 1000t$.
46. Determine the differential voltage gain in the circuit of Figure 1.31 if $R_b = 8$ kΩ, $R_{mirror} = 22$ kΩ, $R_C = 10$ kΩ, $V_{CC} = 18$ V, $V_{EE} = -15$ V, $r_E = 25$ Ω.
47. For the circuit of Problem 46, determine the output at collector 1 if $V_{in1}(t) = -.005 \sin 2\pi 2000t$ and $V_{in2}(t) = .005 \sin 2\pi 2000t$.
48. Determine the tail and emitter currents in the circuit of Figure 1.33 if $R_B = 6$ kΩ, $R_{mirror} = 50$ kΩ, $V_{CC} = 15$ V, $V_{EE} = -15$ V, $r_E = 0$ Ω.

Challenge Problems

49. You would like to use a voltmeter to take dBm readings in a 600 Ω system. What voltage should produce 0 dBm?
50. Assuming that it takes about an 8 dB increase in sound pressure level in order to produce a sound that is subjectively "twice as loud" to the human ear, can a hi-fi using a 100 W amplifier sound twice as loud as one with a 40 W amplifier (assuming the same loudspeakers)?
51. Hi-fi amplifiers are often rated with a "headroom factor" in dB. This indicates how much extra power the amplifier can produce for short periods of time, over and above its nominal rating. What is the maximum output power of a 250 W amplifier with 1.6 dB headroom?
52. If the amplifier of Problem 34 picks up an extraneous signal that is a −10 dBV sine wave at 15 kHz, what is the output?
53. If the amplifier of Problem 39 picks up a high-frequency interference signal at 30 MHz, how much is it attenuated over a normal signal? If this input signal is measured at 2 dBV, what should the output be?
54. If an amplifier has two lag networks and both are critical at 2 MHz, is the resulting f_2 less than, equal to, or greater than 2 MHz?
55. If an amplifier has two lead networks and both are critical at 30 Hz, is the resulting f_1 less than, equal to, or greater than 30 Hz?

Computer Simulation Problems

56. Use a simulator to plot the Bode gain response of the circuit in Problem 39.
57. Use a simulator to plot the Bode phase response of the circuit in Problem 34.
58. Use a simulation program to generate a Bode plot for a lead network comprised of a 1 kΩ resistor and a .1 μF capacitor.

CHAPTER 2

OPERATIONAL AMPLIFIER INTERNALS

CHAPTER OBJECTIVES

After completing this chapter, you should be able to:

- Describe the internal layout of a typical op amp.
- Describe a simple op amp computer simulation model.
- Determine fundamental parameters from an op amp data sheet.
- Describe an op amp-based comparator, and note where it might be used.
- Describe how integrated circuits are constructed.
- Define *monolithic planar* construction.
- Define *hybrid* construction.

2.1 INTRODUCTION

In this chapter we introduce the fundamentals of the operational amplifier, or op amp, as it is commonly known. We will investigate the construction and usage of a typical general-purpose op amp. Specific along with generalized internal circuits and associated block diagrams are examined. An initial op amp data sheet interpretation is given as well. Toward the middle of the chapter, the first op amp circuit examples are presented. The chapter finishes with an explanation of semiconductor integration and construction techniques. After finishing this chapter, you should be familiar with the concepts of what an op amp is composed of, how it is manufactured, and a beginning idea of how it might be used in application circuit design.

FIGURE 2.1

General op amp symbol

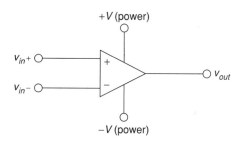

2.2 What Is an Op Amp?

An operational amplifier is, in essence, a multistage high-gain amplifier treated as a single entity. Normally, op amps have a differential input and a single-ended output. In other words, one input produces an inverted output signal, and the other input produces a noninverted output signal. Often, the op amp is driven from a bipolar power supply (i.e., two supplies, one positive and one negative). Just about any sort of active amplifying device may be used for the individual stages. Op amps can be made entirely from vacuum tubes or discrete bipolar transistors (and of course, they were made that way some years ago). The advances in semiconductor manufacture in the late 1960s and early 1970s eventually made it possible to miniaturize the required components and place the whole affair on a single silicon chip (hence the term *integrated circuit*). Through common use, this is what is generally meant by the term *op amp* today.

As seen in Figure 2.1, a typical op amp has at least five distinct connections: an inverting input (labeled "−"), a noninverting input (labeled "+"), an output, and positive and negative power supply inputs. These power supply connections are sometimes referred to as *supply rails*. Note that a ground connection is not directly given. Rather, a ground connection is implied through the other connections. This symbol and its associated connections are typical, but by no means absolute. There is a wide variety of devices available to the designer that offer such features as differential outputs or unipolar power supply operation. In any case, some form of triangle will be used for the schematic symbol.

It is best to think of op amps as general-purpose building blocks. With them, you can create a wide variety of useful circuits. For general-purpose work, designing with op amps is usually much quicker and more economical than an all-discrete approach. Likewise, troubleshooting and packaging constraints may be lessened. For more demanding applications, such as those requiring very low noise, high output current and/or voltage, or wide bandwidth, manufacturers have created specialized op amps. The final word in performance today is still dominated by discrete circuit designs, though. Also, it is very common to see a mix of discrete and integrated devices in a given circuit. There is certainly no law that op amps can only be used with other op amps. Often, a judicious mix of discrete devices and op amps can produce a circuit superior to one made entirely of discretes or op amps alone.

Where might you find op amp circuits? In a word, anywhere. They're probably in use in your home stereo or TV, where they help capture incoming signals; in electronic musical instruments, where they can be used to create and modify tones; in a camera in

FIGURE 2.2

General op amp schematic

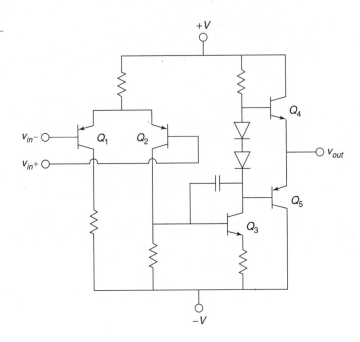

Block Diagram of an Op Amp

At this point you may be asking yourself, "What's inside of the op amp?" The generic op amp consists of three main functional stages. A real op amp may contain more than three distinct stages, but can be reduced to this level for analysis. A generalized discrete representation is given in Figure 2.2. Because the op amp requires a differential input scheme, the first stage is most often a differential amplifier. As seen here, Q_1 and Q_2 comprise a PNP-based differential amplifier. The output of one collector (Q_2 here) is then fed to a high-gain second stage. This stage usually includes a lag network capacitor that plays a major role in setting the op amp's AC characteristics (this is examined further in Chapter 5). Q_3 makes up the second stage in the example. It is set in a common emitter configuration for both current and voltage gain. The aforementioned lag capacitor is positioned across Q_3's base-collector junction in order to take advantage of the Miller effect. The third and final section is a class B or class AB follower for the most effective load drive. Q_4 and Q_5 make up the final stage. The twin diodes compensate for the Q_4 and Q_5 V_{BE} drops and produce a trickle bias current that minimizes distortion. This is a relatively standard class AB stage. Note that the entire circuit is direct-coupled. There are no lead networks, and thus the op amp can amplify down to zero Hertz (DC). There are many possible changes that may be seen in a real-world circuit, including the use of Darlington pairs or FETs for the differential amplifier, multiple high-gain stages, and output current limiting for the class B section.

FIGURE 2.3

LF411 simplified schematic

(Reprinted with permission of National Semiconductor Corporation)

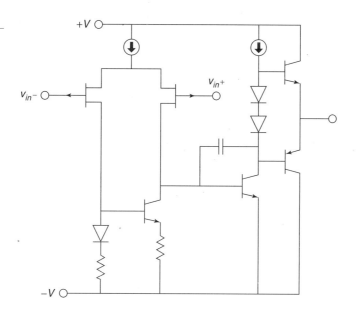

 The example discrete circuit uses only five transistors and two diodes. In contrast, an integrated version may use two to three dozen active devices. Because of the excellent device matching abilities of single-chip integration, certain techniques are used in favor of standard discrete designs. Internal IC current sources are normally made through the use of current mirrors. Current mirror configurations are also employed to create active loads, in order to achieve maximum circuit gain. A typical integrated op amp will contain very few resistors and usually only one or two lag network capacitors. Due to size limitations and other factors, inductors are virtually unseen in these circuits. A simplified equivalent circuit of the LF411 op amp is shown in Figure 2.3. Note that this device uses JFETs for the diff amp with an active load. The diff amp tail current source and the class AB trickle bias source are shown as simple current sources. In reality, they are a bit more complex, utilizing current mirror arrangements.

 One of the most popular op amps over the years has been the 741. The specifications of this device seem rather lackluster by today's standards, but it was one of the first easy-to-use devices produced. As a result, it has found its way into a large number of designs. Indeed, it is still a wise choice for less-demanding applications, or where parts costs are a major consideration. A complete schematic of the μA 741 is shown in Figure 2.4. Several different manufacturers make the 741. This version is manufactured by Signetics and may be somewhat different from a 741 made by another company.[1] The circuit contains 20 active devices and about one dozen resistors.

 At first glance, this circuit may look hopelessly confusing. A closer look reveals many familiar circuit blocks. First, you will notice that a number of devices show a

---

[1] Although the exact internal circuitry may be altered, the various manufacturers' versions will have the same pinouts and very similar performance specifications.

Figure 2.4

Schematic of the µA741
Courtesy of Philips Semiconductors

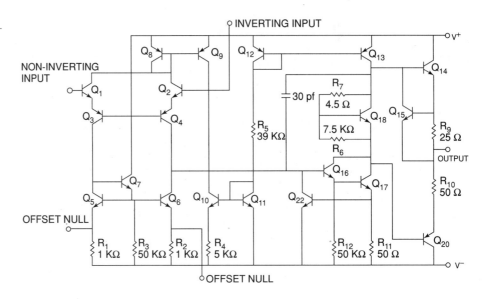

shorting connection between their base and collector terminals, such as Q_8, Q_{11}, and Q_{12}. In essence, these are diodes. (They are drawn this way because they are manufactured as transistor junctions. It is actually easier to make diodes in this fashion.) For the most part, these diodes are part of current mirror biasing networks. The bias setup is found in the very center of the schematic and revolves around Q_9 through Q_{12}. The setup current is found by subtracting two diode drops (Q_{11}, Q_{12}) from the total power supply potential ($V_+ - V_-$) and dividing the result by R_5. For a standard ±15 V power supply, this works out to

$$I_{bias} = \frac{V_+ - V_- - V_{BE-Q11} - V_{BE-Q12}}{R_5}$$

$$I_{bias} = \frac{30V - 1.4\ V}{39\ k\Omega}$$

$$I_{bias} = .733\ mA$$

This current is reflected into Q_{13}. A close look at Q_{10}, Q_{11} reveals that this portion is not a simple current mirror. By including R_4, the voltage drop across the base-emitter of Q_{10} is decreased, thus producing a current less than .733 mA. This configuration is known as a *Widlar current source*. The derivation of the exact current equation is rather involved and beyond the scope of this chapter.[2] This current is reflected into Q_8 via Q_9, and establishes the tail current for the differential amplifier. The diff amp stage uses a total of four amplifying transistors in a common-collector/common-base

---

[2] A complete derivation of the Widlar current source may be found in S. G. Burns and P. R. Bond, *Principles of Electronic Circuits* (Eagan, Minn.: West Publishing Company, 1987).

configuration (Q_1 through Q_4). In essence, Q_1 and Q_2 are configured as emitter followers, thus producing high input impedance and reasonable current gain. Q_3 and Q_4 are configured as common base amplifiers, and as such, produce a large voltage gain. The gain is maximized by the active load comprised of Q_5 through Q_7. The output signal at the collector of Q_4 passes on to a dual transistor high-gain stage (Q_{16} and Q_{17}). Q_{16} is configured as an emitter follower and buffers Q_{17}, which is set as a common emitter voltage amplifier. Resistor R_{11} serves to stabilize both the bias and gain of this stage (i.e., it is an emitter degeneration or swamping resistor). Q_{17} is directly coupled to the class AB output stage (Q_{14} and Q_{20}). Note the use of a V_{BE} multiplier to bias the output transistors. The V_{BE} multiplier is formed from Q_{18} and resistors R_7 and R_8. Note that this section receives its bias current from Q_{13}, which is part of the central current mirror complex.

Some transistors in this circuit are used solely for protection from overloads. A good example of this is Q_{15}. As the output current increases, the voltage across R_9 will increase proportionally. Note that this resistor is in parallel with the base emitter junction of Q_{15}. If this potential gets high enough, Q_{15} will turn on, shunting base drive current around the output device (Q_{14}). In this manner, current gain is reduced, and the maximum output current is limited to a safe value. This limiting value may be found via Ohm's law:

$$I_{limit} = \frac{V_{BE}}{R_9}$$
$$I_{limit} = \frac{.7 \text{ V}}{25 \text{ }\Omega}$$
$$I_{limit} = 28 \text{ mA}$$

In a similar fashion, Q_{20} is protected by R_{10}, R_{11}, and Q_{22}. If the output tries to sink too large of a current, Q_{22} will turn on, shunting current away from the base of Q_{16}. Although individual op amp schematics will vary widely, they generally hold to the basic four-part theme presented here:

1. A central current source/current mirror section to establish proper bias
2. A differential amplifier input stage with active load
3. A high-voltage gain intermediate stage
4. A class B or AB follower output section

Fortunately for the designer or repair technician, intimate knowledge of a particular op amp's internal structure is usually not required for successful application of the device. In fact, a few simple models can be used for the majority of cases. One very useful model is given in Figure 2.5. Here the entire multistage op amp is modeled with a simple resistive input network and a voltage source output. This output source is a dependent source. Specifically, it is a voltage-controlled voltage source. The value of this source is

$$E_{out} = A_v(V_{in+} - V_{in-})$$

The input network is specified as a resistance from each input to ground, as well as an input-to-input isolation resistance. For typical op amps, these values are normally

FIGURE 2.5

Simplified model

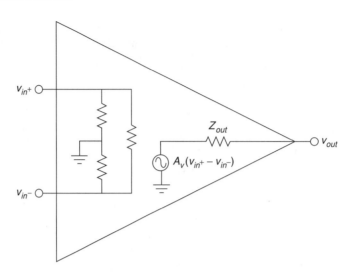

hundreds of kilo-ohms or more at low frequencies. Due to the differential input stage, the difference between the two inputs is multiplied by the system gain. This signal is presented to the output terminal through the final stage's output impedance. The output impedance will most likely be less than 100 Ω. System voltage gains in excess of 80 dB (10,000) are the norm.

A Simple Op Amp Simulation Model

It is possible to create a great variety of simulation models for any given op amp. Generally speaking, the more accurate the model is, the more likely it is to be complex. Due to the nature of most simulators, a more complex model requires a greater amount of time for an analysis to be completed. There is always a trade-off between model complexity and computation time. We can create a very simple model based on the previous section. This model is shown in Figure 2.6. It consists of just five nodes. The input section is modeled as a single resistor, R_{in}, between nodes 1 and 2. These two nodes are the noninverting and inverting inputs of the op amp, respectively. The second half of the model consists of a voltage-controlled voltage source and an output resistor. The value of this dependent source is a function of the differential input voltage and the voltage gain. With a minimum of components, the simulation time for this model is very low. In order to use this model, you need only set three parameters: the input resistance, the output resistance, and the voltage gain. An example is shown in Figure 2.7 using Electronics Workbench MultiSIM.

This model must be used with great care because it is so simplistic. It is useful as a learning tool for investigating the general operation of an op amp, but should never be considered as part of a true-to-life simulation. This model makes no attempt to consider the many limitations of the op amp. Because this model in no way imposes output signal swing limits, the effects of saturation will go unnoticed. Similarly, no attempt has been made at modeling the frequency response of the op amp. This is of great concern, and we will spend considerable time on this subject in later chapters. Many

FIGURE 2.6

Simple SPICE model

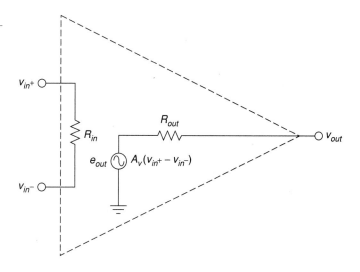

FIGURE 2.7

Simple op amp model in MultiSIM

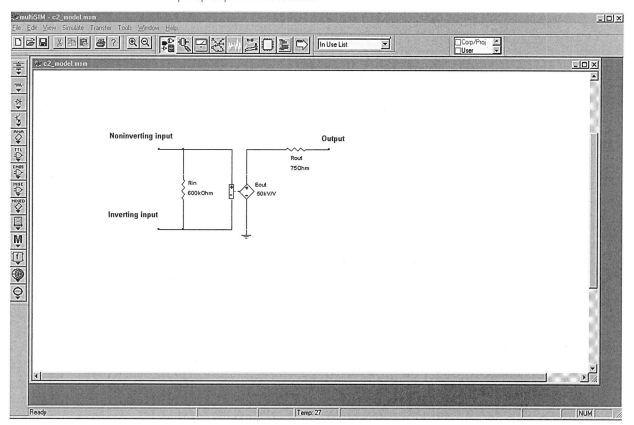

other effects are also ignored. With so many limitations, you might wonder just where such a model may be used. This model is useful for noncritical simulations given low-frequency inputs. You must also recognize the onset of saturation (clipping) yourself. Its primary advantage is that the circuit model is small, and thus computationally fast. Because of this, it is very efficient for students who are new to both op amps and circuit simulation. Perhaps of equal importance is the fact that this model points out the fact that your simulation results can only be as good as the models you use. Many people fall into the trap that "because the simulation came from a computer, it must be correct." Nothing could be further from the truth. Always remember the old axiom: GIGO (Garbage In = Garbage Out). It can be very instructive to simulate a circuit using differing levels of accuracy and complexity and then noting how closely the results match the same circuit built in the laboratory.

Op Amp Data Sheet and Interpretation

Different manufacturers often use special codes and naming conventions to delineate their products from those of other manufacturers, as well as providing quality level and manufacturing information. A manufacturer's code is usually a letter prefix, whereas a quality or construction code is a suffix. Common prefix codes include μA (Fairchild); AD (Analog Devices); CS (Crystal); LM, LH, and LF (National Semiconductor, with M indicating monolithic construction, H indicating hybrid construction, and F indicating an FET device); LT (Linear Technology); MC (Motorola); NE and SE (Signetics); OPA (Burr-Brown); and TL (Texas Instruments).

Many manufacturers make a host of standard parts such as the 741. For example, National Semiconductor makes the LM741, and Fairchild makes the μA741. These parts are generally considered to be interchangeable, although they may vary in some ways. Some manufacturers will use the prefix code of the original developer of a part and reserve their prefix for their own designs. As an example, Signetics produces their version of the 741, which they call μA741 as this op amp was first developed by Fairchild. (Signetics is then referred to as a *second source* for the μA741.)

Suffix codes vary widely between manufacturers. Typical designations for consumer grade parts are C and CN. The suffix N often means *Not Graded*. Interestingly, the lack of a final suffix often indicates a very high quality part, usually with an extended temperature range. Suffix codes are also used to indicate package styles. This practice is particularly popular among voltage regulators and other high-power linear ICs.

Finally, some manufacturers will use a "parallel" numbering system for high-grade parts. For example, the commercial grade device may have a "300 series" part number, with industrial grade given a "200 series" designation, and a military grade part given a "100 series" number. One possible example is the LM318 commercial grade op amp versus its high-grade counterpart, the LM118. Generally, military-specified parts will have a very wide temperature range, with industrial and commercial grades offering progressively less.

The data sheet for the LF411 op amp is shown in Figure 2.8. Let's take a look at some of the basic parameters and descriptions. The values given are typical of a modern op amp. A complete investigation of all parameters will be given in Chapter 5, once we've gotten a little more familiar with the device.

Figure 2.8

Data sheet for the LF411
Reprinted with permission of National Semiconductor Corporation

Absolute Maximum Ratings

If Military/Aerospace specified devices are required, please contact the National Semiconductor Sales Office/Distributors for availability and specifications. (Note 8)

	LF411A	LF411
Supply Voltage	±22V	±18V
Differential Input Voltage	±38V	±30V
Input Voltage Range (Note 1)	±19V	±15V
Output Short Circuit Duration	Continuous	Continuous

	H Package	N Package
Power Dissipation (Notes 2 and 9)	670 mW	670 mW
T_j max	150°C	115°C
$\theta_j A$	162°C/W (Still Air) 65°C/W (400 LF/min Air Flow)	120°C/W
$\theta_j C$	20°C/W	
Operating Temp. Range	(Note 3)	(Note 3)
Storage Temp. Range	$-65°C \leq T_A \leq 150°C$	$-65°C \leq T_A \leq 150°C$
Lead Temp. (Soldering, 10 sec.)	260°C	260°C
ESD Tolerance		Rating to be determined.

DC Electrical Characteristics (Note 4)

Symbol	Parameter	Conditions	LF411A Min	LF411A Typ	LF411A Max	LF411 Min	LF411 Typ	LF411 Max	Units
V_{OS}	Input Offset Voltage	$R_S = 10\,k\Omega$, $T_A = 25°C$		0.3	0.5		0.8	2.0	mV
$\Delta V_{OS}/\Delta T$	Average TC of Input Offset Voltage	$R_S = 10\,k\Omega$ (Note 5)		7	10		7	20 (Note 5)	μV/°C
I_{OS}	Input Offset Current	$V_S = \pm 15V$ (Notes 4,6), $T_j = 25°C$		25	100		25	100	pA
		$T_j = 70°C$			2			2	nA
		$T_j = 125°C$			25			25	nA
I_B	Input Bias Current	$V_S = \pm 15V$ (Notes 4,6), $T_j = 25°C$		50	200		50	200	pA
		$T_j = 70°C$			4			4	nA
		$T_j = 125°C$			50			50	nA
R_{IN}	Input Resistance	$T_j = 25°C$		10^{12}			10^{12}		Ω
A_{VOL}	Large Signal Voltage Gain	$V_S = \pm 15V$, $V_O = \pm 10V$, $R_L = 2K$, $T_A = 25°C$	50	200		25	200		V/mV
		Over Temperature	25	200		15	200		V/mV
V_O	Output Voltage Swing	$V_S = \pm 15V$, $R_L = 10k$	±12	±13.5		±12	±13.5		V
V_{CM}	Input Common-Mode Voltage Range		±16	±19.5		±11	+14.5		V
				−16.5			−11.5		V
CMRR	Common-Mode Rejection Ratio	$R_S \leq 10k$	80	100		70	100		dB
PSRR	Supply Voltage Rejection Ratio	(Note 7)	80	100		70	100		dB
I_S	Supply Current			1.8	2.8		1.8	3.4	mA

AC Electrical Characteristics (Note 4)

Symbol	Parameter	Conditions	LF411A Min	LF411A Typ	LF411A Max	LF411 Min	LF411 Typ	LF411 Max	Units
SR	Slew Rate	$V_S = \pm 15V$, $T_A = 25°C$	10	15		8	15		V/μs
GBW	Gain-Bandwidth Product	$V_S = \pm 15V$, $T_A = 25°C$	3	4		2.7	4		MHz
e_n	Equivalent Input Noise Voltage	$T_A = 25°C$, $R_S = 100\Omega$, $f = 1\,kHz$		25			25		$nV/\sqrt{Hz}$
i_n	Equivalent Input Noise Current	$T_A = 25°C$, $f = 1\,kHz$		0.01			0.01		$pA/\sqrt{Hz}$

Note 1: Unless otherwise specified the absolute maximum negative input voltage is equal to the negative power supply voltage.

Note 2: For operating at elevated temperature, these devices must be derated based on a thermal resistance of $\theta_j A$.

Note 3: These devices are available in both the commercial temprature range $0°C \leq T_A \leq 70°C$ and the military temperature range $-55°C \leq T_A \leq 125°C$. The temperature range is designated by the position just before the package type in the device number. A "C" indicates the commercial temperature range and an "M" indicates the military temperature range. The military temperature range is available in "H" package only.

Note 4: Unless otherwise specified, the specifications apply over the full temperature range and for $V_S = \pm 20V$ for the LF411A and for $V_S = \pm 15V$ for the LF411. V_{OS}, I_B, and I_{OS} are measured at $V_{CM} = 0$.

Note 5: The LF411A is 100% tested to this specification. The LF411 is sample tested to insure at least 90% of the units meet this specification.

Note 6: The input bias currents are junction leakage currents which approximately double for every 10°C increase in the junction temperature, T_j. Due to limited production test time, the input bias currents measured are correlated to junction temperature. In normal operation the junction temperature rises above the ambient temperature as a result of internal power dissipation, P_D. $T_j = T_A + \theta_{jA} P_D$ where θ_{jA} is the thermal resistance from junction to ambient. Use of a heat sink is recommended if input bias current is to be kept to a minimum.

Note 7: Supply voltage rejection ratio is measured for both supply magnitudes increasing or decreasing simultaneously in accordance with common practice, from ±15V to ±5V for the LF411 and from ±20V to ±5V for the LF411A.

Note 8: RETS 411X for LF411MH and LF411MJ military specifications.

Note 9: Max. Power Dissipation is defined by the package characteristics. Operating the part near the Max. Power Dissipation may cause the part to operate outside guaranteed limits.

First, note that two versions of the IC are given. We will examine the LF411 rather than the high-grade LF411A. At the very top of the data sheet is a listing of the absolute maximum ratings. The op amp should never operate at values greater than those presented, as doing so may permanently damage it. Like most general-purpose op amps, the LF411 is powered by a bipolar power supply. The supply rails should never exceed ±18 V DC. Normally, op amps will be used with ±15 V supplies. Maximum power dissipation is given as 670 mW. Obviously, then, this is a small signal device. In keeping with this, the operating temperature range and maximum junction temperatures are relatively low. We also see that the device can withstand differential input signals of up to 30 V and single-ended inputs of up to 15 V without damage. On the output, the LF411 is capable of withstanding a shorted load condition continuously. This makes the op amp a bit more "bulletproof." The remainder of this section details soldering conditions. Excessive heat during soldering may damage the device.

The second section of the data sheet lists the DC characteristics of the op amp. This table is broken down into five major sections:

1. The parameter symbol
2. The parameter name
3. The conditions under which the parameter is measured
4. The parameter values, either typical or min/max
5. The parameter units

We shall examine a few of these parameters right now. The fourth parameter given is I_B, the input bias current. I_B is the current drawn by the bases (or gates) of the input differential amplifier stage. Because the LF411 utilizes a JFET diff amp, we expect this value to be rather small. For an operating temperature of 25°C, a typical LF411 will draw 50 pA, and a worst-case LF411 no more than 200 pA. If we extend the temperature range out a bit, I_B can extend out to 50 nA. This is a sizable jump, but even 50 nA is a very small value for general-purpose work. Because larger bias currents are normally seen as undesirable, the maximum I_B is the worst-case scenario, hence a minimum I_B is not reported. Along with this, we see a very high value for input resistance, some 10^{12} Ω, typically. Op amps utilizing bipolar input devices will show much higher values for I_B and much lower values for R_{in}.

Next in line comes A_{vol}. This is the DC voltage gain. Note the test conditions. The power supply is set to ±15 V, the load is 2 kΩ, and V_{out} is 10 V peak. Normally, we desire as much gain as possible, so the worst-case scenario is the minimum A_{vol}. For 25°C operation, this is specified as 25 V/mV, or 25,000. The average device will produce a gain of 200,000. As is typical, once the temperature range is expanded performance degrades. Over the operating temperature range, the minimum gain may drop to 15,000.

Because the op amp uses a class AB follower for its output stage, we should expect the output compliance to be very close to the power supply rails. The output voltage swing is specified for ±15 V supplies with a 10 kΩ load. The typical device can swing out to ±13.5 V, with a worst-case swing of ±12 V. A reduction in power supply value will naturally cause the maximum output swing to drop. A sizable reduction in the load resistance will also cause a drop in V_o, as we shall see a bit later. These maximum output values are caused by the internal stages reaching their saturation limits. When this

FIGURE 2.9

Comparator (single input)

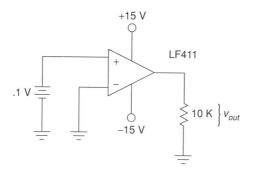

happens, the op amp is said to be clipping or in saturation. As a general rule of thumb, saturation may be approximated as 1.5 V less than the magnitude of the power supplies.

The last item in the list is the standby current draw, I_S. Note how small this is, only 1.8 mA, 3.4 mA worst case. This is the current the op amp draws from the supply under no signal conditions. When producing output signals, the current draw will rise.

The final section of the data sheet lists certain AC characteristics of the op amp that will be of great concern to us in later sections. Many device parameters change a great deal with frequency, temperature, supply voltage, or other factors. Because of this, data sheets also include a large number of graphs that further detail the op amp's performance. Finally, application hints and typical circuits may round out the basic data sheet.

Simple Op Amp Comparator

Now that you have a feel for what an op amp is and what some typical parameters are, let's take a look at an application. The one thing that jumps to most people's attention is the very high gain of the average op amp. The typical LF411 showed A_{vol} at approximately 200,000. With gains this high, it is obvious that even very small input signals can force the output into saturation (clipping). Take a look at Figure 2.9. Here an op amp is being supplied by ±15 V and is driving a 10 kΩ load. As seen in our model of Figure 2.5, V_{out} should equal the differential input voltage times the op amp's gain, A_{vol}.

$$V_{out} = A_{vol}(V_{in+} - V_{in-})$$
$$V_{out} = 200,000 \times (.1 \text{ V} - 0 \text{ V})$$
$$V_{out} = 20,000 \text{ V}$$

The op amp cannot produce 20,000 V. The data sheet lists a maximum output swing of only ±13.5 V when using ±15 V supplies. The output will be truncated at 13.5 V. If the input signal is reduced to only 1 mV, the output will still be clipped at 13.5 V. This holds true even if we apply a signal to the inverting input, as in Figure 2.10.

$$V_{out} = A_{vol}(V_{in+} - V_{in-})$$
$$V_{out} = 200,000 \times (.5 \text{ V} - .3 \text{ V})$$
$$V_{out} = 40,000 \text{ V}$$
$$V_{out} = 13.5 \text{ V, due to clipping}$$

FIGURE 2.10

Comparator (dual input)

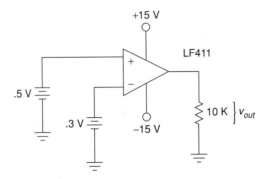

> ### multiSIM COMPUTER SIMULATION
>
> A simulation of Figure 2.10 using Electronics Workbench MultiSIM is shown in Figure 2.11 (see pp. 57–58). The LF411 op amp is selected from the component library, and you need not concern yourself with the model's internal makeup just yet. This particular model includes the effects of power supply limitations (i.e., output saturation), which the very simple dependent-source model presented earlier does not. Individual DC sources are used for the input signals. Although no AC signals are applied, it is perfectly valid to run a transient-mode simulation. The first millisecond of the output voltage is shown. It verifies the manual calculation, indicating a DC level of slightly more than 13.5 V.

For any reasonable set of inputs, as long as the noninverting signal is greater than the inverting signal, the output will be positive saturation. If you trade the input signals so that the inverting signal is the larger, the converse will be true. As long as the inverting signal is greater than the noninverting signal, the output will be negative saturation. If the inverting and noninverting signals are identical, V_{out} should be 0 V. In the real world, this will not happen. Due to minute discrepancies and offsets in the diff amp stage, either positive or negative saturation will result. You have no quick way of knowing in which direction it will go. It is for this reason that it is impractical to amplify a very small signal, say around 10 μV. You might then wonder, "What is the use of this amplifier if it always clips? How can I get it to amplify a simple signal?" Well, for normal amplification uses, we will have to add on some extra components, and through the use of negative feedback (next chapter), we will create some very well controlled, useful amplifiers. This is not to say that our barren op amp circuit is useless. Quite to the contrary; we have just created a comparator.

A comparator has two output states: high and low. In other words, it is a digital, logical output. Our comparator has a high state potential of 13.5 V and a low state potential of -13.5 V. The input signals, in contrast, are continuously variable analog potentials. A comparator, then, is an interface between analog and digital circuitry. One input will be considered the reference, whereas the other input will be considered the sensing line. Note that the differential input signal is the difference between the

What Is an Op Amp?   57

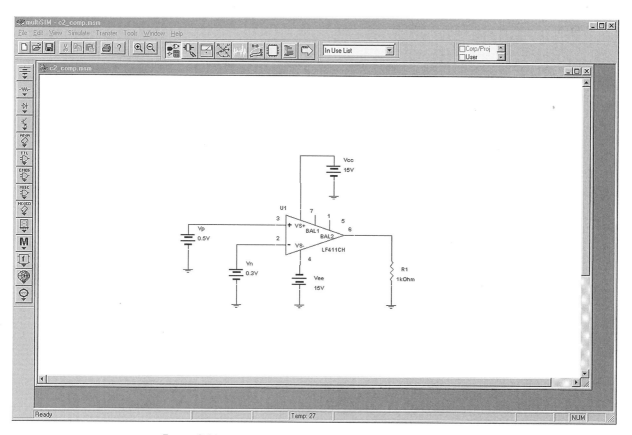

FIGURE 2.11a

Comparator example in MultiSIM

sensing input and the reference input. When the polarity of the differential input signal changes, the logical output of the comparator changes state.

EXAMPLE 2-1

Figure 2.12 shows a light leakage detector that might be used in a photographer's dark room. This circuit utilizes a Cadmium Sulfide (CdS) cell that is used as a light-sensitive resistor. The inverting input of the op amp is being used as the reference input, with a 1 V DC level. The noninverting input is being used as the sensing input. Under normal (no light) conditions, the CdS cell acts as a very high resistance, perhaps 1 MΩ. Under these conditions, a voltage divider is set up with the 10 kΩ resistor, producing about .15 V at the noninverting input. Remember, no loading of the divider occurs because the LF411 utilizes a JFET input. The noninverting input is less than the inverting input, thus the comparator's output is negative saturation, or approximately −13.5 V. If the ambient light level rises, the resistance of the CdS cell drops, thus raising the signal applied to the noninverting input. Eventually, if the light

58 Operational Amplifier Internals

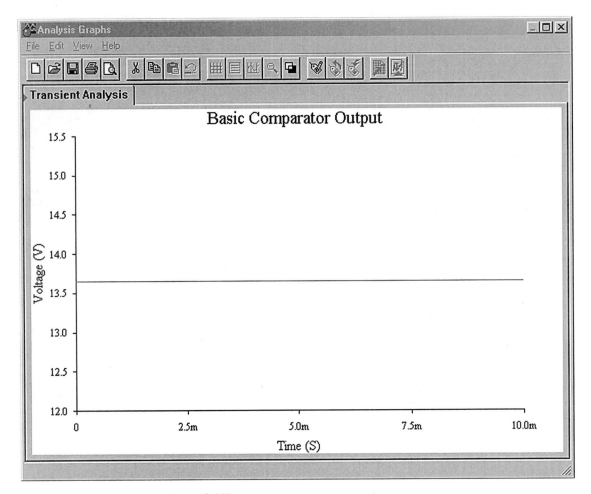

FIGURE 2.11b

Comparator output

FIGURE 2.12

Light alarm

FIGURE 2.13

"Square-up" circuit

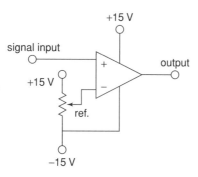

level is high enough, the noninverting input signal will exceed the 1 V reference, and the comparator's output will move to positive saturation, about +13.5 V. This signal could then be used to trigger some form of audible alarm. A real-world circuit would need the flexibility of an adjustable reference in place of the fixed 1 V reference. By swapping the CdS cell and the 10 kΩ resistor and adjusting the reference, an inverse circuit (i.e., an alarm that senses darkness) may be produced.

Circuits of this type can be used to sense a variety of over-level/under-level conditions, including temperature and pressure. All that is needed is an appropriate sensing device. Comparators can also be used with AC input signals.

EXAMPLE 2-2

Sometimes, it is necessary to square up an AC signal for further processing. That is, we must turn it into an equivalent pulse waveform. One example of this might be a frequency counter. A frequency counter works by tallying the number of high-to-low or low-to-high transitions in the input signal over a specific length of time. For accurate counts, good edge transitions are required. A simple sine wave changes relatively slowly compared to an equal frequency square wave, so some inaccuracy may creep into the readings. We can turn the input into a pulse-type output by running it through the comparator of Figure 2.13. Note that the reference signal is adjustable from -15 to $+15$ V. Normally, the reference is set for 0 V. Whenever the input is greater than the reference, the output will be positive saturation. When the input is less than the reference, the output will be negative saturation. By making the reference adjustable, we have control over the output duty cycle and can also compensate for DC offsets on the input signal. A typical input/output signal set is given in Figure 2.14.

There are a few limitations with our simple op amp comparator. For very fast signal changes, a typical op amp will not be able to accurately track its output. Also, the output signal range is rather wide and is bipolar. It is not at all compatible with normal TTL logic circuits. Extra limiting circuitry is required for proper interfacing. To help reduce these problems, a number of circuits have been specially optimized for comparator purposes. We will take a closer look at a few of them in Chapter 7.

FIGURE 2.14

Output of "square-up" circuit

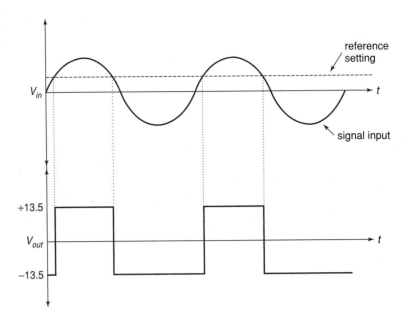

2.3 Op Amp Manufacture

Op amps are generally manufactured in one of two ways: The device is either a *hybrid* or is *monolithic*. In either case, the circuit can contain hundreds of components. The resulting op amp will be packaged in a variety of styles, including plastic and ceramic dual in-line and single in-line types, multilead cans, flat packs, and surface mount forms. Some examples are shown in Figure 2.15. In each type, the circuitry is completely encased and not accessible to the designer or technician. If one of the components should fail, the entire op amp is replaced. The design and layout of the integrated circuit itself is normally carried out with the use of special computer workstations and software tools. These allow the designers to simulate portions of the circuit and to create the outlines and interconnections for the various components to be formed.

FIGURE 2.15

Package styles

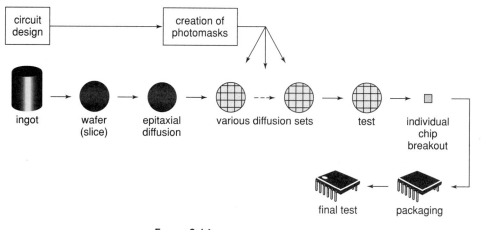

FIGURE 2.16

Chip manufacturing process

Monolithic Construction

The term *monolithic* is from the Greek, meaning, literally, "single stone." In this process, all circuit elements are created and interconnected using a single slab of silicon (or other suitable material). Normally, several op amps are made from a single silicon wafer. Each wafer may be a few inches in diameter, with each op amp circuit chip comprising perhaps a square 1 millimeter by 1 millimeter in area. A single transistor can easily be smaller than 15 micrometers by 20 micrometers. Because the scale of construction is so small, special *clean rooms* are required in order to remove tiny airborne particles of dust and grit that could interfere with the production of these super-small components. Workers in clean rooms are required to wear special suits as well.

Figure 2.16 outlines the major steps in the chip manufacturing process. This process starts with the preparation of a p-type silicon wafer. This is referred to as the *substrate*. After it has been cleaned and polished, an n-type *epitaxial* region is *diffused* into the p-type base. *Epitaxial* is from the Greek, roughly meaning "to arrange upon." It is within this thin epitaxial region that the circuit elements will be formed, the remainder of the substrate lending mechanical support to the structure. The term *diffusion* refers to the manner in which the semiconductor material becomes doped. In essence, the base material is surrounded by a high concentration of doping material, usually gaseous, along with the application of heat. The low-concentration wafer material will be infiltrated by the high-concentration doping material. Diffusion is a relatively accurate and inexpensive means of controlling the semiconductor's properties.

Once the n-type region is produced, the wafer will undergo an oxidizing process that will leave the top surface covered with silicon dioxide. This layer prevents impurities from entering the n-type region. At this point, a series of steps will be used to create *wells* or deposits of alternate p and n material. These deposits will form the various active and passive components. Normally, this is done through a photolithographic process. This involves the use of light sensitive materials and *masks*. Conceptually, the process is not very much different from the way in which printed circuit boards

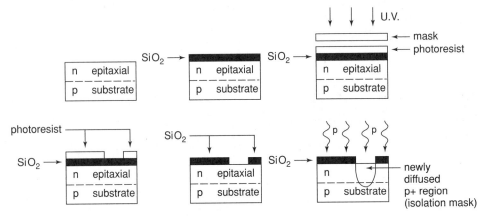

FIGURE 2.17

Diffusion process (one run)

are often made. In essence, specific areas of the silicon dioxide layer will be stripped away, thus exposing the epitaxial region, and allowing diffusion of other acceptor/donor impurities to take place. Because the silicon dioxide serves as an effective barrier to diffusion, only areas cleared of silicon dioxide will be affected by the diffusion process. In this manner, specific areas can be singled out and selectively doped to create specific components. This is detailed below and in Figure 2.17.

In order to selectively remove the silicon dioxide, the top surface is coated with a light-sensitive material called *photoresist*. Above this is placed a mask. This mask is much like a black-and-white negative; some areas are clear, and some areas are opaque. The resulting sandwich is then exposed to ultraviolet light. The clear areas of the mask will allow the light to pass through and cause a chemical change in the photoresist. A solution is then used to wash away the unexposed photoresist. At this point, a second solution is used to wash away the silicon dioxide. This solution will not affect the exposed photoresist, and thus, the silicon dioxide beneath it is unaffected. After this protecting layer of photoresist is removed, all that remains on the top surface of the wafer are alternating patches of silicon dioxide. The wafer can now be lead through another diffusion process.

The process of oxidizing, masking, and diffusing will be repeated several times. The initial run will be produced with an isolation mask. This is used to separate the various components. Normally, a base mask will be used next, followed by the emitter mask. The final masks will be used for contacts and interconnections. In this way, n-type material can be placed next to, or completely within, p-type material. The adjoining areas are, of course, PN junctions. Because all circuit elements are laid out lengthwise on a thin strip, this form of manufacture is referred to as a *planar* process.

Once the final mask is completed, the wafer will be inspected. The individual chips will then be broken out of the wafer and mounted into the desired package. Leads will be connected to the chip with fine *angel-hair* wire, and then the package will be sealed. It is now ready for final test and inspection. Part numbers and date codes will also be imprinted.

Virtually all general-purpose op amps today use a planar monolithic process. Some of the advantages of monolithic construction are its relative simplicity and low per-part cost.

Hybrid Construction

Hybrids are usually used where a complete monolithic solution is impractical. This is usually the case for special-purpose devices, such as those requiring very high output current, very wide bandwidth, or which are very complex or sensitive. Hybrids, as the name suggests, are a collection of smaller circuit elements interconnected. A typical hybrid may contain two or three smaller monolithic chips and assorted miniaturized passive and/or power components. Passive components may be further integrated by using either a thin or thick film chip process. (A discussion of thin and thick film chip techniques is beyond the scope of this text.) Due to the complexity of a hybrid chip, it is normally more expensive than its monolithic cousins. Although the IC itself may be more expensive, the complete application may very well wind up being less costly to produce because the cost of other components are effectively absorbed within the hybrid IC. One place where hybrids are often used is in consumer stereo music systems. A hybrid power amplifier IC offers the convenience of a single IC solution with the capabilities of a discrete transistor approach. As an op amp user, it makes little difference whether the device is hybrid or monolithic when it comes to circuit analysis or design.

SUMMARY

Op amps are presently in wide use in just about every aspect of linear electronics. An op amp is a multistage amplifier treated as a single entity. The first stage usually utilizes a differential amplifier that can be made with either bipolar or FET devices. The following stage(s) create a large voltage gain. The final stage is a class B voltage follower. The resulting op amp typically has a high-input impedance, a low-output impedance, and voltage gains in excess of 10,000. The op amp operates from a bipolar power supply, usually around ±15 V. Externally, it has connections for the inverting and noninverting inputs, the single-ended output, and the power supplies. The op amp may be packaged in a variety of forms, including DIPs, SIPs, cans, flat packs, and surface mount.

The general-purpose op amp is manufactured using a monolithic structure and a photolithographic process. Several chips are created from a single master wafer. Creation is a multistep process involving the selective doping of specific areas on the chip through diffusion. The monolithic technique is relatively inexpensive and accurate. Because integration allows for very tight part matching and consistency, certain circuit design techniques are favored, including the use of current mirrors and active loads.

Finally, with very little supporting circuitry, simple op amps can make effective comparator circuits. A comparator is in essence a bridge between the analog and digital worlds.

REVIEW QUESTIONS

1. What is an op amp?
2. Give several examples of where op amps might be used.
3. What is the typical stage layout of an op amp?
4. What comprises the first stage of a typical op amp?
5. What comprises the final stage of a typical op amp?
6. How does integrated circuit design differ from discrete design?
7. What is a comparator and how might it be used?
8. What is meant by monolithic planar construction?
9. What is a mask and how is it used in the construction of IC op amps?
10. What is the process of diffusion and how does it relate to the construction of IC op amps?
11. What are the advantages of monolithic IC construction?
12. How does hybrid construction differ from monolithic construction?

Problems

1. For the circuit in Figure 2.18, find V_{out} for the following inputs: (a) 0 V, (b) −1 V, (c) +2 V.

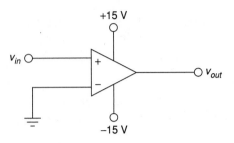

Figure 2.18

2. For the circuit in Figure 2.19, find V_{out} for the following inputs: (a) 0 V, (b) −4 V, (c) +5 V, (d) −.5 V.

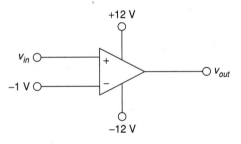

Figure 2.19

3. For the circuit in Figure 2.20, find V_{out} for the following inputs: (a) 0 V, (b) −2 V, (c) +1 V, (d) −.5 V.

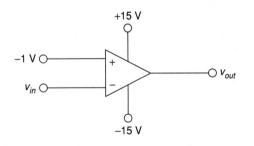

Figure 2.20

4. Sketch V_{out} if $V_{in}(t) = 1 \sin 60t$ in Figure 2.18.

5. Sketch V_{out} if $V_{in}(t) = 2 \sin 20t$ in Figure 2.19.
6. Sketch V_{out} if $V_{in}(t) = 3 \sin 10t$ in Figure 2.20.
7. A Temperature Dependent Resistor is used in the comparator of Figure 2.21. At what temperature will the comparator change state?

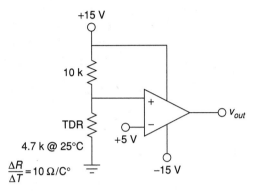

Figure 2.21

8. What is the value of the Light Dependent Resistor at the comparator trip point in Figure 2.22?

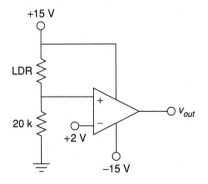

Figure 2.22

9. What reference voltage is required for a 50°C trip point in Figure 2.21?
10. A thermal fuse is device found in such common items as coffeemakers. Normally, its resistance is very low, ideally 0 Ω. With the application of excessive heat, the fuse

opens, presenting a very high resistance, ideally infinite. Explain the operation of the circuit in Figure 2.23. Is the output voltage normally high or low?

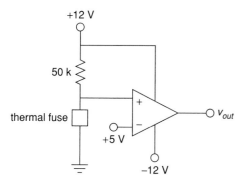

FIGURE 2.23

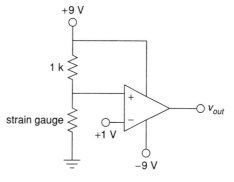

FIGURE 2.24

11. A strain gauge is a device that can be used to measure the amount of bend or deflection in a part it is attached to. It can be connected so that its resistance rises as the bend increases. What resistance is required to trip the comparator of Figure 2.24?

12. Determine the current flowing out of the collector of Q_{13} in Figure 2.4 using ± 15 volt power supplies.

Computer Simulation Problem

13. Alter the simple simulation model presented in the chapter to include saturation effects. (Hint: Consider a device that limits voltage.)

CHAPTER 3

Negative Feedback

CHAPTER OBJECTIVES

After completing this chapter, you should be able to:

- Give examples of how negative feedback is used in everyday life.
- Discuss the four basic feedback connections, detailing their similarities and differences.
- Detail which circuit parameters negative feedback will alter, and how.
- Discuss which circuit parameters are not altered by negative feedback.
- Define the terms *sacrifice factor*, *gain margin*, and *phase margin*, and relate them to a Bode plot.
- Discuss in general, the limits of negative feedback in practical amplifiers.

3.1 Introduction

As we saw in the last chapter, op amps are very useful devices. However, in many applications, the device's gain is simply too large and its bandwidth too narrow for effective use. In this chapter we will explore the concept of negative feedback. This concept is realized by *feeding* a portion of the output signal *back* to the input of the system. The proper use of negative feedback will allow us to exercise fine control over the performance of electronic circuits. As a matter of fact, negative feedback is so useful to us that we will seldom use op amps without it. Negative feedback is not tied solely to op amps, though, as almost any electronic circuit may benefit from its application. As with most things, there are disadvantages as well. A successful design will minimize the disadvantages and capitalize on the positive aspects. We will begin with the basic concepts of what negative feedback is and does, and then fine-tune our viewing by examining its four specific variants. We will look at specific examples of how negative feedback is applied to op amps, and finish off with a discussion of its practical limits.

3.2 What Negative Feedback Is and Why We Use It

People use negative feedback every day of their lives. In fact, we probably couldn't get along without it. Simply put, negative feedback is a very rudimentary part of intelligence. In essence, negative feedback lets something correct for mistakes. It tends to stabilize operations and reduce change. Negative feedback relies on a loop concept. In human terms, it is akin to knowing what you are doing and being able to correct for mistakes as they happen. You are constantly evaluating and correcting your actions in order to achieve a desired goal. This may be stated as letting the input know what the output is doing. A good example of this is your ability to maintain a constant speed while driving along the highway. You have a desired result, or set point, in mind, say 60 MPH. As you drive, you constantly monitor the speedometer. If you glance down and see that you're zipping along at 70 MPH, you think "Ooops, I'm going a bit too fast" and lift your foot slightly off of the gas pedal. On the other hand, if you're only going 40 MPH, you will depress the gas pedal further. The faster and more accurate your updates are, the better you will be at maintaining an exact speed.

In contrast to negative feedback is positive feedback, which reinforces change. If you were to correct your speed by saying "Hmmm, I'm going 70 MPH, I'd better step on the gas," you'd be using positive feedback. Other examples of positive feedback include the "acoustic squeal" often heard over public address systems, and thermal runaway effects seen in discrete devices. When positive feedback is applied to normal amplifiers, they *oscillate*. That is, they produce their own signals without any input applied.

3.3 Basic Concepts

Seeing the usefulness of negative feedback, it would be nice if we could apply the concept to our electronic circuits. The basic idea is quite simple, really. What we will do is sample a piece of the output signal, and then add it to the input signal out of phase (i.e., subtract it). By doing so, the circuit will see the difference between the input and the output. If the output signal is too large, the difference will be negative. Conversely, the difference signal will be positive if the output is too small. This signal is then multiplied by the circuit gain and cancels the output error. Thus, the circuit will be presented with the undesired errors in a way that will force the output to compensate (move in the opposite direction). This process is done continuously; the only time lags involved are the propagation delays of the circuits used. Because the sampled output signal is effectively subtracted from the input signal, negative feedback is sometimes referred to as degenerative or destructive feedback. This subtraction can be achieved in a variety of ways. A differential amplifier is tailor-made for this task because it has one inverting input and one noninverting input. (Note: If the error is presented in phase, the circuit magnifies the errors, and positive feedback results.)

FIGURE 3.1

Negative feedback

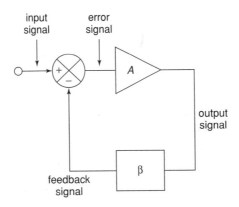

To see an example of how this works, refer to Figure 3.1. The triangle represents an amplifying circuit. It has a gain of A. The output signal is also presented to the input of the feedback network represented by the box. This network scales the output signal by a factor, β. The feedback network ranges from very simple to complex. It may contain several resistors, capacitors, diodes, and what not, or it may be as simple as a single piece of wire. In any case, this scaled output signal is referred to as the *feedback signal* and is effectively subtracted from the input signal. This combination, called the *error signal*, is then fed to the amplifier, where it is boosted and appears at the output. The process repeats like this forever (or at least until the power is switched off).

Let's assume that for some reason (perhaps a temperature change) the amplifier's gain were to rise. This should make the output signal increase by a similar percentage, but it doesn't. Here's why: As the output signal tries to rise, the feedback signal tracks with it. Now that there is a larger feedback signal, the error signal will become smaller (remember, error = input − feedback). This smaller signal is multiplied by the gain of the amplifier, thus producing a smaller output signal that almost completely offsets the original positive change. Note that if the output signal were too small, the error signal would increase, thus bringing the output back up to a normal level. When everything is working right, the feedback and input signals are almost the same size. (Actually, the feedback signal is somewhat smaller in magnitude.)

The Effects of Negative Feedback

Besides smoothing out gain anomalies, negative feedback can reduce the effect of device nonlinearities, thus producing a reduction in static forms of distortion such as THD (Total Harmonic Distortion). Basically, these nonlinearities can be viewed as a string of small gain errors. As such, they produce appropriate error signals and are compensated for in the above manner. Negative feedback can also increase the bandwidth of the system. It can increase the upper cutoff frequency f_2 and decrease the lower cutoff frequency f_1 (assuming the system has one). Also, we can exercise control over the input and output impedances of the circuit. It is possible to increase or decrease the impedances. As you might have guessed, we don't receive these benefits for nothing. The downside to negative feedback is that you lose gain. Effectively, you get to trade off gain for an increase in bandwidth, a decrease in distortion, and control over impedances. The more gain you trade off, the greater your rewards in the other three areas. In the case of our op amp, this is a wise trade-off because we already have more gain than we need for typical applications. This give and take is a very important idea, so remember "Big D." That stands for **B**andwidth, **I**mpedance, **G**ain, and **D**istortion.

At this point, we need to define a few terms. *Closed loop* refers to the characteristics of the system when feedback exists. For example, *closed-loop gain* is the gain of the system with feedback, whereas *closed-loop frequency response* refers to the new system break points. Generalized closed-loop quantities will be shown with the subscript '*cl*.' Similarly, we will denote impedances, gain, and the like for specific feedback variants with a two-letter subscript abbreviating the exact feedback configuration. One possibility for closed-loop gain would be A_{sp}. *Open loop* refers to the characteristics of the amplifier itself. To remember this, think of disconnecting or opening the path through the feedback network. Once the path is broken, the amplifier is on its own. *Open-loop gain*, then, refers to the gain of the amplifier by itself with no feedback. All open-loop quantities will be shown with the subscript '*ol*.' The symbol for open-loop gain would be A_{ol}. The term *loop gain* refers to the ratio between the open- and closed-loop gains. (It may also be computed from their difference in decibels on a Bode plot.) Loop gain indicates how much gain we have given up or sacrificed in order to enhance the operation of the system. Consequently, loop gain is often called *sacrifice factor* and given the symbol S. Generally, trade-offs are proportional to the sacrifice factor. For example, if we cut the gain in half, we will generally double the bandwidth and halve the distortion. An example that illustrates this can be seen in Figure 3.2. Note how the loop gain decreases with increasing frequency. This means that the effects of feedback at higher frequencies are not as great.

FIGURE 3.2

Response with and without feedback

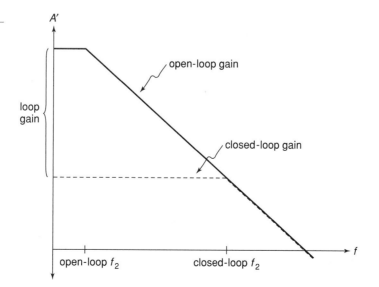

Up to this point we have made one simple assumption about our system—that it exhibits no "extra" phase change beyond the desired inversion. As we saw in the first chapter, however, all circuits do produce phase changes as the input frequency is increased. If this extra phase change were to reach $-180°$ while the gain was greater than unity (0 dB), our negative feedback will turn into positive feedback (the inversion = $-180°$, plus this extra $-180°$, places us at $-360°$. The net result is an in-phase signal). If this were to happen, our amplifier would no longer be stable. In fact, it may very well turn into a high-frequency oscillator. (You will see how to do this on purpose when we cover Chapter 9.) As the input frequency is raised, the phase will eventually exceed $-180°$, and the gain will drop to a fraction (<0 dB). The real key here is making sure that the phase never reaches or exceeds $-180°$ when the gain falls to 1. Stated another way, when the phase hits $-180°$, the gain should be a fraction. Generally, the farther you are from this "danger zone," the better. In other words, you're better off if the extra phase at the unity gain point is $-90°$ rather than $-170°$. Both are stable, but the first one gives you some breathing room.

Two measures of circuit safety are the *gain* and *phase margins*. Phase margin indicates the difference between the actual phase at unity gain and $-180°$. In the example above the first circuit would have a phase margin of $90°$, and the second would have a margin of $10°$. Gain margin is the difference between the actual gain in dB at the $-180°$ phase point and 0 dB. If our gain was -9 dB at $-180°$, the gain margin would be 9 dB (i.e., we have 9 dB "to spare"). Reasonable values for gain and phase margin are >6 dB and >45°. Gain and phase margins are depicted in Figure 3.3. It is possible to guarantee safe margins if the amplifier's open loop response maintains a 20 dB-per-decade rolloff up to the unity-gain frequency, f_{unity}. This means that there is only one dominating lag network that will add a maximum phase shift of $-90°$. Even if the second network coincided with f_{unity}, it would add $-45°$ at most. This would still leave us with a 45° phase margin. (Note that if we had several secondary networks critical at f_{unity} the phase could exceed $-180°$, however the slope would no longer be 20 dB-per-decade in reality.) It is for this reason that the general-purpose op amps examined in Chapter 2 included a compensating capacitor. No matter how much feedback we wish to use, our circuits will always end up being stable. For the best circuit performance, it is possible to use amplifiers that do not have the "constant rolloff" characteristic. The possibility exists that they may go into oscillation or become unstable if you are not careful and ignore the margins.

3.4 The Four Variants of Negative Feedback

Negative feedback can be achieved via four different forms. They differ in how the input and output impedances are changed. We have basically two choices when it comes to connecting the input and output of the amplifier to the output and input of the feedback network. We may produce either a series connection or a parallel connection. This yields four possibilities total. Each connection will produce a specific effect on the input or output impedance of the system. As you might guess, parallel connections decrease the impedance, and series connections increase it. A high input impedance is desirable for maximum voltage transfer, whereas a low impedance is

FIGURE 3.3

Gain and phase margin graphically determined from Bode plot

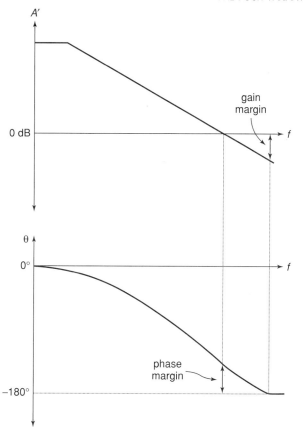

required for maximum current transfer. As a memory aid, think of volt meters and ammeters. For the smallest loading effect, volt meters should exhibit a high impedance and ammeters a very low one. In the case of the output connection, a low source impedance is required for the best voltage transfer and a high source impedance for current transfer. You should think of the ideal voltage source (zero Z_{out}) and the ideal current source (infinite Z_{out}) here. Consequently, if we were to connect our feedback network in series with the amplifier's input and in parallel with its output, we would have an increase in Z_{in} and a decrease in Z_{out}. This means that our system would be very good at sensing an input voltage and ideal for producing a voltage. We will have created a voltage-controlled voltage source (VCVS), the ideal voltage amplifier. So that you can get a good idea of the possibilities, all four types are summarized below.

Type (in-out)	Z_{in}	Z_{out}	Model	Idealization	Transfer Ratio
Series-Parallel	High	Low	VCVS	Voltage Amplifier	V_{out}/V_{in} (Voltage gain)
Series-Series	High	High	VCCS	Voltage to Current Transducer	I_{out}/V_{in} (Transconductance)

FIGURE 3.4

Series-parallel connection

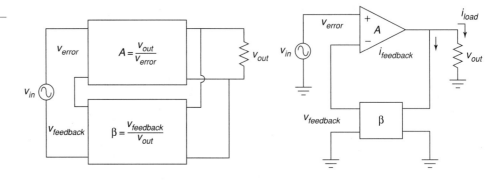

Parallel-Parallel	Low	Low	CCVS	Current to Voltage Transducer	V_{out}/I_{in} (Transresistance)
Parallel-Series	Low	High	CCCS	Current Amplifier	I_{out}/I_{in} (Current Gain)

Generally speaking, the input and output impedances will be raised or lowered from the nonfeedback value by the sacrifice factor. Note that by using the proper form of feedback, we can achieve any of the possible models. This greatly enhances our ability to deal with specific applications. Much work in our field relies on optimal voltage transfer, therefore Series-Parallel (SP) is often used. A variation on Parallel-Parallel (PP) is also used frequently, as we shall see.

Series-Parallel (SP)

The SP connection makes for the ideal voltage amplifier. A generalized block diagram is shown in Figure 3.4. You can tell that it has a series input because there is no input current node. Contrast this with the output: Note that the op amp's output current splits into two paths, one through the load and the other into the feedback network. This output node clearly denotes the feedback's parallel output connection. Let's take a look at exactly how SP negative feedback alters the system gain, impedances, and frequency response. For starters, let's examine the closed-loop gain (A_{sp}). Our amplifier block produces a gain A and could be a diff amp, op amp, or other multistage possibility. The feedback network is typically a voltage divider and produces a loss, β. The signal presented to the inverting input of the amp is the feedback signal and is equal to $V_{out}\beta$. Note that the source's signal, V_{in}, is applied to the noninverting input. Therefore, the differential input voltage (usually referred to as V_{error}), equals $V_{in} - V_{feedback}$. We also know from previous work with differential amplifiers that that $V_{out} = V_{error}A_{ol}$ (A_{ol} is the amplifier's open-loop gain). In other words we know:

$$V_{in} = V_{error} + V_{feedback} \tag{3.1}$$

$$V_{feedback} = V_{out}\beta \tag{3.2}$$

$$V_{out} = V_{error}A_{ol} \tag{3.3}$$

and by definition,

$$A_{sp} = \frac{V_{out}}{V_{in}} \tag{3.4}$$

Substituting Equation 3.3 into Equation 3.2,

$$V_{feedback} = V_{error}\beta A_{ol} \tag{3.5}$$

Substituting Equation 3.5 into Equation 3.1 and simplifying,

$$V_{in} = V_{error}(1 + \beta A_{ol}) \tag{3.6}$$

Finally, substituting Equation 3.6 and Equation 3.3 into Equation 3.4 and simplifying yields

$$A_{sp} = \frac{A_{ol}}{1 + \beta A_{ol}} \tag{3.7}$$

Because the fundamental definition of sacrifice factor, S, is $\frac{A_{ol}}{A_{cl}}$, we may also say $A_{sp} = \frac{A_{ol}}{S}$ and therefore, for SP,

$$S = 1 + \beta A_{ol} \tag{3.8}$$

Equation 3.7 is our general gain equation, but if we can make $\beta A_{ol} \gg 1$, we may ignore the "+1" in the denominator and further simplify this as

$$A_{sp} = \frac{1}{\beta} \tag{3.9}$$

This seemingly innocent equation packs a rather hefty punch. What it is telling us is that the open-loop gain of the amplifier does not play a role in setting the system gain as long as the open-loop gain is very large. In other words, the system gain is controlled solely by the feedback network. Consequently, our amplifier can exhibit large-gain changes in its open-loop response, but the closed-loop response will remain essentially constant. For this reason we will achieve identical closed-loop gains for op amps that exhibit sizable differences in their open-loop gains. Because signal distortion is produced by nonlinearities that can be viewed as dynamic gain changes, our closed-loop distortion drops as well. Also, it is this very effect that extends our closed-loop frequency response. Imagine that our amplifier exhibits a gain of 10,000 at its upper break frequency of 100 Hz. If the feedback factor equals .1, our exact gain is:

$$A_{sp} = \frac{10,000}{1 + .1 \times 10,000} = 9.99$$

If we were to measure the amplifier's open-loop gain one decade up, at 1 kHz, it should be around 1,000 (assuming 20 dB/decade loss). The closed-loop gain now equals:

$$A_{sp} = \frac{1{,}000}{1 + .1 \times 1{,}000} = 9.9$$

As you can see, the closed-loop gain changed only about 1% despite the fact that the open-loop gain dropped by a factor of 10. If we continue to raise the frequency, A_{sp} would equal 9.09 at 10 kHz. Finally, at 100 kHz a sizable drop is seen because the gain falls to 5. At this point, our assumption of $\beta A_{ol} \gg 1$ falls apart. Note, however, that our loss relative to the midband gain is only a few dB. We have effectively stretched out the bandwidth of the system. Actually, this calculation is somewhat oversimplified, as we have ignored the extra phase lag produced by the amplifier above the open-loop break frequency. If we assume that the open-loop response is dominated by a single lag network (and it should be, in order to guarantee stability, remember?), a phase sensitive version of Equation 3.7 would be:

$$A_{sp} = \frac{-jA_{ol}}{1 - jA_{ol}\beta}$$

This extra phase will reach its maximum of −90 degrees approximately one decade above the open-loop break frequency. Consequently, when we find the magnitude of gain at 100 kHz, it's not

$$A_{sp} = \frac{10}{1 + 1}$$

but rather

$$A_{sp} = \frac{10}{\sqrt{1^2 + 1^2}}$$

which equals 7.07, for a −3 dB relative loss.

A simpler way of stating all of this is: The new upper-break frequency is equal to the open-loop upper-break times the sacrifice factor, S. Because S is the loop gain, it is equal to $\frac{A_{ol}}{A_{sp}}$. Note that our low-frequency $S = \frac{10{,}000}{10}$, or 1,000. Therefore, our closed-loop break equals 1,000 × 100 Hz, or 100 kHz. A very important item to notice here is that there is an inverse relation between closed-loop gain and frequency response. Systems with low gains will have high upper-breaks, while high-gain systems will suffer from low upper-breaks. This sort of trade-off is very common. Although most diff amps and op amps do not have lower break frequencies, circuits that do will see an extension of their lower response in a similar manner (i.e., the lower break will be reduced by S). In order to achieve both high gain and wide bandwidth, it may be necessary to cascade multiple low-gain stages.

EXAMPLE 3-1

Assume that you have an amplifier connected as in Figure 3.4. The open-loop gain (A_{ol}) of the amp is 200 and its open-loop upper-break frequency (f_{2-ol}) is 10 kHz. If the feedback factor (β) is .04, what are the closed-loop gain (A_{sp}) and break frequency (f_{2-sp})?

For A_{sp},

$$A_{sp} = \frac{A_{ol}}{1 + \beta A_{ol}}$$

$$A_{sp} = \frac{200}{1 + .04 \times 200}$$

$$A_{sp} = 22.22$$

The approximation says

$$A_{sp} = \frac{1}{\beta}$$

$$A_{sp} = \frac{1}{.04}$$

$$A_{sp} = 25$$

That is reasonably close to the general equation's answer (note that there is no need to include phase effects as we are looking for the midband gain). The approximation is more accurate when A_{ol} is larger.

For f_{2-sp}, first find the sacrifice factor, S.

$$S = \frac{A_{ol}}{A_{sp}}$$

$$S = \frac{200}{22.22}$$

$$S = 9$$

$$f_{2-sp} = f_{2-ol} S$$

$$f_{2-sp} = 10 \text{ kHz} \times 9$$

$$f_{2-sp} = 90 \text{ kHz}$$

One interesting thing to note is that the product of the gain and upper-break frequency will always equal a constant value, assuming a 20 dB per decade rolloff. Our open-loop product is 200 × 10 kHz, or 2 MHz. Our closed-loop product is 22.22 × 90 kHz, which is 2 MHz. If we choose any other feedback factor, the resulting A_{sp} and f_{2-sp} will also produce a product of 2 MHz (try it and see). The reason for this is simple. A 20 dB per decade rolloff means that the gain drops by a factor of 10 when the frequency is increased by a factor of 10. There is a perfect 1:1 inverse relationship between the two parameters. No matter how much you increase one parameter, the other one will decrease by the exact same amount. Thus, the product is a constant.

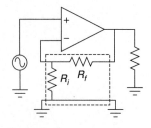

FIGURE 3.5

Simple voltage divider for β

At this point you may be asking yourself, "What exactly is in that feedback network and how do I figure out β?" Usually, the feedback network just needs to produce a loss—it has to scale V_{out} down to $V_{feedback}$. The simplest item for the job would be a resistive voltage divider. (It is possible to have complex frequency dependent or nonlinear elements in the network, as we shall see in the future). An example is presented in Figure 3.5. If you study this diagram for a moment, you will notice that the feedback factor β is really nothing more than the voltage divider loss. V_{out} is the input to the feedback network and appears across $R_f + R_i \| Z_{in}$. The output of the network is $V_{feedback}$, which appears across R_i. Simply put, the ratio is

$$\beta = \frac{R_i \| Z_{in}}{R_f + R_i \| Z_{in}}$$

If the Z_{in} is large enough to ignore, as in most op amps, this simplifies to

$$\beta = \frac{R_i}{R_f + R_i}$$

By substituting this equation into our approximate gain Equation 3.9, we find

$$A_{sp} = \frac{R_f + R_i}{R_i} = \frac{R_f}{R_i} + 1 \quad (3.10)$$

Note that the values of R_f and R_i are not really important, rather, their ratio is. We would arrive at the same gain if $R_f = 10$ kΩ and $R_i = 1$ kΩ, or $R_f = 20$ kΩ and $R_i = 2$ kΩ. We obviously have quite a bit of latitude when designing circuits for a specific gain, but we do face a few practical limits. If the resistors are too small, we will run into problems with op amp output current. On the other hand, if the resistors are too large, excessive noise, offset, drift, and loading effects will result. As a guideline for general-purpose circuits, $R_f + R_i$ is usually in the range of 10 k to 100 kΩ.

multiSIM Computer Simulation

As evidenced earlier, if the open-loop gain is very high, its precise value does not matter. We'll examine this effect using Example 3-1, but with much higher open loop gains. The simulation is shown in Figure 3.6 (see pp. 77–78) using Electronics Workbench MultiSIM and the basic op amp model presented in Chapter 2. The circuit uses $R_f = 24$ kΩ and $R_i = 1$ kΩ for an ideal closed-loop voltage gain of 25. The input signal is set to 40 millivolts. Using the Parameter Sweep option, the open-loop gain is initialized at 10,000 and is progressively doubled to a maximum of 160,000. These are reasonable gain values for a production op amp. The output transient analysis shows that in spite of a 16 to 1 variation in open-loop gain, all output voltages are approximately 1 V, achieving a closed loop gain of nearly 25 in all cases.

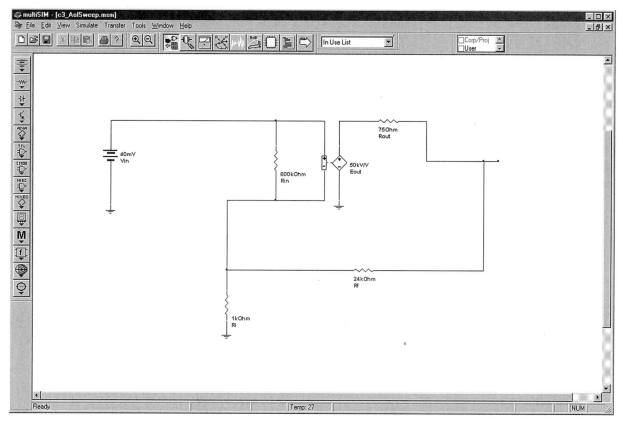

Figure 3.6a

MultiSIM schematic for gain sweep

Example 3-2

Let's say that the microphone you use for acoustic instruments produces a signal that is just too weak for you to record without excessive tape hiss. After a little experimentation in lab, you discover that you need about 20 dB of voltage gain before you can successfully capture a softly picked guitar. Using Figure 3.7 as a guide, design this amplifier with the following device: $A_{ol} = 50,000$, $Z_{in-ol} = 600$ kΩ.

First, note that our A_{ol} and Z_{in-ol} values are more than sufficient for us to use the approximation formulas. Because our formulas all deal with ordinary gain, we must convert 20 dB.

$$A_{sp} = \log^{-1} \frac{A'_{sp}}{20}$$
$$A_{sp} = \log^{-1} \frac{20 \text{ dB}}{20}$$
$$A_{sp} = 10$$

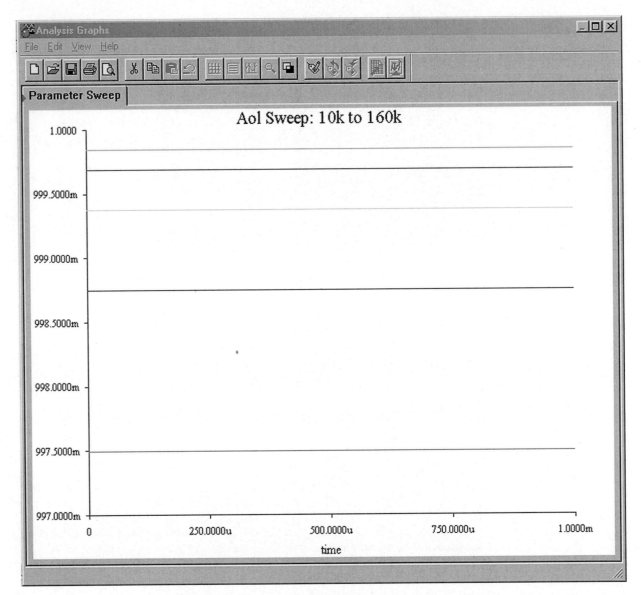

FIGURE 3.6b

Output variation due to open-loop gain change

By rearranging Equation 3.10,

$$\frac{R_f}{R_i} = A_{sp} - 1$$
$$\frac{R_f}{R_i} = 9$$

FIGURE 3.7

Microphone amplifier

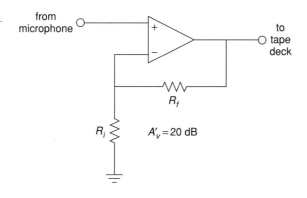

We see that R_f must be 9 times larger than R_i. There is no single right answer here; there are many possibilities.

multiSIM COMPUTER SIMULATION

One viable solution for Example 3-2 is simulated using Electronics Workbench MultiSIM in Figure 3.8 (see page 80). This circuit uses $R_f = 9$ kΩ and $R_i = 1$ kΩ. The op amp model is the simple dependent source version examined in Chapter 2. The input signal is set to .1 V DC for simplicity. Both the output and feedback potentials are presented. The results of this simulation verify our hand calculations. In order to note the sensitivity of the design, you can alter certain parameters of the input file and rerun the simulation. Two of the more interesting areas are the absolute values of the feedback resistors and the open-loop gain of the op amp. You will note that as these quantities are lowered, our approximation formulas become less accurate. With the given values, the approximations deviate from the simulation results by less than 1%.

SP Impedance Effects

As noted earlier, negative feedback affects the closed-loop input and output impedances of our system. Series connections increase the impedance, and parallel connections decrease the impedance.

Let's see exactly how this works in the SP case. First, we must distinguish between the Z_{in} of the amplifier itself and the Z_{in} of the system with feedback. We shall call them Z_{in-ol} and Z_{in-sp} respectively. Figure 3.9 shows this difference by using a simple model of the amplifier. By definition,

$$Z_{in-sp} = \frac{V_{in}}{I_{in}}$$

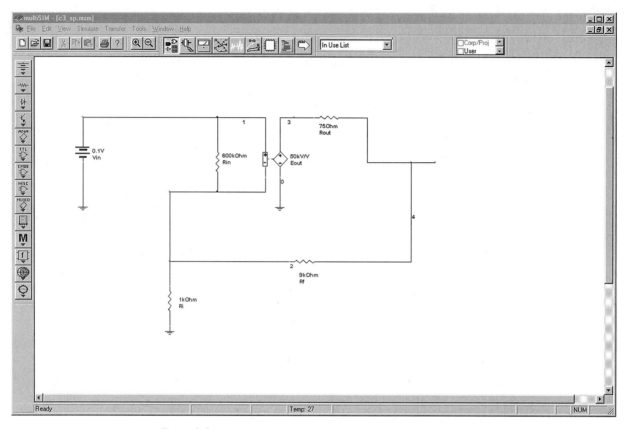

Figure 3.8a

Analysis of the simple op amp model of Example 3-2. MultiSIM schematic

Figure 3.8b

MultiSIM DC outputs

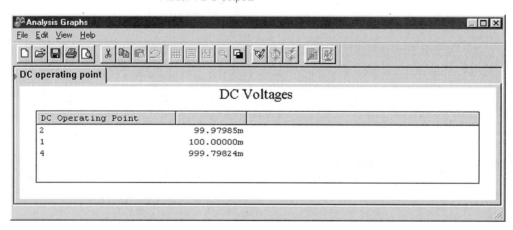

FIGURE 3.9

Series-parallel input impedance

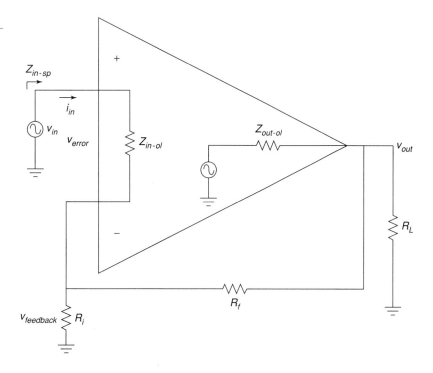

The idea here is to notice that the source only needs to supply enough signal current to develop the V_{error} drop across the amplifier's Z_{in-ol}. As far as the V_{in} signal source is concerned, $V_{feedback}$ is a voltage source, not a voltage drop. Therefore, $I_{in} = \frac{V_{error}}{Z_{in-ol}}$. We may now say

$$Z_{in-sp} = \frac{V_{in}}{\frac{V_{error}}{Z_{in-ol}}} = \frac{Z_{in-ol} V_{in}}{V_{error}}$$

Because V_{error} ideally equals $\frac{V_{out}}{A_{ol}}$,

$$Z_{in-sp} = \frac{Z_{in-ol} A_{ol} V_{in}}{V_{out}} = \frac{Z_{in-ol} A_{ol}}{A_{sp}}$$

Sacrifice factor S is defined as $\frac{A_{ol}}{A_{sp}}$, so

$$Z_{in-sp} = Z_{in-ol} S$$

This is our ideal SP input impedance. Obviously, even moderate open-loop Z_{in}s with moderate sacrifice factors can yield high closed-loop Z_{in}s. The upper limit to this will be the impedance seen from each input to ground. In the case of a typical op

FIGURE 3.10

Common-mode input impedance

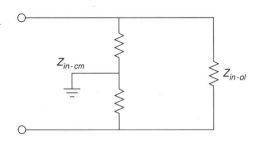

amp, this is sometimes referred to as the common-mode input impedance, Z_{in-cm}, and can be very high (perhaps hundreds of megohms). This is the impedance presented to common-mode signals. This value effectively appears in parallel with our calculated Z_{in-sp}, above. An example is shown in Figure 3.10. Note that because Z_{in-cm} is measured with the inputs of the op amp in parallel, each input has approximately twice the value to ground. In the case of a discrete amplifier, you would be most concerned with the noninverting input's Z_{in}. In any case, because S drops as the frequency increases, Z_{in-sp} decreases as well. At very high frequencies, input and stray capacitances dominate, and the real system input impedance may be a small fraction of the low-frequency value. Negative feedback cannot reduce effects that live outside of the loop.

Now for our Z_{out-sp}. Refer to Figure 3.11. Z_{out-sp} is the Thevenin output impedance. In order to find this, we will drive the amplifier's output with a voltage source and

FIGURE 3.11

Series-parallel output impedance

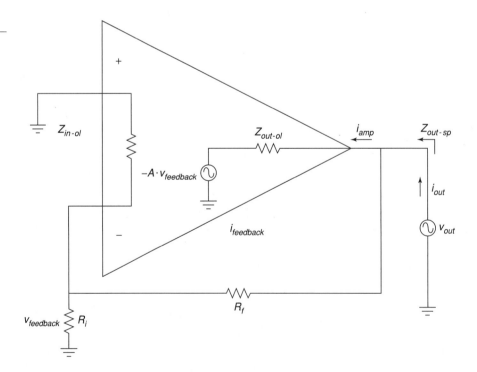

reduce all other independent voltage sources to zero. (We have no current sources. Remember, this is a "paper" analysis technique and may not work in lab due to other factors.) By figuring out the resulting output current, we can find Z_{out-sp} (by definition, $Z_{out-sp} = \frac{V_{out}}{I_{out}}$).

First, notice that I_{out} is made of two pieces, $I_{feedback}$ and I_{amp}. If we can find the two impedances associated with these parts, we can simply perform a parallel equivalent in order to determine Z_{out-sp}. The $I_{feedback}$ portion is very easy to determine. Ignoring the inverting input's loading effects on R_i, this impedance is just $R_f + R_i$. Finding the output impedance of the amplifier itself is a little more involved. I_{amp} is found by taking the drop across Z_{out-ol} and using Ohm's law. The voltage across Z_{out-ol} is the difference between V_{out} and the signal created by the feedback path to the inverting input. This signal is $-A_{ol}V_{feedback}$.

$$I_{amp} = \frac{V_{out} - (-A_{ol}V_{feedback})}{Z_{out-ol}}$$

As $V_{feedback} = V_{out}\beta$

$$I_{amp} = \frac{V_{out} + A_{ol}\beta V_{out}}{Z_{out-ol}}$$

$$I_{amp} = \frac{V_{out}(1 + A_{ol}\beta)}{Z_{out-ol}}$$

By using Equation 3.8, this may be simplified to

$$I_{amp} = \frac{SV_{out}}{Z_{out-ol}}$$

Because

$$Z_{out-amp} = \frac{V_{out}}{I_{amp}}$$

we may say

$$Z_{out-amp} = \frac{Z_{out-ol}}{S}$$

This is the part of Z_{out-sp} contributed by I_{amp}. To find Z_{out-sp}, just combine the two pieces in parallel:

$$Z_{out-sp} = \frac{Z_{out-ol}}{S} \| (R_f + R_i)$$

With op amps, $R_f + R_i$ is much larger than the first part and can be ignored. For example, a typical device may have $Z_{out-ol} = 75\ \Omega$. Even a very modest sacrifice factor will yield a value many times smaller than a typical $R_f + R_i$ combo (generally over 1 kΩ). Discrete circuits using common emitter or common base connections will have larger Z_{out-ol} values, and therefore the feedback path may produce a sizable effect. As in the case of input impedance, Z_{out-sp} is a function of frequency. Because S decreases as the frequency rises, Z_{out-sp} will increase.

Example 3-3

An op amp has the following open-loop specs: $Z_{in} = 300\ k\Omega$, $Z_{out} = 100\ \Omega$ and $A = 50{,}000$. What are the low-frequency system input and output impedances if the closed-loop gain is set to 100?

First we must find S.

$$S = \frac{A_{ol}}{A_{sp}}$$

$$S = \frac{50{,}000}{100} = 500$$

We may now find the approximate solutions.

$$Z_{in-sp} = S\, Z_{in-ol}$$
$$Z_{in-sp} = 500 \times 300\ k\Omega = 150\ M\Omega$$
$$Z_{out-sp} = \frac{Z_{out-ol}}{S}$$
$$Z_{out-sp} = \frac{100}{500} = .2\ \Omega$$

The effects are quite dramatic. Note that with such high Z_{in} values, op amp circuits may be used in place of FETs in some applications. One example would be the front-end amplifier/buffer in an electrometer (electrostatic voltmeter).

Distortion Effects

As noted earlier, negative feedback lowers static forms of distortion, such as THD. The question, as always, is "How much?" Something like harmonic distortion is internally generated, so we may model it as a voltage source in series with the output source. An example of this model can be seen in Figure 3.12. V_{dist} is the distortion

FIGURE 3.12

Distortion model (open-loop)

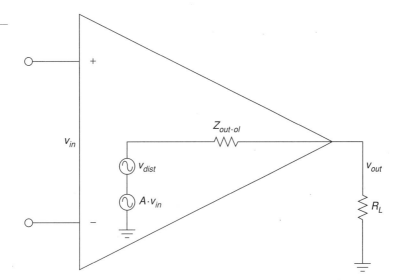

generator. In the case of a simple input sine wave, V_{dist} will contain harmonics at various amplitudes (sine waves at integer multiples of the input frequency). These amplitudes are directly related to the input signal's amplitude. If we assume that Z_{out} is small enough to ignore, V_{dist} will appear at the output of the open-loop circuit. Thus, the total output voltage is the desired AV_{in} plus V_{dist}. When we add feedback, as in Figure 3.13, this distortion signal is fed back to the inverting input, and because it is now out of phase, it partially cancels the internally generated distortion. Thus, the SP distortion signal ($V_{dist-sp}$) is much smaller. The SP output signal is

$$V_{out} = A_{ol}V_{error} + V_{dist}$$
$$V_{out} = A_{ol}(V_{in} - V_{feedback}) + V_{dist}$$
$$V_{out} = A_{ol}(V_{in} - \beta V_{out}) + V_{dist}$$

We now perform some algebra in order to get this into a nicer form and solve for V_{out}:

$$V_{out} = A_{ol}V_{in} - A_{ol}\beta V_{out} + V_{dist}$$
$$V_{out} + A_{ol}\beta V_{out} = A_{ol}V_{in} + V_{dist}$$
$$V_{out}(1 + A_{ol}\beta) = A_{ol}V_{in} + V_{dist}$$

Figure 3.13

Distortion model (closed-loop)

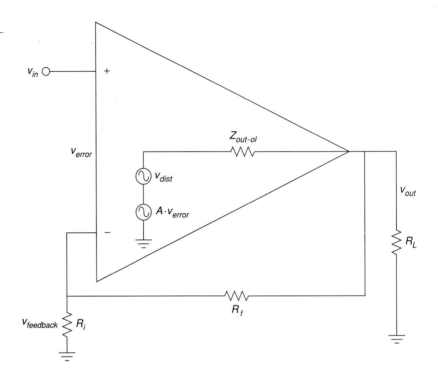

Remember that $1 + A_{ol}\beta = S$, so

$$V_{out} S = A_{ol} V_{in} + V_{dist}$$
$$V_{out} = \frac{A_{ol} V_{in}}{S} + \frac{V_{dist}}{S}$$

Because $\frac{A_{ol}}{S}$ is just A_{sp}, this reduces to

$$V_{out} = A_{sp} V_{in} + \frac{V_{dist}}{S}$$

The internally generated distortion is reduced by the sacrifice factor. As you can see, large sacrifice factors can drastically reduce distortion. An amplifier with 10% THD and a sacrifice factor of 100 produces an effective distortion of only .1%. This analysis does assume that the open-loop distortion is not overly grotesque. If the distortion is large, we cannot use this superposition approach (remember, superposition assumes that the circuit is essentially linear). Also, we are ignoring any additional distortion created by feeding this distortion back into the amplifier. For any reasonably linear amplifier, this extra distortion is a second-order effect, and thus constitutes only a small part of the total output signal.

multiSIM COMPUTER SIMULATION

An example of the reduction of distortion is simulated using the Distortion Analyzer from Electronics Workbench MultiSIM. A basic amplifier is shown in Figure 3.14 (see below and pp. 88–89). The value for R_f is changed from 999 kΩ to 99 kΩ to 9 kΩ. V_{in} is scaled accordingly so that the output of the amplifier remains at approximately 10 volts. The test frequency was set to 1 kHz, and a total of 20 harmonics (up to 20 kHz) were used in the analysis. Using the LF411 op amp, the high-gain version shows a THD of .09%. Reducing the gain by a factor of 10 (and thus increasing sacrifice factor by tenfold) yields a THD of .011%. Finally, a further tenfold reduction of gain yields a THD of .001%. Although the reduction in distortion is not exactly a factor of 10 each time, the trend can be seen clearly. The precise distortion values will depend on the accuracy of the op amp model used, the test frequency, and the number of harmonics kept in the analysis. 20 kHz was chosen here because that represents the upper limit of human hearing, and thus it would be appropriate for an audio amplifier.

Figure 3.14a

MultiSIM distortion analysis

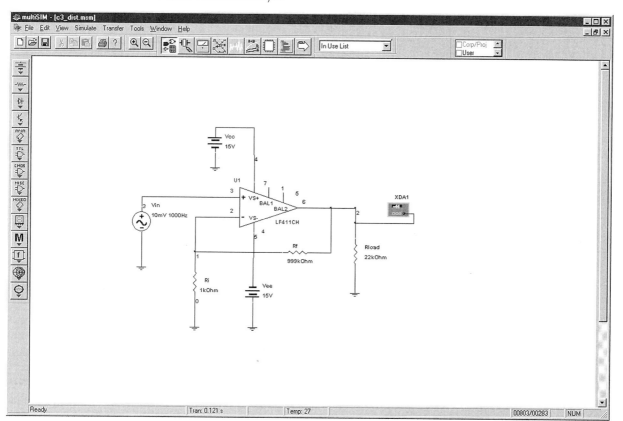

Figure 3.14b

Distortion analyzer results for high-gain version

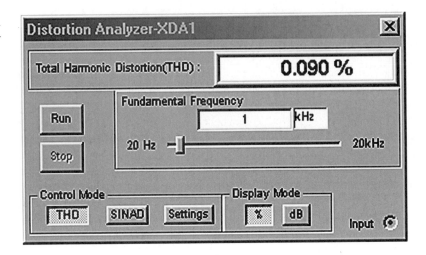

Figure 3.14c

Distortion analyzer results for medium-gain version

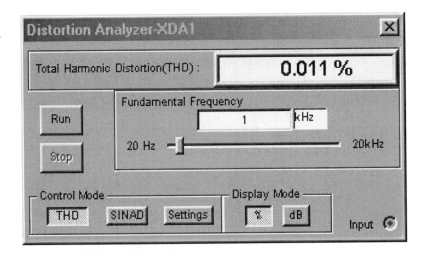

As we have seen, the sacrifice factor is a very useful item. Our gain, distortion, and Z_{out} are all reduced by S, and f_2 and Z_{in} are increased by S.

Noise

It is possible to model noise effects in much the same way as we just modeled distortion effects. By doing so, you will discover that noise can also be decreased by a large amount. Unfortunately, there is one major flaw. Unlike our distortion generator, a noise generator will produce a signal that is not dependent on the input signal. The net result is that although the noise level does drop by the sacrifice factor, so does

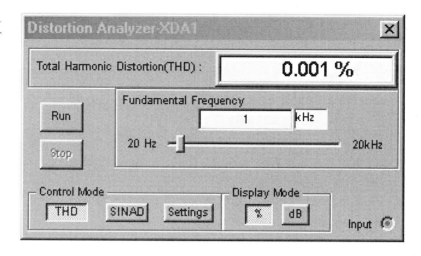

FIGURE 3.14d

Distortion analyzer results for low-gain version

the desired output signal. Thus, the signal-to-noise ratio at the output is unchanged. In contrast, the distortion signal is proportional to the input signal, so that when the desired signal is cut by S, the distortion signal sees a further cut by S (i.e., the distortion drops by S relative to the desired signal). As a matter of fact, it is quite possible that the noise produced in following stages may add up to more noise than the circuit had without feedback. Sad but true, negative feedback doesn't help us when it comes to signal-to-noise ratio.

Parallel-Series (PS)

The Parallel-Series connection is the opposite of the Series-Parallel form. PS negative feedback is used to make the ideal current amplifier. Its gain is dimensionless, but for convenience, it is normally given the units A/A (amps per amp). It produces a low Z_{in} (perfect for sensing I_{in}) and a high Z_{out} (making for an ideal current source). An example of PS is shown in Figure 3.15. The signal source's current splits in two, with part going into the amplifier and part going through the feedback network. This is how you know that you have a parallel input connection. The output current, on the other hand, passes through the load and then enters the feedback network, indicating a series output connection. Note that this general model is an inverting type and that the load is floating (i.e., not ground referenced). It is possible, though, to use ground referenced loads with some additional circuitry. As PS is used for current amplification, let's see how we can find the current gain. Figure 3.16 will help us along. PS current gain is defined as

$$A_{ps} = \frac{I_{out}}{I_{in}} \qquad (3.11)$$

FIGURE 3.15

Parallel-series connection

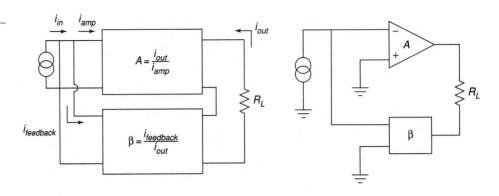

The signal source's current splits into two paths, so

$$I_{in} = I_{amp} + I_{feedback} \qquad (3.12)$$

Because I_{amp} times the amplifier's open-loop current gain is I_{out}, we can also say

$$I_{amp} = \frac{I_{out}}{A_{ol}} \qquad (3.13)$$

The feedback network is nothing more than a current divider, where the feedback network's output ($I_{feedback}$) is β times smaller than its input (I_{out}). Don't let the arbitrary current direction fool you—the feedback "flow" is still from right to left, as always. In this case the amplifier is sinking current instead of sourcing it. As β is just a fraction,

FIGURE 3.16

Parallel-series analysis

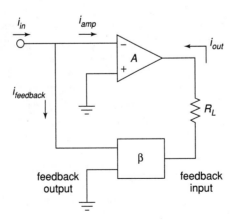

we may say

$$\beta = \frac{I_{feedback}}{I_{out}} \quad \text{or,}$$
$$I_{feedback} = \beta I_{out}$$
$$I_{feedback} = A_{ol}\beta I_{amp} \tag{3.14}$$

By substituting Equation 3.14 into Equation 3.12, we see that

$$I_{in} = I_{amp} + A_{ol}\beta I_{amp} \quad \text{or,}$$
$$I_{in} = I_{amp}(1 + A_{ol}\beta) \tag{3.15}$$

After substituting this into Equation 3.11, we find that

$$A_{ps} = \frac{I_{out}}{I_{amp}(1 + A_{ol}\beta)}$$

or, with the help of Equation 3.13,

$$A_{ps} = \frac{A_{ol}}{1 + A_{ol}\beta} \tag{3.16}$$

To make a long story short, the open-loop gain is reduced by the sacrifice factor. (Where have we seen this before?) The one item that you should note is that we have used only current gains in our derivation (compared to voltage gains in the SP case). It is possible to perform a derivation using the open-loop voltage gain; however, the results are basically the same, as you might have guessed. Once again, our approximation for gain can be expressed as $\frac{1}{\beta}$.

EXAMPLE 3-4

A coworker asks you to measure the output current of a circuit that she's just built. If all is working correctly, this circuit should produce 100 μ amps. Unfortunately, your handheld DMM will only accurately measure down to 1 milliamp. It is 10 times less sensitive than it needs to be. In order to be read accurately, the current will need to be boosted. Does the amplifier of Figure 3.17 have the current gain you need?

If we are using an op amp, we can assume that $A_{ps} = \frac{1}{\beta}$. The question then becomes, "What is β?" β is the current divider ratio. The resistors R_1 and R_2 make up the current

Figure 3.17

Current amplifier for Example 3-4

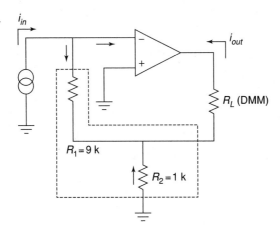

divider. The output current splits between R_1 and R_2, where the R_1 path is $I_{feedback}$. According to the current divider rule,

$$\beta = \frac{R_2}{R_1 + R_2}$$

Consequently,

$$A_{ps} = \frac{R_1 + R_2}{R_2} \quad \text{or,}$$

$$A_{ps} = \frac{R_1}{R_2} + 1$$

This equation looks a lot like the one in the SP example! Solving for A_{ps} yields

$$A_{ps} = \frac{9\,k}{1\,k} + 1$$
$$A_{ps} = 10$$

Yes, this circuit has just the gain you need. All you have to do is connect your coworker's circuit to the input and replace the load resistor with your handheld DMM. Note that most DMMs are floating instruments and are not tied to ground (unlike an oscilloscope), so using it as a floating load presents no problems. The accuracy of the gain (and thus your measurement) depends on the accuracy of the resistors and the relative size of the op amp's input bias current. Therefore, it would be advisable to use a bi-FET type device (i.e., an op amp with an FET diff amp front end) and precision resistors.

FIGURE 3.18

Parallel-series input impedance

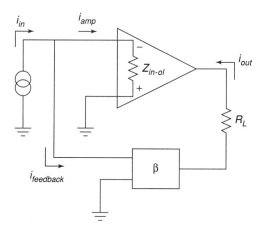

PS Impedance Effects

As you have seen, the gain derivation for PS is similar to that for SP. The same is true for the impedance equations. First, we'll take a look at input impedance with Figure 3.18. As always, we start with our base definition,

$$Z_{in-ps} = \frac{V_{in}}{I_{in}}$$

Recalling Equation 3.15, this can be rewritten as

$$Z_{in-ps} = \frac{V_{in}}{I_{amp}(1 + A_{ol}\beta)} \quad (3.17)$$

V_{in} is merely the voltage that appears from the inverting input to ground. By using Ohm's law, we can say

$$V_{in} = I_{amp} Z_{in-ol}$$

Where Z_{in-ol} is the open-loop input impedance. Finally, substituting this into Equation 3.17 gives

$$Z_{in-sp} = \frac{I_{amp} Z_{in-ol}}{I_{amp}(1 + A_{ol}\beta)}$$

$$Z_{in-sp} = \frac{Z_{in-ol}}{1 + A_{ol}\beta}$$

The input impedance is lowered by the sacrifice factor. If you're starting to wonder whether everything is altered by the sacrifice factor, the answer is yes. This is, of course, an approximation. What do you think is going to happen to the output impedance? At this point, it shouldn't be too surprising. For this proof, refer to Figure 3.19. We shall

Figure 3.19

Parallel-series output impedance

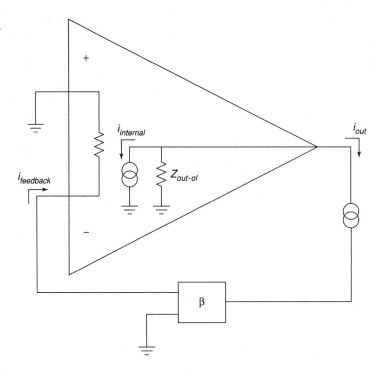

use the same general technique to find Z_{out} as we did with the SP configuration. In this case, we replace the load with a current source, and then determine the resulting output voltage. Note that the input signal current source is opened. The output current drives the feedback network and produces the feedback current. The feedback current is then multiplied by the amplifier's open-loop current gain. Because the feedback current enters the inverting input, the internal source is sinking current. Because we are driving the circuit from the output,

$$Z_{out-ps} = \frac{V_{out}}{I_{out}}$$

V_{out} is found through Ohm's law.

$$V_{out} = Z_{out-ol}(I_{out} + I_{int})$$

We now expand on our currents.

$$I_{int} = I_{feedback} A_{ol}$$
$$I_{int} = I_{out} \beta A_{ol}$$

$$V_{out} = Z_{out-ol}(I_{out} + I_{out}\beta A_{ol})$$
$$V_{out} = Z_{out-ol} I_{out}(1 + A_{ol}\beta)$$
$$V_{out} = Z_{out-ol} I_{out} S$$

Finally, we see that

$$Z_{out-ps} = \frac{Z_{out-ol} I_{out} S}{I_{out}}$$

$$Z_{out-ps} = Z_{out-ol} S$$

As expected, the series output connection increased Z_{out} by the sacrifice factor. The remainder of the PS equations are essentially those used in our earlier SP work. Once again, the bandwidth will be increased by S, and the distortion will be reduced by S. This is also true for the Parallel-Parallel and Series-Series connections. As a matter of fact, the Z_{in} and Z_{out} relations are just what you might expect. The proofs are basically the same as those already presented, so we won't go into them. Suffice to say that parallel connections reduce the impedance by S, and series connections increase it by S.

Parallel-Parallel (PP) and Series-Series (SS)

Unlike our two earlier examples, the concept and modeling of gain is not quite as straightforward in the Parallel-Parallel and Series-Series cases. These forms do not produce gain, so to speak. They are neither voltage amplifiers nor current amplifiers. Instead, these connections are used as *transducers*. Parallel-Parallel turns an input current into an output voltage. It is shown in Figure 3.20. Series-Series turns an input voltage into an output current. It is shown in Figure 3.21. Our normal gain and feedback factors now have units associated with them. Because our output versus input quantities are measured in volts-per-amp (PP) or amps-per-volt (SS), the appropriate units for our factors are ohms and siemens. To be specific, we refer to our PP gain as a transresistance value and the SS gain as a transconductance value. Although it is quite possible (and useful) to derive formulas for gain based on these, it is often done only for discrete designs. Quite simply, you cannot find transresistance or transconductance values on typical op amp data sheets.

Fortunately, we can make a few approximations and create some very useful circuits utilizing PP and SS feedback with op amps. These design and analysis shortcuts are presented in the next chapter, along with practical applications.

FIGURE 3.20

Parallel-parallel connection

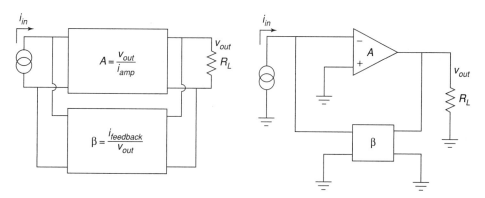

FIGURE 3.21

Series-series connection

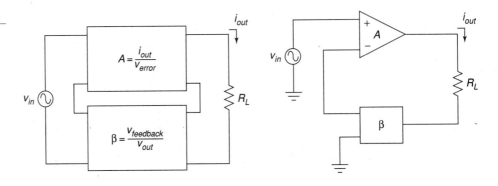

3.5 Limitations on the Use of Negative Feedback

From the foregoing discussion, you may well think that negative feedback can do just about anything, short of curing a rainy day. Such is not the case. Yes, negative feedback can drastically lower distortion and increase bandwidth. Yes, it can have a very profound effect on input and output impedance. And yes, it certainly does stabilize our gains. What, then, is the problem? Like all things, negative feedback has specific limitations. The first thing that you should note is that S is a function of frequency. This was graphically depicted back in Figure 3.2. The amount of change seen in impedances and distortion is a function of S, thus it follows that these changes must be a function of frequency. Because S drops as the frequency increases, the effects of negative feedback diminish as well. For example, if an SP amplifier has an open-loop Z_{in} of 200 kΩ and the low-frequency S is 500, the resulting Z_{in} with feedback is 100 MΩ. If we increase the input frequency past the open-loop f_2, the open-loop gain drops and thus, S drops. One decade up, S will only be 50, so the Z_{in} with feedback will only be 10 MΩ. If this amplifier has a lower break frequency (f_1), S will also drop as the frequency is reduced (below f_1). The same sort of thing occurs with distortion; however, the harmonics each see a different S, so the calculation is a bit more involved. Along with the reduction in gain, there is also a change in phase. If the phase around the feedback loop varies from $-180°$, incomplete cancellation takes place, and thus, the effects of feedback are lessened. The bottom line is that the effects of negative feedback weaken as we approach the frequency extremes.

The other item that must be kept in mind is the fact that negative feedback does not change specific fundamental characteristics of the amplifier. Negative feedback cannot get a circuit to do something beyond its operational parameters. For example, feedback has no effect on clipping level (saturation point). Further, feedback has no effect on *slew rate* (the maximum rate of output signal change, and an item that we will examine in a later chapter). Actually, when an amplifier slews, feedback is effectively blocked. An accurate output signal is no longer sent back to the input, but the amplifier can't correct for the errors any faster than it already is. In a similar manner, even though feedback can be used to lower the output impedance of a system, this does not imply that the system can produce more output current.

One possible "problem" with negative feedback is really the fault of the designer. It can be very tempting to sloppily design an amplifier with poor characteristics and then correct for them with large amounts of feedback. No matter how much feedback you use, the result will never be as good as a system that was designed carefully from the start. An example of this effect can be seen with TIMD (Transient Inter-Modulation Distortion). TIMD is a function of nonlinearities in the first stages of an amplifier, and the excessive application of negative feedback will not remove it. On the other hand, if the initial stages of the system are properly designed, TIMD is not likely to be a problem.[1]

SUMMARY

We have seen that negative feedback can enhance the performance of amplifier circuits. This is done by sampling a portion of the output signal and summing it out of phase with the input signal. In order to maintain stability, the gain of the amplifier must be less than unity by the time its phase reaches $-180°$. There are four basic variants of negative feedback: Series-Parallel, Parallel-Series, Parallel-Parallel, and Series-Series. In all cases, gain and distortion are lowered by the sacrifice factor S, and the bandwidth is increased by S. Parallel connections reduce impedance by S, whereas series connections increase the impedance by S. At the input, parallel connections are current sensing, and series connections are voltage sensing. At the output, parallel connections produce a voltage-source model, and series connections produce a current-source model. The sacrifice factor is the ratio between the open- and closed-loop gains. It is a function of frequency, and therefore, the effects of negative feedback lessen at the frequency extremes.

REVIEW QUESTIONS

1. Give two examples of how negative feedback is used in everyday life.
2. What circuit parameters will negative feedback alter, and to what extent?
3. What is meant by the term Sacrifice Factor?
4. What is the usage of gain and phase margin?
5. Name the different negative feedback connections (i.e., variants or forms).
6. How might negative feedback accidentally turn into positive feedback?
7. What circuit parameters won't negative feedback affect?
8. In practical amplifiers, when does negative feedback "stop working," and why?

PROBLEMS

Analysis Problems

1. An amplifier's open-loop gain plot is given in Figure 3.22. If the amplifier is set up for a closed-loop gain of 100, what is the sacrifice factor (S) at low frequencies? What is S at 1 kHz?
2. If the amplifier in Problem 1 has an open-loop THD of 5%, what is the closed-loop THD at low frequencies? Assuming that the open-loop THD doesn't change with frequency, what is the closed-loop THD at 1 kHz?
3. If an amplifier has an open-loop response as given in Figure 3.22, and a feedback factor (β) of .05 is used, what is the exact low-frequency closed-loop gain? What

---

[1] See E. M. Cherry and K. P. Dabke, "Transient Intermodulation Distortion—Part 2: Soft Nonlinearity," *Journal of the Audio Engineering Society* 34, no. 1/2 (1986):19–35.

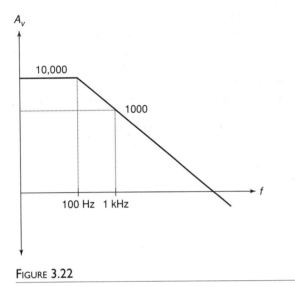

FIGURE 3.22

6. What is the maximum allowable phase shift at 500 kHz for stable usage of negative feedback in Figure 3.23?
7. Determine the gain and phase margins for the amplifier response given in Figure 3.24. Is this amplifier a good candidate for negative feedback?

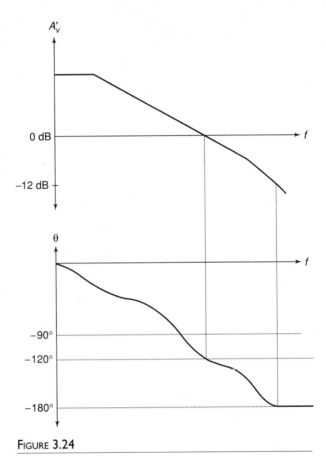

FIGURE 3.24

is the approximate low-frequency gain? What is the approximate gain at 1 kHz?

4. Using the open-loop response curve in Figure 3.23, determine exact and approximate values of β for a closed-loop gain of 26 dB.

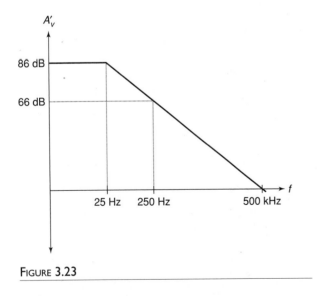

FIGURE 3.23

5. Determine the closed-loop f_2 for the circuit of Problem 4.

8. Determine the closed-loop (midband) gain in Figure 3.25.
9. What is the closed-loop Z_{in} in Figure 3.25? What is Z_{out}?
10. What is the low-frequency sacrifice factor in Figure 3.25?
11. How much distortion reduction can we hope for in Figure 3.25?
12. How much of a signal/noise improvement can we expect in Figure 3.25?
13. Assuming $V_{in}(t) = .1 \sin 2\pi 500 t$ in Figure 3.25, what is $V_{out}(t)$?

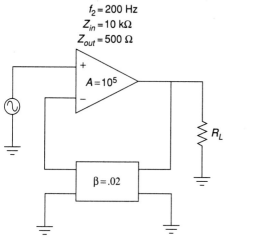

FIGURE 3.25

14. If the circuit in Figure 3.25 had an open-loop f_1 of 10 Hz, what would the closed-loop f_1 be?
15. Determine an appropriate pair of resistors to set β to .1 in Figure 3.25.

Challenge Problems

16. If the feedback network of Figure 3.25 produces a phase shift of $-200°$ at 4 kHz, what effect will this have on circuit operation in Problem 10?
17. Consider the circuit of Figure 3.7. In general, what effect will the following alterations to the feedback network have on the closed-loop system response: (a) Placing a capacitor across R_f, (b) Placing a capacitor across R_i, (c) Placing a rectifying diode across R_f (both polarities)?

Computer Simulation Problems

18. Rerun the simulation of Figure 3.8 using the following open-loop gains: 1 k, 10 k, and 100 k. What can you conclude from the results?
19. Verify the results of Problem 13 using a circuit simulator. It is possible to extend the basic op amp model with a lag network in order to mimic f_{2-ol}.
20. Verify the stability of the circuit used in Problem 13 with regard to the open-loop gain. Run simulations with the given A_{ol} and with values one decade above and below. Compare the resulting simulations and determine the maximum deviation of A_{sp}.

CHAPTER 4

BASIC OP AMP CIRCUITS

CHAPTER OBJECTIVES

After completing this chapter, you should be able to:

- Relate each op amp circuit back to its general feedback form.
- Detail the general op amp circuit analysis idealizations.
- Solve inverting and noninverting voltage amplifier circuits for a variety of parameters, including gain and input impedance.
- Solve voltage/current transducer circuits for a variety of parameters.
- Solve current amplifier circuits for a variety of parameters.
- Define the term *virtual ground*.
- Analyze and design differential amplifiers.
- Analyze and design inverting and noninverting summing amplifiers.
- Discuss how output current capability may be increased.
- Outline the circuit modifications required for operation from a single polarity power supply.

4.1 INTRODUCTION

In this chapter we will be examining some common uses of op amps. Where Chapter 3 focused on the theory of negative feedback in a more abstract way, this chapter zeroes in on the practical results of using negative feedback with op amps. Now that you know what negative feedback is and how it works, you can reap the benefits. The main theme is in the design and analysis of simple, small-signal, linear amplifiers. We will also be looking at some convenient approximation methods of analysis that can prove quite efficient. By the end of the chapter, you should be able to design single and multistage voltage amplifiers, voltage followers, and even amplifiers that sense current and/or produce constant current output. The voltage amplifiers offer the possibilities of inverting or not inverting the signal. Simple differential amplifiers are examined too. The chapter wraps up with a section on using op amps with a single-polarity power supply and how to increase the available output current.

This is an introductory design section, so the details of high frequency response, noise, offsets, and other important criteria are ignored. These items await a detailed analysis in Chapter 5. For the most part, then, all calculations and circuit operations are assumed to be in the midband region.

4.2 Inverting and Noninverting Amplifiers

As noted in our earlier work, negative feedback can be applied in one of four ways. The parallel input form inverts the input signal, and the series input form doesn't. Because these forms were presented as current-sensing and voltage-sensing respectively, you might get the initial impression that all voltage amplifiers must be noninverting. This is not the case. With the simple inclusion of one or two resistors, for example, we can make inverting voltage amplifiers or noninverting current amplifiers. Virtually all topologies are realizable. We will look at the controlled-voltage source forms first (those using SP and PP negative feedback).

For analysis, you can use the classic treatment given in Chapter 3; however, due to some rather nice characteristics of the typical op amp, approximations will be shown. These approximations are only valid in the midband and say nothing of the high-frequency performance of the circuit. Therefore, they are not suitable for general-purpose discrete work. The idealizations for the approximations are:

1. The input current is virtually zero (i.e., Z_{in} is infinite).
2. The potential difference between the inverting and noninverting inputs is virtually zero (i.e., loop gain is infinite). This signal is also called the error signal.

Also, note for clarity that the power supply connections are not shown in most of the diagrams.

The Noninverting Voltage Amplifier

The noninverting voltage amplifier is based on SP negative feedback. An example is given in Figure 4.1. Note the similarity to the generic SP circuits of Chapter 3. Recalling the basic action of SP negative feedback, we expect a very high Z_{in}, a very low Z_{out}, and a reduction in voltage gain. Idealization 1 states that Z_{in} must be infinite. We already know that op amps have low Z_{out}s, so the second item is taken care of. Now let's take a look at voltage gain.

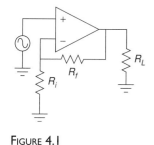

Figure 4.1

Noninverting voltage amplifier

$$A_v = \frac{V_{out}}{V_{in}}$$

Because ideally $V_{error} = 0$,

$$V_{in} = V_{R_i}$$

Also,

$$V_{out} = V_{R_i} + V_{R_f}$$
$$A_v = \frac{V_{R_i} + V_{R_f}}{V_{R_i}}$$

Expansion gives

$$A_v = \frac{R_i I_{R_i} + R_f I_{R_f}}{R_i I_{R_i}}$$

Because $I_{in} = 0$,

$$I_{R_f} = I_{R_i}$$

So, finally we come to

$$A_v = \frac{R_i + R_f}{R_i} \quad \text{or}$$
$$A_v = 1 + \frac{R_f}{R_i} \tag{4.1}$$

Now that's convenient. The gain of this amplifier is set by the ratio of two resistors. The larger R_f is relative to R_i, the more gain you get. Remember, this is an approximation. The closed-loop gain can never exceed the open-loop gain, and eventually, A_v will fall off as frequency increases. Note that the calculation ignores the effect of the load impedance. Obviously, if R_l is too small, the excessive current draw will cause the op amp to clip.

EXAMPLE 4-1

What are the input impedance and gain of the circuit in Figure 4.2?
First off, Z_{in} is ideally infinite. Now for the gain:

$$A_v = 1 + \frac{R_f}{R_i}$$
$$A_v = 1 + \frac{10\,k}{1\,k}$$
$$A_v = 11$$

The opposite process of amplifier design is just as straightforward.

FIGURE 4.2

Noninverting circuit for Example 4-1

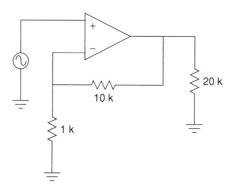

EXAMPLE 4-2

Design an amplifier with a gain of 26 dB and an input impedance of 47 kΩ.
For the gain, first turn 26 dB into ordinary form. This is a voltage gain of about 20.

$$A_v = 1 + \frac{R_f}{R_i}$$

$$\frac{R_f}{R_i} = A_v - 1$$

$$\frac{R_f}{R_i} = 19$$

At this point, choose a value for one of the resistors and solve for the other one. For example, the following would all be valid:

$$R_i = 1 \text{ k}\Omega, \quad R_f = 19 \text{ k}\Omega$$
$$R_i = 2 \text{ k}\Omega, \quad R_f = 38 \text{ k}\Omega$$
$$R_i = 500 \text{ }\Omega, \quad R_f = 9.5 \text{ k}\Omega$$

A reasonable range is 100 kΩ > $R_i + R_f$ > 10 kΩ. The accuracy of this gain will depend on the accuracy of the resistors. Now for the Z_{in} requirement. This is deceptively simple. Z_{in} is assumed to be infinite, so all you need to do is place a 47 kΩ in parallel with the input. The resulting circuit is shown in Figure 4.3. If a specific Z_{in} is not required, a resistor in this

FIGURE 4.3

Noninverting design for Example 4-2

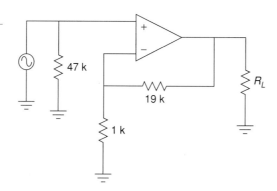

position is not required. There is one exception to this rule. If the driving source is not directly coupled to the op amp input (e.g., it is capacitively coupled), a resistor will be required to establish a DC return path to ground. Without a DC return path, the input section's diff amp stage will not be properly biased. This point is worth remembering, as it can save you a great deal in future headaches. For example, in the lab a circuit like the one in Figure 4.2 may work fine with one function generator, but not with another. This would be the case if the second generator used an output coupling capacitor and the first one didn't.

EXAMPLE 4-3

Design a voltage follower (i.e., ideally infinite Z_{in} and a voltage gain of 1).

The Z_{in} part is straightforward enough. As for the second part, what ratio of R_f to R_i will yield a gain of 1?

$$A_v = 1 + \frac{R_f}{R_i}$$

$$\frac{R_f}{R_i} = A_v - 1$$

$$\frac{R_f}{R_i} = 0$$

This says that R_f must be 0 Ω. Practically speaking, that means that R_f is replaced with a shorting wire. What about R_i? Theoretically, almost any value will do. As long as there's a choice, consider infinite. Zero divided by infinite is certainly zero. The practical benefit of choosing $R_i = \infty$ is that you may delete R_i. The resulting circuit is shown in Figure 4.4. Remember, if the source is not directly coupled, a DC return resistor will be needed. The value of this resistor has to be large enough to avoid loading the source.

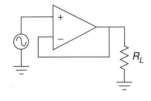

FIGURE 4.4

Voltage follower for Example 4-3

As you can see, designing with op amps can be much quicker than its discrete counterpart. As a result, your efficiency as a designer or repair technician can improve greatly. You are now free to concentrate on the system, rather than on the specifics of an individual biasing resistor. In order to make multistage amplifiers, just link individual stages together.

EXAMPLE 4-4

What is the input impedance of the circuit in Figure 4.5? What is V'_{out}?

As in any multistage amplifier, the input impedance to the first stage is the system Z_{in}. The DC return resistor sets this at 100 kΩ.

To find V_{out}, we need to find the gain (in dB).

$$A_{v1} = 1 + \frac{R_f}{R_i}$$

$$A_{v1} = 1 + \frac{14 \text{ k}}{2 \text{ k}}$$

$$A_{v1} = 8$$

$$A'_{v1} = 18 \text{ dB}$$

Figure 4.5

Multistage circuit for Example 4-4

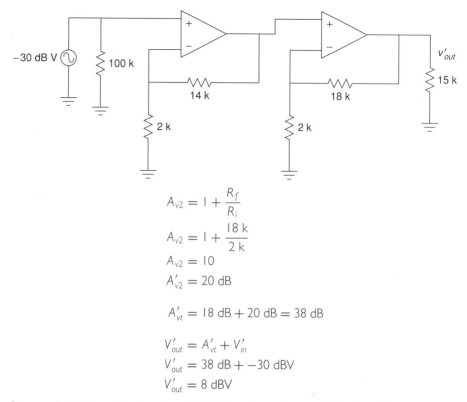

$$A_{v2} = 1 + \frac{R_f}{R_i}$$

$$A_{v2} = 1 + \frac{18\,k}{2\,k}$$

$$A_{v2} = 10$$

$$A'_{v2} = 20\text{ dB}$$

$$A'_{vt} = 18\text{ dB} + 20\text{ dB} = 38\text{ dB}$$

$$V'_{out} = A'_{vt} + V'_{in}$$

$$V'_{out} = 38\text{ dB} + -30\text{ dBV}$$

$$V'_{out} = 8\text{ dBV}$$

Because 8 dBV translates to about 2.5 V, there is no danger of clipping either.

Inverting Voltage Amplifier

The inverting amplifier is based on the PP negative feedback model. The base form is shown in Figure 4.6. By itself, this form is current sensing, not voltage sensing. In order to achieve voltage sensing, an input resistor, R_i, is added. See Figure 4.7. Here's how the circuit works: V_{error} is virtually zero, so the inverting input potential must equal the noninverting input potential. This means that the inverting input is at a *virtual ground*. The signal here is so small that it is negligible. Because of this, we may also say that the impedance seen looking into this point is zero. This last point may cause a bit of confusion. You may ask, "How can the impedance be zero if the current into the op amp is zero?" The answer lies in the fact that all of the entering current will be drawn through R_f, thus bypassing the inverting input.

Refer to Figure 4.8 for the detailed explanation. The right end of R_i is at virtual ground, so all of the input voltage drops across it, creating I_{in}, the input current. This current cannot enter the op amp and instead will pass through R_f. Because a positive signal is presented to the inverting input, the op amp will sink output current, thus drawing I_{in} through R_f. The resulting voltage drop across R_f is the same magnitude as the load voltage. This is true because R_f is effectively in parallel with the load. Note that both elements are tied to the op amp's output and to (virtual) ground. There is a

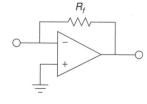

Figure 4.6

A basic parallel-parallel amplifier

FIGURE 4.7

Inverting voltage amplifier

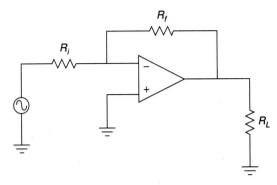

change in polarity because we reference the output signal to ground. In short, V_{out} is the voltage across R_f, inverted.

$$A_v = \frac{V_{out}}{V_{in}}$$
$$V_{in} = I_{in} R_i$$
$$V_{out} = -V_{R_f}$$
$$V_{R_f} = I_{in} R_f$$

Substitution yields

$$A_v = -\frac{I_{in} R_f}{I_{in} R_i}$$
$$A_v = -\frac{R_f}{R_i} \tag{4.2}$$

Again, we see that the voltage gain is set by resistor ratio. Again, there is an allowable range of values.

The foregoing discussion points up the derivation of input impedance. Because all of the input signal drops across R_i, it follows that all the driving source "sees" is R_i. Quite simply, R_i sets the input impedance. Unlike the noninverting voltage amp, there

FIGURE 4.8

Analysis of the inverting amplifier from Figure 4.7

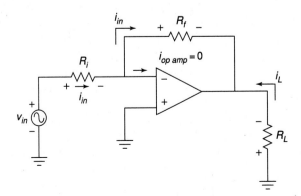

FIGURE 4.9

Inverting amplifier for Example 4-5

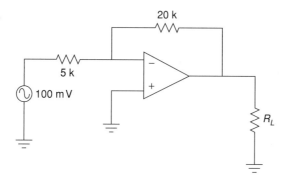

is a definite interrelation between Z_{in} (R_i) and A_v ($\frac{-R_f}{R_i}$). This indicates that it is very hard to achieve both high gain and high Z_{in} with this circuit.

EXAMPLE 4-5

Determine the input impedance and output voltage for the circuit in Figure 4.9.
The input impedance is set by R_i. $R_i = 5$ kΩ, therefore $Z_{in} = 5$ kΩ.

$$V_{out} = V_{in} A_v$$
$$A_v = -\frac{R_f}{R_i}$$
$$A_v = -\frac{20\ k}{5\ k}$$
$$A_v = -4$$

$$V_{out} = 100\ mV \times (-4)$$
$$V_{out} = -400\ mV,\ (\text{i.e., inverted})$$

EXAMPLE 4-6

Design an inverting amplifier with a gain of 10 and an input impedance of 15 kΩ.
The input impedance tells us what R_i must be.

$$Z_{in} = R_i$$
$$R_i = 15\ k$$

Knowing R_i, solve for R_f:

$$A_v = -\frac{R_f}{R_i}$$
$$R_f = R_i(-A_v)$$
$$R_f = 15\ k \times (-(-10))$$
$$R_f = 150\ k$$

multiSIM COMPUTER SIMULATION

An Electronics Workbench MultiSIM simulation of the result of Example 4-6 is shown in Figure 4.10 (see below and page 109), along with its schematic. This simulation uses the simple dependent source model presented in Chapter 2. The input is set at .1 V DC for simplicity. Note that the output potential is negative, indicating the inverting action of the amplifier. Also, note that the virtual ground approximation is borne out quite well, with the inverting input potential measuring in the μV region.

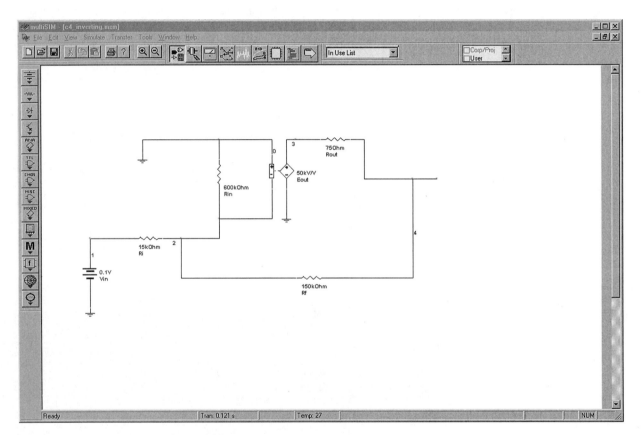

FIGURE 4.10a

MultiSIM simulation of the simple op amp model for Example 4-6. a. MultiSIM schematic

EXAMPLE 4-7

The circuit of Figure 4.11 is a preamplifier stage for an electronic music keyboard. Like most musicians' preamplifiers, this one offers adjustable gain. This is achieved by following the amplifier with a pot. What are the maximum and minimum gain values?

FIGURE 4.10b
MultiSIM output listing

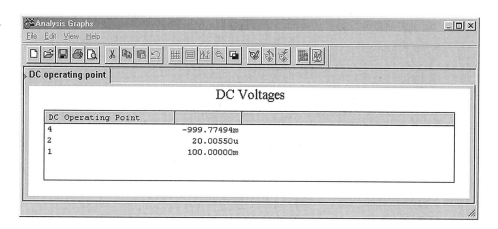

Note that the gain for the preamp is the product of the op amp gain and the voltage divider ratio produced by the pot. For maximum gain, use the pot in its uppermost position. Because the pot acts as a voltage divider, the uppermost position provides no divider action (i.e., its gain is unity). For midband frequencies, the 20 pF may be ignored.

$$A_{v-max} = -\frac{R_f}{R_i}$$
$$A_{v-max} = -\frac{200\ k}{15\ k}$$
$$A_{v-max} = -13.33$$
$$A'_{v-max} = 22.5\ dB$$

FIGURE 4.11
Musical instrument preamplifier for Example 4-7

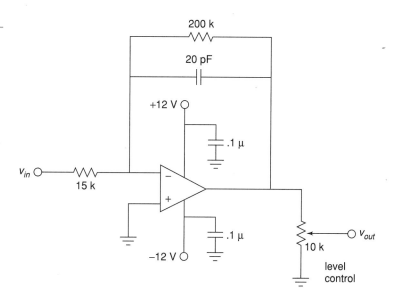

For minimum gain, the pot is dialed to ground. At this point, the divider action is infinite, and thus the minimum gain is 0 (resulting in silence).

Z_{in} for the system is about 15 kΩ. As far as the extra components are concerned, the 20 pF capacitor is used to decrease high frequency gain. The two .1 μF bypass capacitors across the power supply lines are very mportant. Virtually all op amp circuits use bypass capacitors. Due to the high-gain nature of op amps, it is essential to have good AC grounds at the power supply pins. At higher frequencies the inductance of power supply wiring may produce a sizable impedance. This impedance may create a positive feedback loop that wouldn't exist otherwise. Without the bypass capacitors, the circuit may oscillate or produce spurious output signals. The precise values for the capacitors are usually not critical, with .1 to 1 μF being typical.

Inverting Current-to-Voltage Transducer

As previously mentioned, the inverting voltage amplifier is based on PP negative feedback, with an extra input resistor used to turn the input voltage into a current. What happens if that extra resistor is left out, and a circuit such as Figure 4.6 is used? Without the extra resistor, the input is at virtual ground, thus setting Z_{in} to 0 Ω. This is ideal for sensing current. This input current will pass through R_f and produce an output voltage as outlined above. The characteristic of transforming a current to a voltage is measured by the parameter transresistance. By definition, the transresistance of this circuit is the value of R_f. To find V_{out}, multiply the input current by the transresistance. This circuit inverts polarity as well.

$$V_{out} = I_{in} R_f \tag{4.3}$$

Example 4-8

Design a circuit based on Figure 4.6 if an input current of 50 μA should produce an output of 4 V.

The transresistance of the circuit is R_f.

$$R_f = \frac{V_{out}}{I_{in}}$$

$$R_f = \frac{4 \text{ V}}{50 \text{ }\mu\text{A}}$$

$$R_f = 80 \text{ k}$$

The input impedance is assumed to be zero. At first glance, the circuit applications of this topology seem very limited. In reality, there are a number of linear integrated circuits that produce their output in current form.[1] In many cases, this signal must be turned into a voltage in order to properly interface with other circuit elements. The current-to-voltage transducer is widely used for this purpose.

---

[1] Most notably, operational transconductance amplifiers and digital-to-analog convertors, which we shall examine in Chapters 6 and 12, respectively.

INVERTING AND NONINVERTING AMPLIFIERS   111

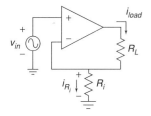

FIGURE 4.12

Voltage-to-current transducer

Noninverting Voltage-to-Current Transducer

This circuit topology utilizes SS negative feedback. It senses an input voltage and produces a current. A conceptual comparison can be made to the FET (a voltage-controlled current source). Instead of circuit gain, we are interested in transconductance. In other words, how much input voltage is required to produce a given output current? The op amp circuit presented here drives a floating load. That is, the load is not referenced to ground. This can be convenient in some cases and a real pain in others. With some added circuitry, it is possible to produce a grounded-load version, although space precludes us from examining it here.

A typical voltage-to-current circuit is shown in Figure 4.12. Because this uses series-input type feedback, we may immediately assume that Z_{in} is infinite. The voltage to current ratio is set by feedback resistor R_i. As V_{error} is assumed to be zero, all of V_{in} drops across R_i, creating current I_{R_i}. The op amp is assumed to have zero input current, so all of I_{R_i} flows through the load resistor, R_l. By adjusting R_i, the load current may be varied.

$$I_{load} = I_{R_i}$$
$$I_{R_i} = \frac{V_{in}}{R_i}$$
$$I_{load} = \frac{V_{in}}{R_i}$$

By definition,

$$\text{Transconductance } (gm) = \frac{I_{load}}{V_{in}}$$
$$gm = \frac{1}{R_i} \tag{4.4}$$

So, the transconductance of the circuit is set by the feedback resistor. As usual, there are practical limits to the size of R_i. If R_i and R_l are too small, the possibility exists that the op amp will "run out" of output current and go into saturation. At the other extreme, the product of the two resistors and the I_{load} cannot exceed the power supply rails. As an example, if R_i plus R_l is 10 kΩ, I_{load} cannot exceed about 1.5 mA if standard ±15 V supplies are used.

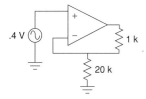

FIGURE 4.13

Voltage-to-current transducer for Example 4-9

EXAMPLE 4-9

Given an input voltage of .4 V in the circuit of Figure 4.13, what is the load current?

$$gm = \frac{1}{R_i}$$
$$gm = \frac{1}{20 \text{ k}}$$
$$gm = 50 \ \mu S$$

$$I_{load} = gm V_{in}$$
$$I_{load} = 50\ \mu S \times .4\ V$$
$$I_{load} = 20\ \mu A$$

There is no danger of current overload here, as the average op amp can produce about 20 mA, maximum. The output current will be 20 μA regardless of the value of R_l, up to clipping. There is no danger of clipping in this situation either. The voltage seen at the output of the op amp to ground is

$$V_{max} = (R_i + R_l)I_{load}$$
$$V_{max} = (20\ k + 1\ k) \times 20\ \mu A$$
$$V_{max} = 420\ mV$$

That's well below clipping level.

multiSIM Computer Simulation

A simulation of the circuit of Example 4-9 is shown in Figure 4.14 (see pp. 113–114). MultiSIM's ideal op amp model has been chosen to simplify the layout. The load current is exactly as calculated at 20 μA. An interesting trick is used here to plot the load current, as many simulators only offer plotting of node voltages. Using MultiSIM's Post Processor, the load current is computed by taking the difference between the node voltages on either side of the load resistor and then dividing the result by the load resistance.

Example 4-10

The circuit of Figure 4.15 can be used to make a high-input-impedance DC voltmeter. The load in this case is a simple meter movement. This particular meter requires 100 μA for full-scale deflection. If we want to measure voltages up to 10 V, what must R_i be?

First, we must find the transconductance.

$$gm = \frac{I_{load}}{V_{in}}$$
$$gm = \frac{100\ \mu A}{10\ V}$$
$$gm = 10\ \mu S$$

$$R_i = \frac{1}{gm}$$
$$R_i = \frac{1}{10\ \mu S}$$
$$R_i = 100\ k$$

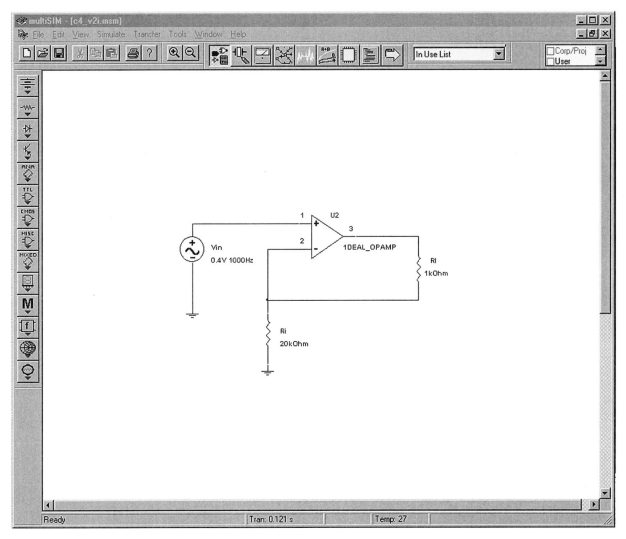

FIGURE 4.14a

Voltage to current transducer simulation schematic

The meter deflection is assumed to be linear. For example, if the input signal is only 5 V, the current produced is halved to 50 μA. 50 μA should produce half-scale deflection. The accuracy of this electronic voltmeter depends on the accuracy of R_i and the linearity of the meter movement. Note that this little circuit can be quite convenient in a lab, being powered by batteries. In order to change scales, new values of R_i can be swapped in with a rotary switch. For a 1 V scale, R_i equals 10 kΩ. Note that for higher input ranges, some form of input attenuator is needed. This is due to the fact that most op amps may be damaged if input signals larger than the supply rails are used.

FIGURE 4.14b

Simulation results

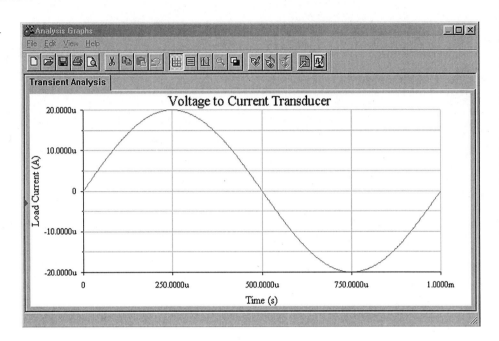

Inverting Current Amplifier

The inverting current amplifier uses PS negative feedback. As in the voltage-to-current transducer, the load is floating. The basic circuit is shown in Figure 4.16. Due to the parallel negative feedback connection at the input, the circuit input impedance is assumed to be zero. This means that the input point is at virtual ground. The current into the op amp is negligible, so all input current flows through R_i to node A. Effectively, R_i and R_f are in parallel (they both share node A and ground—actually virtual ground for R_i). Therefore, V_{R_i} and V_{R_f} are the same value. This means that a current is flowing through R_f from ground to node A. These two currents join to form the load current. In this manner, current gain is achieved. The larger I_{R_f} is relative to I_{in}, the more current gain there is.

$$A_i = \frac{I_{out}}{I_{in}}$$

FIGURE 4.15

DC voltmeter for Example 4-10

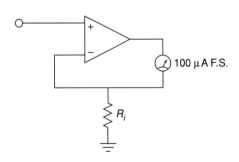

FIGURE 4.16

Inverting current amplifier

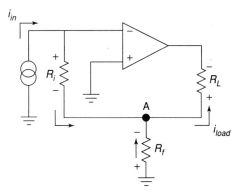

$$I_{out} = I_{R_f} + I_{R_i} \quad (4.5)$$
$$I_{R_i} = I_{in}$$
$$I_{R_f} = \frac{V_{R_f}}{R_f}$$

Because V_{R_f} is the same value as V_{R_i},

$$I_{R_f} = \frac{V_{R_i}}{R_f} \quad (4.6)$$
$$V_{R_i} = I_{in} R_i \quad (4.7)$$

Substitution of 4.7 into 4.6 yields

$$I_{R_f} = \frac{I_{in} R_i}{R_f}$$

Substituting into 4.5 produces

$$I_{out} = I_{in} + \frac{I_{in} R_i}{R_f}$$
$$I_{out} = I_{in}\left(1 + \frac{R_i}{R_f}\right)$$
$$A_i = 1 + \frac{R_i}{R_f} \quad (4.8)$$

As you might expect, the gain is a function of the two feedback resistors. Note the similarity of this result with that of the noninverting voltage amplifier.

FIGURE 4.17

Current amlifier for Example 4-11

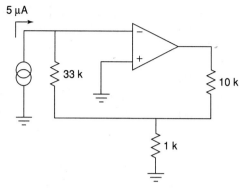

EXAMPLE 4-11

What is the load current in Figure 4.17?

$$I_{out} = A_i I_{in}$$
$$A_i = 1 + \frac{R_i}{R_f}$$
$$A_i = 1 + \frac{33\text{ k}}{1\text{ k}}$$
$$A_i = 34$$
$$I_{out} = 34 \times 5\ \mu A$$
$$I_{out} = 170\ \mu A$$

We need to check to make sure that this current doesn't cause output clipping. A simple Ohm's law check is all that's needed.

$$V_{max} = I_{out} R_{load} + I_{in} R_i$$
$$V_{max} = 170\ \mu A \times 10\text{ k} + 5\ \mu A \times 33\text{ k}$$
$$V_{max} = 1.7\text{ V} + .165\text{ V}$$
$$V_{max} = 1.865\text{ V (no problem)}$$

EXAMPLE 4-12

Design an amplifier with a current gain of 50. The load is approximately 200 kΩ. Assuming a typical op amp ($I_{out-max} = 20$ mA with ± 15 V supplies), what is the maximum load current obtainable?

$$A_i = 1 + \frac{R_i}{R_f}$$
$$\frac{R_i}{R_f} = A_i - 1$$
$$\frac{R_i}{R_f} = 50 - 1$$
$$\frac{R_i}{R_f} = 49$$

FIGURE 4.18

Current amplifier design for Example 4-12

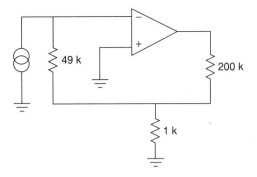

Therefore, R_i must be 49 times larger than R_f. Possible solutions include:

$$R_i = 49 \text{ k}\Omega, \qquad R_f = 1 \text{ k}\Omega$$
$$R_i = 98 \text{ k}\Omega, \qquad R_f = 2 \text{ k}\Omega$$
$$R_i = 24.5 \text{ k}\Omega, \qquad R_f = 500 \text{ }\Omega$$

As far as maximum load current is concerned, it can be no greater than the op amp's maximum output of 20 mA, but it may be less. We need to determine the current at clipping. Due to the large size of the load resistance, virtually all of the output potential will drop across it. Ignoring the extra drop across the feedback resistors will introduce a maximum of 1% error (that's worst case, assuming resistor set number 2).

With 15 V rails, a typical op amp will clip at 13.5 V. The resulting current is found through Ohm's law:

$$I_{max} = \frac{13.5 \text{ V}}{200 \text{ k}}$$
$$I_{max} = 67.5 \text{ }\mu\text{A}$$

Another way of looking at this is to say that the maximum allowable input current is $\frac{67.5 \text{ }\mu\text{A}}{50}$, or 1.35 μA. One possible solution is shown in Figure 4.18.

Summing Amplifiers

It is very common in circuit design to combine several signals into a single common signal. One good example of this is in the broadcast and recording industries. The typical modern music recording will require the use of perhaps dozens of microphones, yet the final product consists of typically two output signals (stereo left and right). If signals are joined haphazardly, excessive interference, noise, and distortion may result. The ideal summing amplifier would present each input signal with an isolated load not affected by other channels.

The most common form of summing amplifier is really nothing more than an extension of the inverting voltage amplifier. Because the input to the op amp is at virtual ground, it makes an ideal current summing node. Instead of placing a single input resistor at this point, several input resistors may be used. Each input source drives its own resistor, and there is very little effect from neighboring inputs. The

FIGURE 4.19

Summing amplifier

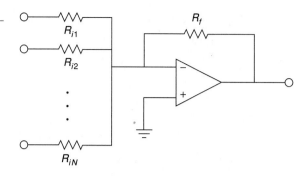

virtual ground is the key. A general summing amplifier is shown in Figure 4.19. The input impedance for the first channel is R_{i1}, and its voltage gain is $\frac{-R_f}{R_{i1}}$. For channel 2, the input impedance is R_{i2}, with a gain of $\frac{-R_f}{R_{i2}}$. In general, then, for channel N we have

$$Z_{inN} = R_{iN}$$
$$A_{vN} = -\frac{R_f}{R_{iN}}$$

The output signal is the sum of all inputs multiplied by their associated gains.

$$V_{out} = V_{in1} A_{v1} + V_{in2} A_{v2} + \cdots + V_{inN} A_{vN}$$

which is written more conveniently as

$$V_{out} = \sum_{i=1}^{n} V_{in_i} A_{v_i} \qquad (4.9)$$

A summing amplifier may have equal gain for each input channel. This is referred to as an *equal-weighted configuration*.

EXAMPLE 4-13

What is the output of the summing amplifier in Figure 4.20, with the given DC input voltages?

The easy way to approach this is to just treat the circuit as three inverting voltage amplifiers, and then add the results to get the final output.

FIGURE 4.20

Summing amplifier for Example 4-13

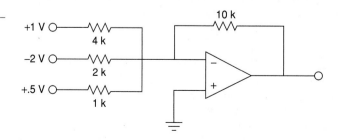

Channel 1:
$$A_v = -\frac{R_f}{R_i}$$
$$A_v = -\frac{10\,k}{4\,k}$$
$$A_v = -2.5$$
$$V_{out} = -2.5 \times 1\,V$$
$$V_{out} = -2.5\,V$$

Channel 2:
$$A_v = -\frac{R_f}{R_i}$$
$$A_v = -\frac{10\,k}{2\,k}$$
$$A_v = -5$$
$$V_{out} = -5 \times -2\,V$$
$$V_{out} = 10\,V$$

Channel 3:
$$A_v = -\frac{R_f}{R_i}$$
$$A_v = -\frac{10\,k}{1\,k}$$
$$A_v = -10$$
$$V_{out} = -10 \times .5\,V$$
$$V_{out} = -5\,V$$

The final output is found via summation:
$$V_{out} = -2.5\,V + 10\,V + (-5\,V)$$
$$V_{out} = 2.5\,V$$

If the inputs were AC signals, the summation is not quite so straightforward. Remember, AC signals of differing frequency and phase do not add coherently. You can perform a calculation similar to the preceding to find the peak value, however, an RMS calculation is needed for the effective value (i.e., square root of the sum of the squares).

For use in the broadcast and recording industries, summing amplifiers will also require some form of volume control for each input channel and a master volume control as well. This allows the levels of various microphones or instruments to be properly balanced. Theoretically, individual-channel gain control may be produced by replacing each input resistor with a potentiometer. By adjusting R_i, the gain may be directly varied. In practice, there are a few problems with this arrangement. First of all, it is impossible to turn a channel completely off. The required value for R_i would be

FIGURE 4.21

Audio mixer

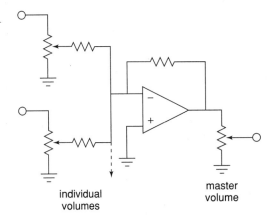

individual volumes

master volume

infinite. Second, because R_i sets the input impedance, a variation in gain will produce a Z_{in} change. This change may overload or alter the characteristics of the driving source. One possible solution is to keep R_i at a fixed value and place a potentiometer before it, as in Figure 4.21. The pot produces a gain from 1 through 0. The R_f/R_i combo is then set for maximum gain. As long as R_i is several times larger than the pot's value, the channel's input impedance will stay relatively constant. The effective Z_{in} for the channel is R_{pot} in parallel with R_i, at a minimum, up to R_{pot}.

As far as a master volume control is concerned, it is possible to use a pot for R_f. Without a limiting resistor though, a very low master gain runs the risk of overdriving the op amp due to the small effective R_f value. This technique also causes variations in offset potentials and circuit bandwidth. A technique that achieves higher performance involves using a stage with a fixed R_f value, followed with a pot, as in Figure 4.21.

Still another application of the summing amplifier is the *level shifter*. A level shifter is a two-input summing amplifier. One input is the desired AC signal, and the second input is a DC value. The proper selection of DC value lets you place the AC signal at a desired DC offset. There are many uses for such a circuit. One possible application is the DC offset control available on many signal generators.

Noninverting Summing Amplifier

Besides the inverting form, summing amplifiers may also be produced in a noninverting form. Noninverting summers generally exhibit superior high-frequency performance when compared to the inverting type. One possible circuit is shown in Figure 4.22. In this example, three inputs are shown, although more could be added. Each input has an associated input resistor. Note that it is not possible to simply wire several sources together in hopes of summing their respective signals. This is because each source will try to bring its output to a desired value that will be different from the values created by the other sources. The resulting imbalance may cause excessive (and possibly damaging) source currents. Consequently, each source must be isolated from the others through a resistor.

In order to understand the operation of this circuit, it is best to break it into two parts: the input source/resistor section and the noninverting amplifier section. The

FIGURE 4.22

Noninverting summing amplifier

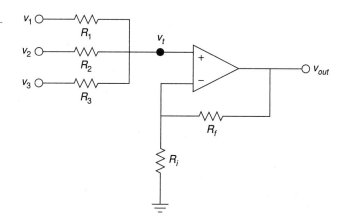

input signals will combine to create a total input voltage, V_t. By inspection, you should see that the output voltage of the circuit will equal V_t times the noninverting gain, or

$$V_{out} = V_t \left(1 + \frac{R_f}{R_i}\right)$$

All that remains is to determine V_t. Each of the input channels contributes to V_t in a similar manner, so the derivation of the contribution from a single channel will be sufficient.

Unlike the inverting summer, the noninverting summer does not take advantage of the virtual-ground summing node. The result is that individual channels will affect each other. The equivalent circuit for channel 1 is redrawn in Figure 4.23. Using superposition, we would first replace the input generators of channels 2 and 3 with short circuits. The result is a simple voltage divider between V_1 and V_{t1}.

$$V_{t1} = V_1 \frac{R_2 \| R_3}{R_1 + R_2 \| R_3}$$

FIGURE 4.23

Channel 1 input equivalent circuit

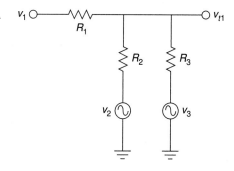

In a similar manner, we can derive the portions of V_t due to channel 2

$$V_{t2} = V_2 \frac{R_1 \| R_3}{R_2 + R_1 \| R_3}$$

and due to channel 3.

$$V_{t3} = V_3 \frac{R_1 \| R_2}{R_3 + R_1 \| R_2}$$

V_t is the summation of these three portions.

$$V_t = V_{t1} + V_{t2} + V_{t3}$$

Thus, by combining these elements, we find that the output voltage is

$$V_{out} = \left(1 + \frac{R_f}{R_i}\right)\left(V_1 \frac{R_2 \| R_3}{R_1 + R_2 \| R_3} + V_2 \frac{R_1 \| R_3}{R_2 + R_1 \| R_3} + V_3 \frac{R_1 \| R_2}{R_3 + R_1 \| R_2}\right)$$

For convenience and equal weighting, the input resistors are often all set to the same value. This results in a circuit that averages together all of the inputs. Doing so simplifies the equation to

$$V_{out} = \left(1 + \frac{R_f}{R_i}\right)\frac{V_1 + V_2 + V_3}{3}$$

or in a more general sense,

$$V_{out} = \left(1 + \frac{R_f}{R_i}\right)\frac{\sum_{i=1}^{n} V_n}{n} \quad (4.10)$$

where n is the number of channels.

One problem still remains with this circuit, and that is interchannel isolation or *crosstalk*. This can be eliminated by individually buffering each input, as shown in Figure 4.24.

Example 4-14

A noninverting summer such as the one shown in Figure 4.22 is used to combine three signals. $V_1 = 1$ V DC, $V_2 = -.2$ V DC, and V_3 is a 2 V peak 100 Hz sine wave. Determine the output voltage if $R_1 = R_2 = R_3 = R_f = 20$ kΩ and $R_i = 5$ kΩ.

Because all of the input resistors are equal, we can use the general form of the summing equation.

$$V_{out} = \left(1 + \frac{R_f}{R_i}\right)\frac{V_1 + V_2 + \ldots + V_n}{\text{Number of channels}}$$

FIGURE 4.24

Buffered and isolated noninverting summing amplifier

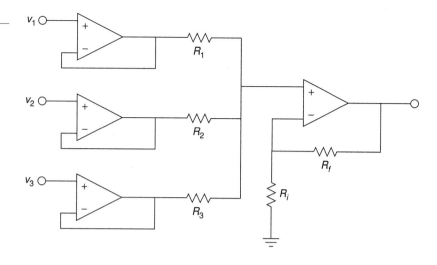

$$V_{out} = \left(1 + \frac{20\ k}{5\ k}\right) \frac{1\ V\ DC + -.2\ V\ DC + 2 \sin 2\pi 100t}{3}$$

$$V_{out} = 5 \frac{.8\ V\ DC + 2 \sin 2\pi 100t}{3}$$

$$V_{out} = 1.33\ V\ DC + 3.33\ \sin 2\pi 100t$$

So we see that the output is a 3.33 V peak sine wave riding on a 1.33 V DC offset.

Differential Amplifier

As long as the op amp is based on a differential input stage, there is nothing preventing you from making a diff amp with it. The applications of an op amp based unit are the same as the discrete version examined in Chapter 1. In essence, the differential amplifier configuration is a combination of the inverting and noninverting voltage amplifiers. A candidate is seen in Figure 4.25. The analysis is identical to that of the two base types, and Superposition is used to combine the results. The obvious problem for this circuit is that there is a large mismatch between the gains if lower values are used. Remember, for the inverting input the gain magnitude is $\frac{R_f}{R_i}$, whereas the noninverting input sees $\frac{R_f}{R_i} + 1$. For proper operation, the gains of the two halves should be identical. The noninverting input has a slightly higher gain, so a simple voltage divider may be used to compensate. This is shown in Figure 4.26. The ratio should be the same as the $\frac{R_f}{R_i}$ ratio. The target gain is $\frac{R_f}{R_i}$, the present gain is $1 + \frac{R_f}{R_i}$, which may be written as $\frac{R_f + R_i}{R_i}$. To compensate, a gain of $\frac{R_f}{R_f + R_i}$ is used.

$$A_{v+} = \frac{R_f + R_i}{R_i} \cdot \frac{R_f}{R_f + R_i}$$

FIGURE 4.25

Differential amplifier candidate

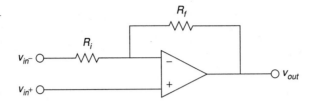

$$A_{v+} = \frac{R_f(R_f + R_i)}{R_i(R_f + R_i)}$$

$$A_{v+} = \frac{R_f}{R_i} \tag{4.11}$$

For a true differential amplifier, R'_i is set to R_i, and R'_f is set to R_f. A small potentiometer is typically placed in series with R'_f in order to compensate for slight gain imbalances due to component tolerances. This makes it possible for the circuit's common-mode rejection ratio to reach its maximum value. Another option for a simple difference amplifier is to set R'_i plus R'_f equal to R_i. Doing so will maintain roughly equal input impedance between the two halves if two different input sources are used.

Once the divider is added, the output voltage is found by multiplying the differential input signal by $\frac{R_f}{R_i}$.

EXAMPLE 4-15

Design a simple difference amplifier with an input impedance of 10 kΩ per leg and a voltage gain of 26 dB.

First of all, converting 26 dB into ordinary form yields 20. Because R_i sets Z_{in}, set $R_i = $ 10 kΩ from the specifications.

$$A_v = \frac{R_f}{R_i}$$

$$R_f = A_v R_i$$

FIGURE 4.26

Differential amplifier with compensation for mismatched gains

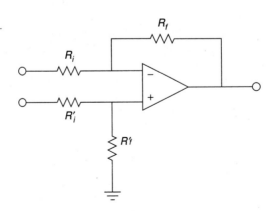

FIGURE 4.27

Difference amplifier for Example 4-15

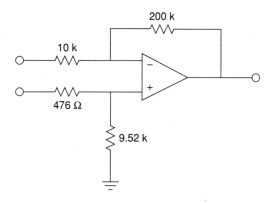

$$R_f = 20 \times 10\,k$$
$$R_f = 200\,k$$

For equivalent inputs,

$$R'_i + R'_f = R_i$$
$$R'_i + R'_f = 10\,k$$

Given that $A_v = 20$,

$$R'_f = 20 \times R'_i$$

Therefore,

$$21 \times R'_i = 10\,k$$
$$R'_i = 476$$
$$R'_f = 20 \times R'_i$$
$$R'_f = 9.52\,k$$

The final result is shown in Figure 4.27. As you will see later in Chapter 6, the differential amplifier figures prominently in another useful circuit, the *instrumentation amplifier*.

Adder/Subtractor

If inverting and noninverting summing amplifiers are combined using the differential amplifier topology, an adder/subtractor results. Normally, all resistors in an adder/subtractor are the same value. A typical adder/subtractor is shown in Figure 4.28. The inverting inputs number from 1 through m, and the noninverting inputs number from $m + 1$ through n. The circuit can be analyzed by combining the preceding proofs of Equations 4.9 through 4.11 via the superposition theorem. The details are left as an exercise (Problem 4.45). When all resistors are equal, the input weightings are unity,

Figure 4.28

Adder-subtractor

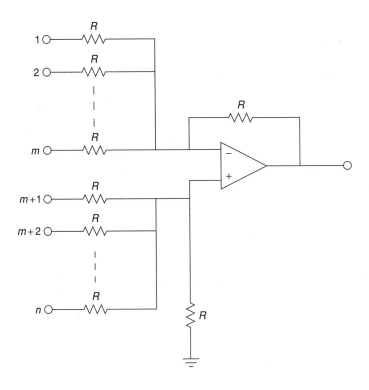

and the output is found by:

$$V_{out} = \sum_{i=m+1}^{n} V_{in_i} - \sum_{j=1}^{m} V_{in_j} \qquad (4.12)$$

In essence, you can think of the output voltage in terms of subtracting the inverting input summation from the noninverting input summation.

Adjustable Inverter/Noninverter

A unique adjustable gain amplifier is shown in Figure 4.29. What makes this circuit interesting is that the gain is continuously variable between an inverting and noninverting maximum. For example, the gain might be set for a maximum of 10. A full turn of the potentiometer would swing the gain from +10 through −10. The exact middle setting would produce a gain of 0. In this way, a single knob controls both the phase and magnitude of the gain. For the analysis of the circuit, refer to Figure 4.30.

As you would expect, the gain of the circuit is defined as the ratio of the output to input voltages. It is important to note that unlike a normal inverting amplifier, the magnitude of the output voltage is not necessarily equal to the voltage across R_2. This is because the inverting terminal of the op amp is not normally a virtual ground. Instead, the voltage across R_3 must also be considered. Because the two inputs of the op amp must be at approximately the same potential (i.e., V_{error} must be 0), the voltage at the

FIGURE 4.29

Adjustable inverter/noninverter

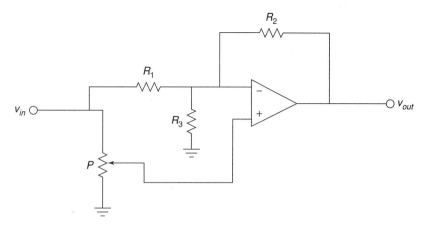

inverting terminal must be the same as the voltage tapped off of the potentiometer. Representing the potentiometer's voltage divider factor as k, we find:

$$V_{out} = kV_{in} - V_{R_2} \qquad (4.13)$$

The drop across R_2 is simply $I_2 R_2$. I_2 is found by Kirchhoff's current law and the appropriate voltage-resistor substitutions:

$$I_2 = I_1 - I_3$$
$$I_2 = \frac{V_{in} - kV_{in}}{R_1} - \frac{kV_{in}}{R_3}$$
$$I_2 = V_{in}\left(\frac{1-k}{R_1} - \frac{k}{R_3}\right)$$

FIGURE 4.30

Inverter/noninverter analysis

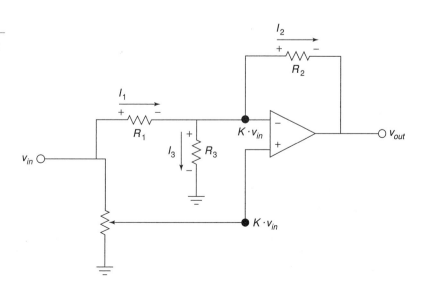

Thus, V_{R_2} is found to be

$$V_{R_2} = V_{in}\left((1-k)\frac{R_2}{R_1} - k\frac{R_2}{R_3}\right) \tag{4.14}$$

Combining Equations 4.13 and 4.14 and then solving for gain, we find

$$A_v = k - \left((1-k)\frac{R_2}{R_1} - k\frac{R_2}{R_3}\right)$$

$$A_v = k - \left(\frac{R_2}{R_1} - k\frac{R_2}{R_1} - k\frac{R_2}{R_3}\right)$$

$$A_v = -\frac{R_2}{R_1} + k\left(1 + \frac{R_2}{R_1} + k\frac{R_2}{R_3}\right)$$

The value of R_3 is chosen so that $R_1 = R_2 \| R_3$. This means that

$$R_3 = \frac{1}{\frac{1}{R_1} - \frac{1}{R_2}}$$

Substituting this into our gain equation and simplifying yields

$$A_v = \frac{R_2}{R_1}(2k - 1)$$

In essence, the resistors R_1 and R_2 set the maximum gain. The potentiometer sets k from 0 through 1. If $k = 1$, then $A_v = \frac{R_2}{R_1}$, or maximum noninverting gain. When $k = 0$, then $A_v = -\frac{R_2}{R_1}$, or maximum inverting gain. Finally, when the potentiometer is set to the midpoint, $k = .5$ and $A_v = 0$.

4.3 SINGLE-SUPPLY BIASING

Up to this point, all of the example circuits have used a bipolar power supply, usually ± 15 V. Sometimes this is not practical. For example, a small amount of analog circuitry may be used along with a predominantly digital circuit that runs off a unipolar supply. It may not be economical to create an entire negative supply just to run one or two op amps. Although it is possible to buy op amps that have been specially designed to work with unipolar supplies,[2] the addition of simple bias circuitry will allow almost any op amp to run from a unipolar supply. This supply can be up to twice as large as the bipolar counterpart. In other words, a circuit that normally runs off of a ± 15 V supply can be configured to run off a $+30$ V unipolar supply, producing similar performance. We will look at examples using both the noninverting and inverting voltage amplifiers.

The idea is to bias the input at one-half of the total supply potential. This can be done with a simple voltage divider. A coupling capacitor may be used to isolate this

---

[2] Examples include the LM324 and TLC270 series.

FIGURE 4.31

Single-supply bias in a noninverting amplifier

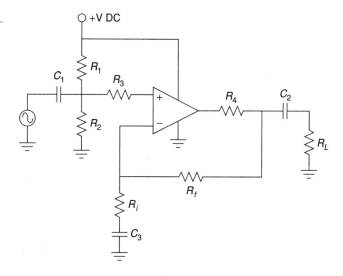

DC potential from the driving stage. For proper operation, the op amp's output should also be sitting at one-half of the supply. This fact implies that the circuit gain must be unity. This may appear to be a very limiting factor, but in reality, it isn't. The thing to remember is that the gain need only be unity for DC. The AC gain can be just about any gain you'd like.

An example using the noninverting voltage amplifier is shown in Figure 4.31. In order to set DC gain to unity without affecting the AC gain, capacitor C_3 is placed in series with R_i. R_1 and R_2 establish the 50% bias point. Their parallel combination sets the input impedance too. Resistors R_3 and R_4 are used to prevent destructive discharge of the coupling capacitors C_1 and C_2 into the op amp. They may not be required, but if present, typically run around 1 kΩ and 100 Ω respectively. The inclusion of the capacitors produces three lead networks. A standard frequency analysis and circuit simplification shows that the approximate critical frequencies are

$$f_{in} = \frac{1}{2\pi C_1 R_1 \| R_2}$$

$$f_{out} = \frac{1}{2\pi C_2 R_{load}}$$

$$f_{fdbk} = \frac{1}{2\pi C_3 R_i}$$

The input bias network can be improved by using the circuit of Figure 4.32. This reduces the hum and noise transmitted from the power supply into the op amp's input. It does so by creating a low impedance at node A. This, of course, does not affect the DC potential. R_5 now sets the input impedance of the circuit.

The important points to remember here are that voltage gain is still $1 + \frac{R_f}{R_i}$ in the midband, Z_{in} is now set by the biasing resistors R_1 and R_2, or R_5 (if used), and that frequency response is no longer flat down to zero Hertz.

FIGURE 4.32

Improved bias for the circuit in Figure 4.31

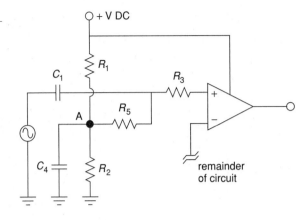

A single-supply version of the inverting voltage amplifier is shown in Figure 4.33. It uses the same basic techniques as the noninverting form. The bias setup uses the optimized low-noise form. Note that there is no change in input impedance; it is still set by R_i. The approximate lead network critical frequencies are found through

$$f_{in} = \frac{1}{2\pi C_1 R_i}$$

$$f_{out} = \frac{1}{2\pi C_2 R_{load}}$$

$$f_{bias} = \frac{1}{2\pi C_3 R_1 \| R_2}$$

Note the general similarity between the circuits of Figures 4.33 and 4.31. A simple redirection of the input signal creates one form from the other.

FIGURE 4.33

Single-supply inverting amplifier

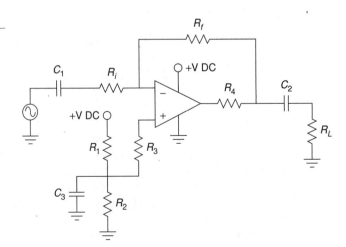

4.4 Current Boosting

As previously noted, general-purpose op amps produce a maximum output current of around 20 mA. This is sufficient for a wide variety of uses. If the load is less than about 1 kΩ, the op amp will start to clip on the higher output signals. The average op amp cannot drive low-impedance loads. A few examples of applications whose loads are inappropriate include distribution amplifiers, small audio power amplifiers such as a headphone amplifier, and small motors. This is most unfortunate, as we have already seen how useful these devices can be. There is a way out, though. It is possible to include a current-gain stage right after the op amp. All that is needed is a simple class B or class AB push-pull follower. This follower will be able to produce the higher current required by low-impedance loads. The op amp only needs to drive the follower stage. In order to increase system linearity and lower distortion, the follower can be placed inside of the op amp's feedback loop. Because the follower is noninverting, there is no problem with maintaining correct feedback (this assumes that the power devices used have a wider bandwidth than the op amp). An example of this is shown in Figure 4.34. This circuit is typical of an electronic crossover or distribution amplifier. (An electronic crossover design example is given in Chapter 11.) This output circuit needs to drive relatively low impedances through long cable runs of perhaps several hundred feet. The excessive capacitance resulting from long cable runs increases the current demand above that of a purely resistive load.

Circuits like the one in Figure 4.34 can produce currents of several hundred milliamps or more. Many times, small resistors are placed in the emitter or collector as a means of limiting maximum current or reducing distortion. The maximum output current limitation is a function of the class B devices. Some manufacturers offer current-boosting ICs to further simplify the design. The current booster is a drop-in replacement for the class B follower. For very high current demands, Darlingtons or multistage designs may be required. It is even possible to provide voltage gain stages. Indeed, several consumer audio power amplifiers have been designed in exactly this way. In essence, the designers produce a discrete power amplifier and then "wrap it" within an op amp feedback loop.

multiSIM Computer Simulation

A basic current booster is simulated using MultiSIM in Figure 4.35 (see pp. 133–134). In order to see secondary effects, the LF411 model has been chosen instead of the ideal device model. The circuit is configured for a voltage gain of unity, thus the 5-volt input signal should yield a 5-volt output. According to its data sheet, the short-circuit current of the LF411 is approximately 25 mA at room temperature. It is not capable of driving a 75 Ω load to 5 volts by itself. The transient analysis shows a full 5-volt output signal, indicating the effectiveness of the current boosting stage. Also, a close inspection of the output waveform shows no obvious forms of distortion such as the crossover distortion typical of simple class B stages. This shows that keeping the class B stage within the feedback loop does indeed minimize distortion.

Figure 4.34

Current boosting

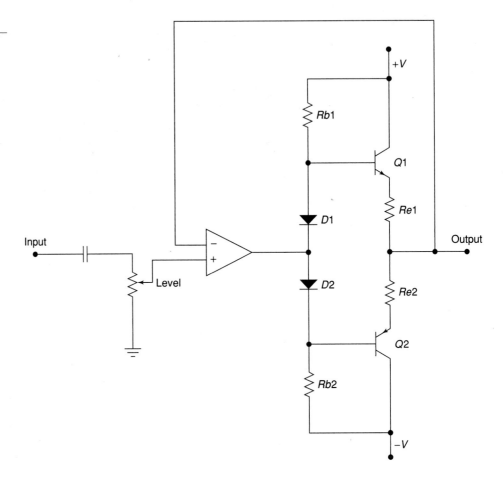

Summary

In this chapter we have explored a variety of basic op amp circuits and learned a few analysis shortcuts. The basic assumptions are that the error voltage (differential input voltage) is zero and that the op amp's input current is zero. In all circuits, the gain or transfer parameter is a function of just one or two resistors. Circuits can be made that produce voltage gain, current gain, voltage-to-current conversion, or current-to-voltage conversion. The most popular op amp circuits are the noninverting voltage amplifier and the inverting voltage amplifier. These are based on SP and PP negative feedback, respectively. The noninverting type shows an ideally infinite input impedance, whereas the inverting type has its input impedance set by one of the feedback resistors. A variation of the inverting voltage amplifier is the summing amplifier. This adds its several input channels together in order to arrive at its single output signal. The input node is at virtual ground. The differential amplifier is basically the simultaneous use of both the inverting and noninverting voltage amplifier forms.

The voltage-to-current transducer is based on SS feedback. Its transconductance is set by a single feedback resistor. In a similar manner, the current-to-voltage transducer is based on PP feedback and has a single feedback resistor to set its transresistance. The current amplifier is based on PS feedback.

Although op amps are designed to run off bipolar power supplies, they can be used with unipolar supplies. Extra circuitry is needed for the proper bias. There is no restriction on AC gain; however, DC gain must be set to unity. Because lead networks are introduced, the system gain cannot be flat down to zero Hertz.

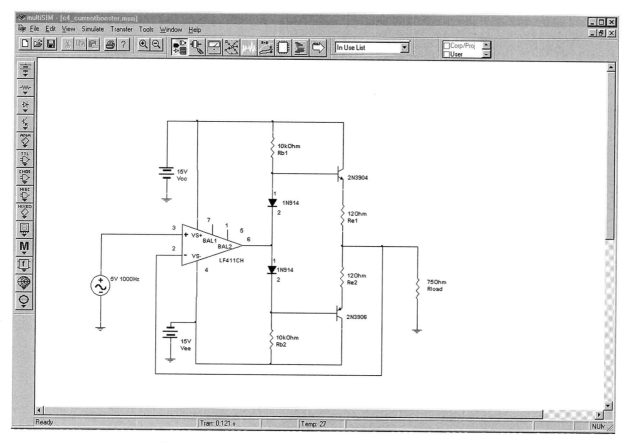

FIGURE 4.35a

Current booster simulation schematic

Finally, if higher output current requirements need to be met, it is possible to boost the op amp's capabilities with a discrete output stage. This stage is typically a class B or class AB push-pull follower. In order to lower system distortion, the follower is kept within the op amp's feedback loop.

REVIEW QUESTIONS

1. What forms of feedback are used for the inverting and noninverting voltage amplifiers?
2. What forms of feedback are used for the current-to-voltage and voltage-to-current transducers?
3. What form of feedback is used for the inverting current amplifier?
4. What are the op amp analysis idealizations?
5. What is virtual ground?
6. What is a summing amplifier?
7. How can output current be increased?
8. What circuit changes are needed in order to bias an op amp with a unipolar supply?
9. What operational parameters change when a circuit is set up for single-supply biasing?
10. How might a circuit's gain be controlled externally?
11. What is meant by the term *floating load*?

FIGURE 4.35b

Simulation of output waveform

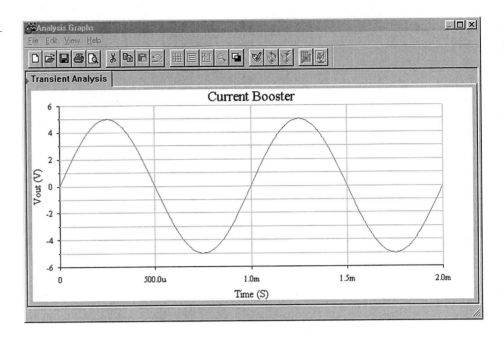

PROBLEMS

Analysis Problems

1. What is the voltage gain in Figure 4.36? What is the input impedance?

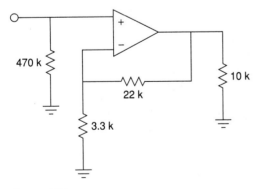

FIGURE 4.36

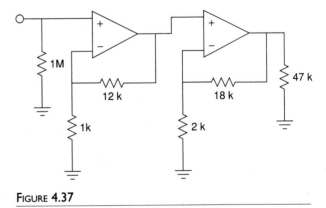

FIGURE 4.37

2. What is the voltage gain for the first stage of Figure 4.37? What is the input impedance?
3. What is the voltage gain for the second stage of Figure 4.37? What is the input impedance?
4. What is the system voltage gain in Figure 4.37? What is the input impedance?
5. If the input to Figure 4.37 is −52 dB V, what is V'_{out}?
6. What is the voltage gain in Figure 4.38? What is the input impedance?
7. If the input voltage to the circuit of Figure 4.38 is 100 mV, what is V_{out}?

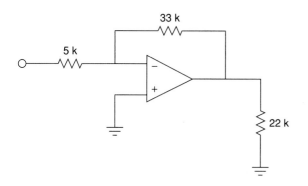

FIGURE 4.38

8. What is the system input impedance in Figure 4.39? What is the system gain?

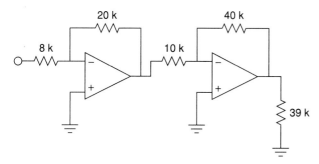

FIGURE 4.39

9. Redesign Figure 4.39 for an input impedance of 20 kΩ.
10. Given an input current of 2 μA, what is the output voltage in Figure 4.40?

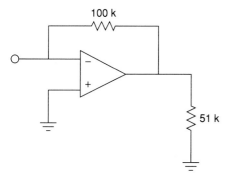

FIGURE 4.40

11. What is the meter deflection in Figure 4.41 if the input voltage is 1 V?

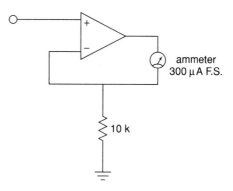

FIGURE 4.41

12. What input voltage will produce full-scale deflection in Figure 4.41?
13. Determine a new value for the 10 kΩ resistor in Figure 4.41 such that a .1 V input will produce full-scale deflection.
14. What is the current gain in Figure 4.42?

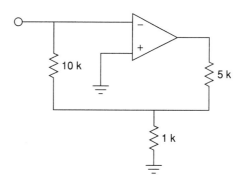

FIGURE 4.42

15. What is the maximum input current in Figure 4.42, assuming the circuit is running off of ±15 V supplies and the op amp has a maximum output current of 25 mA?
16. If the differential input signal is 300 mV in Figure 4.43, what is V_{out}?
17. Determine new values for the voltage divider resistors in Figure 4.43, such that the resulting input impedance is balanced.

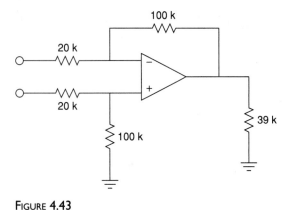

FIGURE 4.43

Design Problems

18. Design a noninverting amplifier with a voltage gain of 32 dB and an input impedance of 200 kΩ.
19. Design a voltage follower with a gain of 0 dB.
20. Design an inverting amplifier with a voltage gain of 14 dB and an input impedance of 15 kΩ.
21. Design a current-to-voltage transducer such that a 20 μA input current will produce a 1 V output.
22. Design a voltage-to-current transducer such that a 100 mV input will produce a 1 mA output.
23. Design a current amplifier with a gain of 20.
24. Design a differential amplifier with a gain of 18 dB and a balanced input impedance of 25 kΩ per input.
25. Design a voltage-to-current transducer with a transconductance of 1 mS. If V_{in} is 200 mV, what is I_{out}?
26. Design a current-to-voltage transducer with a transresistance of 10 kΩ. If the input current is .5 mA, what is V_{out}?
27. Redesign the circuit of Figure 4.36 for single-supply operation (don't bother calculating capacitor values).
28. Redesign the circuit of Figure 4.38 for single-supply operation (don't bother calculating capacitor values).
29. Design a summing amplifier such that channel 1 has a gain of 10, channel 2 has a gain of 15, and channel 3 has a gain of 5. The minimum channel input impedance should be 1 kΩ.
30. Determine capacitor values for Problem 27 if the lower break frequency f_1 is set to 20 Hz.
31. Determine capacitor values for Problem 28 if the lower break frequency f_1 is set to 10 Hz.

Challenge Problems

32. Design a three-channel summing amplifier such that: channel 1 $Z_{in} \geq 10$ kΩ, $A_v = 6$ dB; channel 2 $Z_{in} \geq 22$ k, $A_v = 10$ dB; and channel 3 $Z_{in} \geq 5$ k, $A_v = 16$ dB.
33. Assuming 10% resistor values, determine the production gain range for Figure 4.36.
34. Assuming 5% resistor values, determine the highest gain produced in Figure 4.37.
35. Design an inverting amplification circuit with a gain of at least 40 dB and an input impedance of at least 100 kΩ. No resistor used may be greater than 500 kΩ. Multiple stages are allowed.
36. Redesign the circuit of Figure 4.38 as a voltmeter with 500 mV, 2 V, 5 V, 20 V, and 50 V ranges.
37. Assuming 1% precision resistors and a meter accuracy of 5%, what range of input values may produce a full-scale reading of 2 V for the circuit of Problem 36?
38. Design an amplifier with a gain range from −10 dB to +20 dB, with an input impedance of at least 10 kΩ.
39. Design an amplifier with a gain range from 0 to 20. The input impedance should be at least 5 kΩ.
40. What is the input impedance in Figure 4.44? What is A_v?

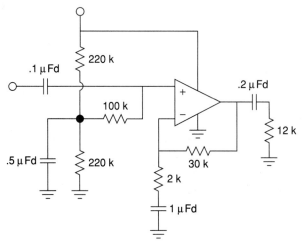

FIGURE 4.44

41. How much power supply ripple attenuation does the input biasing network of Figure 4.44 produce (assume $F_{ripple} = 120$ Hz)?
42. Assume that the circuit of Figure 4.45 utilizes a standard 20 mA output op amp. If the output devices are rated for

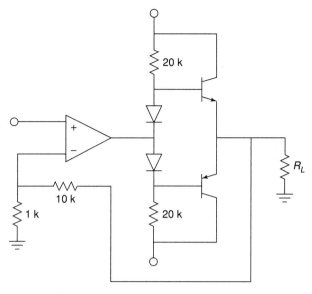

FIGURE 4.45

a maximum collector current of 5 amps and a Beta of 50, what is the maximum load current obtainable?

43. What are the voltage gain and input impedance in Figure 4.45?
44. Given a summer based on Figure 4.22, sketch the output waveform if $R_1 = R_2 = 10$ kΩ, $R_3 = R_f = 30$ kΩ, $R_i = 15$ kΩ, $V_1 = .3$ V DC, $V_2 = .1 \sin 2\pi 50t$ and $V_3 = -.2 \sin 2\pi 200t$.
45. Prove Equation 4.12 for the case when all resistors are of equal value.

Computer Simulation Problems

46. Simulate the operation of the circuit in Figure 4.9. Verify the output voltage and the virtual ground at the inverting input.
47. Use a simulator to verify the maximum and minimum gains of the circuit in Figure 4.11.
48. Use a simulator to verify the load current and the voltage of the circuit in Figure 4.17.
49. Verify the output potential of the circuit in Figure 4.20.
50. Simulate the output voltage of the circuit of Figure 4.39 for the following inputs: (a) $V_{in} = .1$ V DC, (b) $V_{in}(t) = 1 \sin 2\pi 10t$, (c) $V_{in} = 5$ V DC. Also, note the potential at the output of the first stage. How might your op amp model affect the results?
51. Simulate the circuit in Figure 4.27. Determine the output potential for the following inputs: (a) $V_{in+}(t) = .1 \sin 2\pi 10t$, $V_{in-}(t) = .1 \sin 2\pi 10t$, (b) $V_{in+}(t) = .1 \sin 2\pi 10t$, $V_{in-}(t) = -.1 \sin 2\pi 10t$.
52. Simulate the circuit of Figure 4.45, and determine the output of the circuit and op amp for inputs of .1 V DC and 1 V DC.

CHAPTER 5

Practical Limitations of Op Amp Circuits

CHAPTER OBJECTIVES

After completing this chapter, you should be able to:

- Define *gain-bandwidth product* and describe its use in circuit design and analysis.
- Determine upper and lower break frequencies in a multistage circuit.
- Define *slew rate* and *power bandwidth*, and calculate their effect on circuit performance.
- Understand the difference between *power bandwidth* and *small-signal bandwidth*.
- Detail the differences between compensated, noncompensated, and decompensated op amps.
- Calculate the DC offset of an op amp circuit and understand how to minimize it.
- Calculate the DC drift of an op amp circuit and understand how to minimize it.
- Discuss which factors affect the noise performance of an op amp circuit.
- Calculate the noise voltage of an op amp circuit.
- Analyze the CMRR, PSRR, and S/N performance of an op amp circuit.

5.1 Introduction

Up to now, the op amp has been treated as an ideal device. Although these idealizations are very useful in their place, closer examination must follow. Without this knowledge, it will be impossible to accurately predict a circuit's performance for very high or low frequencies, to judge its noise characteristics, or to determine its stability with temperature or power supply variations. With this information, you will be able to optimize circuit performance for given applications. A major part of this is determining the most desirable op amp for the job. The function of this chapter, then, is to delve deeper into the specifics of individual op amps and to present methods for determining system parameters such as frequency response, noise level, offsets, and drift. The primary interest is in investigating the popular inverting and noninverting voltage amplifier topologies.

5.2 FREQUENCY RESPONSE

In Chapter 4, a number of equations were presented for the various amplifier topologies. These enabled you to find the circuit gain, among other things. These equations are, of course, only valid in the midband region of the amplifier. They say nothing of the amplifier response at the frequency extremes. Chapter 1 showed that all amplifiers eventually roll off their gain as the input frequency increases. Some amplifiers exhibit a rolloff as the input frequency is decreased as well. Op amp circuits are no exception. There are two things we can say about the average op amp circuit's frequency response: (1) if there are no coupling or other lead network capacitors, the circuit gain will be flat from midband down to DC; and (2) there will eventually be a well-controlled high-frequency rolloff that is usually very easy to find. Item one should come as no great surprise, but you may well wonder about the second. For general-purpose op amps, the high frequency response may be determined with a parameter called the *gain-bandwidth product*, often abbreviated GBW.

5.3 GAIN-BANDWIDTH PRODUCT

The open-loop frequency response of a general-purpose op amp is shown in Figure 5.1a. Although the exact frequency and gain values will differ from model to model, all devices will exhibit this same general shape and 20 dB per decade rolloff slope. This is because the lag break frequency (f_c) is determined by a single capacitor called the *compensation capacitor*. This capacitor is usually in the Miller position (i.e., straddling input and output) of an intermediate stage, such as C in Figure 5.1b. Although this capacitor is rather small, the Miller effect drastically increases its apparent value. The resulting critical frequency is very low, often in the range of 10 to 100 Hz. The other circuit lag networks caused by stray or load capacitances are much higher, usually over 1 MHz. As a result, a constant 20 dB per decade rolloff is maintained from f_c up to very high frequencies. The remaining lag networks will not affect the open-loop response until the gain has already dropped below zero dB.

This type of frequency response curve has two benefits: (1) The most important benefit is that it allows you to set almost any gain you desire with stability. Only a single network is active, thus satisfactory gain and phase margins will be maintained. Therefore, your negative feedback never turns into positive feedback (as noted in Chapter 3). (2) The product of any break frequency and its corresponding gain is a constant. In other words, the gain decreases at the same rate at which the frequency increases. In Figure 5.1, the product is 1 MHz. As you might have guessed, this parameter is the gain-bandwidth product of the op amp (GBW). GBW is also referred to as f_{unity} (the

FIGURE 5.1

Gain curve

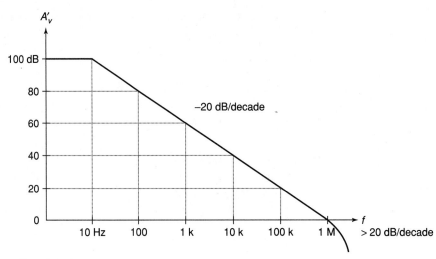

a. Open-loop frequency response

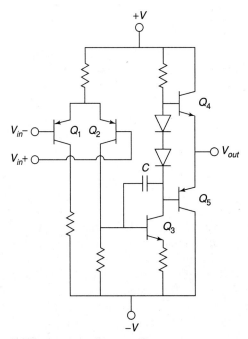

b. Miller compensation capacitor

frequency at which the open-loop gain equals 1). You will find both terms used on manufacturers' spec sheets.

As you already know, operating an op amp with negative feedback lowers the midband gain. To a first approximation, this gain will continue until it reaches the open-loop response. At this point, the closed-loop response will follow the open-loop rolloff. Remember, this is due to the reduction in loop gain, as seen in Chapter 3.

FIGURE 5.2

Comparison of open-loop and closed-loop responses

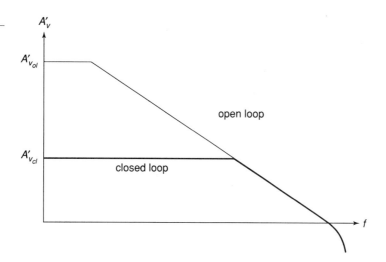

This effect is shown in Figure 5.2. By knowing GBW and the gain, the associated break frequency can be quickly determined. For the inverting and noninverting voltage amplifiers,

$$f_2 = \frac{GBW}{A_{noise}} \qquad (5.1)$$

The use of *noise gain* versus ordinary voltage gain simplifies things and actually makes the results a bit more accurate. Noise gain is the same for both the inverting and noninverting voltage amplifiers. The use of noise gain helps us to take into account the true (nonideal) feedback effects and circuit imperfections. An example of these limitations is that the open-loop gain of an op amp is never infinite. To find the noise gain for any circuit, short all voltage sources and open all current sources. The only item remaining for each source should be its internal resistance. At this point, simplify the circuit as required, and find the gain from the noninverting input to the output of the op amp. This gain is the noise gain. For the standard inverting and noninverting voltage amplifiers, we find

$$A_{noise} = 1 + \frac{R_f}{R_i}$$

Noise gain is the same as ordinary voltage gain for the noninverting voltage amplifier, but is one unit larger than the inverting amplifier's ordinary gain ($\frac{R_f}{R_i}$). The deviation is only noticeable at lower gains. This does imply though, that for the same gain, noninverting amplifiers will exhibit a higher break frequency than inverting types. Thus, for maximum bandwidth with low-gain circuits, the noninverting form is generally preferred. The worst case occurs with an ordinary voltage gain of 1. For the noninverting configuration, the noise gain will also equal 1, and the closed-loop bandwidth will equal f_{unity}. On the other hand, an inverting amplifier with a voltage gain of 1 will produce a noise gain of 2 and will exhibit a small-signal bandwidth of $\frac{f_{unity}}{2}$. Never use the gain in dB form for this calculation.

Example 5-1

Using a 741 op amp, what is the upper break frequency for a noninverting amplifier with a gain of 20 dB?

A 741 data sheet shows a typical GBW of 1 MHz. The noise gain for a noninverting amplifier is the same as its ordinary gain. Converting 20 dB into ordinary form yields a gain of 10.

$$f_2 = \frac{GBW}{A_{noise}}$$

$$f_2 = \frac{1\ MHz}{10}$$

$$f_2 = 100\ kHz$$

Thus, the gain is constant at 10 up to 100 kHz. Above this frequency the gain rolls off at 20 dB per decade.

Example 5-2

Sketch the frequency response of the circuit in Figure 5.3.

This is an inverting voltage amplifier. The gain is

$$A_v = -\frac{R_f}{R_i}$$

$$A_v = -\frac{10\ k}{2\ k}$$

$$A_v = -5$$

$$A'_v = 14\ dB$$

For noise gain,

$$A_{noise} = 1 + \frac{R_f}{R_i}$$

$$A_{noise} = 1 + \frac{10\ k}{2\ k}$$

$$A_{noise} = 6$$

From a data sheet, GBW for a 741 is found to be 1 MHz.

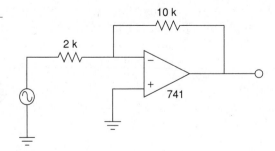

Figure 5.3 Circuit for Example 5-2

multiSIM COMPUTER SIMULATION

The simulation results for Example 5-2 are also shown in Figure 5.4 (see below and pp. 144–145). The low frequency gain agrees with the hand calculation of approximately 14 dB. The 3 dB down point (f_2) also agrees with the calculated break of approximately 167 kHz. It is important to note that the simple dependent source model presented in Chapter 2 cannot be used because it does not have the proper frequency response. Instead, manufacturers offer accurate models for their op amps. There are several variations on the theme. The model presented here is typical. It is fairly complex and is quite accurate. The op amp model is comprised of two basic parts, a differential amplifier input portion and a dependent source output section. The input portion utilizes a pair of NPN transistors with simple resistors for the loads (R_{C_1} and R_{C_2}). Resistors R_{E_1} and R_{E_2} serve as swamping or emitter-degeneration resistors. The tail-current source is set by the independent source I_{EE}. The nonideal internal impedance and frequency limitations of this current source are taken into account by R_E and C_E, whereas C_I helps to model the high-frequency loading of the diff amp's output. The output portion revolves around a series of voltage-controlled current sources. G_{CM} models common mode gain, G_A models the ordinary gain, and R_2 serves as the combined internal impedance of these sources. C_2 is the system compensation capacitor and has a value of 30 pF. R_{O_1} and R_{O_2} serve to model the output impedance of the op amp. Diodes D_1 through D_4 and voltage sources E_C, V_C, and V_E model the limits of the op amp's class AB output stage.

In spite of their accuracy, models such as this are time consuming and tedious to recreate. Fortunately, many manufacturers offer simulation models for their components in library form. To use these models, all you need to do is reference the appropriate part number from the library. The op amp models normally use typical rather than worst-case values.

FIGURE 5.4a

Analysis for the inverting 741 op amp of Figure 5.3

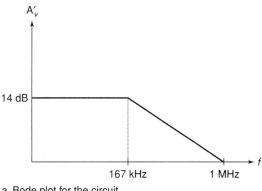

a. Bode plot for the circuit

$$f_2 = \frac{GBW}{A_{noise}}$$

$$f_2 = \frac{1 \text{ MHz}}{6}$$

$$f_2 = 167 \text{ kHz}$$

The resulting gain Bode plot is shown in Figure 5.4. Note that if a "faster" op amp is used (i.e., one with a higher GBW, such as the LF411), the response will extend further. As you might guess, faster op amps are more expensive.

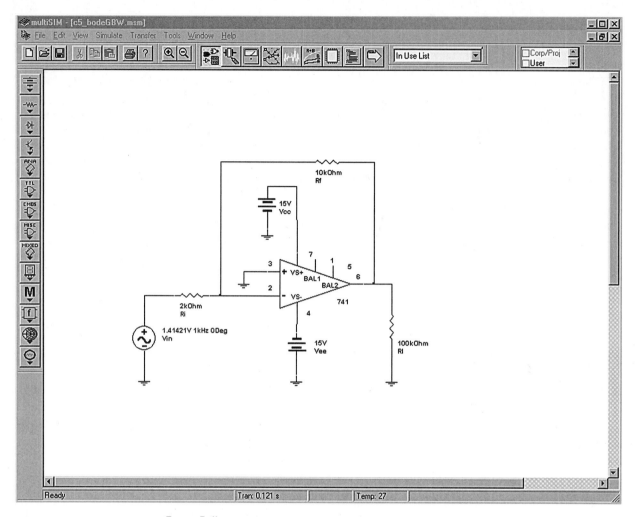

Figure 5.4b

MultiSIM schematic for Bode plot simulation

Example 5-3

Determine the minimum acceptable f_{unity} for the circuit of Figure 5.5 if response should extend to at least 50 kHz.

$$A_{noise} = 1 + \frac{R_f}{R_i}$$

$$A_{noise} = 1 + \frac{20\text{ k}}{500}$$

$$A_{noise} = 41$$

$$f_{unity} = A_{noise} f_2$$
$$f_{unity} = 41 \times 50 \text{ kHz}$$
$$f_{unity} = 2.05 \text{ MHz}$$

GAIN-BANDWIDTH PRODUCT   145

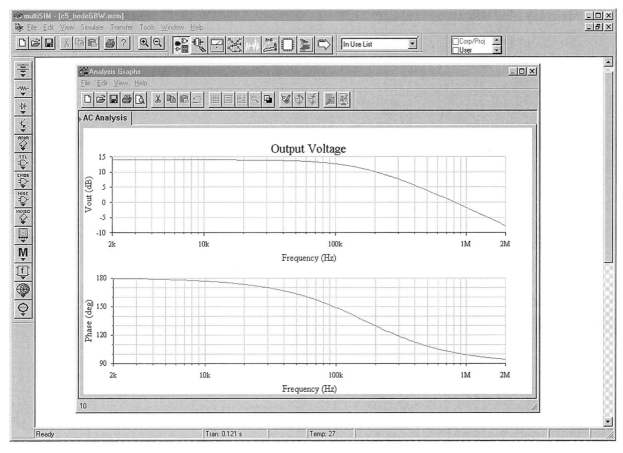

FIGURE 5.4c

Simulation of Bode plot

FIGURE 5.4d

Typical op amp model

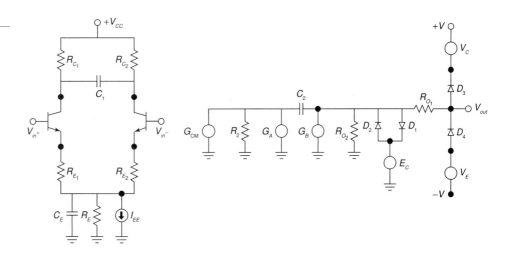

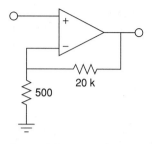

FIGURE 5.5

Circuit for Example 5-3

For this application, a stock 741 would not be fast enough; however, a 411 would be fine. From the foregoing, it is apparent that there is a direct trade-off between circuit gain and high-frequency performance for a given device. For an application requiring both high gain and wide bandwidth, a multistage approach should be considered.

Multistage Considerations

By combining two or more wide-bandwidth, low-gain stages, a single high-gain, wide-bandwidth system may be produced. Although the overall system gain will simply be the combination of the individual stage gains, the upper break frequency calculation can be a little tricky. Chances are, in a multistage op amp design, all stages will not exhibit the same upper break frequency. In this case the system's upper break is approximately equal to the lowest of the stage f_cs. In other words, the system is treated as though it was a discrete stage with multiple lag networks. On the other hand, if the break frequencies are close, this approximation can lead to a sizable error. This is best illustrated with a quick example. Imagine two stages exhibiting a 100 kHz break. If each stage produces a 3 dB loss at 100 kHz, it is obvious that the cascaded system must be producing a 6 dB loss at 100 kHz. Therefore, the system's critical frequency (i.e., −3 dB point) must be somewhat lower than 100 kHz (to be exact, it is the frequency at which each stage produces a 1.5 dB loss). Taking this a step further, if we cascade three identical stages, the total loss at 100 kHz will be 9 dB. The system break will be the point at which each of the three stages produces a 1 dB loss. The more identical stages that are added, the lower the effective break becomes. If we make a few assumptions about the exact shape of the rolloff curve, we can reduce this to a simple equation. In Chapter 1, we derived the general equation describing the amplitude response of a lead network (1.7). In a similar vein, the response for a lag network may be determined to be

$$A_V = \frac{1}{\sqrt{1 + \frac{f^2}{f_c^2}}} \qquad (5.2)$$

where f is the frequency of interest and f_c is the critical frequency.

It is more convenient to write this equation in terms of a *normalized frequency of interest*. Instead of being expressed in Hertz, the frequency of interest is represented as a factor relative to f_c. If we call this normalized frequency k_n, we may rewrite the amplitude response equation.

$$A_V = \frac{1}{\sqrt{1 + k_n^2}} \qquad (5.3)$$

We now solve for k_n

$$\frac{1}{A_V} = \sqrt{1 + k_n^2}$$

$$k_n^2 + 1 = \frac{1}{A_V^2}$$

$$k_n^2 = \frac{1}{A_v^2} - 1$$

$$k_n = \sqrt{\frac{1}{A_v^2} - 1} \tag{5.4}$$

We will now find the gain contribution of each stage. If all stages are critical at the same frequency, each stage must produce the same gain as the other stages at any other frequency. Because the combined gain of all stages must, by definition, be −3 dB or .707 at the system's break frequency, we may find the gain of each stage at this new frequency.

$$A_v^n = .707 \tag{5.5}$$

where n is the number of stages involved.

We may rewrite this as

$$A_v = .707^{\frac{1}{n}} \tag{5.6}$$

Combining 5.6 with 5.4 yields

$$k_n = \sqrt{\frac{1}{(.707^{\frac{1}{n}})^2} - 1}$$

$$k_n = \sqrt{2^{\frac{1}{n}} - 1} \tag{5.7}$$

As k_n is nothing more than a factor, this may be rewritten into a final convenient form.

$$f_{2\text{-system}} = f_2 k_n$$

$$f_{2\text{-system}} = f_2 \sqrt{2^{\frac{1}{n}} - 1} \tag{5.8}$$

where n is the number of identical stages.

Example 5-4

Assuming that all stages in Figure 5.6 use 741s, what is the system gain and upper break frequency?

Stage 1:

$$A_v = 1 + \frac{R_f}{R_i}$$

$$A_v = 1 + \frac{14\,k}{2\,k}$$

$$A_v = 8$$

Figure 5.6

Multistage circuit for Example 5-4

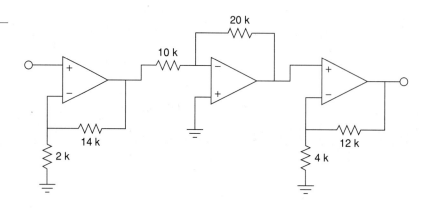

$$A_{noise} = 1 + \frac{R_f}{R_i}$$
$$A_{noise} = 8$$

$$f_2 = \frac{GBW}{A_{noise}}$$
$$f_2 = \frac{1 \text{ MHz}}{8}$$
$$f_2 = 125 \text{ kHz}$$

Stage 2:

$$A_v = -\frac{R_f}{R_i}$$
$$A_v = -\frac{20 \text{ k}}{10 \text{ k}}$$
$$A_v = -2$$

$$A_{noise} = 1 + \frac{R_f}{R_i}$$
$$A_{noise} = 3$$

$$f_2 = \frac{GBW}{A_{noise}}$$
$$f_2 = \frac{1 \text{ MHz}}{3}$$
$$f_2 = 333 \text{ kHz}$$

Stage 3:

$$A_v = 1 + \frac{R_f}{R_i}$$
$$A_v = 1 + \frac{12 \text{ k}}{4 \text{ k}}$$
$$A_v = 4$$

$$A_{noise} = 1 + \frac{R_f}{R_i}$$
$$A_{noise} = 4$$

$$f_2 = \frac{GBW}{A_{noise}}$$
$$f_2 = \frac{1 \text{ MHz}}{4}$$
$$f_2 = 250 \text{ kHz}$$

System:

$$A_v = 8 \times (-2) \times 4$$
$$A_v = -64$$

f_2 = dominant stage. The dominant break here is 125 kHz (stage 1).
The system has a gain of 64 and an upper break of 125 kHz. If this level of performance is to be achieved with a single op amp, it would need a gain-bandwidth product of 125 kHz × 64, or 8 MHz.

EXAMPLE 5-5

A three-stage amplifier uses identical noninverting voltage stages with gains of 10 each. If the op amps used have an f_{unity} of 4 MHz, what is the system gain and upper break?

Because these are noninverting amplifiers, the noise gain equals the signal gain. The break frequency for each stage is:

$$f_2 = \frac{f_{unity}}{A_{noise}}$$
$$f_2 = \frac{4 \text{ MHz}}{10}$$
$$f_2 = 400 \text{ kHz}$$

Because the three stages are identical, the system will roll off before 400 kHz.

$$f_{2-system} = f_2 \sqrt{2^{\frac{1}{n}} - 1}$$
$$f_{2-system} = 400 \text{ kHz} \sqrt{2^{\frac{1}{3}} - 1}$$
$$f_{2-system} = 203.9 \text{ kHz}$$

Note that the system response in this case is reduced about an octave from the single-stage response.

Example 5-6

Using only LF411 op amps, design a circuit with an upper break frequency of 500 kHz and a gain of 26 dB.

A gain of 26 dB translates to an ordinary gain of 20. Assuming a noninverting voltage stage, a single op amp would require an f_{unity} of:

$$f_{unity} = f_2 A_v \ (A_{noise} = A_v \text{ for noninverting form})$$
$$f_{unity} = 500 \text{ kHz} \times 20$$
$$f_{unity} = 10 \text{ MHz}$$

Because the 411 has a typical f_{unity} of 4 MHz, at least two stages are required. There are many possibilities. One option is to set one stage as the dominant stage and set its gain to produce the desired f_2. The second stage will then be used to make up the difference in gain to the desired system gain.

Stage 1:

$$f_2 = \frac{f_{unity}}{A_v}$$
$$A_v = \frac{f_{unity}}{f_2}$$
$$A_v = \frac{4 \text{ MHz}}{500 \text{ kHz}}$$
$$A_v = 8$$

Stage 2:
To achieve a final gain of 20, stage two requires a gain of 2.5. Its f_2 is:

$$f_2 = \frac{f_{unity}}{A_v}$$
$$f_2 = \frac{4 \text{ MHz}}{2.5}$$
$$f_2 = 1.6 \text{ MHz}$$

Note that if this frequency worked out to less than 500 kHz, three or more stages would be needed. To set the resistor values, the rules of thumb presented in Chapter 4 may be used. One possible solution is shown in Figure 5.7.

Low-Frequency Limitations

As mentioned earlier, standard op amps are direct-coupled. That is, their gain response extends down to 0 Hz. Consequently, many op amp circuits have no lower frequency limit. They will amplify DC signals just as easily as AC signals. Sometimes it is desirable to introduce a low-frequency rolloff. Two cases of this are single-supply biasing (Chapter 4) and interference rejection (the removal of undesired signals, such as low-frequency rumble). In both cases, the circuit designer produces a low-frequency rolloff (lead network) by introducing coupling capacitors. For single-supply circuits, these capacitors are a necessary evil. Without them, stages would quickly overload from

FIGURE 5.7

Completed design for Example 5-6

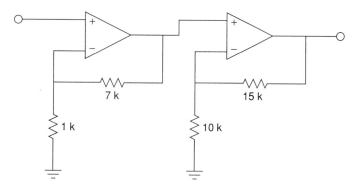

the large DC input. Also, signal sources and loads may be very intolerant of the DC bias potential. The result could be gross distortion or component failure. Even if a circuit uses a normal bipolar supply, a lead network may be used to reduce interference signals. For example, a well-chosen coupling capacitor can reduce 60 Hz hum interference while hardly affecting the quality of a voice transmission. Generally, these coupling capacitors can be simplified into the straightforward lead networks discussed back in Chapter 1. (Remember, for lead networks, the highest critical frequency is the dominant one.) Also, if multiple networks are dominant, the resulting critical frequency will be higher than the individual break frequency. The relationship is the mirror image of Equation 5.8. The proof for the following equation is very similar to that of Equation 5.8 and is left as an exercise.

$$f_{1\text{-system}} = \frac{f_1}{\sqrt{2^{\frac{1}{n}} - 1}} \qquad (5.9)$$

If you decide to add coupling capacitors in order to reduce interference, remember that the op amp will need a DC return resistor. An example is shown in Figure 5.8. The 100 kΩ resistor is needed so that the inverting input's half of the diff amp stage is properly biased. Note that for a typical op amp, this 100 kΩ also ends up setting the input impedance. Assuming a relatively low source impedance, the lead network simplification boils down to the .1 μF capacitor along with the 100 kΩ. The critical

FIGURE 5.8

DC-return resistor (100 k)

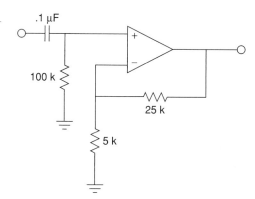

frequency is:

$$f_c = \frac{1}{2\pi RC}$$
$$f_c = \frac{1}{2\pi \times 100\text{ k} \times .1\ \mu\text{F}}$$
$$f_c = 15.9 \text{ Hz}$$

To sum up, then, when using general-purpose op amps, if no signal-coupling capacitors are being used, the gain response extends back to 0 Hz. If coupling capacitors are used, general lead network analysis techniques can be used to find the critical frequencies.

5.4 SLEW RATE AND POWER BANDWIDTH

As noted in the previous section, general-purpose op amps contain a compensation capacitor that is used to control the open-loop frequency response. The signal developed across this capacitor will be amplified in order to create the final output signal. In essence, this capacitor serves as the load for the preceding stage inside of the op amp. Like all stages, this one has a finite current output capability. Due to this, the compensation capacitor can be charged no faster than a rate determined by the standard capacitor charge equation:

$$i = C\frac{dv}{dt}$$
$$\frac{dv}{dt} = \frac{i}{C}$$

The rate of change of voltage versus time is $\frac{dv}{dt}$. By definition, this parameter is called *slew rate* (SR). The base unit for slew rate is volts per second; however, given the speed of typical devices, slew rate is normally specified in volts per microsecond. Slew rate is very important in that it helps determine whether or not a circuit can accurately amplify high-frequency or pulse-type waveforms. In order to create a fast op amp, either the charging current i must be large, or the compensation capacitor C must be very small. Because C also plays a role in determining the gain-bandwidth product, there is a lower limit to its size. A typical op amp might use a 30 pF compensation capacitor, and the driving stage may effectively produce a charging current of 100 μA. The resulting slew rate would be:

$$SR = \frac{dv}{dt} = \frac{i}{C}$$
$$SR = \frac{100\ \mu\text{A}}{30\text{ pF}}$$
$$SR = 3.33 \text{ Megavolts/Second}$$
$$SR = 3.33 \text{ V/}\mu\text{S}$$

TABLE 5.1

Device	Slew Rate
µA741	.5 V/µS
LF411	15 V/µS
OPA134	20 V/µS
LM318	70 V/µS
LM6364	300 V/µS
LT1363	1000 V/µS

This means that the output of the op amp can change no faster than 3.33 V over the course of one microsecond. It would take this op amp about 3 microseconds for its output signal to change a total of 10 V. It can go no faster than this. The ideal op amp would have an infinite slew rate. Although this is a practical impossibility, it is possible to find special high-speed devices that exhibit slew rates in the range of several thousand volts per microsecond. Comparative slew rates for a few selected devices are found in Table 5.1.

Slew rate is always output-referred. This way, the circuit gain need not be taken into account. Slew rate is normally the same regardless of whether the signal is positive or negative going. There are a few devices that exhibit an asymmetrical slew rate. One example is the 3900. It has a slew rate of .5 V/µS for positive swings, but shows 20 V/µS for negative swings.

The Effect of Slew Rate on Pulse Signals

An ideal pulse waveform will shift from one level to the other instantaneously, as shown in Figure 5.9. In reality, the rising and falling edges are limited by the slew rate. If this signal is fed into a 741 op amp, the output pulse would be decidedly trapezoidal, as shown in Figure 5.10. The 741 has a slew rate of .5 V/µS. Because the voltage change is 2 V, it takes the 741 4 µS to traverse from low to high, or from high to low. The resulting waveform is still recognizable as a pulse, however. Gross distortion of the pulse occurs if the pulse width is decreased, as in Figure 5.11. Here, the pulse width is only 3 microseconds, so the 741 doesn't even have enough time to reach the high level. In 3 microseconds, the 741 can only change 1.5 V. By the time the 741 gets to 1.5 V, the input signal is already swinging low, so the 741 attempts to track it. The result is a triangular waveform of reduced amplitude as shown in Figure 5.12. This same effect can occur if the amplitude of the pulse is increased. Obviously, then, pulses that are both fast and large require high slew rate devices. Note that a 411 op amp would produce a nice output in this example. Because its slew rate is 15 V/µS, it requires only .134 microseconds for the 2 V output swing. Its output waveform is shown in Figure 5.13.

The Effect of Slew Rate on Sinusoidal Signals and Power Bandwidth

Slew rate limiting produces an obvious effect on pulse signals. Slew rate limiting can also affect sinusoidal signals. All that is required for slewing to take place is that the

FIGURE 5.9

Ideal output

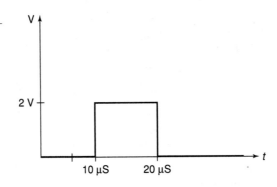

FIGURE 5.10

Slewed output of the 741

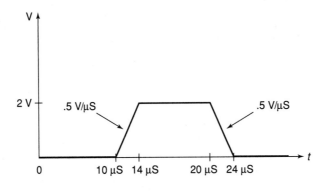

FIGURE 5.11

Ideal output

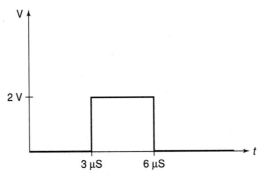

FIGURE 5.12

Slewed output of the 741

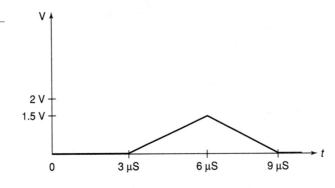

FIGURE 5.13

Slewed output of the 411

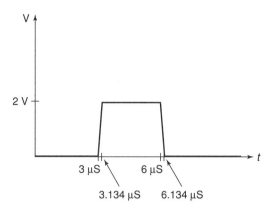

signal change faster than the device's slew rate. If the rate of change of the signal is never greater than the slew rate, slewing will never occur. To find out just how fast a given sine wave does change, we need to find the first derivative with respect to time. Assume that the input sine wave has a frequency f, and a peak amplitude K.

$$v(t) = K \sin 2\pi f t$$
$$\frac{dv}{dt} = 2\pi f K \cos 2\pi f t$$

The rate of change of the signal with respect to time is $\frac{dv}{dt}$. The maximum rate of change will occur when the sine wave passes through zero (i.e., at $t = 0$). To find this maximum value, substitute 0 in for t, and solve the equation.

$$\frac{dv}{dt} = 2\pi f K \tag{5.10}$$

So, the rate of change of the signal is directly proportional to the signal's frequency (f), and its amplitude (K). From this, it is apparent that high-amplitude, high-frequency signals require high slew rate op amps in order to prevent slewing. We can rewrite our equation in a more convenient form:

$$\text{Slew Rate required} = 2\pi V_p f_{max} \tag{5.11}$$

where V_p is the peak voltage swing required and f_{max} is the highest frequency sine wave reproduced. Often, it is desirable to know just how "fast" a given op amp is. A further rearranging yields

$$f_{max} = \frac{\text{Slew Rate}}{2\pi V_p}$$

In this case, f_{max} represents the highest frequency sine wave that the op amp can reproduce without producing **S**lewing **I**nduced **D**istortion (SID). This frequency is

FIGURE 5.14

Sine wave disorted by heavy slewing

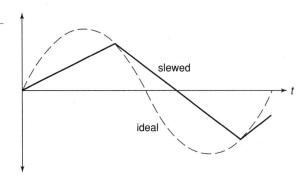

commonly referred to as the *power bandwidth*. To be on the conservative side, set V_p to the op amp's clipping level. Note that slew rate calculations are not dependent on either the circuit gain or small-signal bandwidth. Power bandwidth and small-signal bandwidth (f_2) are **not** the same thing. This is a very important point!

The effects of slewing can be either subtle or dramatic. Small amounts of SID are very difficult to see directly on an oscilloscope and require the use of a distortion analyzer or a spectrum analyzer for verification. Heavy slewing will turn a sine wave into a triangular wave. An example of this is shown in Figure 5.14.

EXAMPLE 5-7

A 741 is used as part of a motor control system. If the highest reproducible frequency is 3 kHz and the maximum output level is 12 V peak, does slewing ever occur?

Another way of stating the problem is to ask "Is the 741's power bandwidth at least 3 kHz?"

$$f_{max} = \frac{Slew\ Rate}{2\pi V_p}$$

$$f_{max} = \frac{.5\ V/\mu S}{2\pi \times 12\ V}$$

$$f_{max} = \frac{.5\ MV/S}{\pi \times 24\ V}$$

$$f_{max} = 6631\ Hz$$

For this application, the 741 is twice as fast as it needs to be. Note in the calculation how the slew rate is transferred from V/μS into V/S and how the volts units cancel between denominator and numerator. This leaves units of "1/Seconds," which is another way of saying "Hertz." If the calculation produced a smaller value, say 2 kHz, then slewing is a possibility for certain signals.

Example 5-8

An audio preamplifier needs to reproduce signals as high as 20 kHz. The maximum output swing is 10 V peak. What is the minimum acceptable slew rate for the op amp used?

$$Slew\ Rate = 2\pi V_p f_{max}$$
$$Slew\ Rate = 2\pi \times 10\ V \times 20\ kHz$$
$$Slew\ Rate = 1.257\ MV/S$$
$$Slew\ Rate = 1.257\ V/\mu S$$

For this design, a 741 would not be fast enough. The aforementioned 411 through 318 would certainly be satisfactory, whereas the 1363 would probably be overkill.

multiSIM Computer Simulation

To verify the results of Example 5-8, a simple noninverting voltage amplifier may be used with differing op amp models. The simulation is shown Figure 5.15 (see pp. 158–160) using MultiSIM. The circuit is configured with a 2-volt input at 20 kHz and a gain of 5. This will yield the worst-case output of 10 volts at 20 kHz. For the first Transient Analysis, a 741 is used. Note how the output waveform is essentially triangular. It is also below the expected peak output level. Clearly, this waveform is severely slewed and results in undesired distortion and a reduction in audio quality. The second simulation is performed using the faster LF411. In this case, the simulation shows a full 10-volt peak output with no discernable distortion. The LF411 would certainly meet the circuit requirements.

Design Hint

There is a convenient way of graphically determining whether the output of an op amp will be distorted. It involves graphing output levels versus frequency. The two major distortion causes are clipping and slewing. We start with a grid measuring frequency on the horizontal axis, and output voltage on the vertical axis. The first step involves plotting the output level limit imposed by clipping. Clipping is dependent on the circuit's power supply and is independent of frequency. Therefore, a horizontal line is drawn across the graph at the clipping level (see Figure 5.16). If we assume a standard ±15 V power supply, this level will be around ±13 V. The output level cannot swing above this line because clipping will be the result. Everything below this line represents unclipped signals. The second step is to plot the slewing line. To do this, a point needs to be calculated for f_{max}. In Example 5-7 a 741 was used and 12 V produced an f_{max} of 6631 Hz. Plot this point on the graph. Now, as the slew rate is directly proportional to f_{max} and V_p, it follows that doubling f_{max}, while halving V_p, results in the same slew rate. This new point lets you graphically determine the slope of the slew limiting line. Plot this new point and connect the two points with a straight line (see Figure 5.17). Everything above this line represents slewed signals, and everything below this line represents nonslewed signals. As long as the desired output signal falls within the

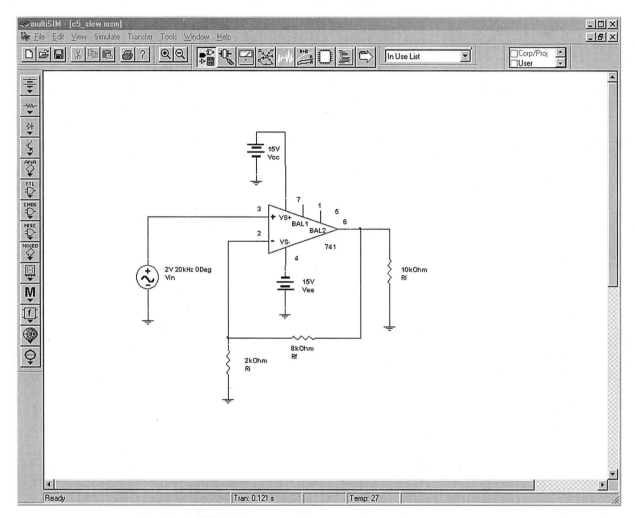

FIGURE 5.15a

MultiSIM schematic for slew simulation

lower intersection area of the two lines, the signal will not experience either slewing or clipping. A quick glance at the graph allows you to tell what forms of distortion may affect a given signal.

Slew Rate and Multiple Stages

Consider the three-stage circuit shown in Figure 5.18. The slew rates for each device are found in Table 5.1. What is the effective slew rate of the system? You might think that it is set by the slowest device (741 at .5 V/μS) or perhaps by the final device (318 at 70 V/μS). The fact is that the system slew rate could be set by any of the devices, and it depends on the gains of the stages.

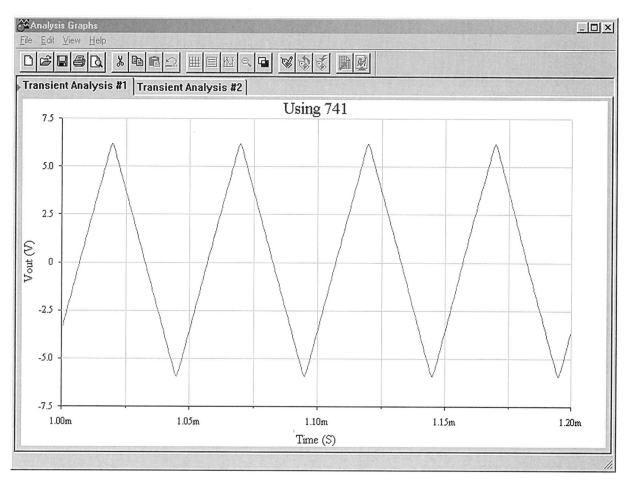

FIGURE 5.15b

Output with 741

The one thing that you can say immediately is that the system slew rate will never be faster than the final device. In this example, the slew rate cannot be greater than 70 V/μS. It may be less than this, however. The trick to finding the effective system slew rate is to start at the output of the first stage and then determine the maximum rate of change for the following stages in sequence. Looking at stage 1, its maximum output rate is .5 V/μS. This is the maximum rate of change going into stage 2. Because stage 2 has a gain of 32, it will attempt to increase this rate to 16 V/μS. This cannot happen, however, because the 411 has a slew rate of only 15 V/μS. Therefore, the 411 is the limiting factor at this point. The maximum rate of change out of stage 2 is 15 V/μS. This signal is then applied to stage 3, which has a gain of 3. So, the 318 triples its input signal to 45 V/μS. Because the 318 is capable of changing as fast as 70 V/μS, 45 V/μS becomes the limiting output factor. The system slew rate is 45 V/μS. This is the value used to calculate the system power bandwidth, if needed. The first op amp to slew in

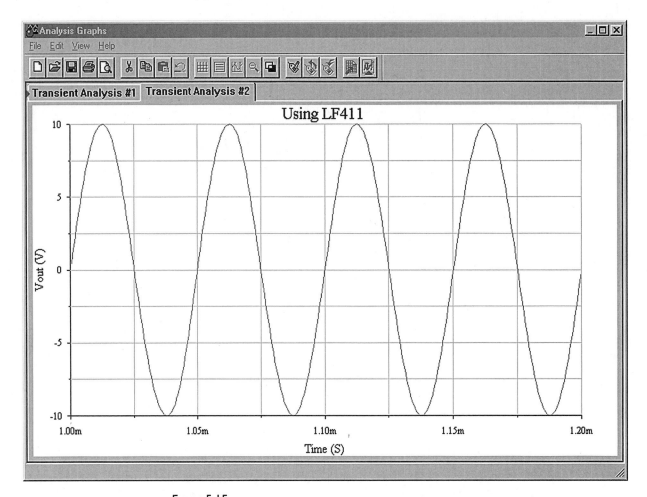

FIGURE 5.15c

Output with LF411

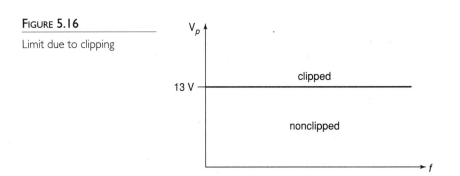

FIGURE 5.16

Limit due to clipping

FIGURE 5.17

Limits due to clipping and slewing

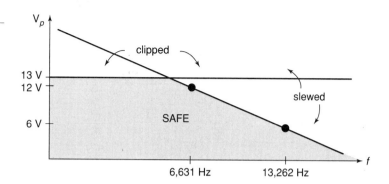

this circuit is the 411, even though it is about 30 times faster than the 741 used in stage 1. The reason for this is that it must handle signals 32 times as large. Note that if the final stage had a larger gain, say 5, the 318 would become the limiting factor. The important thing to remember is that the front-end stages of a system don't need to be as fast as the final stages, as they handle smaller signals.

Noncompensated Devices

As discussed in Chapter 3, all op amps need some form of frequency compensation in order to ensure that their closed-loop response is stable. The most straightforward way to do this is to add a compensation capacitor, which forces a 20 dB/decade rolloff to f_{unity}. In this way, no matter what gain you choose, the circuit will be stable. Although this is very convenient, it is not the most efficient form of compensation for every circuit. High-gain circuits need less compensation capacitance than a low-gain circuit does. The advantage of using a smaller compensation capacitor is that slew rate is increased. Also, available loop gain at higher frequencies is increased. This allows the resulting circuit to have a wider small-signal bandwidth (the effect is as if f_{unity} increased). Therefore, if you are designing a high-gain circuit, you are not producing the maximum slew rate and small-signal bandwidth that you might. The compensation capacitor is large enough to achieve unity gain stability, but your circuit is a high-gain design. The bottom line is that you are paying for unity gain stability with slew rate and bandwidth.

FIGURE 5.18

Multistage circuit incorporating the 741, 411, and 318

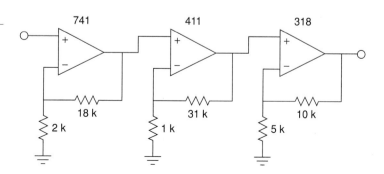

FIGURE 5.19

Bode gain plots for the 301

Reprinted with permission of National Semiconductor Corporation

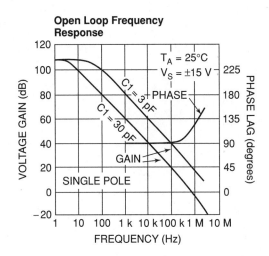

To get around this, manufacturers offer noncompensated op amps. No internal capacitor is used. Instead, connections are brought out to the IC package so that you may add your own capacitor. This way, the op amp may be tailored to your application. There is no set way of determining the values for the external compensation circuit (it may be more complex than a single capacitor). Compensation details are given on manufacturers' data sheets. One example of a noncompensated op amp is the 301. You can think of a 301 as a 741 without a compensation capacitor. If a 33 pF compensation capacitor is used, the 301 will be unity gain stable and produce an f_{unity} of 1 MHz, and a slew rate of .5 V/μS. For higher gains, a smaller capacitor may be used. A 10 pF unit will produce an effective f_{unity} of 3 MHz and a slew rate of 1.5 V/μS. Comparative Bode gain plots are shown in Figure 5.19 for the 301.

Feedforward Compensation

Besides the ordinary form of compensation, noncompensated devices like the 301 may utilize other methods that can make them even faster, such as *feedforward compensation*. The concept of feedforward is in direct contrast to feedback. As its name suggests, feedforward involves adding a portion of the input signal to the output, thus bypassing certain sections of the system. Compared to ordinary feedback, feedforward is seldom used as part of the design of an amplifier.[1]

Ordinarily, a compensation capacitor must be large enough to maintain sufficient gain and phase margin for the slowest stage inside of an op amp. Quite often, the slowest stage in an op amp is one of the first stages, such as a level shifter. Here is where feedforward comes into play. If the high-frequency content of the input signal can be shunted around this slow stage, the effective bandwidth and speed of the op amp

---

[1] It can offer similar advantages though, such as a reduction in distortion. Also, feedforward and feedback techniques may be combined in order to achieve complementary increases in performance. For an example, see M. J. Hawksford, "Reduction of Transistor Slope Impedance Dependent Distortion in Large Signal Amplifiers," *Journal of the Audio Engineering Society*, 36, no. 4 (1988): 213–22

may be increased. This is the key behind the technique of feedforward compensation. Normally, manufacturers will provide details of specific feedforward realizations for their op amps. (It would be very difficult to create successful feedforward designs without detailed knowledge of the internal design of the specific op amp being used.) Not all op amps lend themselves to feedforward techniques.

In summary, noncompensated op amps require a bit more work to configure than fully compensated devices, but offer higher performance. This means faster slew rates and higher upper break frequencies.

Decompensated Devices

Straddling the worlds of compensated and uncompensated op amps is the *decompensated* device. Decompensated devices are also known as *partially compensated* devices. They include some compensation capacitance, but not enough to make them unity gain stable. Usually these devices are stable for gains above 3 to 5. Because the majority of applications require gains in this area or above, decompensated devices offer the ease of use of compensated devices and the increased performance of customized noncompensated units. If required, extra capacitance can usually be added to make the circuit unity gain stable. One example of this type is the 5534. Its gain Bode plot is shown in Figure 5.20. Note how the addition of an extra 22 pF reduces the open-loop gain. This 22 pF is enough to make the device unity gain stable. It also has a dramatic effect on the slew rate, as seen in the 5534's spec sheet. Without the 22 pF capacitor, the slew rate is 13 V/μS, but with it, the slew rate drops to 6 V/μS. The performance that you give up in order to achieve unity gain stability is obvious here.

There is one more interesting item to note about the 5534. A close look at the Bode gain curve shows a hump in the rolloff region. Some other devices, such as the 318, exhibit this hump too. This is a nice extra. In essence, the manufacturer has been able

FIGURE 5.20

Bode gain plot for the 5534
Courtesy of Philips Semiconductors

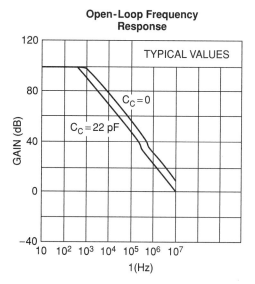

to achieve a slightly higher open-loop gain than a normal device would allow. This means that your circuits will have higher loop gains, and therefore, the nice effects of negative feedback will remain active to higher frequencies. This also means that very high gain circuits will be able to achieve a higher f_2 than is predicted by the gain-bandwidth calculation.

5.5 Offsets

Offsets are undesirable DC levels appearing at the output of a circuit. If op amps were perfect, there would be no such thing as an offset. Even though part matching is very close when ICs are made, the parts will not be identical. One possible example is the fact that the transistors used for the differential amplifier stage will not have identical characteristics. Because of this, their DC bias points are slightly different. This difference, or unbalance, is amplified by the remaining stages and will eventually produce a DC voltage at the output. Because all op amps are slightly different, you never know what the exact output offset will be. For measurement applications, this offset creates uncertainty in readings. For example, if the circuit output measures 100 mV, the signal might be 99 mV with 1 mV of offset. It might also be 101 mV with −1 mV offset. In other applications, offsets can harm following stages or loads. Dynamic loudspeakers and headphones are two loads that should not be fed DC signals. This will reduce their maximum volume and increase their distortion. In short, offsets are not desired. Let's see what the causes are and how we can reduce or eliminate their effect.

Offset Sources and Compensation

For bipolar input sections the major cause of input current mismatch is the variation of beta. Base-emitter junction voltage variation is the major cause of input voltage deviation. For field effect devices, current variation is much less of a problem, as the magnitude of input current is very low to begin with. Unfortunately, FETs do suffer from larger input voltage variations due to transconductance curve mismatches.

As mentioned in Chapter 2, the input current into the bases (or gates, in the case of an FET) of the first stage is called I_B, the input bias current. In reality, this is an average of the two input currents, I_{B_+} and I_{B_-}. The magnitude of their difference is called the *input offset current*, I_{OS} (some manufacturers use the symbol I_{IO}). Note that the actual direction of I_B is normally not specified, but can usually be determined from the manufacturer's circuit diagram. I_B flows into the op amp if the input devices are NPN and out of the op amp if the input devices are PNP.

$$I_B = \frac{I_{B_+} + I_{B_-}}{2}$$

$$I_{OS} = |I_{B_+} - I_{B_-}| \tag{5.12}$$

The voltage difference for the input stage is referred to as the *input offset voltage*, V_{OS} (some manufacturers use the symbol V_{IO}). This is the potential required between the two inputs to null the output, that is, to realign the output to 0 V DC. Both I_{OS} and

TABLE 5.2

Device	I_B	I_{OS}	V_{OS}
5534	800 nA	10 nA	.5 mV
411	50 pA	25 pA	.8 mV
318	150 nA	30 nA	4 mV
741	80 nA	20 nA	1 mV

V_{OS} are available on data sheets. The absolute magnitude of these offsets generally gets worse at temperature extremes. Table 5.2 shows some typical values. Note the low I_B and I_{OS} values for the FET input 411.

Remember, these numbers are absolutes, so when I_{OS} is specified as 10 nA, it means that the actual I_{OS} can be anywhere between -10 nA and $+10$ nA. I_B, I_{OS}, and V_{OS} combine with other circuit elements to produce an *output offset voltage*. As this is a linear circuit, superposition may be used to separately calculate their effects. The model in Figure 5.21 will be used. R_i and R_f are the standard feedback components, and R_{off} is called the *offset compensation resistor* (in some cases it may be zero). Because the input signal is grounded, this model is valid for both inverting and noninverting amplifiers.

V_{OS} is seen as a small input voltage and is multiplied by the circuit's noise gain in order to find its contribution to the output offset. Offsets are by nature DC, so it is important to use the DC noise gain. Consequently, any capacitors found within the feedback loop should be mathematically "opened" for this calculation (for example, when working with the filter circuits presented in Chapter 11).

$$V_{out-offset1} = A_{noise} V_{OS} \tag{5.13}$$

where

$$A_{noise} = 1 + \frac{R_f}{R_i} = \frac{R_i + R_f}{R_i}$$

FIGURE 5.21

Offset model

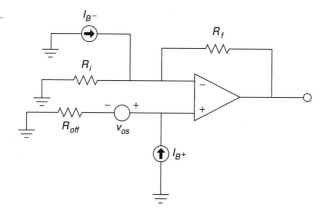

I_B and I_{OS} pass through input and feedback resistors to produce their output contributions. First, consider the effect of I_{B_+}. This creates a voltage across R_{off}. This voltage is then multiplied by the circuit noise gain to yield its portion of the output offset.

$$V_{out\text{-}offset2} = I_{B_+} R_{off} A_{noise} \tag{5.14}$$

For I_{B_-}, recall that the inverting input is at virtual ground. This implies that the voltage across R_i must be zero, and therefore, the current through R_i must be zero. Consequently, all of I_{B_-} flows through R_f. This creates a relative negative potential at the output.

$$V_{out\text{-}offset3} = -I_{B_-} R_f \tag{5.15}$$

So the combination of the input bias current effects is

$$V_{out\text{-}offset(I_B)} = \left| I_{B_+} R_{off} A_{noise} - I_{B_-} R_f \right| \tag{5.16}$$

Expanding this produces

$$V_{out\text{-}offset(I_B)} = I_{B_+} R_{off} \frac{R_i + R_f}{R_i} - I_{B_-} R_f$$

$$V_{out\text{-}offset(I_B)} = \left(I_{B_+} R_{off} \frac{R_i + R_f}{R_i R_f} - I_{B_-} \right) R_f$$

By noting the product-sum rule for resistor combination R_i, R_f, this can be further simplified to

$$V_{out\text{-}offset(I_B)} = \left(\frac{I_{B_+} R_{off}}{R_i \| R_f} - I_{B_-} \right) R_f \tag{5.17}$$

If R_{off} is set to equal $R_i \| R_f$, this reduces to

$$V_{out\text{-}offset(I_B)} = \left(I_{B_+} - I_{B_-} \right) R_f$$

By definition,

$$I_{OS} = \left| I_{B_+} - I_{B_-} \right|$$

so we finally come to

$$V_{out\text{-}offset(I_B)} = I_{OS} R_f \tag{5.18}$$

If it is possible, R_{off} should be set to $R_i \| R_f$. This drastically reduces the effect of the input bias current. Note that the value of R_{off} includes the driving source internal

FIGURE 5.22

Offset compensation for a follower

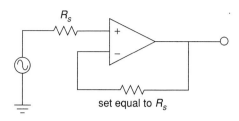

set equal to R_s

resistance. If $R_i \| R_f = 2$ kΩ and the driving source resistance is 100 Ω, the required resistance value would be 1.9 kΩ. If setting R_{off} to the optimum value is not possible, you can at least reduce the effect of I_B by using a partial value. Also, note that it is possible to determine the polarity of the offset caused by I_{B_+} and I_{B_-} (Equation 5.17) if actual currents and the type of device used in the diff amp stage are known. The circuit of Figure 5.21 assumes that NPN devices are being used, hence the currents are drawn as entering the op amp. PNP input devices would produce the opposite polarity. Typically, though, we don't have precise values for I_{B_+} and I_{B_-}, and thus can only compute the worst-case magnitude.

For a final result, we may combine our components:

$$V_{out\text{-}offset} = V_{OS} A_{noise} + I_{OS} R_f \quad (5.19)$$

if $R_{off} = R_i \| R_f$, and

$$V_{out\text{-}offset} = V_{OS} A_{noise} + \left| I_{B_+} R_{off} A_{noise} - I_{B_-} R_f \right| \quad (5.20)$$

if $R_{off} \neq R_i \| R_f$.

There is one special case involving the selection of R_{off}, and that deals with a voltage follower. Normally for a follower, $R_f = 0\,\Omega$. What if the driving source resistance is perhaps 50 Ω? The calculation would require an R_{off} of 0 Ω and thus a -50 Ω resistor to compensate for the source resistance. This is, of course, impossible! To compensate for the 50 Ω source, use 50 Ω for R_f. The circuit gain will still be unity, but I_B will now be compensated for. This is shown in Figure 5.22.

EXAMPLE 5-9

Determine the typical output offset voltage for the circuit of Figure 5.23 if R_{off} is 0 Ω. Then determine an optimum size for R_{off} and calculate the new offset.

From the data sheet for the 5534, we find $V_{OS} = .5$ mV, $I_{OS} = 10$ nA, and $I_B = 800$ nA. Because this is an approximation, assume $I_{B_+} = I_{B_-} = I_B$.

$$A_{noise} = 1 + \frac{R_f}{R_i}$$

$$A_{noise} = 1 + \frac{10\,k}{1\,k}$$

$$A_{noise} = 11$$

FIGURE 5.23

Circuit for Example 5-9

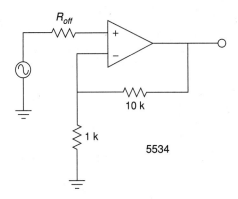

$$V_{out-offset} = V_{OS} A_{noise} + |I_{B+} R_{off} A_{noise} - I_{B-} R_f|$$
$$V_{out-offset} = .5 \text{ mV} \times 11 + |800 \text{ nA} \times 0 \times 11 - 800 \text{ nA} \times 10 \text{ k}|$$
$$V_{out-offset} = 5.5 \text{ mV} + 8 \text{ mV}$$
$$V_{out-offset} = 13.5 \text{ mV}$$

Remember, this is the magnitude of the offset, which could be anywhere within ±13.5 mV. It might be worse if this is not a typical device. Now we find the optimum offset compensating resistor:

$$R_{off} = R_i \| R_f$$
$$R_{off} = 1 \text{ k} \| 10 \text{ k}$$
$$R_{off} = 909 \text{ }\Omega$$

For this case, the offset equation reduces to

$$V_{out-offset} = V_{OS} A_{noise} + I_{OS} R_f$$
$$V_{out-offset} = .5 \text{ mV} \times 11 + 10 \text{ nA} \times 10 \text{ k}$$
$$V_{out-offset} = 5.5 \text{ mV} + 100 \text{ }\mu\text{V}$$
$$V_{out-offset} = 5.6 \text{ mV}$$

By adding R_{off}, the output offset voltage is more than halved. This may lead you to think that it is always wise to add R_{off}. Such is not the case. There are two times when you may prefer to leave it out: (1) to optimize noise characteristics, as we shall see shortly, and (2) when using FET input op amps. FET input devices have very small input bias and offset currents to begin with, so their effect is negligible when using typical resistor values.

EXAMPLE 5-10

The circuit of Figure 5.24 is used as part of a measurement system. Assuming that the DC input signal is 3 mV, how much uncertainty is there in the output voltage typically?

The desired output from the amplifier is

FIGURE 5.24

Circuit for Example 5-10

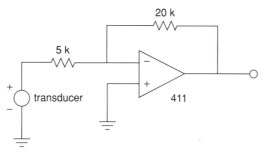

$$V_{out} = A_v V_{in}$$
$$V_{out} = -\frac{R_f}{R_i} V_{in}$$
$$V_{out} = -\frac{20\text{ k}}{5\text{ k}} \times 3\text{ mV}$$
$$V_{out} = -12\text{ mV}$$

The typical specs for the 411 are $V_{OS} = .8$ mV, $I_{OS} = 25$ pA, $I_B = 50$ pA

$$A_{noise} = 1 + \frac{R_f}{R_i}$$
$$A_{noise} = 1 + \frac{20\text{ k}}{5\text{ k}}$$
$$A_{noise} = 5$$

$$V_{out-offset} = V_{OS} A_{noise} + |I_{B+} R_{off} A_{noise} - I_{B-} R_f|$$
$$V_{out-offset} = .8\text{ mV} \times 5 + |50\text{ pA} \times 0 \times 11 - 50\text{ pA} \times 20\text{ k}|$$
$$V_{out-offset} = 4\text{ mV} + 1\ \mu\text{V}$$
$$V_{out-offset} = 4\text{ mV}$$

The output can vary by as much as ±4 mV. As this a DC measurement system, the results are devastating. The output can be anywhere from -12 mV $- 4$ mV $= -16$ mV, to -12 mV $+ 4$ mV $= -8$ mV. That's a 2:1 spread, and it's caused solely by the op amp. Note that the addition of R_{off} would have little effect here. Because the 411 uses a FET input, its I_B contribution is only 1 μV.

So, then, how do you keep output offsets to a minimum? First and foremost, make sure that the op amp chosen has low I_{OS} and V_{OS} ratings. Second, use the offset compensation resistor, R_{off}. Third, keep the circuit resistances as low as possible. Finally, if the output offset is still too large, it can be reduced by manually nulling the circuit.

Nulling involves summing in a small signal that is of opposite polarity to the existing offset. By doing this, the new signal will completely cancel the offset, and the output will show 0 V DC. This is much easier than it sounds. Most op amps have connections for null circuits. These are specified by the manufacturer and usually consist of a single potentiometer and perhaps one or two resistors. An example of a nulling connection is shown in Figure 5.25. Usually the potentiometer is a multi-turn

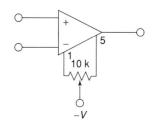

FIGURE 5.25

A typical nulling connection

trim type to allow for fine adjustment. To null the circuit, the technician monitors the output with a very sensitive DC voltmeter. The input is grounded (or perhaps tied to ground through a resistor equal to the driving source resistance if it's large). The potentiometer is then adjusted until the meter reads zero.

The drawback to this procedure is that it requires someone (or perhaps some thing) to perform the nulling. Also, the unit will require periodic adjustment to compensate for aging and environmental effects.

---

multiSIM COMPUTER SIMULATION

As stated, it is very important to match the input resistors in order to keep offsets low. This can be seen clearly in the simulation shown in Figure 5.26 (see pp. 171–172). A noninverting voltage amplifier is modeled here using the 741. In order to focus on the offset current effect, the contribution of V_{OS} is kept small by keeping the voltage gain low. The simulation is run twice for DC Operating Point. In the case where the resistances are matched, the DC output voltage is less than 1 mV. For the unmatched case, R_{off} is set to a fraction of an ohm. The resulting DC output voltage is much larger at approximately 17 mV. It is worthy to note that this is approximately equal to I_{bias} times R_f (80 nA × 200 kΩ). This simulation also points out the poor performance caused by excessively large resistor values. If the simulation is rerun with the resistors scaled down in size, the offsets will be lessened.

5.6 DRIFT

Drift is a variation in the output offset voltage. Often, it is temperature induced. Even if a circuit has been manually nulled, an output offset can be produced if the temperature changes. This is because V_{OS} and I_{OS} are temperature sensitive. The only way around this is to keep the circuit in a constant temperature environment. This can be very costly. If the drift can be kept within an acceptable range, the added cost of cooling and heating equipment may be removed. As you might expect, the magnitude of the drift depends on the size of the temperature change. It also depends on the I_{OS} and V_{OS} sensitivities. These items are $\frac{\Delta V_{OS}}{\Delta T}$, the change in V_{OS} with respect to temperature, and $\frac{\Delta I_{OS}}{\Delta T}$, the change in I_{OS} with respect to temperature. Drift rates are specified in terms of change per centigrade degree. These parameters are usually specified as worst-case values and can produce either a positive or negative potential.

The development of the drift equation pretty much follows that of the equation for offsets. The only difference is that offset parameters are replaced by their temperature coefficients and the temperature change. The products of the coefficients and the change in temperature produce an input offset voltage and current.

$$V_{drift} = \frac{\Delta V_{OS}}{\Delta T} \Delta T \, A_{noise} + \frac{\Delta I_{OS}}{\Delta T} \Delta T \, R_f \qquad (5.21)$$

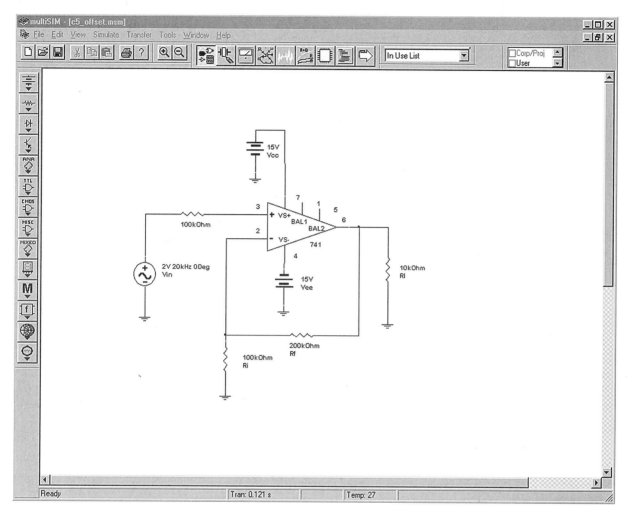

FIGURE 5.26a

MultiSIM schematic for offset simulation

As with the offset calculation, the drift result may be either positive or negative. Also, because I_{OS} is so small for FET input devices, $\frac{\Delta I_{OS}}{\Delta T}$ is often not listed, as it is almost always small enough to ignore. For lowest drift, it is assumed that the op amp uses the offset compensation resistor R_{off}, and that the circuit has been nulled.

EXAMPLE 5-11

Determine the output drift for the circuit of Figure 5.23 for a target temperature of 80°C. Assume that $R_{off} = 909\ \Omega$ and that the circuit has been nulled at 25°C.

The parameters for the 5534 are $\frac{\Delta V_{OS}}{\Delta T} = 5\ \mu V/C°$, $\frac{\Delta I_{OS}}{\Delta T} = 200\ pA/C°$.

FIGURE 5.26b

Results with offset resistor

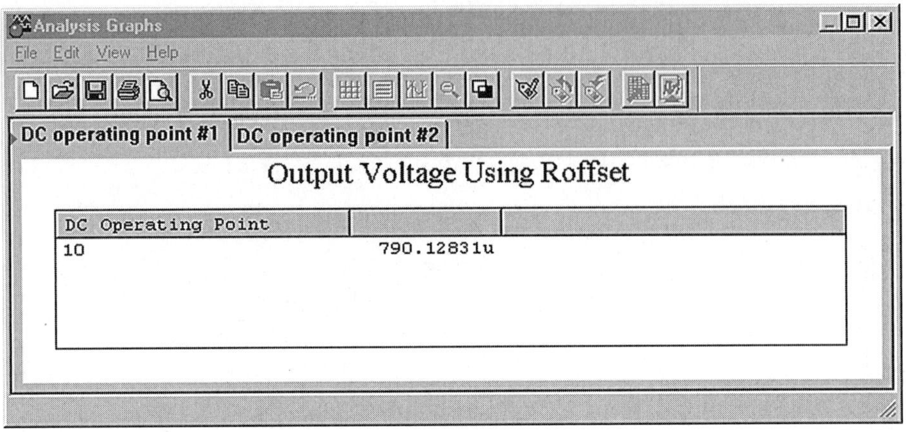

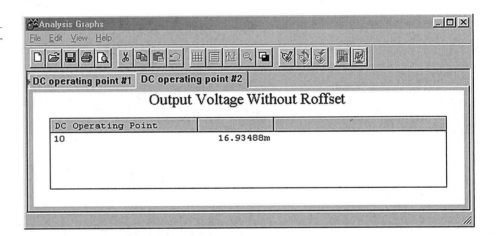

FIGURE 5.26c

Results without offset resistor

The noise gain was already determined to be 11 in Example 5-9. The total temperature change is from 25°C to 80°C, or 55C°.

$$V_{drift} = \frac{\Delta V_{OS}}{\Delta T}\Delta TA_{noise} + \frac{\Delta I_{OS}}{\Delta T}\Delta TR_f$$
$$V_{drift} = 5\ \mu V/C° \times 55C° \times 11 + 200\ pA/C° \times 55C° \times 10\ k$$
$$V_{drift} = 3.025\ mV + .11\ mV$$
$$V_{drift} = 3.135\ mV$$

Note that for this circuit the V_{OS} drift is the major source of error. At 80°C, the output of the circuit may have up to ±3.135 mV of DC error. As with offsets, drift is partially a function of the circuit gain. Therefore high-gain circuits often appear to have excessive drift. In order to compare different amplifiers, *input referred drift* is often used. To find input referred drift, just divide the output drift by the **signal gain** of the amplifier. Don't use the noise gain! In this way, both inverting and noninverting amplifiers can be compared on an equal footing. For this circuit, the input referred drift is

$$V_{drift(input)} = \frac{V_{drift}}{A_v}$$

$$V_{drift(input)} = \frac{3.135 \text{ mV}}{11}$$

$$V_{drift(input)} = 285 \ \mu V$$

For many applications, particularly those primarily concerned with AC performance, drift specification is not very important. A communications amplifier, for example, might use an output coupling capacitor to block any drift or offset from reaching the output if it had to. Drift is usually important for applications involving DC or very low frequencies where coupling capacitors are not practical.

5.7 CMRR AND PSRR

CMRR stands for **C**ommon **M**ode **R**ejection **R**atio. It is a measure of how well the two halves of the input differential amplifier stage are matched. A common-mode signal is a signal that is present on both inputs of the diff amp. Ideally, a differential amplifier completely suppresses or rejects common-mode signals. Common-mode signals should not appear at the circuit output. Due to the nonperfect matching of transistors, some portion of the common-mode signal will make it to the output. Exactly how much this signal is reduced relative to desired signals is measured by the CMRR.

Ideally, CMRR is infinite. A typical value for CMRR would be 100 dB. In other words, if an op amp had both desired (i.e., differential) and common-mode signals at its input that were the same size, the common-mode signal would be 100 dB smaller than the desired signal at the output.

CMRR is particularly important when using the op amp in differential mode (Chapter 4) or when making an instrumentation amplifier (Chapter 6). There are two broad uses for these circuits. First, the amplifier may be receiving a low-level, balanced signal over a considerable distance.[2] Good examples of this are a microphone cable in a recording studio and an instrumentation cable on a factory floor. Interference signals tend to be induced into the cable in-phase (i.e., common-mode). Because the desired signal is presented out-of-phase (i.e., differential), a high CMRR will effectively remove the interference signal. Second, the op amp may be used as part of a bridge-type measurement system. Here, the desired signal is seen as a small variation between two DC potentials. The op amp must amplify the difference signal, but suppress the DC outputs of the bridge circuit.

---

[2] A balanced system uses two signal-carrying conductors and a shield (ground). The two signals are the reference, or in-phase signal, and the inverted, or out-of-phase signal.

Example 5-12

An amplifier has a closed-loop voltage gain of 20 dB and a CMRR of 90 dB. If a common-mode signal is applied to the input at −60 dBV, what is the output?

If the input signal were differential instead of common-mode, the output would be

$$V'_{out} = A'_v + V'_{in}$$
$$V'_{out} = 20 \text{ dB} + -60 \text{ dBV}$$
$$V'_{out} = -40 \text{ dBV}$$

Because this is a common-mode signal, it is reduced by the CMRR

$$V'_{out} = -40 \text{ dBV} - 90 \text{ dB}$$
$$V'_{out} = -130 \text{ dBV}$$

This signal is so small that it is probably overshadowed by the circuit noise.

One final note concerning CMRR is that it is specified for DC. In truth, CMRR is frequency dependent. The shape of its curve is reminiscent of the open-loop gain curve. The stated CMRR may remain at its DC level up to perhaps 100 or 1000 Hz, and then fall off as frequency increases. For example, the 741 data sheet found in the Appendix states a typical CMRR of 90 dB. By looking at the CMRR graph, though, you can see that it starts to roll off noticeably around 1 kHz. By the time it hits 1 MHz, only 20 dB of rejection remains. A more gentle rolloff is exhibited by the 411. At 1 MHz, almost 60 dB of rejection remains. What this means is that the op amp cannot suppress high-frequency interference signals as well as it suppresses low-frequency interference.

Similar to CMRR is PSRR, or **P**ower **S**upply **R**ejection **R**atio. Ideally, all ripple, hum, and noise from the power supply will be prevented from reaching the output of the op amp. PSRR is a measure of exactly how well the op amp reaches this ideal. Typical values for PSRR are in the 100 dB range. Like CMRR, PSRR is frequency-dependent and shows a rolloff as frequency increases. If an op amp is powered by a 60 Hz source, the ripple frequency from a standard full-wave rectifier will be 120 Hz. At the output, this ripple will be reduced by the PSRR. Higher-frequency noise components on the power supply line are not reduced as much because PSRR rolls off. Normally, PSRR is consistent between power rails. Sometimes there is a marked performance difference between the positive and negative PSRR. One good example of this is the 411. The positive rail exhibits about a 30 dB improvement over the negative rail. Note that PSRR is only a few decibels for the negative rail by the time it reaches 1 MHz.

Example 5-13

An op amp is operated off of a power supply that has a peak-to-peak ripple voltage of .5 V. If the op amp's PSRR is 86 dB, how much of this ripple is seen at the circuit output?

First, determine the PSRR as an ordinary value.

$$PSRR = \log_{10}^{-1} \frac{PSRR'}{20}$$

$$PSRR = \log_{10}^{-1} \frac{86 \text{ dB}}{20}$$

$$PSRR = 20,000$$

Divide the ripple voltage by the PSRR to find the amount that is seen at the output.

$$V_{out-ripple} = \frac{V_{ripple}}{PSRR}$$

$$V_{out-ripple} = \frac{.5 \text{ V}_{pp}}{20,000}$$

$$V_{out-ripple} = 25 \text{ } \mu V_{pp}$$

5.8 Noise

Generally speaking, *noise* refers to undesired output signals. The background hiss found on audiotape is a good example of noise. If noise levels get too high, the desired signals are lost. We will narrow our definition down a bit by only considering noise signals that are created by the op amp circuit. Noise comes from a variety of places. First of all, all resistors have *thermal*, or *Johnson noise*. This is due to the random effects that thermal energy produces on the electrons. Thermal noise is also called *white noise*, because it is equally distributed across the frequency spectrum. Semiconductors exhibit other forms of noise. *Shot noise* is caused by the fact the charges move as discrete particles (electrons). It is also white. *Popcorn noise* is dominant at lower frequencies and is caused by manufacturing imperfections. Finally, *flicker noise* has a $\frac{1}{f}$ spectral density. This means that it increases as the frequency drops. It is sometimes referred to as $\frac{1}{f}$ noise.

Any op amp circuit exhibits noise from all of these sources. Because we are not designing the op amps, we don't really need to distinguish the exact sources of the noise; rather, we'd just like to find out how much total noise arrives at the output. This will allow us to determine the *signal-to-noise ratio* (S/N) of the circuit, or how quiet the amplifier is.

If noise performance for a particular design is not paramount, it can be quickly estimated from manufacturers' data sheets. Some manufacturers will specify RMS noise voltages for specified signal bandwidths and source impedances. A typical plot is found in Figure 5.27. To use this, simply find the source impedance of your circuit on the horizontal axis, and by using the appropriate signal bandwidth curve, find the noise voltage on the vertical axis. This noise voltage is input-referred. In order to find the output noise voltage, multiply this number by the noise gain of the circuit. The result will not be very exact, but it will put you in the ballpark.

A more exact approach involves the use of two op amp parameters, *input noise voltage density*, v_{ind}, and *input noise current density*, i_{ind}. Nanovolts per root hertz are used to specify v_{ind}. Picoamps per root hertz are used to specify i_{ind}. Refer to the 5534 data sheet for example specifications. (Some manufacturers square these values and give units of volts squared per hertz and amps squared per hertz. To translate to the more common form, just take the square root of the values given. The data sheet for

Figure 5.27

Noise voltage for given bandwidth versus source resistance

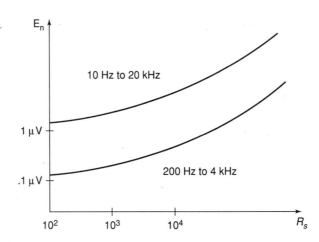

the 741 is typical of this form.) These two parameters take into account noise from all internal sources. As a result, these parameters are frequency dependent. Due to the flicker noise component, the curves tend to be rather flat at higher frequencies and then suddenly start to increase at lower frequencies. The point at which the graphs start to rise is called the *noise corner frequency*. Generally, the lower this frequency, the better.

In order to find the output noise we will combine the noise from three sources:

1. v_{ind}
2. i_{ind}
3. The thermal noise of the input and feedback resistors

Before we start, there are a few points to note. First, the strength of the noise is dependent on the *noise bandwidth* of the circuit. For general-purpose circuits with 20 dB/decade rolloffs, the noise-bandwidth, BW_{noise}, is 1.57 times larger than the small-signal bandwidth. It is larger because some noise still exists in the rolloff region. The noise-bandwidth would only equal the small-signal bandwidth if the rolloff rate were infinitely fast. Second, because the noise-bandwidth factor is common to all three sources, instead of calculating its effect three times, we will combine the partial results of the three sources first, and then apply the noise-bandwidth effect. This will make the calculation faster, as we have effectively factored out BW_{noise}. Each of the three sources will be noise voltage densities, all having units of volts per root hertz. Because v_{ind} is already in this form, we only need to calculate the thermal and i_{ind} effects before doing the summation. Finally, because noise signals are random, they do not add coherently. In order to find the effective sum we must perform an RMS summation, that is, a *square root of the sum of the squares*. This will result in the input noise voltage. To find the output noise voltage we will then multiply by A_{noise}.

The first thing that must be done is to determine the *input noise resistance*. R_{noise} is the combination of the resistance seen from the inverting input to ground and from the noninverting input to ground. To do this, short the voltage source and ground the

Figure 5.28

Equivalent noise analysis circuit

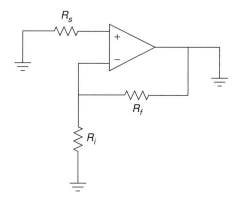

output. You will end up with circuit like Figure 5.28. Note that R_i and R_f are effectively in parallel. Therefore,

$$R_{noise} = R_s + R_i \| R_f \tag{5.22}$$

You might note that R_s is in the same position as R_{off} for the offset calculations. For absolute minimum noise, the offset compensating resistor is not used. R_{noise} is used to find the thermal noise and the contribution of i_{ind}. As always, Ohm's law still applies, so as you might expect, $i_{ind} R_{noise}$ produces a noise voltage density with our desired units of volts per root hertz. The only source left is the thermal noise.

The general equation for thermal noise is:

$$e_{th} = \sqrt{4\, KTBW_{noise}\, R_{noise}} \tag{5.23}$$

Where

e_{th} is the thermal noise.
K is Boltzmann's constant, 1.38×10^{-23} Joules/Kelvin degree.
T is the temperature in Kelvin degrees (Celsius + 273).
BW_{noise} is the effective noise bandwidth.
R_{noise} is the equivalent noise resistance.

Because we are interested in finding the noise density, we can pull out the BW_{noise} factor. Also, when we perform the RMS summation, this quantity will need to be squared. Instead of taking the square root and then squaring it again, we can just leave it as e_{th}^2. These two considerations leave us with the *mean squared thermal noise voltage density*, or

$$e_{th}^2 = 4\, KTR_{noise} \tag{5.24}$$

Now that we have the components, we may perform the summation.

$$e_{total} = \sqrt{v_{ind}^2 + (i_{ind} \times R_{noise})^2 + e_{th}^2} \tag{5.25}$$

The total input noise voltage density is e_{total}. Its units are in volts per root hertz. At this point, we may now include the noise bandwidth effect. In order to find BW_{noise}, multiply the small-signal bandwidth by 1.57. If the amplifier is DC coupled, the small-signal bandwidth is equal to f_2, otherwise it is equal to $f_2 - f_1$. For most applications setting the bandwidth to f_2 is sufficient.

$$BW_{noise} = 1.57 f_2 \qquad (5.26)$$

For the final step, e_{total} is multiplied by the square root of BW_{noise}. Note that the units for e_{total} are volts per root hertz. Consequently, we need a *root hertz* bandwidth. We are performing a mathematical shortcut here. You could square e_{total} in order to get units of volts squared per hertz, multiply by BW_{noise}, and then take the square root of the result to get back to units of volts, but the first way is quicker. Anyway, we end up with the input referred RMS noise voltage.

$$e_n = e_{total}\sqrt{BW_{noise}} \qquad (5.27)$$

EXAMPLE 5-14

Determine the output noise voltage for the circuit of Figure 5.29. For a nominal output of 1 V RMS, what is the signal-to-noise ratio? Assume T = 300° K (room temperature).

The 5534 shows the following specs, $v_{ind} = 4$ nV/$\sqrt{Hz}$, $i_{ind} = .6$ pA/$\sqrt{Hz}$. These values do rise at lower frequencies, but we shall ignore this effect for now. Also, f_{unity} is 10 MHz.

$$A_v = 1 + \frac{R_f}{R_i}$$
$$A_v = 1 + \frac{99\,k}{1\,k}$$
$$A_v = 100 = 40 \text{ dB}$$

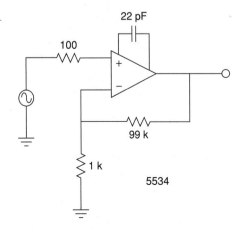

FIGURE 5.29

Circuit for Example 5-14

For the noninverting amplifier, $A_v = A_{noise}$, so

$$A_{noise} = 100 = 40 \text{ dB}$$
$$R_{noise} = R_s + R_f \| R_i$$
$$R_{noise} = 100 + 99 \text{ k} \| 1 \text{ k}$$
$$R_{noise} = 1090 \text{ }\Omega$$

$$e_{th}^2 = 4KTR_{noise}$$
$$e_{th}^2 = 4 \times 1.38 \times 10^{-23} \times 300 \times 1090$$
$$e_{th}^2 = 1.805 \times 10^{-17} \text{ volts squared per hertz}$$

$$e_{total} = \sqrt{v_{ind}^2 + (i_{ind} R_{noise})^2 + e_{th}^2}$$
$$e_{total} = \sqrt{(4 \text{ nV}/\sqrt{\text{Hz}})^2 + (.6 \text{ pA}/\sqrt{\text{Hz}} \times 1090)^2 + 1.805 \times 10^{-17} \text{V}^2/\text{Hz}}$$
$$e_{total} = \sqrt{1.6 \times 10^{-17} + 4.277 \times 10^{-19} + 1.805 \times 10^{-17}}$$
$$e_{total} = 5.87 \text{ nV}/\sqrt{\text{Hz}}$$

Note that the major noise contributors are v_{ind} and e_{th}. Now to find BW_{noise},

$$f_2 = \frac{f_{unity}}{A_{noise}}$$
$$f_2 = \frac{10 \text{ MHz}}{100}$$
$$f_2 = 100 \text{ kHz}$$

$$BW_{noise} = f_2 1.57$$
$$BW_{noise} = 100 \text{ kHz} \times 1.57$$
$$BW_{noise} = 157 \text{ kHz}$$

$$e_n = e_{total} \sqrt{BW_{noise}}$$
$$e_n = 5.87 \text{ nV}/\sqrt{\text{Hz}} \sqrt{157 \text{ kHz}}$$
$$e_n = 2.33 \text{ }\mu\text{V RMS}$$

To find the output noise, multiply by the noise gain.

$$e_{n-out} = e_n A_{noise}$$
$$e_{n-out} = 2.33 \text{ }\mu\text{V} \times 100$$
$$e_{n-out} = 233 \text{ }\mu\text{V RMS}$$

For a nominal output signal of 1 V RMS, the signal-to-noise ratio is

$$S/N = \frac{\text{Signal}}{\text{Noise}}$$
$$S/N = \frac{1 \text{ V}}{233 \text{ }\mu\text{V}}$$
$$S/N = 4290$$

Normally S/N is given in dB.

$$S/N' = 20 \log_{10} S/N$$
$$S/N' = 20 \log_{10} 4290$$
$$S/N' = 72.6 \text{ dB}$$

As was noted earlier, the noise curves increase at lower frequencies. How do you take care of this effect? First of all, when using a wide-band design with a noise corner frequency that is relatively low (as in Example 5-14), it can be safely ignored. If the low-frequency portion takes up a sizable chunk of the signal frequency range, it is possible to split the calculation into two or more segments. One segment would be for the constant part of the curves. Other segments can be made for the lower frequency portions. In these regions, an averaged value would be used for v_{ind} and i_{ind}. This is a rather advanced treatment, and we will not pursue it here.

Finally, it is common to use the parameter *input referred noise voltage*. Input referred noise voltage is the output noise voltage divided by the circuit signal gain.

$$e_{in-ref} = \frac{e_{n-out}}{A_v} \tag{5.28}$$

This value is the same as e_n for noninverting amplifiers, but varies a bit for inverting amplifiers because A_{noise} does not equal A_v for inverting amplifiers.

SUMMARY

In this chapter we have taken a closer look at op amp characteristics. First of all, we find that the upper frequency limit is a function of the op amp parameter f_{unity}, also known as the gain-bandwidth product, and the circuit's noise gain. The higher the gain is, the lower the upper break frequency will be. Op amps are capable of flat response down to DC. If coupling capacitors are used, the lower break frequency may be found by using standard lead network analysis. When stages are cascaded, the results echo those of cascaded discrete stages. The lowest f_2 is dominant and becomes the system f_2. The highest f_1 is dominant and sets the system f_1. If more than one stage exhibits the dominant critical frequency, the actual critical frequency will be somewhat lower for f_2 and somewhat higher for f_1.

In order to make the op amp unconditionally stable, a compensation capacitor is used to tailor the open-loop frequency response. Besides setting f_{unity}, this capacitor also sets the slew rate. Slew rate is the maximum rate of change of output voltage with respect to time. Slewing slows down the edges of pulse signals and distorts sinusoidal signals. The highest frequency that an amplifier can produce without slewing is called the power bandwidth. In order to optimize f_{unity} and slew rate, some amplifiers are available without the compensation capacitor. The designer then adds just enough capacitance to make the design stable.

Due to slight imperfections between the input transistors, op amps may produce small DC output voltages called offsets. Offsets may be reduced through proper resistor selection. Simple nulling circuits may be used to completely remove the offset. A variable offset due to temperature variation is called drift. The larger the temperature variation, the larger the drift will be. The transistor mismatch also means that common-mode signals will not be completely suppressed. Just how well common-mode signals are suppressed is measured by the common-mode rejection ratio, CMRR. Similar to CMRR is PSRR, the power-supply rejection ratio. PSRR measures how well power-supply noise and ripple are suppressed by the op amp. Both PSRR and CMRR are frequency dependent. Their maximum values are found at DC and then they decrease as frequency increases.

Finally, noise is characterized as an undesired random output signal. The noise in op amp circuits may be characterized by three components: the thermal noise of the input and feedback resistors, the op amp's input noise voltage density, v_{ind}, and its input noise current density, i_{ind}. The combination of these elements requires an

RMS summation. In order to find the output noise voltage, the input noise voltage is multiplied by the circuit's noise gain. The ratio of the desired output signal and the noise voltage is called, appropriately enough, the signal-to-noise ratio. Normally, signal-to-noise ratio is specified in decibels.

Review Questions

1. Define gain-bandwidth product. What is its use?
2. How do you determine f_2 and f_1 for a multistage circuit?
3. What happens if two or more stages share the same break frequency?
4. What is slew rate?
5. How is power bandwidth determined?
6. How do power bandwidth and small-signal bandwidth differ?
7. What are the advantages and disadvantages of noncompensated op amps?
8. What are decompensated op amps?
9. What causes DC offset voltage?
10. What causes DC drift voltage?
11. What is CMRR?
12. What is PSRR?
13. What parameters describe an op amp's noise performance?
14. What is S/N?

Problems

Analysis Problems

1. Determine f_2 for the circuit in Figure 5.3 if a 411 op amp is used.
2. Determine f_2 for the circuit of Figure 5.5 if a 318 op amp is used.
3. What is the minimum acceptable f_{unity} for the op amp in Figure 5.3 if the desired f_2 is 250 kHz?
4. What is the minimum acceptable f_{unity} for the op amp in Figure 5.5 if the desired f_2 is 20 kHz?
5. Determine the power bandwidth for Problem 5.1. Assume $V_p = 10$ V.
6. Determine the power bandwidth for Problem 5.3. Assume $V_p = 12$ V.
7. What is the minimum acceptable slew rate for the circuit of Figure 5.3 if the desired power bandwidth is 20 kHz with a V_p of 10 V?
8. What is the minimum acceptable slew rate for the circuit of Figure 5.5 if the desired power bandwidth is 40 kHz with a V_p of 5 V?
9. A circuit has the following specifications: ±15 V power supply, voltage gain equals 26 dB, desired power bandwidth equals 80 kHz at clipping. Determine the minimum acceptable slew rate for the op amp.
10. Determine the system f_2 in Figure 5.6 if all three devices are 318s.
11. Determine the system slew rate for Figure 5.7. The first device is a 741 and the second unit is a 411.
12. Find the output offset voltage for Figure 5.3.
13. If $R_s = R_f = 100\Omega$ in Figure 5.22, find the output offset voltage using a 318 op amp.
14. Assume that the circuit of Figure 5.23 is nulled at 25°C and that an optimum value for R_{off} is used. Determine the drift at 75°C.
15. Assuming that the 120 Hz power supply ripple in Figure 5.30 is 50 mV, how large is its contribution to the output?

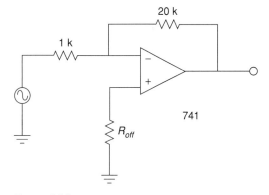

FIGURE 5.30

16. Utilizing a 5534 op amp, what is the approximate input noise voltage for a source resistance of 1 kΩ and a bandwidth from 10 Hz to 20 kHz?

17. Assuming that the op amp of Problem 5.16 is connected like Figure 5.30, what is the approximate output noise voltage? What is the approximate input referred noise voltage?
18. Assume that $R_s = 0\ \Omega$, $R_i = 500\ \Omega$, and $R_f = 10\ \text{k}\Omega$ in Figure 5.28. Find the input noise voltage if the op amp is a 411.
19. For a nominal output voltage of 2 V RMS, determine the signal-to-noise ratio for Problem 5.18.
20. Assume that the input to Figure 5.22 is a 5 V peak 50 kHz square wave. Draw the output waveform if a 741 is used.
21. Repeat Problem 5.20 using a 10 V peak, 100 kHz sine wave.

Design Problems

22. Determine an optimum value for R_{off} in Figure 5.3 and determine the resulting offset voltage.
23. Determine the optimum value for R_{off} in Figure 5.30. Assuming that the circuit has been nulled at 25°C, find the drift at 60°C.
24. Determine a new value for the 100 kΩ resistor in Figure 5.8 in order to minimize the output offset.
25. Using the optimum resistor found in Problem 5.24, determine a new value for the input capacitor that will maintain the original f_1.
26. Design a circuit with a gain of 32 dB and an f_2 of at least 100 kHz. You may use any of the following: 741, 411, or 318.
27. Design a circuit with a gain of 50 and an f_{max} of at least 50 kHz, given a maximum output swing of 10 V peak. You may use any of the following: 741, 411, or 318.
28. Design a circuit with a gain of 12 dB, a small-signal bandwidth of at least 100 kHz, and an f_{max} of at least 100 kHz for a peak output swing of 12 V.
29. Utilizing two or more stages, design a circuit with a gain of 150 and a small-signal bandwidth of at least 600 kHz.

Challenge Problems

30. Determine the system f_2 in Figure 5.18.
31. Determine the input noise voltage for the circuit of Figure 5.30. Assume $R_{off} = 950\ \Omega$.
32. Determine the output noise voltage and the input-referred noise voltage for Problem 5.31.
33. Assuming that driving source resistance in Figure 5.7 is $0\ \Omega$, how much offset voltage is produced at the output of the circuit? Assume that both devices are 411s.
34. Assume that you have one each of: 411, 741, and 318. Determine the combination that will yield the highest system slew rate in Figure 5.31.

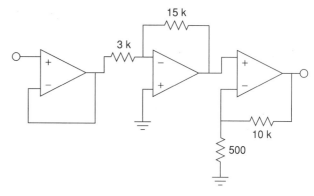

FIGURE 5.31

35. Repeat Problem 5.34 in order to produce the highest system f_2.
36. Assuming that the driving source resistance is $0\ \Omega$ in Figure 5.7, determine the output noise voltage if both devices are 411s.
37. Derive Equation 5.9 from the text.

Computer Simulation Problems

38. Use a simulator to create a Bode plot for Problem 5.10. If a macro model for the LM318 is not available, use 741s instead.
39. Create a time domain representation of the output voltage of Problem 5.20 using a simulator.
40. Create a time domain representation of the output voltage of Problem 5.21 using a simulator. Repeat the simulation using an LM318 op amp in place of the 741. What do the results indicate?
41. Simulate the circuit designed in Problem 5.26. Verify f_2 and A_V through a Bode plot.
42. Generate a Bode plot using a simulator and verify f_2 and A_V for the circuit designed in Problem 5.29.
43. Generate a Bode plot using a simulator and verify f_2 for the circuit designed in Problem 5.35.

CHAPTER 6

Specialized Op Amps

CHAPTER OBJECTIVES

After completing this chapter, you should be able to:

- Analyze *instrumentation amplifiers*.
- Discuss the advantages of using instrumentation amplifiers versus simple op amp differential amplifiers.
- Analyze and detail the advantages of *programmable* op amps and give possible applications.
- Detail the advantages of high-speed and high-power op amps and give possible applications.
- Describe an *OTA*, noting how it differs from a programmable op amp, and give possible applications.
- Describe the operation of a *Norton amplifier*, compare how it is used relative to ordinary op amps, and analyze circuitry that makes use of Norton amplifiers.
- Describe the operation of *current feedback* op amps, and explain what their primary advantages are.
- Describe the need and usage of application-specific integrated circuits.

6.1 Introduction

So far, you have seen how to analyze and design op amp circuits for a variety of general-purpose applications. There are many applications in which the ease of op amp circuit design would be welcome, but in which the average op amp's performance is not suitable. Over the years, manufacturers have extended the performance of op amps outside of their original low-power, low-frequency realm. A variety of special-purpose op amps and op amp derivatives now exist for the designer's convenience. This chapter takes a look at a number of the areas where specialized op amps may now be used.

Perhaps the two most noticeable areas of extended device performance are in power handling and speed. At one time, op amps were considered to be suitable for use below 1 MHz. Today it is possible to find high-speed devices designed for applications such as video where bandwidths are measured in the tens of MHz. In the area of power, the general-purpose op amp typically operates from a ±15 V supply and can produce output currents in the neighborhood of 25 milliamps. New devices can produce output currents measured in amperes, while other devices can produce signals in the hundreds of volts. This makes it possible to directly connect low-impedance loads to the op amp, which is a very useful commodity.

There are a variety of other useful variations on the basic op amp theme. These include devices optimized for single-supply operation, dedicated voltage followers, devices that allow you to trade off speed for power consumption, and application-specific items like low-noise audio preamplifiers and voltage-controlled amplifiers. This chapter will introduce a selection of these devices and take a look at a few of their applications. Although these devices may not find as wide an appeal as, say, a 411 or 741, they allow a user to extend designs without "reinventing the wheel" each time.

6.2 Instrumentation Amplifiers

There are numerous applications where a differential signal needs to be amplified. These include low-level bridge measurements, balanced microphone lines, communications equipment, thermocouple amplifiers, and the like. The immediate answer to these applications is the differential op amp configuration noted in Chapter 4. There are limitations to this form, unfortunately. For starters, it is practically impossible to achieve matched high-impedance inputs while maintaining high gain and satisfactory offset and noise performance. For that matter, the input impedances are not isolated; indeed, the impedance of one input may very well be a function of the signal present on the other input. Simply put, this is an unacceptable situation when a precision amplifier is needed, particularly if the source impedance is not very low.

An *instrumentation amplifier* overcomes these problems. Instrumentation amplifiers offer very high impedance, isolated inputs along with high gain, and excellent CMRR performance. Some people like to think of instrumentation amplifiers as a form of "souped up" differential amplifier. Instrumentation amplifiers can be fashioned from separate op amps. They are also available on a single IC for highest performance.

Instrumentation amplifiers are, in essence, a three-amplifier design. To understand how they work, it is best to start with a differential amplifier based on a single op amp, as seen in Figure 6.1a. One way to increase the input impedances and also maintain input isolation is to place a voltage follower in front of each input. This is shown in Figure 6.1b. The source now drives the very high input impedance followers. The followers exhibit very low output impedance and have no trouble driving the differential stage. In this circuit, op amp 3 is used for common-mode rejection as well as for voltage gain. Note that the gain-bandwidth requirement for op amp 3 is considerably higher than for the input followers. An enhancement to this circuit is

FIGURE 6.1

Basic differential amplifier (left) and with buffers (right)

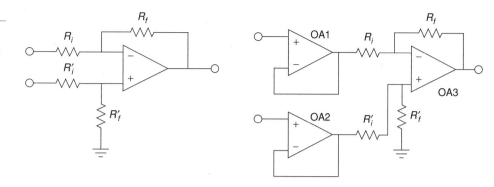

shown in Figure 6.2. Here, op amps 1 and 2 are used for signal gain along with their previous duty of input buffering. The major problem with this configuration is that it requires very close matching of resistors in order to keep the gains the same. Any mismatch in gain between the two inputs will result in a degradation of CMRR. For optimum common-mode rejection, the undesired signal must be identical at the inputs to op amp 3. For high CMRR systems, resistor matching may need to be better than .01%. This is an expensive requirement if discrete resistors are used. Another way around this would be to allow some adjustment of the gain to compensate for gain mismatches, perhaps by using a resistor/potentiometer combination in place of R'_f for op amp 3. For reasons of cost and time, adjustments are frowned upon.

Fortunately, there is a very slight modification to Figure 6.2 that will remove the problem of mismatched gains. This modification involves joining the R_i values of op amps 1 and 2 into a single resistor. This is shown in Figure 6.3. In order to prevent

FIGURE 6.2

Basic instrumentation amplifier

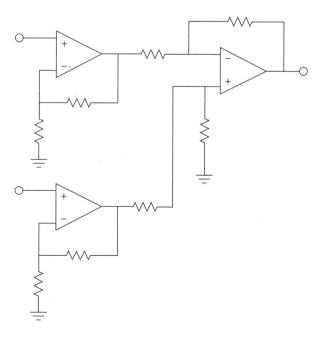

FIGURE 6.3

Improved instrumentation amplifier

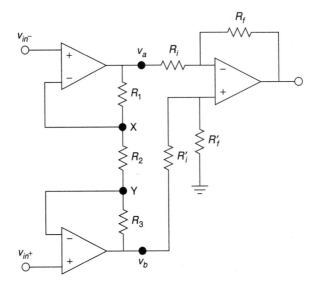

possible confusion with op amp 3, the three resistors used for the input section have been labeled as R_1, R_2, and R_3. In order to analyze this circuit, we shall use the approximation techniques examined earlier.

First of all, from preceding work we already know the gain equation for the differential section is

$$V_{out} = \frac{R_f}{R_i}(V_b - V_a) \qquad (6.1)$$

Our goal, then, is to find equations for V_b and V_a. Let's examine op amp 1 for V_a. First, due to the approximation that the error voltage of the op amps is zero, the inverting and noninverting inputs of each op amp are the same. We can therefore say that V_x must equal V_{in_-}, and that V_y must equal V_{in_+}.

$$V_x = V_{in_-} \qquad (6.2)$$
$$V_y = V_{in_+} \qquad (6.3)$$

The output voltage V_a must equal V_x plus the drop across R_1.

$$V_a = V_x + V_{R_1} \qquad (6.4)$$

The "positive" potential of V_{R_1} is arbitrary. We are assuming that the current is flowing down through R_1. If the actual numbers work out opposite, V_{R_1} will come up negative. The voltage drop V_{R_1} is found by Ohm's law:

$$V_{R_1} = R_1 I_{R_1} \qquad (6.5)$$

To expand,

$$I_{R_1} = I_{R_2} + I_{op\,amp1}$$

$I_{op\,amp1}$ is approximately 0, therefore,

$$V_{R_1} = R_1 I_{R_2}$$

The current through R_2 is set by its value and the drop across it.

$$I_{R_2} = \frac{V_x - V_y}{R_2} \tag{6.6}$$

Substituting 6.6 and 6.5 into 6.4 yields

$$V_a = V_x + \frac{R_1(V_x - V_y)}{R_2} \tag{6.7}$$

Substituting 6.2 and 6.3 into 6.7 yields

$$V_a = V_{in_-} + \frac{R_1(V_{in_-} - V_{in_+})}{R_2}$$

$$V_a = V_{in_-} + \frac{R_1}{R_2}(V_{in_-} - V_{in_+})$$

$$V_a = V_{in_-} + V_{in_-}\frac{R_1}{R_2} - V_{in_+}\frac{R_1}{R_2}$$

$$V_a = V_{in_-}\left(1 + \frac{R_1}{R_2}\right) - V_{in_+}\frac{R_1}{R_2} \tag{6.8}$$

A close look at Equation 6.8 reveals that it is made up of two terms. The first term is V_{in_-} times the noninverting gain of op amp 1, as you may have expected. The second term is V_{in_+} times the inverting gain of op amp 1. This output potential is important to note. Even though it may not appear as though op amp 1 will clip a given signal, it might if the input to the second op amp is large enough and of the proper polarity.

By a similar derivation, the equation for V_b is found.

$$V_b = V_{in_+}\left(1 + \frac{R_3}{R_2}\right) - V_{in_-}\frac{R_3}{R_2} \tag{6.9}$$

For gain matching, R_3 is set equal to R_1. R_2 can then be used to control the gain of the input pair, in tandem.

Finally, substituting 6.8 and 6.9 into 6.1 yields,

$$V_{out} = \frac{R_f}{R_i}\left(\left(V_{in_+}\left(1 + \frac{R_1}{R_2}\right) - V_{in_-}\frac{R_1}{R_2}\right) - \left(V_{in_-}\left(1 + \frac{R_1}{R_2}\right) - V_{in_+}\frac{R_1}{R_2}\right)\right) \tag{6.10}$$

Combining terms produces

$$V_{out} = \frac{R_f}{R_i}\left((V_{in_+} - V_{in_-})\left(1 + \frac{R_1}{R_2}\right) + (V_{in_+} - V_{in_-})\frac{R_1}{R_2}\right)$$

$$V_{out} = (V_{in_+} - V_{in_-})\left(\frac{R_f}{R_i}\right)\left(1 + 2\frac{R_1}{R_2}\right) \qquad (6.11)$$

The first term is the differential input voltage. The second term is the gain produced by op amp 3, and the third term is the gain produced by op amps 1 and 2. Note that the system common-mode rejection is no longer solely dependent on op amp 3. A fair amount of common-mode rejection is produced by the first section, as evidenced by equations 6.8 and 6.9. Because the inverting and noninverting gains are almost the same for very high values, high input gains tend to optimize the system CMRR. The remaining common-mode signals can then be dealt with by op amp 3.

EXAMPLE 6-1

The instrumentation amplifier of Figure 6.4 is used to amplify the output of a balanced microphone. The output of the microphone is 6 mV peak (12 m differential), and a common-mode hum signal is induced into the lines at 10 mV peak (0 mV differential). If the system has a CMRR of 100 dB, what is the output signal?

First, let's check the outputs of the first section to make sure that no clipping is occurring. We shall use superposition and consider the desired signal and hum signal separately.

$$V_a = V_{in_-}\left(1 + \frac{R_1}{R_2}\right) - V_{in_+}\frac{R_1}{R_2}$$

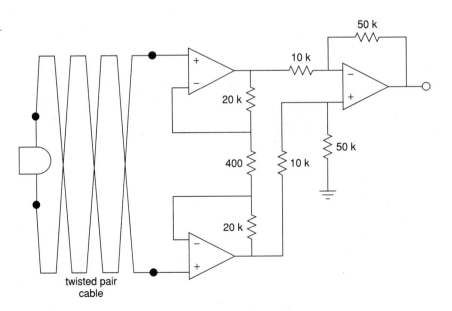

FIGURE 6.4

Instrumentation amplifier for Example 6-1

$$V_a = -6 \text{ mV}\left(1 + \frac{20 \text{ k}}{400}\right) - 6 \text{ mV}\frac{20 \text{ k}}{400}$$
$$V_a = -306 \text{ mV} - 300 \text{ mV}$$
$$V_a = -606 \text{ mV}$$

Performing the same calculation on the hum signal produces a contribution of

$$V_a = 10 \text{ mV}\left(1 + \frac{20 \text{ k}}{400}\right) - 10 \text{ mV}\frac{20 \text{ k}}{400}$$
$$V_a = 10 \text{ mV}$$

For worst case, these two components' magnitudes add, yielding 616 mV, which is far below clipping. The same results are produced for V_b, except that the desired signal is positive.

Now for the output voltage. The second section has a gain of

$$A_v = \frac{R_f}{R_i}$$
$$A_v = \frac{50 \text{ k}}{10 \text{ k}}$$
$$A_v = 5$$

The desired differential input signal is $V_b - V_a$, so

$$V_{out} = A_v(V_b - V_a)$$
$$V_{out} = 5(606 \text{ mV} - -606 \text{ mV})$$
$$V_{out} = 6.06 \text{ V}$$

This result may also be produced in one step by using Equation 6.11.

$$V_{out} = (V_{in_+} - V_{in_-})\frac{R_f}{R_i}\left(1 + 2\frac{R_1}{R_2}\right)$$
$$V_{out} = 12 \text{ mV}\frac{50 \text{ k}}{10 \text{ k}}\left(1 + 2\frac{20 \text{ k}}{400}\right)$$
$$V_{out} = 6.06 \text{ V}$$

Note that the total gain is 505. Because this amplifier is not perfect, some common-mode signal gets through. It is suppressed by 100 dB over a desired signal. Thus, 100 dB translates to a factor of 10^5 in voltage gain. To find the hum signal at the output, multiply the hum by the ordinary signal gain, and then divide it by the CMRR.

$$V_{hum} = V_{out(cm)}$$
$$V_{out(cm)} = V_{in(cm)}\frac{A_v}{CMRR}$$
$$V_{out(cm)} = 10 \text{ mV}\frac{505}{10^5}$$
$$V_{out(cm)} = 50.5 \text{ } \mu\text{V}$$

Notice how the hum signal started out just as large as the desired signal, but is now many times smaller. The very high CMRR of the instrumentation amplifier is what makes this possible.

multiSIM COMPUTER SIMULATION

In Figure 6.5 (below and on p. 191), Electronics Workbench MultiSIM is used to simulate the amplifier of Example 6-1. The simple three-terminal op amp model is used here. In order to clearly see the common-mode rejection, the desired differential input signal is set to a 1 mV sine wave, and the common-mode signal is set to 1 V DC. The initial bias solution shows that op amps 1 and 2 amplify both the AC and DC portions of the input, while common-mode rejection is left up to op amp 3. This is evidenced by the fact that the input nodes of the final op amp both see the same DC potential. At the output, the sine wave has been amplified by approximately 500 as expected. There is no DC offset at the output, indicating rejection of the common-mode DC signal.

Having seen how useful instrumentation amplifiers are, it should come as no surprise to find that manufacturers have produced these devices on a single IC. One such device is the LT1167 from Linear Technology. This amplifier is suited for a variety of applications including bridge amplifiers and differential to single-ended converters.

FIGURE 6.5a

Instrumentation amplifier in MultiSIM

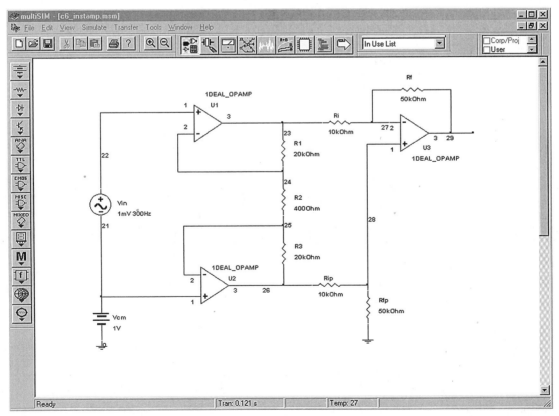

INSTRUMENTATION AMPLIFIERS 191

FIGURE 6.5b
DC voltages for instrumentation amplifier

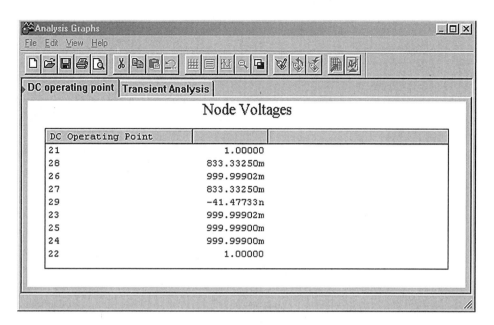

FIGURE 6.5c
AC output voltage for instrumentation amplifier

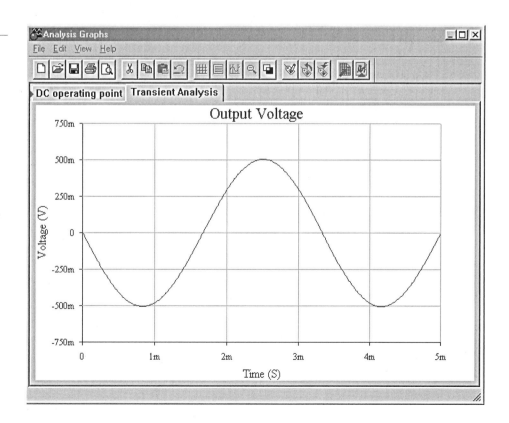

FIGURE 6.6

Guard drive internal connections

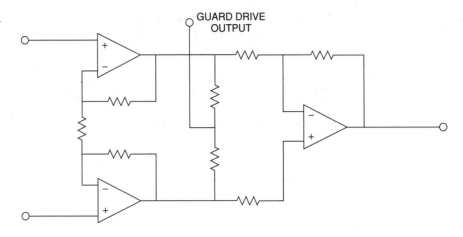

It is very similar in layout to Figure 6.3, using 24.7 kΩ resistors for R_1 and R_3, and 10 kΩ resistors for R_i and R_f. It features a typical CMRR of 115 dB for gains of 10, and up to 140 dB for gains of 1000. Its input resistance is 200 GΩ minimum. The gain is set by placing an appropriate resistor between the *gain set pins*. The gain may be approximated as

$$A_v = 1 + \frac{49.4 \text{ k}\Omega}{R_g}$$

For superior high-frequency performance, some instrumentation amplifiers include connections for a *guard drive*. This is shown in Figure 6.6. This signal is derived from the outputs of the first section. By using two equal-valued resistors, the differential signals cancel, and thus the guard-drive signal is equal to the common-mode signal. This signal will then be buffered and used to drive the shields of the input wires as shown in Figure 6.7.

To understand how performance is improved, refer to Figure 6.8. Here, the cables are replaced with a simple model. *R* represents the cable resistance and *C* represents the cable capacitance. The cable model is little more than a lag network. As you know, this causes high-frequency rolloff and phase change. If the two sections are not identical, the rolloffs and phase changes will not be the same in the two lines. These changes will affect the common-mode signal and can lead to a degradation of the system CMRR.

FIGURE 6.7

Guard drive

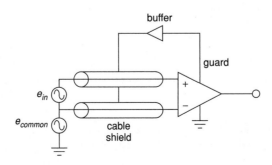

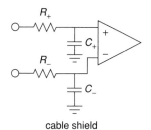

FIGURE 6.8

Unbalanced lag network equivalents

In the guard-drive circuit, the cable shields are not connected to ground at the source signal. They are only connected to the guard-buffer output. This is an important point. By driving the shield with a signal equal to the common-mode signal, the voltage developed across C will be zero. Both ends of the capacitor see the same potential; therefore, the drop is nonexistent. This reduces the cable effects considerably.

EXAMPLE 6-2

The signal produced by a transducer in a factory automation system provides a nominal level of .1 V. For proper use, the signal needs to be amplified up to 1 V. Because this signal must pass through the relatively noisy (electrically speaking) environment of the production floor, a balanced cable with an instrumentation amplifier is appropriate. Using the LT1167, design a circuit to meet these requirements.

For the power supply, a standard ±15 V unit will suffice. R_g is used to set the desired gain of ten.

$$A_v = 1 + \frac{49.4\,k}{R_g}$$

$$R_g = \frac{49.4\,k}{A_v - 1}$$

$$R_g = \frac{49.4\,k}{10 - 1}$$

$$R_g = 5.489\,k$$

The completed circuit is shown in Figure 6.9.

FIGURE 6.9

Completed instrumentation amplifier for Example 6-2

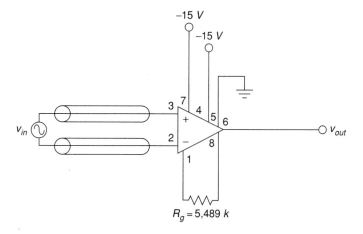

6.3 PROGRAMMABLE OP AMPS

As noted in Chapter 5, there is always a trade-off between the speed of an op amp and its power consumption. In order to make an op amp fast (i.e., high slew rate and

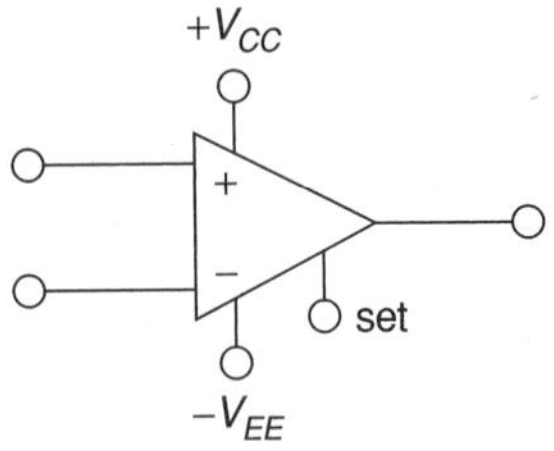

FIGURE 6.10

Basic connections of the LM4250

f_{unity}), the charging current for the compensation capacitor needs to be fairly high. Other requirements may also increase the current draw of the device. This situation is not unlike that of an automobile where high fuel mileage and fast acceleration tend to be mutually exclusive. Op amps are available that can be set for predetermined performance levels. If the device needs to be very fast, the designer can adjust an external voltage or resistance and optimize the device for speed. At the other end of the spectrum, the device can be optimized for low power draw. Because of their ability to change operational parameters, these devices are commonly called *programmable op amps*.

Generally, programmable op amps are just like general-purpose op amps, except that they also include a programming input. The current into this pin is called I_{set}. I_{set} controls a host of parameters including slew rate, f_{unity}, input bias current, standby supply current, and others. A higher value of I_{set} increases f_{unity}, slew rate, standby current, and input bias current. The value of I_{set} can be controlled by a simple transistor current source, or even a single resistor in most cases.

One example of a programmable op amp is the National Semiconductor LM4250. The outline of the LM4250 is shown in Figure 6.10. Two possible ways of adjusting I_{set} are shown in Figures 6.11 and 6.12. A variety of graphs showing the variation of device parameters with respect to I_{set} are shown in Figure 6.13. As an example, assume that $I_{set} = 10$ μA and that we're using standard ±15 V supplies. At this current, we find that f_{unity} is 230 kHz, slew rate is .21 V/μS, and the standby current draw is 50 μA. The current draw is decidedly lower than that of a 741. This makes the device very attractive for applications where power draw must be kept to a minimum. Battery operated systems generally fall into the low power consumption category. By keeping current drain down, battery life is extended.

Perhaps the strongest feature of this class of device is its ability to go into a *power down* mode. When the device is sitting at idle and not being used, I_{set} can be lowered, and thus the standby current drops considerably. For example, if I_{set} for the LM4250 is lowered to .1 μA, standby current drops to under 1 μA. By doing this, the device may only dissipate a few microwatts of power during idle periods. When the op amp again needs to be used for amplification, I_{set} can be brought back up to a normal level. This form of real-time adjustable performance can be instituted with a simple controlling voltage and a transistor current source arrangement as in Figure 6.12.

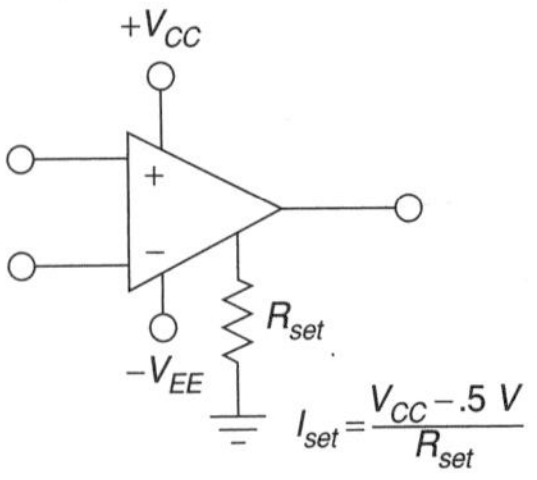

FIGURE 6.11

Programming via resistor

EXAMPLE 6-3

Determine the power bandwidth for $V_p = 10$ V for the circuit of Figure 6.14. Also find f_2. ±15 V power supplies are used.

In order to find these quantities, we need the slew rate and f_{unity}. These are controlled by I_{set}, so finding I_{set} is the first order of business.

$$I_{set} = \frac{V_{CC} - .5}{R_{set}}$$

PROGRAMMABLE OP AMPS 195

FIGURE 6.12

Programming via voltage

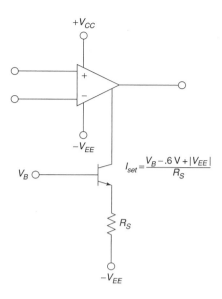

FIGURE 6.13

LM4250 device curves
Reprinted with permission of National Semiconductor Corporation

Typical Performance Characteristics

Input Bias Current vs I_{SET}

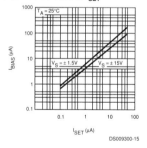

Input Bias Current vs Temperature

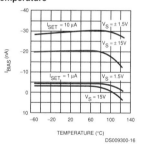

Input Offset Current vs Temperature

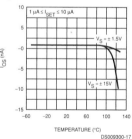

Unnulled Input Offset Voltage Change vs I_{SET}

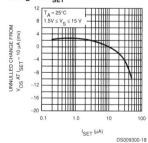

Unnulled Input Offset Voltage Change vs Temperature

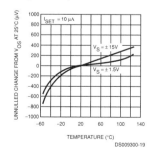

Peak to Peak Output Voltage Swing vs Load Resistance

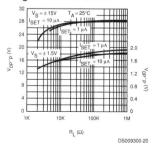

Typical Performance Characteristics (Continued)

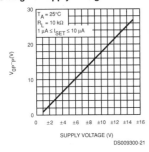
Peak to Peak Output Voltage Swing vs Supply Voltage

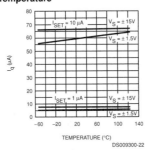
Quiescent Current (I_q) vs Temperature

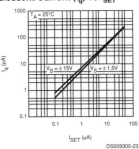
Quiescent Current (I_q) vs I_{SET}

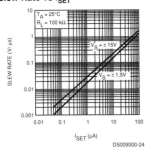
Slew Rate vs I_{SET}

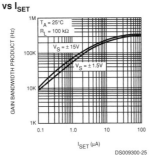
Gain Bandwidth Product vs I_{SET}

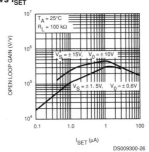
Open Loop Voltage Gain vs I_{SET}

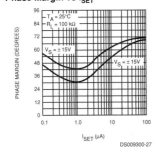
Phase Margin vs I_{SET}

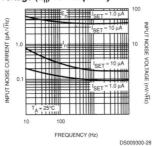
Input Noise Current (I_n) and Voltage (E_n) vs Frequency

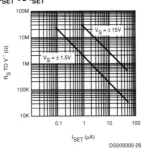
R_{SET} vs I_{SET}

FIGURE 6.13

(Cont.)

$$I_{set} = \frac{15 - .5}{3\,M\Omega}$$

$$I_{set} = 4.83\,\mu A$$

From the graph "Gain-Bandwidth Product versus Set Current" we find that 4.83 μA yields an f_{unity} of approximately 200 kHz. The graph "Slew Rate versus Set Current" yields approximately .11 V/μS for the slew rate.

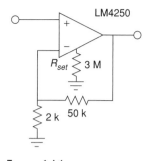

FIGURE 6.14

Circuit for Example 6-3

$$A_{noise} = 1 + \frac{R_f}{R_i}$$

$$A_{noise} = 1 + \frac{50\text{ k}}{2\text{ k}}$$

$$A_{noise} = 26$$

$$f_2 = \frac{f_{unity}}{A_{noise}}$$

$$f_2 = \frac{200\text{ kHz}}{26}$$

$$f_2 = 7.69\text{ kHz}$$

$$f_{max} = \frac{SR}{2\pi V_p}$$

$$f_{max} = \frac{.11\text{ V}/\mu S}{2\pi \times 10\text{ V}}$$

$$f_{max} = 1.75\text{ kHz}$$

As you can see, once the set current and associated parameters are found, analysis proceeds as in any general-purpose op amp.

6.4 Op Amps for High Current, Power, and Voltage Applications

General-purpose op amps normally run on no more than ±15 V power rails and typically produce less than 40 milliamps of output current. This makes it impossible to directly connect them to a low impedance load such as a loudspeaker or motor. The lack of high voltage capability hurts in many places, including many forms of display technology. In short, general-purpose op amps are low-power devices. There are a few ways around these limitations. One way of boosting the output current capability by using a discrete follower was shown in Chapter 4. In recent years, manufacturers have produced a variety of op amps with extended power performance.

High-Current Devices

Perhaps the most immediate desire for higher current capacity op amps came from the audio community. If an op amp could be directly connected to a loudspeaker, a great deal of time and money could be saved for general-purpose audio design work. Instead of a collection of perhaps half a dozen transistors and a handful of required biasing resistors and capacitors, an audio amplifier could be produced with a single op amp and just a few resistors and capacitors. Indeed, some of the first high-output devices were aimed squarely at the audio market. By 1980 it was possible to choose from a number of amplifiers designed to drive loudspeakers with power levels of 1 to 5 watts. In an effort to make designs ever easier, the strict op amp form was modified, and devices with preset and programmable gains were produced. Due to the increased dissipation requirements, the standard plastic dual inline package was abandoned in favor of multilead TO-220 style cases. Devices have been created that can produce output currents in excess of 10 amps. Besides their use in the audio/communications area, these high-current devices find use in direct motor drive applications, power supply and regulation design, and in other areas.

Let's take a closer look at two representative devices. First, on the low end of the scale is the LM386. This is a low-voltage device ideally suited to battery-powered designs. It operates from a single power supply between 4 and 12 V and is capable of producing .8 W into a 16 Ω load with 3% THD. The gain is set internally to 20, but can be increased up to 200 with the addition of a few extra components. A minimum configuration layout requires only one external component, as shown in Figure 6.15. In the minimum configuration the PSRR bypass capacitor is omitted. This produces a PSRR of less than 10 dB. A more respectable PSRR of 50 dB (rolling off below 100 Hz) can be achieved by adding a 50 μF capacitor from the PSRR bypass pin to ground. The small number of required external components is due to the fact that an internal feedback path has already been established by the manufacturer. Normally, this device would not be set up in the forms we have already examined. Configured as in Figure 6.15, the circuit exhibits a 300 kHz bandwidth and an input impedance of 50 kΩ. The LM386 is housed in an 8 pin mini DIP.

A typical application using the LM386 is shown in Figure 6.16. This is a personal amplifier called The Pocket Rockit. It is designed for musicians, so that they can plug a guitar, keyboard, or other electronic instrument into the unit and hear what they play through a pair of headphones. It is in essence, a personal practice amp. In keeping with its small portable design, The Pocket Rockit runs off of a single 9 V radio battery. Note how the single-supply biasing techniques explained in Chapter 4 are used to bias op amps 1a and 1b. The circuit is laid out in three main blocks. The first block serves as a preamplifier and distortion circuit. Because the load for the first stage consists of a pair of diodes, D_2 and D_3, large output signals will be clipped at the diode forward voltage. Now normally, clipping is not a desired result in linear circuit design. This circuit is used by musicians, and a great number of them find this effect to be very rewarding, particularly when applied to guitar. (This is one case where subjective valuations can make a designer do rather odd things.) The second stage comprises a bass/mid/treble

FIGURE 6.15

Simple power amp

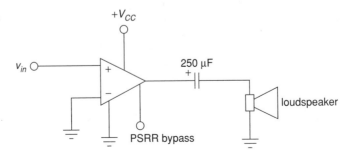

Figure 6.16

Pocket Rockit schematic

From Electronic Musician, *Vol. 3, No. 6. Reprinted with permission*

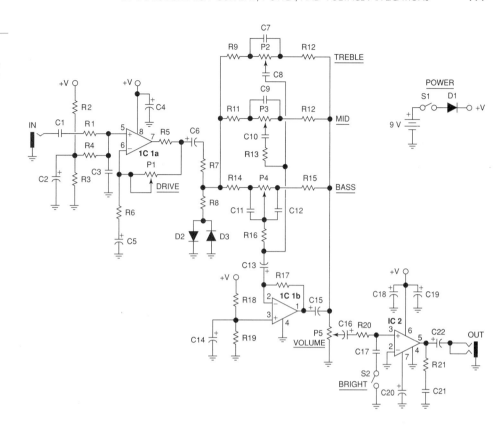

control. This is little more than a frequency selective amplifier, a subject that will be covered in detail in Chapter 11. From the musician's view, it allows control over the tone color, or timbre, of the sound. The final stage utilizes an LM386 to drive a pair of mini headphones. As you can see, the power amplifier section is the smallest of the three. It utilizes the default internal gain of 20, and includes C_{20} for optimum PSRR. This is an important detail, as this circuit may also be powered from 9 V AC adapters, which are not known for their low noise. The volume control is merely a potentiometer that acts as a voltage divider to the input signal. Coupling capacitors are used throughout to prevent DC bias signals from affecting the load or adjacent stages. They are also used to make sure that DC potentials don't appear across the potentiometers, which can increase adjustment noise.

At the other end of the spectrum is the LM12. The LM12 can produce 150 watts of "sine wave power" into a 4 ohm load. The device may be operated from ±40 V supplies and delivers up to ±10 amps to its load. The device comes in a TO-3 type power case with five connections: $\pm V_{supply}$, ± input, and output. To make the designer's life easier, the LM12 features an array of useful characteristics: thermal limiting, overvoltage shutdown, output current limiting, and dynamic output-transistor safe-area protection. Unlike many power op amps, the LM12 is unity-gain stable. It shows an f_{unity} of 700 kHz and a slew rate of 9 V/µS. For even higher output currents, it is possible

FIGURE 6.17

High-power amplifier

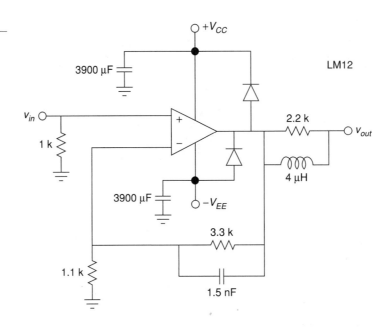

to connect LM12's in parallel. Due to the high-power nature of this device, extra care needs to be exercised in the areas of power supply bypassing, ground loops, and reactive loads. Whereas general-purpose op amps may bypass their power supplies with .1 μF to 1 μF capacitors, the LM12 requires 20 μF bypass capacitors. Also, the device may go into oscillation when driving capacitive loads. In this case some form of load isolation/compensation is used. This usually takes the form of a resistor/inductor combination. Generally, all high power op amps require a bit more attention than general-purpose devices do.

As an example, Figure 6.17 shows a low distortion (<.01% THD) audio amplifier. Note the large (3900 μF) bypass capacitors and the output clamping diodes. As with discrete power amplifiers, the clamp diodes are used to prevent an overvoltage condition occurring when driving inductive loads to the clipping level. The stored energy in the inductor can raise the output of the op amp above the power supply rail while the output transistors are in saturation. This condition will damage the amplifier. The diodes are used to shunt the stored energy to the power supply. The 4 μH/2.2 Ω combination is used to stabilize the system in the event of adversely reactive loads. Obviously, with the sort of power dissipation this device is capable of, appropriate cooling methods, such as heat sinks, cannot be overlooked.

Besides driving a loudspeaker, the LM12 can also drive motors. If only a single-polarity supply is available, a pair of devices may be hooked up in bridge mode in order to achieve bi-directional load current. Figure 6.18 shows a single-supply bridge configuration driving a servo motor. Either input may be grounded. This system utilizes current drive output (i.e., it is a voltage-to-current transducer). Current output is sometimes preferred with servo motors, as it helps stabilize the system in the presence of the sizable inductance of the servo motor.

FIGURE 6.18

Motor drive circuit

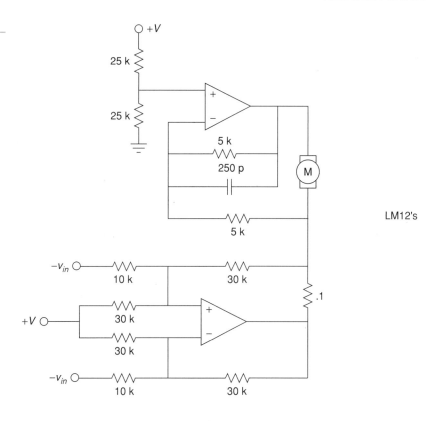

LM12's

High-Voltage Devices

Sometimes an amplifier needs to deliver output voltages in excess of the 12 to 13 V limits presented by most op amps. In this case, you may turn to a number of high-voltage op amps. These devices do not necessarily have a higher output-current capability as well. Basically, they may be treated as general-purpose op amps that may be operated from higher supply voltages. Because of this, their output compliance is increased. Devices are available that may operate from power supplies in excess of ±100 V.

One example of this breed is the Burr-Brown 3584. The 3584 runs from power supplies in the range of ±70 to ±150 V. The output current limit is typically 15 mA; not much different than an ordinary op amp. The minimum gain-bandwidth product is 20 MHz and the typical slew rate is 150 V/μS. The high output voltage capability makes this IC ideal for driving piezo-electric or electrostatic transducers.

6.5 High-Speed Amplifiers

There are many applications where amplification in the 1 MHz to 100 MHz region is desired. In this range falls the broadcast spectrums of FM radio, VHF television, citizen's band radio, and video processing in general. The bulk of general-purpose op amps exhibit f_{unity}s in the 1 MHz to 10 MHz range. Also, slew rates tend to be below

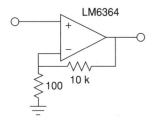

FIGURE 6.19

40 dB amplifier

20 V/μS. These characteristics make them wholly unsuitable for higher frequency applications. Even standard "fast" op amps like the LM318 and LF357 offer f_{unity}s of no more than 20 MHz with 50 V/μS slew rates.

In contrast to general-purpose devices, "ultra-fast" op amps exhibit gain bandwidth products in excess of 50 MHz and slew rates over 100 V/μS. Devices are available with slew rates greater than 1000 V/μS and others boast gain bandwidth products around 1 GHz. In order to optimize performance, these devices are normally of the uncompensated variety. In this way, the largest possible bandwidth is produced for any given gain.

Of course, when working with such fast devices, circuit layout is far more critical than normal. A very good circuit design may be crippled by a careless layout. First of all, PC-board ground planes are advised. A ground plane offers isolation along with low resistance and inductance pathways. These parasitic and stray impedance effects, which are often ignored at lower frequencies, can reduce circuit performance at high frequencies. Shielding around input and output paths may also be required for some applications, but care must be exercised, as the resulting capacitance may also produce high-frequency attenuation. In general, the circuit layout should minimize stray capacitance between the output and feedback point, from the feedback-summing node to ground, and between the circuit input and output. One common technique involves extending a ground plane in order to isolate the input and output traces. Also, the traces should be kept short to minimize coupling effects, and thus, compact layouts are generally preferred. IC sockets are generally avoided due to the increased lead inductance and capacitance they present. Finally, when bread-boarding high-speed circuits, the popular multirow socket boards used for general-purpose work are normally unacceptable due to high lead inductance and capacitance effects.

As with power op amps, some high-speed op amps have been optimized for specific applications and are no longer interchangeable with the generic op amp model. For example, specific gains may be selected by grounding or opening certain pins on the IC.

The LM6364 is one example of an ultra-fast op amp. It boasts typical specs of 300 V/μS for slew rate and 175 MHz for gain-bandwidth product. The LM6364 is stable for gains greater than 5 and requires no compensation. Other members of the family range from the LM6361 with unity-gain stability and 50 MHz gain-bandwidth, to the LM6365 with a 725 MHz gain-bandwidth (stable for gains as low as 25). As with any high-frequency device, power supply bypass capacitors are required.

Creating a 40 dB amplifier with the LM6364 is a straightforward affair. Figure 6.19 shows this connection. A circuit such as this could be used to amplify signals in the AM broadcast band. Due to the high gain chosen, an LM6365 could also be used and would exhibit increased bandwidth. For very low gains, the LM6361 or LM6362 would be used, with some reduction in bandwidth. All members of the family exhibit similar slew rates.

6.6 VOLTAGE FOLLOWERS AND BUFFERS

Unity gain noninverting buffers (voltage followers) are used in wide variety of applications. Any time a signal source needs to be isolated, a buffer is needed. As you have

FIGURE 6.20

Video RF cable driver

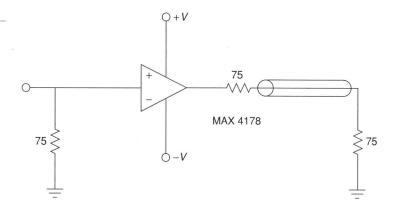

seen, connecting op amps in a follower configuration is a very straightforward exercise. This is not always the best choice. By optimizing the amplifier design for buffer operation, the manufacturer can increase performance, or in some cases, reduce the case size. This is possible because followers need very few connections: input, output, and power supply. Also, as the devices must, by nature, be unity gain stable, an external compensation connection is normally not desired.

One buffer amplifier optimized for high performance is the Maxim MAX4178. This device is designed with video applications, high-speed drivers, and analog-to-digital convertors in mind. Outside of the standard power supply bypass capacitors and perhaps a line termination resistor, no other parts are needed. Figure 6.20 shows a minimum configuration for driving 75 Ω coaxial cable. This buffer is capable of supplying ±100 mA to its load, so direct 75 Ω connection poses no problem. Slew rate is typically 1300 V/μS, and circuit bandwidth is 330 MHz.

EXAMPLE 6-4

Determine the approximate bandwidth required for a high-resolution video-display amplifier. The display has a resolution of 1024 pixels wide by 1024 pixels high, with a refresh rate of 60 frames per second.[1] This specification is typical of a workstation or a high-quality personal computer.

In its simplest form (monochrome or black and white), each pixel is either on or off. As a controlling voltage, this means either a high or a low. An updated grid of 1024 by 1024 dots must be drawn on the display monitor 60 times each second. From this, we can determine the pixel, or dot, rate.

$$\text{Pixel Rate} = Height \times Width \times Refresh$$
$$\text{Pixel Rate} = 1024 \times 1024 \times 60$$
$$\text{Pixel Rate} = 62,914,560 \text{ pixels per second}$$

---

[1] The term *pixel* is a contraction of *picture element*. The display is considered to be nothing more than a large grid of dots. The term *refresh rate* refers to how quickly the display is updated or redrawn.

This indicates that each pixel requires $\frac{1}{62,914,560}$ seconds, or about 15.9 nS, to reproduce. For proper pulse shape, the risetime should be no more than 30% of the pulse width (preferably less). From the risetime/bandwidth relationship established in Chapter 1,

$$f_2 = \frac{.35}{T_r}$$
$$f_2 = \frac{.35}{.3 \times 15.9 \text{ nS}}$$
$$f_2 = \frac{.35}{4.77 \text{ nS}}$$
$$f_2 = 73.4 \text{ MHz}$$

Theoretically, a 73.4 MHz bandwidth is required. Once overhead such as vertical retrace is added, a practical circuit will require a bandwidth on the order of 100 MHz.

6.7 Operational Transconductance Amplifier

The Operational Transconductance Amplifier, or *OTA* as it is normally abbreviated, is primarily used as a controlled-gain block. This means that an external controlling signal, either a current or a voltage, will be used to set a key parameter of the circuit, such as closed-loop gain or f_2. This is not the same thing as the programmable op amps discussed earlier.

OTAs exhibit high output impedances. They are intended to be used as current sources. Therefore, the output voltage is a function of the output current and the load resistance. The output current is set by the differential input signal and the OTA's transconductance. Transconductance is, in turn, controlled by a setup current, I_{abc}. What this all means is that the output signal is a function of I_{abc}. If I_{abc} is increased, transconductance rises and with it, output voltage. Normally, OTAs are operated without feedback. One major consequence of this is that the differential input signal must remain relatively small in order to avoid distortion. Typically, the input signal remains below 100 mV peak to peak.

One of the first popular OTAs was the 3080. An improved version of this chip is the LM13700. It features two amplifiers in a single package. Because OTAs are modeled as output current sources, the LM13700 includes simple emitter follower buffers for each OTA. Connection to the buffers is optional. Also included are input-linearizing diodes that allow higher input levels, and thus, superior signal-to-noise performance. The schematic of the LM13700 is shown in Figure 6.21. The differential amplifier stage comprised of Q_4 and Q_5 is biased by Q_2. The current set by Q_2 eastablishes the transconductance of this stage. Q_2 is part of a current mirror made along with Q_1 and D_1. Q_2 mirrors the current flowing through Q_1. The current flowing through Q_1 is I_{abc}, and is set externally. Thus, the differential amplifier is directly controlled by the external current.

FIGURE 6.21

LM13700 equivalent circuit
Reprinted with permission of National Semiconductor Corporation

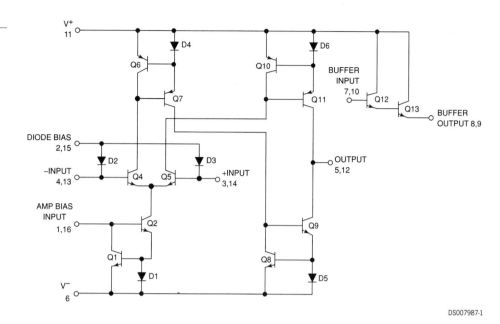

EXAMPLE 6-5

One example of OTA use is shown in Figure 6.22. This circuit is a simple voltage-controlled amplifier (VCA). Let's walk through the circuit to see how the component values were derived. The gain of the system is set by $V_{control}$. The 22 kΩ resistor is used to turn $V_{control}$ into I_{abc}. The maximum value for I_{abc} is specified as 1 mA. By referring to Figure 6.13, an expression for the maximum I_{abc} may be determined from Ohm's law.

$$I_{abc} = \frac{V_+ - V_- - V_{D_1}}{R_{control}}$$

$$I_{abc} = \frac{5 - (-5) - .7}{22\,k}$$

$$I_{abc} = .423\text{ mA}$$

This places us at about half the allowed maximum, giving a twofold safety margin. The LM13700 transconductance graph indicates that we may expect a maximum value of about 10 milliSiemens with our maximum I_{abc}. Note how the input signal must be attenuated in order to prevent overload. For reasonable distortion levels the input to the LM13700 should be below 50 mV peak to peak. This is accomplished by the 33 kΩ/470 Ω voltage divider. This divider produces an input impedance of approximately 33 kΩ, and a loss factor of

$$Loss = \frac{470}{33\,k + 470}$$

$$Loss = .014$$

This means that the circuit may amplify peak to peak signals as large as

$$50\text{ mV} = V_{in-max} Loss$$

Figure 6.22

Voltage-controlled amplifier

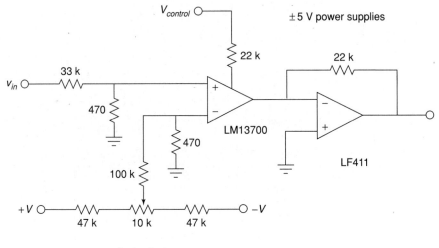

$$V_{in-max} = \frac{50 \text{ mV}}{Loss}$$
$$V_{in-max} = 3.56 \text{ V}$$

The current out of the LM13700 drives an LF411 configured as a current-to-voltage converter. The transresistance of this portion is 22 kΩ. Using the maximum 50 mV input, the maximum output signal will be

$$I_{out-13700} = V_{in-13700} \, gm$$
$$I_{out-13700} = 50 \text{ mV} \times 10 \text{ mS}$$
$$I_{out-13700} = .5 \text{ mA}$$

This current will drive the LF411 to a maximum output of

$$V_{out} = I_{out-13700} R_f$$
$$V_{out} = .5 \text{ mA} \times 22 \text{ k}$$
$$V_{out} = 11 \text{ V}$$

With the signals shown, this circuit acts as a simple amplitude modulator. The $V_{control}$ signal is impressed upon the input signal. These are the audio and carrier signals, respectively. This setup can also be used as part of an AGC (automatic gain control) loop. Also, note the adjustable DC signal applied to the inverting input. This is used to null the amplifier. This is an important point. Any DC offset is also affected by the gain change, in effect causing the control signal to "leak" into the output. This can produce many undesirable side effects, including clicking and popping noises if the circuit is used as an audio VCA. By nulling the output offset, these effects are eliminated. In order to keep the inputs balanced, a 470 Ω is added to the inverting input. Now, each input sees approximately 470 Ω to ground.

When designing a VCA such as this, you usually start with known maximum input and output voltages. The feedback resistor of the current-to-voltage converter is chosen to yield the desired maximum output potential when the OTA is producing its maximum output current. The OTA's maximum input potential times its transconductance will yield the OTA's maximum output current. Once this current has been established, the value of $R_{control}$ may be determined given the range of $V_{control}$. The maximum allowable OTA input potential is determined from the OTA distortion graph, and in association with the desired maximum input voltage, dictates the ratio of the input voltage divider.

OTAs may also be used to create voltage-controlled filters (VCF), where the cutoff frequency is a function of $V_{control}$. Discussion of filters in general is found in Chapter 11. OTAs lend themselves to several unique applications. The downside is that they have limited dynamic range. That is, the range of signal amplitudes that they can handle without excessive distortion or noise is not as great as it is in many other devices.

6.8 NORTON AMPLIFIER

It is possible to create an input-differencing function without using a differential amplifier. One alternative is to use a current mirror arrangement to form a *current differencing amplifier*. Because the input function deals with a difference of current instead of voltage, amplifiers of this type are often referred to as *Norton amplifiers*. Norton amplifiers have the distinct advantages of low cost and the ability to operate from a single-polarity power supply. Perhaps the most popular Norton amplifier is the LM3900. The LM3900 is a quad device, meaning that four amplifiers are combined in a single package.

The internal circuitry of one Norton amplifier is shown in Figure 6.23a. Norton amplifiers also use a slightly modified schematic symbol, shown in Figure 6.23b, to distinguish them from ordinary op amps. The amplifier is comprised of two main

FIGURE 6.23

LM3900 Norton schematic (left) and schematic symbol (right)

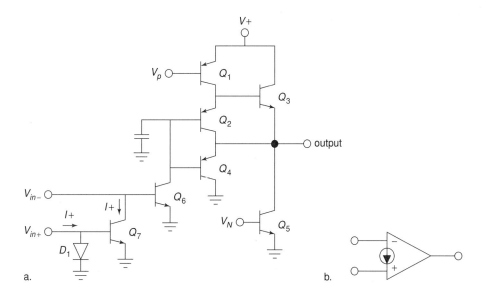

FIGURE 6.24

Norton equivalent circuit

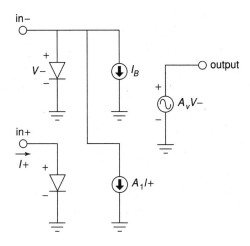

sections: a current-differencing input portion consisting of Q_6, Q_7, and D_1, and a gain stage comprising Q_1 through Q_5. The current-differencing portion is of greatest interest to us and is based on the current mirror concept. D_1 and Q_7 make up a current mirror that is fed by the noninverting input's source current, I_+. This current flows down through D_1 and is mirrored into Q_7. Therefore, the collector current of Q_7 is equal to I_+. This current is effectively subtracted from the current that enters the inverting input. The differential current, $I_- - I_+$, is the net input current that feeds Q_6 and is amplified by the following stage. Note how the schematic symbol of Figure 6.23b echoes the current mirror subtraction process.

Due to its unique input configuration, the equivalent model of the Norton amplifier is rather different from the standard op amp. This is illustrated in Figure 6.24. This clearly shows that the Norton amplifier is a current-sensing device. Also, note that the input impedance for the noninverting input is little more than the dynamic resistance of D_+. This resistance is dependent on the input current, and may be found using the standard diode resistance equation

$$Z_{in_+} = \frac{.026}{I_+} \qquad (6.12)$$

Another important point is that both the inverting and noninverting inputs are locked at approximately one diode drop above ground. This can be beneficial for a current summing node, but it does require that some form of input resistor be used if a voltage input is expected. A side benefit of this is that given a large enough input resistor, there is virtually no limit to the input common-mode voltage range, as the potential will drop across the input resistor. For proper AC operation, an appropriate biasing current must be used. A typical inverting amplifier is shown in Figure 6.25. Note the input and output coupling capacitors, as well as R_B, which serves to convert the bias voltage into a bias current. A generalized model of this circuit is shown in

FIGURE 6.25

Typical inverting amplifier

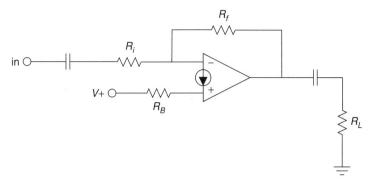

Figure 6.26. From this, we will be able to determine a voltage gain expression and a technique for determining a proper value for R_B. We'll start by setting up an expression for the output voltage for the bias equivalent.

$$V_{out} = V_{D_-} + R_f(I_+ + I_B)$$

FIGURE 6.26

Generalized model of the amplifier in Figure 6.25

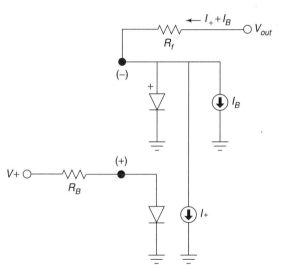

a. Bias equivalent

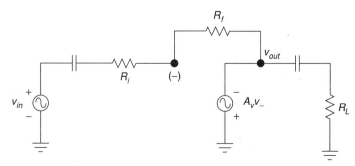

b. AC equivalent

and,

$$I_+ = \frac{V_{supply} - V_{D_+}}{R_B}$$

Combining the above two equations yields

$$V_{out} = V_{D_-} + R_f \left(\frac{V_{supply} - V_{D_+}}{R_B} + I_B \right) \quad (6.13)$$

Normally, $V_{supply} \gg V_D$, and I_B is relatively small. Given these factors, 6.13 may be approximated as

$$V_{out} = V_{supply} \frac{R_f}{R_B} \quad (6.14)$$

Therefore, for a midpoint bias,

$$R_B = 2R_f \quad (6.15)$$

For the AC equivalent circuit, the gain equation may be derived in the same fashion as it was in Chapter 4 for a standard inverting op amp configuration.

$$A_v = -\frac{R_f}{R_i} \quad (6.16)$$

As in the standard op amp inverting amplifier, the input impedance may be approximated as R_i.

EXAMPLE 6-6

Using the LM3900, design an amplifier that operates from a single +9 V battery, has an inverting voltage gain of 20, an input impedance of at least 50 kΩ, and a lower break frequency of no more than 100 Hz.

First, R_i may be established from the input impedance spec at 50 kΩ. From here, both R_f and R_B may be determined.

$$A_v = -\frac{R_f}{R_i}$$
$$R_f = -A_v R_i$$
$$R_f = -(-20) \times 50 \text{ k}\Omega$$
$$R_f = 1 \text{ M}\Omega$$

$$R_B = 2 R_f$$
$$R_B = 2 \times 1 \text{ M}\Omega$$
$$R_B = 2 \text{ M}\Omega$$

FIGURE 6.27
Completed design

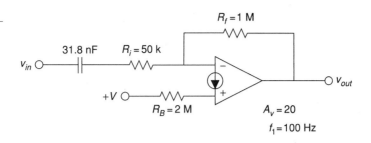

The input coupling capacitor will be used to set the lower break frequency in association with R_i. This is basically unchanged from the single-supply bias calculations presented in Chapter 4.

$$f_c = \frac{1}{2\pi RC}$$

$$C = \frac{1}{2\pi R f_c}$$

$$C = \frac{1}{2\pi \; 50 \text{ k} \; 100 \text{ Hz}}$$

$$C = 31.8 \text{ nF}$$

The completed circuit is shown in Figure 6.27. Although this circuit does not detail the output coupling capacitor and load resistance, it would be handled in a similar vein.

multiSIM COMPUTER SIMULATION

A MultiSIM simulation of the amplifier of Example 6-6 is shown in Figure 6.28 (pp. 212–213). A simple coupling network has been added to the output. The critical frequency of the output network is approximately 1.6 Hz, well below the input network. In this way, the input network is clearly dominant, and its effect on the system lower frequency response can be seen clearly. The AC analysis is plotted from 1 Hz to 10 kHz. The midband gain is approximately 26 dB, yielding an ordinary gain of 20 as desired. The lower end of the frequency response begins to roll off below 200 Hz, reaching approximately 23 dB (i.e., 3 dB down) at the target frequency of 100 Hz. The response continues its constant descent to the lower limit of the graph. Only a slight increase in slope may be noted in the 1 to 2 Hz region. If the frequency range is extended another decade lower, the effect of the output coupling network will be more obvious.

One possible problem with the circuit presented in Example 6-6 is that any power supply ripple present on V_{supply} will be passed into the amplifier via R_B, and appear at the output. In order to minimize this problem, a power supply decoupling circuit such as that presented in Chapter 4 may be used. Figure 6.29 shows a typical inverting amplifier with a decoupling circuit added. Note that R_B is not set to twice R_f, but

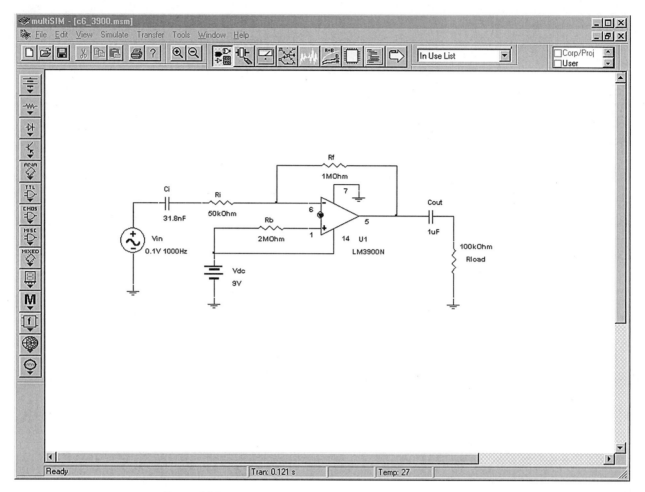

FIGURE 6.28a

Simulation of LM3900 amplifier

rather, is set equal to it. This is because R_B is only being driven by one-half of the supply voltage due to the lower resistance decoupling divider.

Norton amplifiers may be used in a variety of other configurations as well, including noninverting and differential amplifiers. A simple noninverting amplifier is shown in Figure 6.30. Unlike the ordinary op amp version, the Norton amplifier requires an input resistor. Remembering that the input impedance of the noninverting input may be quite low (Equation 6.12), we can derive equations for both circuit input impedance and voltage gain.

The input impedance is the series combination of R_i and the impedance looking into the noninverting input, Z_{in+}.

$$Z_{in} = R_i + Z_{in+} \qquad (6.17)$$

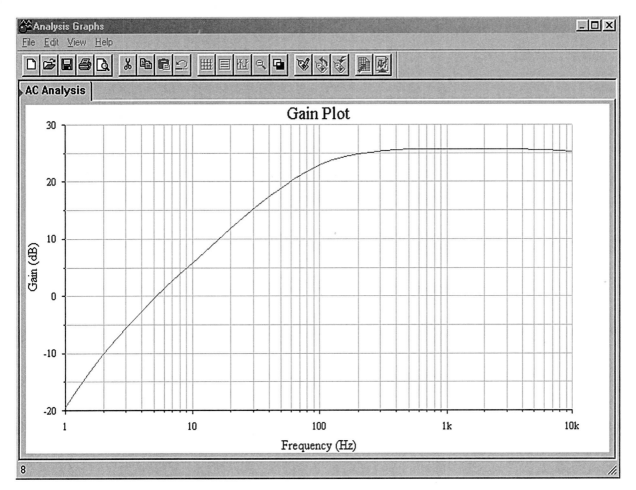

FIGURE 6.28b

Gain plot for LM3900 amplifier

FIGURE 6.29

Optimized bias circuitry

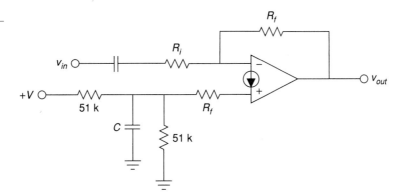

FIGURE 6.30

Noninverting amplifier

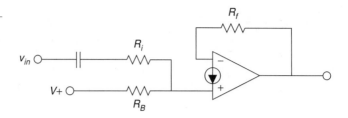

In order to find voltage gain, note that due to the virtual ground at the inverting input, the drop across R_f is equivalent to V_{out}.

$$V_{out} = V_{R_f} = R_f I_- \qquad (6.18)$$

V_{in} drops across the system input impedance. Given Equation 6.17, we may say

$$V_{in} = I_+(R_i + Z_{in+}) \qquad (6.19)$$

Because, by definition, $A_v = \frac{V_{out}}{V_{in}}$, we may say

$$A_v = \frac{R_f I_-}{I_+(R_i + Z_{in+})}$$

The two currents are identical in the AC equivalent circuit, so this simplifies to

$$A_v = \frac{R_f}{R_i + Z_{in+}} \qquad (6.20)$$

In summary, there are a few things that you should remember about Norton amplifiers:

1. The inputs are current-sensing, and thus, require some form of input-current limiting resistance.
2. As with an ordinary op amp, voltage gain is set by the two feedback resistors.
3. A DC bias potential must be fed into the noninverting input via a current limiting resistor, R_B. This resistor is normally twice the size of R_f.

6.9 CURRENT FEEDBACK AMPLIFIERS

The current feedback amplifier is a unique device in the world of linear integrated circuits. Its prime advantage is that it does not suffer from the strict gain/bandwidth trade-off typical of normal op amps. This means that it is possible to increase its gain without incurring an equal decrease in bandwidth. This has obvious advantages for high-speed applications. Also, these devices tend to be free of slewing effects.

An equivalent model of a noninverting current feedback amplifier is shown in Figure 6.31. This device uses a transimpedance source (i.e., a current-controlled voltage source) to translate the feedback current into an output voltage. Instead of relying

FIGURE 6.31

Current feedback amplifier equivalent

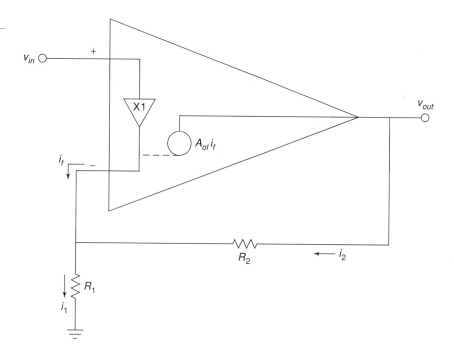

on negative feedback to keep the inverting and noninverting terminals at the same potential, the current feedback amplifier utilizes a unity-gain buffer placed between the two inputs. In this manner, the potential that drives the noninverting input must also appear at the inverting input. Also, the buffer exhibits a low output impedance into or out of which the feedback current may flow. From this diagram we may derive the general voltage-gain expression.

By inspection

$$V_{in} = V_+ = V_- \tag{6.21}$$

$$I_1 = \frac{V_{in}}{R_1} \tag{6.22}$$

$$I_2 = \frac{V_{out} - V_{in}}{R_2} \tag{6.23}$$

$$I_f = I_1 - I_2 \tag{6.24}$$

$$V_{out} = A_{ol} I_f \tag{6.25}$$

Combining 6.22, 6.23, and 6.24 with 6.25 yields

$$V_{out} = A_{ol}\left(\frac{V_{in}}{R_1} - \frac{V_{out} - V_{in}}{R_2}\right) \tag{6.26}$$

Simplification follows

$$\frac{V_{out}}{A_{ol}} = \frac{V_{in}}{R_1} - \frac{V_{out} - V_{in}}{R_2}$$

$$\frac{V_{out}}{A_{ol}} = \frac{V_{in} R_2 - V_{out} R_1 + V_{in} R_1}{R_1 R_2}$$

$$\frac{V_{out} R_1 R_2}{A_{ol}} = V_{in} R_2 - V_{out} R_1 + V_{in} R_1$$

$$\frac{V_{out} R_1 R_2}{A_{ol}} + V_{out} R_1 = V_{in}(R_2 + R_1)$$

$$V_{out}\left(\frac{R_1 R_2}{A_{ol}} + R_1\right) = V_{in}(R_2 + R_1) \tag{6.27}$$

We now solve 6.27 in terms of $\frac{V_{out}}{V_{in}}$, which is A_v.

$$\frac{V_{out}}{V_{in}} = \frac{R_1 + R_2}{\frac{R_1 R_2}{A_{ol}} + R_1}$$

$$A_v = \frac{R_1 + R_2}{\frac{R_1 R_2}{A_{ol}} + R_1}$$

$$A_v = \frac{\frac{R_2}{R_1} + 1}{\frac{R_2}{A_{ol}} + 1} \tag{6.28}$$

If, for convenience, we define $\frac{R_2}{R_1} + 1$ as G, Equation 6.28 becomes

$$A_v = \frac{G}{\frac{R_2}{A_{ol}} + 1} \tag{6.29}$$

If A_{ol} is sufficiently large, this may be approximated as

$$A_v = G$$
$$A_v = \frac{R_2}{R_1} + 1 \tag{6.30}$$

Equation 6.30 is identical to the approximate gain equation for an ordinary non-inverting op amp stage. A closer look at the exact equation (6.29) reveals an important difference with the exact series-parallel expression examined in Chapter 3.

$$A_v = \frac{A_{ol}}{1 + \beta A_{ol}}$$

As you may recall, β is the feedback factor and determines the closed loop gain. To a first approximation, $\beta A_{ol} \gg 1$, and thus, the series-parallel gain is ideally $\frac{1}{\beta}$. As frequency is raised, however, the value of A_{ol} drops, and soon this approximation no longer holds. It is very important to note that **the point at which the idealization** ceases to be accurate is a function of β. For higher closed-loop gains, β is smaller, and thus, the βA_{ol} product reaches unity at a lower frequency. It is this very interdependency that causes the gain/bandwidth trade-off. A close look at Equation 6.29 shows no such interdependency. In this case, the denominator is not influenced by the feedback factor. In its place is the fixed resistor value R_2. Normally, R_2 is set by the manufacturer of the device, and is usually around 1 kΩ to 2 kΩ in size. Closed-loop gain is set by altering the value of R_1. This means that the closed-loop upper-break frequency is not dependent on the closed-loop gain. This is the ideal case. In reality, secondary effects will cause some decrease of f_2 with increasing A_v. This effect is noticeable at medium to high gains, but is still considerably less than what an ordinary op amp would produce. For example, a jump in gain from 10 to 100 may drop f_2 from 150 MHz to only 50 MHz, whereas an ordinary device would drop to 15 MHz.

The OPA658 from Burr-Brown is one example of a current feedback amplifier. It boasts a slew rate of 1700 V/μS, a bandwidth of 900 MHz, and an output current capability of 80 mA. This makes it ideal for applications such as high-speed line drivers or video display drivers. The features and typical circuit are shown in Figure 6.32. Figure 6.33 shows the AC specifications. Note the small variation in f_2 as the gain is changed from 1 to 10.

Example 6-7

Design a noninverting 100 MHz amplifier with a voltage gain of 20 dB, utilizing the OPA658.

By design, the OPA623 will meet our bandwidth requirement, as long as the desired gain is not excessive. A voltage gain of 20 dB translates to an ordinary gain of 10, which is not excessive. Assuming an R_f of 1.5 kΩ, we find

$$A_v = 1 + \frac{1.5\,k}{R}$$
$$R = \frac{1.5\,k}{A_v - 1}$$
$$R = \frac{1.5\,k}{10 - 1}$$
$$R = 167\,\Omega$$

As with any high-speed amplifier, care must be taken when laying out this circuit. Failure to do so may seriously degrade amplifier performance.

6.10 Other Specialized Devices

Op amps have been further refined into other specialized devices. Along with these, a wide range of specialized linear integrated circuits have emerged. In the area of simple amplifiers, devices are available that are designed to work primarily from single-polarity power supplies. Also, a large array of specialty amplifiers exist for such applications as low-noise phono and tape preamplifiers. For very low offset applications, Chopper Auto

OPA658

Wideband, Low Power Current Feedback OPERATIONAL AMPLIFIER

FEATURES
- UNITY GAIN STABLE BRANDWIDTH: 900 MHz
- LOW POWER: 50mW
- LOW DIFFERENTIAL GAIN/PHASE ERRORS: 0.025%/0.02°
- HIGH SLEW RATE: 1700V/μs
- GAIN FLATNESS: 0.1dB to 135MHz
- HIGH OUTPUT CURRENT (80mA)

APPLICATIONS
- MEDICAL IMAGING
- HIGH-RESOLUTION VIDEO
- HIGH-SPEED SIGNAL PROCESSING
- COMMUNICATIONS
- PULSE AMPLIFIERS
- ADC/DAC GAIN AMPLIFIER
- MONITOR PREAMPLIFIER
- CCD IMAGING AMPLIFIER

DESCRIPTION

The OPA658 is an ultra-wideband, low power current feedback video operational amplifier featuring high slew rate and low differential gain/phase error. The current feedback design allows for superior large signal bandwidth, even at high gains. The low differential gain/phase errors, wide bandwidth and low quiescent current make the OPA658 a perfect choice for numerous video, imaging and communications applications.

The OPA658 is optimized for low gain operation and is also available in dual (OPA2658) and quad (OPA4658) configurations.

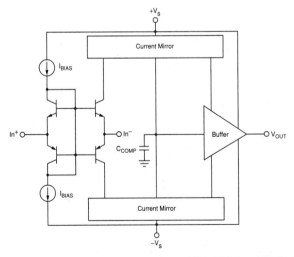

International Airport Industrial Park • Mailing Address: PO Box 11400, Tucson, AZ 85734 • Street Address: 6730 S.Tucson Blvd., Tucson, AZ 85706 • Tel: (520) 746-1111 • Twx: 910-952-1111
Internet: http://www.burr-brown.com/ • FAXLine: (800) 548-6133 (US/Canada Only) • Cable: BBRCORP • Telex: 066-6491 • FAX: (520) 889-1510 • Immediate Product Info: (800) 548-6132

© 1994 Burr-Brown Corporation   PDS-1268F   Printed in U.S.A. March, 1998

FIGURE 6.32

OPA658 current feedback amplifier
(Courtesy of Burr-Brown)

SPECIFICATIONS

At $T_A = +25°C$, $V_S = ±5V$, $R_L = 100Ω$, and $R_{FB} = 402Ω$, unless otherwise noted.

PARAMETER	CONDITION	OPA658P, U, N MIN	OPA658P, U, N TYP	OPA658P, U, N MAX	OPA658UB, NB MIN	OPA658UB, NB TYP	OPA658UB, NB MAX	UNITS
FREQUENCY RESPONSE								
Closed-Loop Bandwidth(2)	G = +1(4)		900			*(1)		MHz
	G = +2		680		400	*		MHz
	G = +5		370			*		MHz
	G = +10		200			*		MHz
Slew Rate(3)	G = +2, 2V Step		1700		1000	*		V/μs
At Minimum Specified Temperature			1500		900	*		V/μs
Settling Time: 0.01%	G = +2, 2V Step		15			*		ns
0.1%	G = +2, 2V Step		11.5			*		ns
1%	G = +2, 2V Step		6			*		ns
Spurious Free Dynamic Range	f = 5MHz, G = +2, V_O = 2Vp-p		68			*		dBc
	f = 20MHz, G = +2, V_O = 2Vp-p		56			*		dBc
Third Order Intercept Point	f = 10MHz, 4dBm Each Tone		40			*		dBm
Differential Gain	G = +2, NTSC, V_O = 1.4Vp-p, R_L = 150Ω		0.025			*		%
Differential Phase	G = +2, NTSC, V_O = 1.4Vp-p, R_L = 150Ω		0.02			*		degrees
Bandwidth for 0.1dB Flatness	G = +2		135(5)			*		MHz
OFFSET VOLTAGE								
Input Offset Voltage	V_{CM} = 0V		±3	±5.5		±2	±4.5	mV
Over Temperature Range			±5	±8		±4	±7	mV
Power Supply Rejection Ratio	V_S = ±4.7 to ±5.5V	55	64		58	67		dB
INPUT BIAS CURRENT								
Non-Inverting	V_{CM} = 0V		±5.7	±30		±18		μA
Over Temperature Range			±10	±80		±35		μA
Inverting	V_{CM} = 0V		±1.1	±35		*	*	μA
Over Temperature Range			±30	±75		*	*	μA
NOISE								
Input Voltage Noise Density								
f = 100Hz			16			*		nV/√Hz
f = 2kHz			4.9			*		nV/√Hz
f = 10kHz			3.2			*		nV/√Hz
f = 1MHz			3.2			*		nV/√Hz
f_B = 100Hz to 200MHz			45.3			*		μVrms
Input Bias Current Noise Density								
Inverting: f = 1MHz			32			*		pA/√Hz
Non-Inverting: f = 1MHz			11.9			*		pA/√Hz
INPUT VOLTAGE RANGE								
Common-Mode Input Range								
Over Temperature Range		±2.5	±2.9		*	*		V
Common-Mode Rejection	V_{CM} = ±1V	45	50		*	*		dB
INPUT IMPEDANCE								
Non-Inverting			500 ‖ 1			*		KΩ ‖ pF
Inverting			50			*		Ω
OPEN-LOOP TRANSRESISTANCE								
Open-Loop Transresistance	V_O = ±2V, R_L = 100Ω	150	190		200	250		KΩ
Over Temperature Range	V_O = ±2V, R_L = 100Ω	100			150			KΩ
OUTPUT								
Voltage Output	No Load	±2.7	±2.9		*	*		V
Over Temperature Range		±2.5	±2.75		*	*		V
Voltage Output	R_L = 250Ω	±2.7	±2.9		*	*		V
Over Temperature Range		±2.5	±2.7		*	*		V
Voltage Output	R_L = 100Ω	±2.2	±2.8		*	*		V
Over Temperature Range		±2.0	±2.5		*	*		V
Output Current, Sourcing		80	120		*	*		mA
Over Temperature		70			*			mA
Output Current, Sinking		60	80		*	*		mA
Over Temperature		35			*			mA
Short Circuit Current			150			*		mA
Output Resistance	0.1MHz, G = +2		0.02			*		Ω
POWER SUPPLY								
Specified Operating Voltage			±5			*		V
Operating Voltage Range		±4.5		±5.5	*		*	V
Quiescent Current	V_S = ±5V		±5	±7.75		±4.5	±5.75	mA
Over Temperature Range			±5.5	±8.5		±4.7	±6.5	mA
TEMPERATURE RANGE								
Specification: P,U, N, UB, NB		−40		+85	*		*	°C
Thermal Resistance, $θ_{JA}$								
P 8-Pin DIP			100			*		°C/W
U SO-8			125			*		°C/W
N SOT23-5			150			*		°C/W

NOTES: (1) An asterisk (*) specifies the same value as the grade to the left. (2) Frequency response can be strongly influenced by PC board parasitics. The demonstration boards show low parasitic layouts for this part. Refer to the demonstration board layout for details. (3) Slew rate is rate of change from 10% to 90% of output voltage step. (4) At G = +1, R_{FB} = 560Ω for PDIP and 402Ω for SCO-8. (5) This specification is PC board layout dependent.

OPA658

FIGURE 6.33

OPA658 AC specifications
(Courtesy of Burr-Brown)

Zero or Commutating Auto Zero (CAZ) amplifiers are available. CAZ amplifiers actually monitor their DC offset, and actively correct it.

Moving away from the strict confines of op amps, linear ICs are available for most popular functions, making the designer's job even easier. Some of these items will be explored in later chapters. Other devices are for very specific applications and will not be covered here. A quick sampling will give you an idea of the level of integration now in wide use. Some of the applications now covered by single linear ICs include: TV video modulator, automotive fuel injector drive controller, wide-band RMS-DC converter block, dynamic noise reduction, LED bar graph meter, AM/FM radio system, and CRT driver.

Some manufacturers develop ICs solely for specific market segments. As an example, both Analog Devices and THAT Corporation make very specialized ICs for audio and musical equipment manufacturers. The Analog Devices SSM line includes very high precision voltage-controlled amplifiers, line drivers, noise-reduction ICs, and similar items. In like fashion, THAT Corporation makes voltage-controlled amplifiers, RMS level detectors and other items aimed at the audio industry. Of course, nothing prevents a designer from using one of these ICs in a nonmusical product.

Summary

In this chapter we have examined operational amplifiers that exhibit extended performance, and those that have been tailored to more specific applications. In the area of precision differential amplification comes the instrumentation amplifier. Instrumentation amplifiers may be formed from three separate op amps, or they may be purchased as single hybrid or monolithic ICs. Instrumentation amplifiers offer isolated high impedance inputs and excellent common-mode rejection characteristics.

Programmable op amps allow the designer to set desired performance characteristics. In this way, an optimum mix of parameters such f_{unity} and slew rate versus power consumption is achieved. Programmable op amps can also be set to a very low power consumption standby level. This is ideal for battery powered circuits. Generally, programming is performed by either a resistor for static applications, or via an external current or voltage for dynamic applications.

The output drive capabilities of the standard op amp have been pushed to high levels of current and voltage. Power op amps may be directly connected to low impedance loads such as servo motors or loudspeakers. Moderate power devices exist that have been designed for line drivers and audio applications. Due to the higher dissipation requirements these applications produce, power op amps are often packaged in TO-220 and TO-3 type cases.

Along with higher power devices, still other devices show increased bandwidth and slewing performance. These fast devices are particularly useful in video applications. Perhaps the fastest amplifiers are those that rely on current feedback and utilize a transimpedance output stage. These amplifiers are a significant departure from the ordinary op amp. They do not suffer from strict gain-bandwidth limitations, and can achieve very wide bandwidth with moderate gain.

Operational transconductance amplifiers, or OTAs, may be used as building blocks for larger circuits such as voltage-controlled amplifiers or filters. Norton amplifiers rely on a current mirror to perform a current-differencing operation. They are relatively inexpensive and operate directly from single-polarity power supplies with little support circuitry. Because they are current-sensing, input limiting resistors are required.

Review Questions

1. What are the advantages of using an instrumentation amplifier versus a simple op amp differential amplifier?
2. How might an instrumentation amplifier be constructed from general-purpose op amps?
3. What are the advantages of using programmable op amps?
4. What are the results of altering the programming current in a programmable op amp?

5. Give at least two applications for a high-power op amp.
6. Give at least two applications for a high-speed op amp.
7. What is an OTA?
8. How is an OTA different from a programmable op amp?
9. Give an application that might use an OTA.
10. Give at least three applications that could benefit from specialty linear integrated circuits.
11. Describe how a Norton amplifier achieves input differencing.
12. List a few of the major design differences that must be considered when working with Norton amplifiers versus ordinary op amps.
13. Explain why a current feedback amplifier does not suffer from the same gain-bandwidth limitations that ordinary op amps do.

Problems

Analysis Problems

1. An instrumentation amplifier has a differential input signal of 5 mV and a common-mode hum input of 2 mV. If the amplifier has a differential gain of 32 dB and a CMRR of 85 dB, what are the output levels of the desired signal and the hum signal?
2. Determine V_{out} in Figure 6.34 if $V_{in_+} = +20$ mV DC and $V_{in_-} = -10$ mV DC.

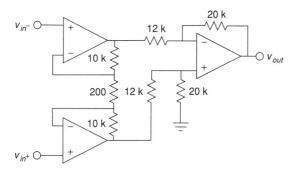

FIGURE 6.34

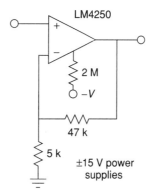

FIGURE 6.35

3. Repeat Problem 2 for a differential input signal of 10 mV peak to peak.
4. Repeat Problem 3 for a differential input signal of 20 mV peak to peak and a common-mode signal of 5 mV peak to peak. Assume the system CMRR is 75 dB.
5. Determine the programming current in Figure 6.11 if $R_{set} = 1$ MΩ. Assume standard ±15 V supplies.
6. Using an LM4250 programmable amplifier, determine the following parameters if $I_{set} = 5$ μA: slew rate, f_{unity}, input noise voltage, input bias current, standby supply current, and open-loop gain.
7. Determine the voltage gain, f_2, and power bandwidth (assume $V_p = 10$ V) for the circuit of Figure 6.35.
8. The circuit of Figure 6.36 uses a light-dependent resistor to enable or disable the amplifier. Under full light, this LDR exhibits a resistance of 1 kΩ, but under no light conditions, the resistance is 50 MΩ. What are the standby current and f_{unity} values under full light and no light conditions?

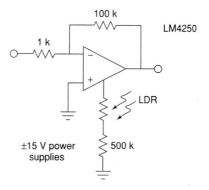

FIGURE 6.36

9. Determine the required capacitance to set PSRR to at least 20 dB at 100 Hz for an LM386 power amp.
10. What is the power bandwidth for a 5 V peak signal in Figure 6.20?
11. Determine the device dissipation for an LM386 delivering .5 W into an 8 ohm load. Assume a 12 V power supply is used.
12. Determine the input resistance, output resistance, and transconductance for an LM13700 OTA with $I_{abc} = 100\ \mu A$.
13. Determine the value of I_{abc} in Figure 6.22 if $V_{control}$ .5 V DC.
14. Recalculate the value of R_i for the circuit of Figure 6.26 if a voltage gain of 45 is desired.
15. Recalculate the input capacitance value for the circuit of Figure 6.26 if a lower break frequency of 15 Hz is desired.

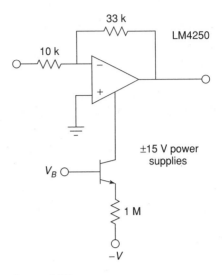

FIGURE 6.37

Design Problems

16. Design an instrumentation amplifier with a gain of 20 dB using the LT1167.
17. Utilizing the LM3900, design an inverting amplifier with a gain of 12 dB, an input impedance of at least 100 kΩ, and a lower break frequency no greater than 25 Hz.
18. Utilizing the LM3900, design a noninverting amplifier with a gain of 24 dB, an input impedance of at least 40 kΩ, and a lower break frequency no greater than 50 Hz.
19. Utilizing the OPA623, design a noninverting amplifier with a gain of 32 dB.
20. Utilizing the OPA623, design an inverting amplifier with a gain of 16 dB.

Challenge Problems

21. Using Figure 6.34 as a guide, design an instrumentation amplifier with a differential gain of 40 dB. The system bandwidth should be at least 50 kHz. Indicate which op amps you intend to use.
22. Determine the gain, f_2, power bandwidth ($V_p = 10$ V), and standby current for the circuit of Figure 6.37 if $V_b = 0$ V, and if $V_b = -15$ V.

23. Design a power amplifier with a voltage gain of 32 dB using the LM386.
24. Determine V_{out} in Figure 6.22 if $V_{control} = 1$ V DC and $V_{in} = 100$ mV sine wave.
25. Sketch V_{out} for Figure 6.22 if $V_{control}(t) = 1 \sin 2\pi\ 1000t$ and $V_{in}(t) = .5 \sin 2\pi\ 300,000t$.
26. Alter Figure 6.22 such that a 100 mV input signal will yield a 1 V output signal when $V_{control}$ is $+2$ V.
27. Derive Equation 6.9 from the text.

Computer Simulation Problems

28. Use a simulator to verify the design produced in Problem 21.
29. Simulate the output of the circuit shown in Figure 6.26. Use a +15 V DC power supply with $V_{in}(t) = .01 \sin 2\pi\ 100t$.
30. Use a simulator to verify the gain of the design produced in Problem 18.
31. Device mismatching can adversely affect the CMRR of an instrumentation amplifier. Rerun the simulation of the instrumentation amplifier (Figure 6.5) with a $\pm 5\%$ tolerance applied to the value of R'_i (Rip). Based on these simulations, what do you think would happen if all of the circuit resistors had this same tolerance?

CHAPTER 7

Nonlinear Circuits

CHAPTER OBJECTIVES

After completing this chapter, you should be able to:

- Understand the advantages and disadvantages of active rectifiers versus passive rectifiers.
- Analyze the operation of various half- and full-wave precision rectifiers.
- Analyze the operation of peak detectors, limiters, and clampers.
- Utilize function generation circuits to alter the amplitude characteristics of an input signal.
- Detail the usefulness of a Schmitt trigger and compare its performance to an ordinary open loop comparator.
- Utilize dedicated comparator ICs, such as the LM311.
- Analyze log and antilog amplifiers and multipliers.

7.1 Introduction

Besides their use as an amplification device, op amps may be used for a variety of other purposes. In this chapter, we will look into a number of circuits that fall under the general classification of nonlinear applications. The term *nonlinear* is used because the input/output transfer characteristic of the circuit is no longer a straight line. Indeed, the characteristic may take on a wide variety of shapes, depending on the application. It might seem odd to be studying nonlinear circuits when op amps are generally considered to be linear devices. Rest assured that nonlinear elements are often essential tools in a larger system. The general usage of nonlinear elements is that of wave shaping. Wave shaping is the process of reforming signals in a desired fashion. You might think of it as signal sculpting.

Wave shaping runs the gamut from simple half-wave rectification up through transfer-function generation. Although many of the applications presented in this chapter can be realized without op amps, the inclusion of the op amp lends the circuit a higher level of precision and stability. This is not to say that the op amp versions are always superior to a passive counterpart. Due to bandwidth and slew rate limitations, the circuits presented perform best at lower frequencies—primarily in the audio range and below. Also, there are definite power handling and signal level limitations. In spite of these factors, the circuits do see quite a bit of use in the appropriate areas. Many of the applications fall into the instrumentation and measurement areas.

Besides the wave-shaping functions, the other uses noted here are comparators and logarithmic amplifiers. General-purpose comparators made from simple op amps were explained in Chapter 2. In this chapter, specialized devices are examined, and their superiority to the basic op amp comparator is noted. Logarithmic amplifiers have the unique characteristic of compressing an input signal. That is, the range of signal variation is reduced. The mirror image of the log amplifier is the antilog amplifier. It increases the range of input signal variation.

7.2 Precision Rectifiers

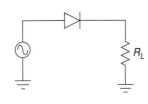

FIGURE 7.1

Passive rectifier

Imagine for a moment that you would like to half-wave rectify the output of an oscillator. Probably the first thing that pops into your head is the use of a diode, as in Figure 7.1. As shown, the diode passes positive half-waves and blocks negative half-waves. But, what happens if the input signal is only .5 V peak? Rectification never occurs because the diode requires .6 to .7 V to turn on. Even if a germanium device is used with a forward drop of .3 V, a sizable portion of the signal will be lost. Not only that, the circuit of Figure 7.1 exhibits vastly different impedances to the driving source. Even if the signal is large enough to avoid the forward voltage drop difficulty, the source impedance must be relatively low. At first glance it seems as though it is impossible to rectify a small AC signal with any hope of accuracy.

One of the items noted in Chapter 3 about negative feedback was the fact that it tended to compensate for errors. Negative feedback tends to reduce errors by an amount equal to the loop gain. This being the case, it should be possible to reduce the diode's forward voltage drop by a very large factor by placing it inside of a feedback loop. This is shown in Figure 7.2 and is called a *precision half-wave rectifier*. To a first approximation, when the input is positive, the diode is forward-biased. In essence, the

FIGURE 7.2

Precision half-wave rectifier

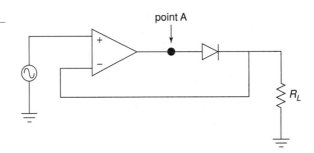

FIGURE 7.3

Output signal

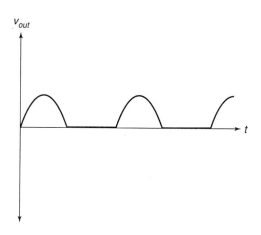

circuit reduces to a simple voltage follower with a high input impedance and a voltage gain of one, so the output looks just like the input. On the other hand, when the input is negative, the diode is reverse-biased, opening up the feedback loop. No signal current is allowed to the load, so the output voltage is zero. Thanks to the op amp, though, the driving source still sees a high impedance. The output waveform consists of just the positive portions of the input signal, as shown in Figure 7.3. Due to the effect of negative feedback, even small signals may be properly rectified. The resulting transfer characteristic is presented in Figure 7.4. A perfect one-to-one input/output curve is seen for positive input signals, whereas negative input signals produce an output potential of zero.

In order to create the circuit output waveform, the op amp creates an entirely different waveform at its output pin. For positive portions of the input, the op amp must produce a signal that is approximately .6 to .7 V greater than the final circuit output. This extra signal effectively compensates for the diode's forward drop. Because the feedback signal is derived after the diode, the compensation is as close as the available loop

FIGURE 7.4

Transfer characteristic

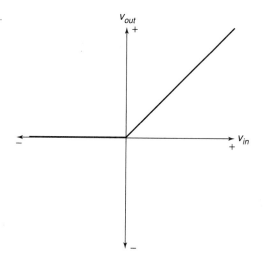

FIGURE 7.5

Output of op amp

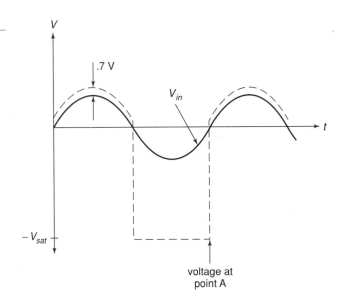

gain allows. At low frequencies where the loop gain is high, the compensation is almost exact, producing a near perfect copy of positive signals. When the input signal swings negative, the op amp tries to sink current in response. As it does so, the diode becomes reverse-biased, and current flow is halted. At this point the op amp's noninverting input will see a large negative potential relative to the inverting input. The resulting negative error signal forces the op amp's output to go to negative saturation. Because the diode remains reverse-biased, the circuit output stays at 0 V. The op amp is no longer able to drive the load. This condition will persist until the input signal goes positive again, at which point the error signal becomes positive, forward-biasing the diode and allowing load current to flow. The op amp and circuit output waveforms are shown in Figure 7.5.

One item to note about Figure 7.5 is the amount of time it takes for the op amp to swing in and out of negative saturation. This time is determined by the device's slew rate. Along with the decrease of loop gain at higher frequencies, slew rate determines how accurate the rectification will be. Suppose that the op amp is in negative saturation and that a quick positive input pulse occurs. In order to track this, the op amp must climb out of negative saturation first. Using a 741 op amp with ±15 V supplies, it will take about 26 μS to go from negative saturation (−13 V) to zero. If the aforementioned pulse is only 20 μS wide, the circuit doesn't have enough time to produce the pulse. The input pulse will have gone negative again, before the op amp has a chance to "climb out of its hole." If the positive pulse were a bit longer, say 50 μS, the op amp would be able to track a portion of it. The result would be a distorted signal as shown in Figure 7.6. In order to accurately rectify fast-moving signals, op amps with high f_{unity} and slew rate are required. If only slow signals are to be rectified, it is possible to configure the circuit with moderate gain if needed, as a cost-saving measure. This is shown in Figure 7.7. Finally, for negative half-wave output, the only modification required is the reversal of the diode.

FIGURE 7.6
High-frequency errors

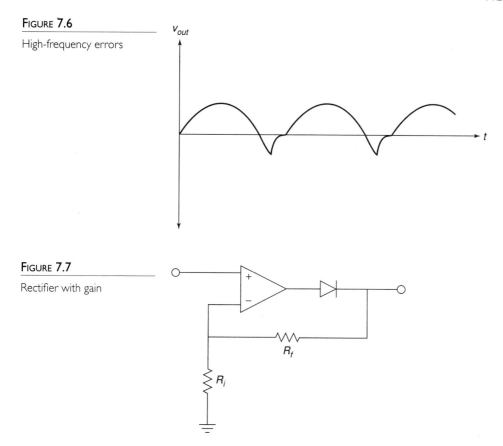

FIGURE 7.7
Rectifier with gain

multiSIM COMPUTER SIMULATION

An Electronics Workbench MultiSIM simulation of the circuit shown in Figure 7.2 is presented in Figure 7.8 (pp. 228–229). The circuit is shown redrawn with the nodes labeled. This example utilizes the 741 op amp model examined earlier. The output waveform is also shown in Figure 7.8. Note the accuracy of the rectification. The output of the op amp is also shown so that the effects of negative feedback illustrated in 7.5 are clearly visible. Because this circuit utilizes an accurate op amp model, it is very instructive to rerun the simulation for higher input frequencies. In this way, the inherent speed limitations of the op amp are shown, and effects such as those presented in Figure 7.6 may be noted.

Peak Detector

One variation on the basic half-wave rectifier is the *peak detector*. This circuit will produce an output that is equal to the peak value of the input signal. This can be configured for either positive or negative peaks. The output of a peak detector can be

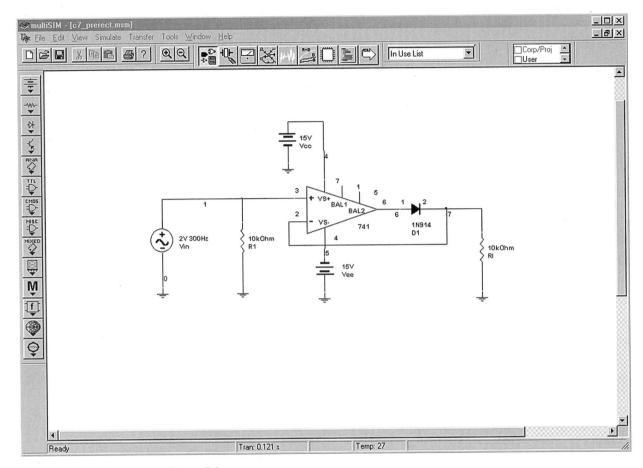

FIGURE 7.8a

Precision rectifier simulation schematic

used for instrumentation or measurement applications. It can also be thought of as an analog pulse stretcher.

A simple positive peak detector is shown in Figure 7.9. Here is how it works: The first portion of the circuit is a precision positive half-wave rectifier. When its output is rising, the capacitor, C, is being charged. This voltage is presented to the second op amp that serves as a buffer for the final load. The output impedance of the first op amp is low, so the charge time constant is very fast, and thus the signal across C is very close to the input signal. When the input signal starts to swing back toward ground, the output of the first op amp starts to drop along with it. Due to the capacitor voltage, the diode ends up in reverse-bias, thus opening the drive to C. C starts to discharge, but the discharge time constant will be much longer than the charge time constant. The discharge resistance is a function of R, the impedance looking into the noninverting input of op amp 2, and the impedance looking into the inverting input of op amp 1,

FIGURE 7.8b

Output waveforms of precision rectifier

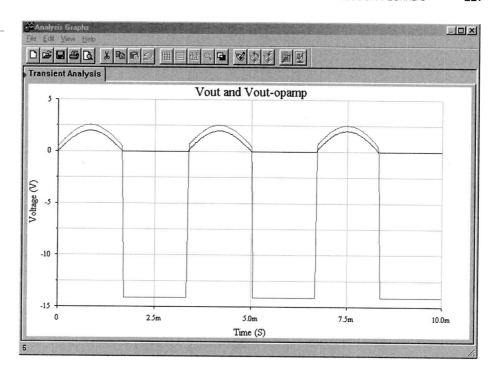

all in parallel. Normally, FET input devices are used, so from a practical standpoint, R sets the discharge rate.

The capacitor will continue to discharge toward zero until the input signal rises enough to overtake it again. If the discharge time constant is much longer than the input period, the circuit output will be a DC value equal to the peak value of the input. If the discharge time constant is somewhat shorter, it has the effect of lengthening the pulse time. It also has the effect of producing the overall contour, or envelope, of complex signals, so it is sometimes called an *envelope detector*. Possible output signals are shown in Figure 7.10. For very long discharge times, large capacitors must be used. Larger capacitors will, of course, produce a lengthening of the charge time (i.e., the rise time will suffer). Large capacitors can also degrade slewing performance. C can only

FIGURE 7.9

Peak detector

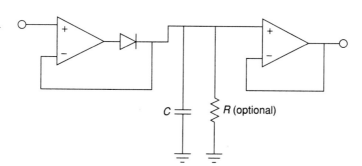

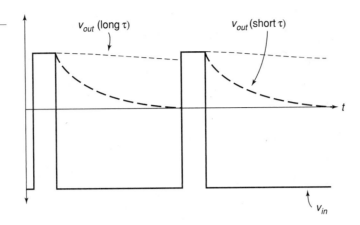

FIGURE 7.10

Effect of τ on pulse shape

be charged so fast because a given op amp can only produce a finite current. This is no different from the case presented with compensation capacitors back in Chapter 5. As an example, if C is 10 μF, and the maximum output current of the op amp is 25 mA,

$$i = C\frac{dv}{dt}$$
$$\frac{dv}{dt} = \frac{i}{C}$$
$$\frac{dv}{dt} = \frac{25 \text{ mA}}{10 \text{ }\mu\text{F}}$$
$$\frac{dv}{dt} = 2500 \text{ V/S} = 2.5 \text{ mV}/\mu\text{S}$$

This is a very slow slew rate. If FET input devices are used, the effective discharge resistance can be very high, thus lowering the requirement for C. For typical applications, C would be many times smaller than the value used here. For long discharge times, high-quality capacitors must be used, as their internal leakage will place the upper limit on discharge resistance.

EXAMPLE 7-1

A positive peak detector is used along with a simple comparator in Figure 7.11 to monitor input levels and warn of possible overload. The LF412 is a dual-package version of the LF411. Explain how it works and determine the point at which the LED lights.

The basic problem when trying to visually monitor a signal for overloads is that the overloading peak may come and go faster than the human eye can detect it. For example, the signal might be sent to a comparator that could light an LED when a preset threshold is exceeded. When the input signal falls, the comparator and LED will go into the off state. Even though the LED does light at the peak, it remains on for such a short time that humans won't notice it. This sort of result is quite possible in the communications industry, where the output of a radio station's microphone will produce very dynamic waves with a great many

FIGURE 7.11

Detector for Example 7-1

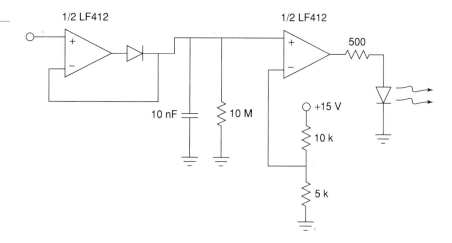

peaks. These peaks can cause havoc in other pieces of equipment down the line. The LED needs to remain on for longer periods.

The circuit of Figure 7.11 uses a peak detector to stretch out the positive pulses. These stretched pulses are then fed to a comparator, which drives an LED. The input pulses are expanded, so the LED will remain on for longer periods.

The discharge time constant is set by R and C. Because FET input devices are used, their impedance is high enough to ignore.

$$T = RC$$
$$T = 10\ M\Omega \times 10\ nF$$
$$T = .1\ S$$

The 10 nF capacitor is small enough to maintain a reasonable slew rate. You may wish to verify this as an exercise. The comparator trip point is set by the 10 kΩ/5 kΩ voltage divider at 5 V. When the input signal rises above 5 V, the comparator output goes high. Assuming that the LED forward drop is about 2.5 V, the 500 Ω resistor limits the output current to

$$I_{LED} = \frac{V_{sat} - V_{LED}}{500}$$
$$I_{LED} = \frac{13\ V - 2.5\ V}{500}$$
$$I_{LED} = \frac{10.5\ V}{500}$$
$$I_{LED} = 21\ mA$$

The LF412 should be able to deliver this current.

In summary, then, the input pulses are stretched by the peak detector. If any of the resulting pulses are greater than 5 V, the comparator trips, and lights the LED. An example input/output wave is shown in Figure 7.12. The one problem with this is that only positive peaks are detected. If large negative peaks exist, they will not cause the LED to light. It is possible to use a similar circuit to detect negative peaks and use that output to drive a

FIGURE 7.12

Waveforms for the circuit of Figure 7.11

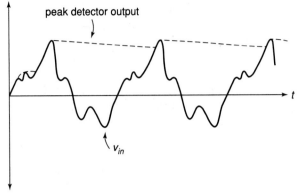

a. Output of detector portion

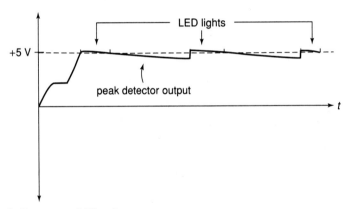

b. Comparator/LED action

common LED along with the positive peak detector. Another way to accomplish this is to utilize a full-wave rectifier/detector.

Precision Full-Wave Rectifier

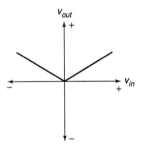

FIGURE 7.13

Transfer characteristic for full-wave rectification

A full-wave rectifier has the input/output characteristic shown in Figure 7.13. No matter what the input polarity is, the output is always positive. For this reason, this circuit is often referred to as an *absolute value* circuit. The design of a precision full-wave rectifier is a little more involved than the single-polarity types. One way of achieving this design is to combine the outputs of negative and positive half-wave circuits with a differential amplifier. Another way is shown in Figure 7.14.

The precision rectifier of circuit 7.14 is convenient in that it only requires two op amps and that all resistors (save one) are the same value. This circuit is comprised of two parts: an inverting half-wave rectifier and a weighted summing amplifier. The rectifier portion is redrawn in Figure 7.15. Let's start the analysis with this portion.

First, note that the circuit is based on an inverting voltage amplifier, with the diodes D_1 and D_2 added. For positive input signals, the input current will attempt to flow

FIGURE 7.14

Precision full-wave rectifier

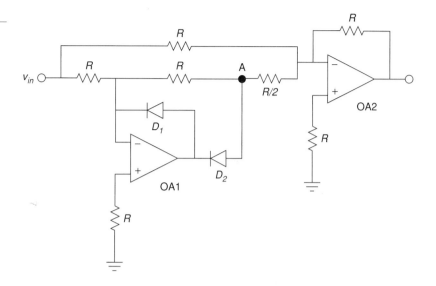

through R_f, to create an inverted output signal with a gain of $\frac{R_f}{R_i}$. (Normally, gain is set to unity.) Because the op amp's inverting input is more positive than its noninverting input, the op amp tries to sink output current. This forces D_2 on, completing the feedback loop, while also forcing D_1 off. As D_2 is inside the feedback loop, its forward drop is compensated for. Thus, positive input signals are amplified and inverted as in a normal inverting amplifier.

If the input signal is negative, the op amp will try to source current. This turns D_1 on, creating a path for current flow. Because the inverting input is at virtual ground, the output voltage of the op amp is limited to the .6 to .7 V drop of D_1. In this way, the op amp does not saturate; rather, it delivers the current required to satisfy the source demand. The op amp's output polarity also forces D_2 off, leaving the circuit output at an approximate ground. Therefore, for negative input signals, the circuit output is zero. The combination of the positive and negative input swings creates an inverted, half-wave rectified output signal, as shown in Figure 7.16. This circuit can be used on its own as a half-wave rectifier if need be. Its major drawback is a somewhat limited input impedance. On the plus side, because the circuit is nonsaturating, it may prove to be faster than the half-wave rectifier first discussed.

FIGURE 7.15

Inverting half-wave rectifier

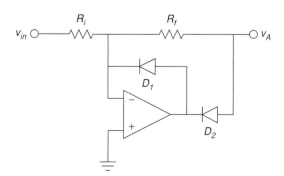

FIGURE 7.16

Output of half-wave rectifier

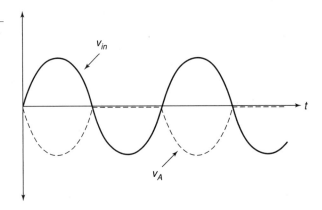

The voltage at point A in Figure 7.14 is the output of the half-wave rectifier and is shown in Figure 7.16. This is one of two signals applied to the summer configured around op amp 2. The other input to the summer is the main circuit's input signal. This signal is given a gain of unity, and the half-wave signal is given a gain of two. These two signals will combine as shown in Figure 7.17 to create a positive full-wave output. Mathematically,

For the first 180 degrees,

$$V_{out} = -K \sin \omega t + 2K \sin \omega t$$
$$V_{out} = K \sin \omega t$$

For the second 180 degrees,

$$V_{out} = K \sin \omega t + 0$$
$$V_{out} = K \sin \omega t$$

In order to produce a negative full-wave rectifier, simply reverse the polarity of D_1 and D_2.

An example application of an op amp-based rectifier is shown in Figure 7.18. This circuit is used detect dangerous overloads and faults in an audio power amplifier. Short-term signal clipping may not be a severe problem in certain applications; however, long-term clipping may create very stressful conditions for the loudspeakers.

FIGURE 7.17

Combination of signals produces output

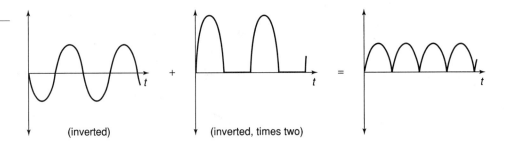

FIGURE 7.18

Power amplifier overload detector

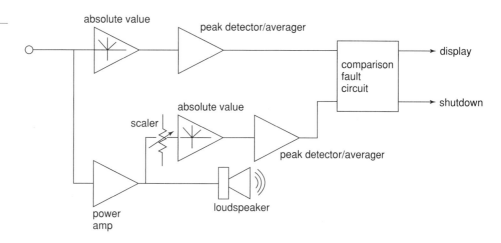

This would also be the case if an improperly functioning power amplifier produced a DC offset. Unfortunately, a simple scaled comparison of the input and output signals of the power amplifier may be misleading. In order to compare long-term averages, the input and scaled output signals are precision full-wave rectified and then passed through a peak-detecting or averaging stage. This might be as simple as a single RC network. These signals are then compared by the fault stage. If there is a substantial difference between the two signals, the amplifier is most likely clipping the signal considerably or producing an unwanted DC offset. The fault stage can then light a warning LED, or in severe cases, trip system shutdown circuitry to prevent damage to other components.

7.3 Wave Shaping

Active Clampers

Clampers are used to add a specific amount of DC to a signal. Generally, the amount of DC will be equal to the peak value of the signal. Clampers are intelligent in that they can adjust the amount of DC if the peak value of the input signal changes. Ideally, clampers will not change the shape of the input signal; rather, the output signal is simply a "vertically shifted" version of the input, as shown in Figure 7.19. One common application of the clamper is in television receivers. Here, a clamper is referred to as a *DC restorer*. Certain portions of the video signal, such as sync pulses, are required to be at specific levels. After the video signal has been amplified by AC coupled gain stages, the DC restorer returns the video to its normal orientation. Without the clamping action, the various parts of the signal cannot be decoded properly. Clampers are commonly made with passive components, just as rectifiers are. Like rectifiers, simple clampers produce errors caused by the diode's forward voltage drop. Active clampers remove this error.

Conceptually, a clamper needs to sense the amplitude of the input signal and create a DC signal of equal value. This DC signal is then added to the input, as shown

FIGURE 7.19

Effect of clampers on input signal

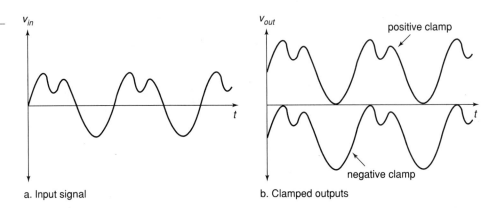

a. Input signal

b. Clamped outputs

in Figure 7.20. If the peak value is known, and does not change, it is possible to create this function with a summing amplifier. If the signal level is dynamic, a different course needs to be taken. The trick is in getting the DC source to properly track input level changes. Obviously, a simple DC supply is not appropriate; however, a charged capacitor will fit the bill nicely—as long as its discharge time constant is much longer than the period of the input wave. All that needs to be done is to have the capacitor charge to the peak level of the input. When the capacitor voltage is added to the input signal, the appropriate DC shift will result.

Figure 7.21 shows an active positive clamper. On the first negative going peak of the input, the source will attempt to pull current through the capacitor in the direction shown. This forces the inverting input of the op amp to go slightly negative, thus creating a positive op amp output. The result of this action will be to forward-bias the diode and supply charging current to the capacitor. The capacitor will charge to the negative peak value of the input. The output resistance of the op amp is very low, so charging is relatively fast, only being limited by the maximum output current of the op amp. When the source changes direction, the op amp will produce a negative output, thus turning off the diode and effectively removing the op amp from the circuit. Because the discharge time for C is much longer than the input period, its potential stays at roughly V_{p-}. It now acts like a voltage source. These two sources add, so we can see that the output must be

$$V_{out} = V_{in} + |V_{p-}| \qquad (7.1)$$

The op amp will remain in saturation until the next negative peak, at which point the capacitor will be recharged. During the charging period, the feedback loop is closed, and thus, the diode's forward drop is compensated for by the op amp. In other words, the op amp's output will be approximately .6 to .7 V above the inverting input's potential. The discharge time of the circuit is set by the load resistor, R_l. If particularly long time constants are required, a buffer stage may be used, along with an FET input clamping op amp. The resistor R is used to prevent possible damage to the op amp from capacitor discharge. This value is normally in the low kilohm region. In order to make a negative clamper, just reverse the polarity of the diode.

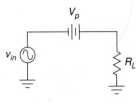

FIGURE 7.20

Simple clamper model

FIGURE 7.21

Active positive clamper

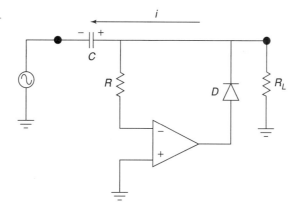

Sometimes, it is desirable to clamp a signal and add a fixed offset as well. An example output waveform of this type is shown in Figure 7.22. This function is relatively easy to add to the basic clamper. In order to include the offset, all that needs to be done is to change the op amp's reference point. In the basic clamper, the noninverting input is tied to ground. Consequently, this establishes the point at which the charge/discharge cycle starts. If this reference is altered, the charge/discharge point is altered too. To create a positive offset, a DC signal equal to the offset is applied to the noninverting input, as shown in Figure 7.23. A similar arrangement may be used for negative clampers.

EXAMPLE 7-2

An active clamper is shown in Figure 7.24. An LF412 op amp is used (a dual LF411). Determine the capacitor's voltage and verify that the time constant is appropriate for the input waveform. Also, sketch the output waveform and determine the maximum differential input voltage for the first op amp.

FIGURE 7.22

Clamped output with offset

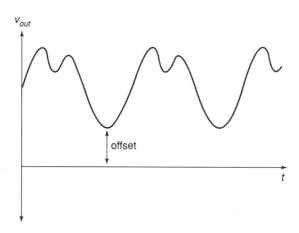

FIGURE 7.23

Active positive clamper with offset

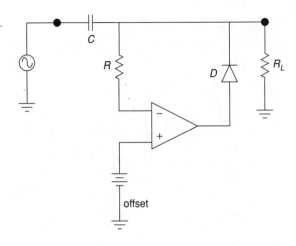

FIGURE 7.24

Active clamper and waveform for Example 7-2

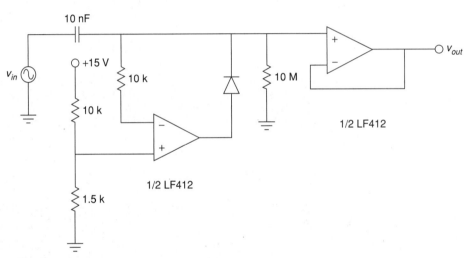

a. Circuit

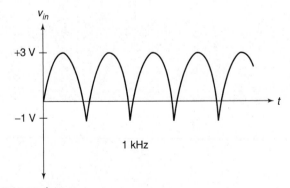

b. Input wave form

FIGURE 7.25

Output of clamper for Example 7-2

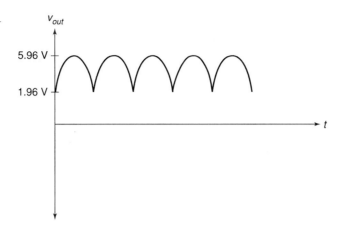

Because this is a positive clamper, the capacitor's voltage is the sum of the offset potential and the negative peak potential of the input waveform. V_{offset} is set by the 10 kΩ/1.5 kΩ divider.

$$V_{offset} = V_{CC} \frac{R_2}{R_1 + R_2}$$
$$V_{offset} = 15\,V \frac{1.5\,k}{10\,k + 1.5\,k}$$
$$V_{offset} = 1.96\,V$$

$$V_c = V_{offset} + |V_{p-}|$$
$$V_c = 1.96\,V + |-1\,V|$$
$$V_c = 2.96\,V$$

This is the amount of DC added to the input signal. The output waveform will look just like the input, except that it will be shifted up by 2.96 V. This is shown in Figure 7.25.

The discharge time constant should be much larger than the period of the input wave. The input wave has a frequency of 1 kHz, and thus, a period of 1 mS. The capacitor discharges through the load resistor. As FET input op amps are used, their effect on the discharge rate is minimal.

$$T = R_l C$$
$$T = 10\,M \times 10\,nF$$
$$T = .1\,S$$

The discharge rate is 100 times longer than the input period, so the capacitor droop will be satisfactory.

The maximum differential input signal is the worst-case difference between the inverting and noninverting input potentials. The noninverting input is tied directly to 1.96 V. The inverting input sees the output waveform. The maximum value of the output is 5.96 V. So the difference is

$$V_{in-diff} = V_{in_+} - V_{in_-}$$
$$V_{in-diff} = 1.96 \text{ V} - 5.96 \text{ V}$$
$$V_{in-diff} = -4 \text{ V}$$

The LF412 will have no problem with a differential input of this size.

---

multiSIM COMPUTER SIMULATION

A simulation of the clamper circuit of Example 7-2 is shown in Figure 7.26 (pp. 241–242). In order to simplify the simulation somewhat, the input signal has been altered modestly. Instead of the shifted 4 volt peak-to-peak full-wave rectified signal originally used, a shifted 4 volt peak-to-peak sine wave is used. Both the input and output waveforms are plotted in the Transient Analysis.

The circuit requires about 1 cycle of the waveform before the output stabilizes. After that point the clamping action and shift of nearly 2 volts is quite evident in the output waveform. There is no noticeable droop, scaling error, or obvious distortion in the output either, although some minor aberrations can be seen at the negative peaks on occasion.

Active Limiters

A *limiter* is a circuit that places a maximum restriction on its output level. The output of a limiter can never be above a specific preset level. In one sense, all active circuits are limiters, in that they will all eventually clip the signal by going into saturation. A true limiter, though, prohibits the signal at levels considerably lower than saturation. The exact level is fairly easy to set. Limiters can be used to protect following stages from excessive input levels. They can also be seen as a type of wave shaper. One form of limiter was shown in Chapter 6, in the Pocket Rockit's schematic. In that circuit, a limiter was used to purposely clip the music signal for artistic effect. The limitation of that form is that the limit potential is locked at $\pm.7$ V by the parallel signal diodes. For a more general form, an arbitrary limit potential is desired. Instead of using parallel signal diodes, a series combination of Zener diodes will prove useful.

An example limiter is shown in Figure 7.27. It is based upon the inverting voltage amplifier form. As long as the output signal is below the Zener potential, the output equals the input times the voltage gain. If the output voltage tries to rise above the Zener potential, one of the diodes will go into Zener conduction, and the other diode will go into forward bias. Once this happens, the low dynamic resistance of the diodes will disallow any further increase in output potential. The output will not be allowed to move outside of the Zener potential (plus the .7 V turn-on for the second diode):

$$|V_{out}| <= V_{zener} + .7 \text{ V} \tag{7.2}$$

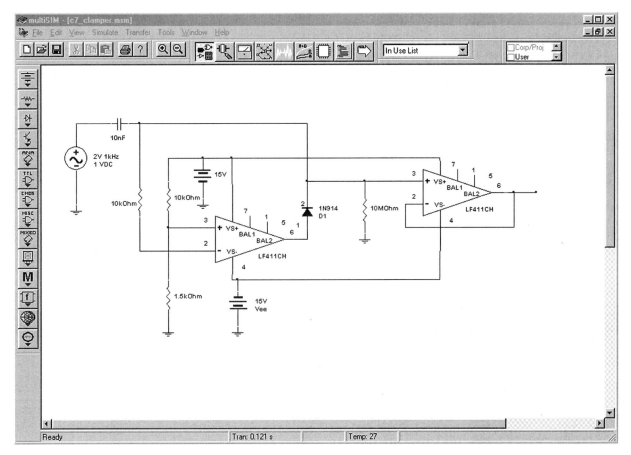

FIGURE 7.26a

Clamper schematic in MultiSIM

Or we might reword this as

$$V_{limit} = \pm(V_{zener} + .7 \text{ V}) \tag{7.3}$$

This effect is shown graphically in Figure 7.28, the limiter's transfer characteristic.

EXAMPLE 7-3

If the input to the circuit of Figure 7.29 is a 4 V peak triangle wave, sketch V_{out}.

Without the Zener limit diodes, this amplifier would simply multiply the input by a gain of two and invert it.

$$V_{out} = A_v V_{in}$$
$$V_{out} = -2 \times 4 \text{ V (peak)}$$
$$V_{out} = -8 \text{ V (peak)}$$

Figure 7.26b

Clamper waveforms

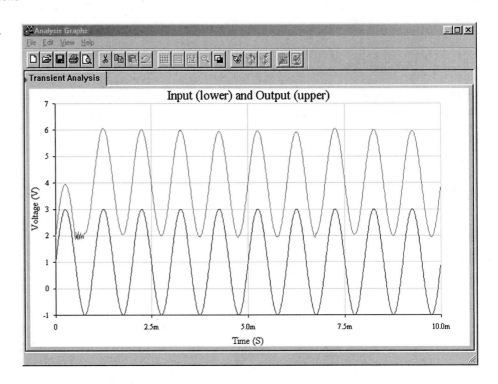

So, an inverted 8 V peak triangle would be the output. With the inclusion of the diodes, the maximum output is

$$V_{limit} = \pm(V_{zener} + .7\text{ V})$$
$$V_{limit} = \pm(5.1\text{ V} + .7\text{ V})$$
$$V_{limit} = \pm 5.8\text{ V}$$

Consequently, the output wave is clipped at 5.8 V, as shown in Figure 7.30.

By using different Zeners, it is also possible to produce asymmetrical limiting. For example, positive signals might be limited to 10 V, whereas negative signals could be limited to −5 V.

Figure 7.27

Active limiter

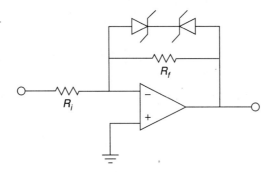

WAVE SHAPING   243

FIGURE 7.28

Transfer characteristic of limiter

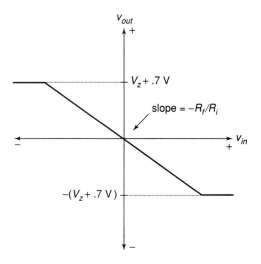

FIGURE 7.29

Limiter for Example 7-3

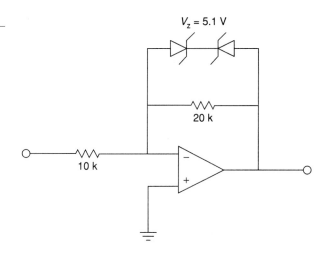

FIGURE 7.30

Input/output signals of limiter for Example 7-3

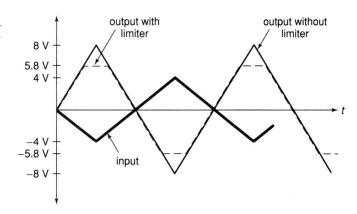

7.4 Function Generation

Function generation circuits are used to create arbitrary transfer characteristics. These can be used for a variety of purposes, including transducer linearization and sine shaping. Indeed, many modern laboratory signal generators do not directly create a sine wave; rather, they generate a triangle wave and pass it through a function circuit, which will then produce the desired sine wave. In essence, a function circuit gives different gains to different parts of the input signal. It impresses its shape on the input waveform. If a "straight-sided" waveform such as a ramp or triangle is fed into a function circuit, the resulting output waveform will bear a striking resemblance to the circuit's transfer curve. (This effect can be put to good use when investigating function circuits via computer simulation.)

Basically, there are two ways in which to create a function circuit with an op amp. The first way is an extension of the Zener limiter circuit. The second form relies on a biased-diode network. In both cases, the base circuit form is that of an inverting-voltage amplifier. Also, both techniques allow the resulting transfer curve's slope to increase or decrease at specified break points and rates.

A simple function circuit is shown in Figure 7.31. Note that it is very similar to the limiter of Figure 7.27. The only difference is the inclusion of R_a. Below the Zener potential, the circuit gain is still set by R_f. When the Zener potential is eventually reached, the diodes can no longer force the output to an unchanging value, as R_a is in series with them. Instead of an ideal shunt across R_f as in the limiter, R_f is effectively paralleled by R_a. In other words, once the Zener potential is reached, the gain drops to $\frac{R_f \| R_a}{R_i}$. This is shown in Figure 7.32. To be a bit more precise, the gain change is not quite so abrupt as this approximation. In reality, the transfer curve is smoother and somewhat delayed, as indicated by the dashed line.

In order to achieve an increasing slope, the Zener/resistor combination is placed in parallel with R_i. When the Zener potential is reached, R_a will effectively parallel R_i, reducing gain. This has the unfortunate side effect of reducing the input impedance, so it may be necessary to include an input buffer. An example circuit and resulting transfer curve are shown in Figure 7.33.

If more than two slopes are required, it is possible to parallel multiple Zener/resistor combinations. This technique can be used to create a piecewise linear approximation of a desired transfer characteristic. A two-section circuit is shown in Figure 7.34. Note

Figure 7.31

Simple function generator

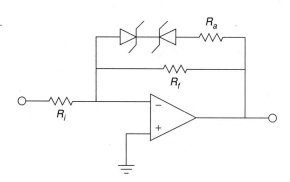

FIGURE 7.32

Transfer characteristic (decreasing slope)

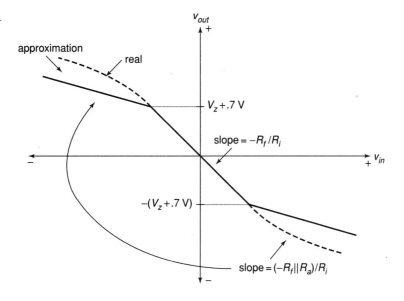

how the resistors are constantly placed in parallel, thus reducing gain. If this multiple-section technique is used with a circuit such as the one in Figure 7.33, the gain will increase.

EXAMPLE 7-4

Given the circuit in Figure 7.35, sketch its transfer curve.

The first step is to note the break points in the curve. Because this circuit uses a decreasing gain scheme with diodes across R_f, the output break points are set by the Zener potentials. Also note that the break points are symmetrical about zero, as identical diodes are used.

$$V_{break} = \pm(V_z + .7\text{ V})$$
$$V_{break} = \pm(3.9\text{ V} + .7\text{ V})$$
$$V_{break} = \pm 4.6\text{ V}$$

The next item to determine is the base voltage gain. This is the gain without the Zener/resistor effect.

$$A_v = -\frac{R_f}{R_i}$$
$$A_v = -\frac{20\text{ k}}{5\text{ k}}$$
$$A_v = -4$$

FIGURE 7.33

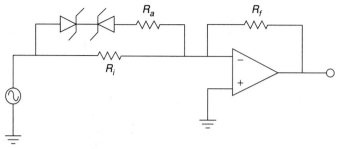

a. Simple function generator (increasing gain)

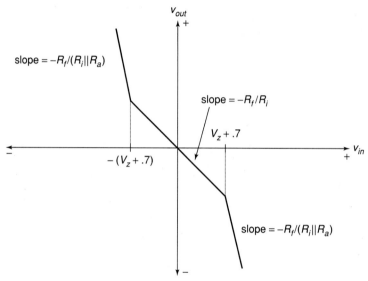

b. Transfer characteristic (increasing slope)

The second level gain occurs when the diodes are on, placing R_a in parallel with R_f.

$$A_{v2} = -\frac{R_f \| R_a}{R_i}$$

$$A_{v2} = -\frac{20\,k \| 10\,k}{5\,k}$$

$$A_{v2} = -1.33$$

The resulting plot is shown in Figure 7.36. In order to find the break point for the input signal (in this case only the output break is found), you need to divide the output break by the slope of the transfer curve. For an output of 4.6 V, the input signal will be 4 times less at 1.15 V. If multiple sections are used, the changes due to each section are added in order to find the corresponding input signal. If the circuit is an increasing-gain type (as in Figure 7.33),

FUNCTION GENERATION   247

FIGURE 7.34

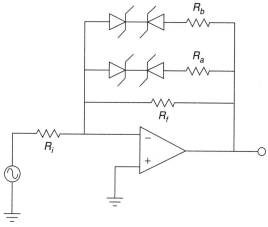

a. Multiple section function generator

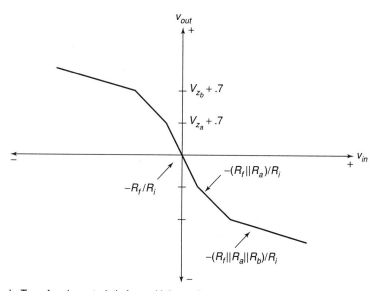

b. Transfer characteristic for multiple section generator

FIGURE 7.35

Function generator for Example 7-4

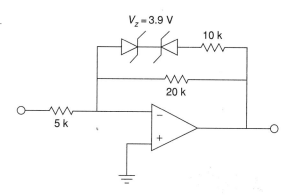

FIGURE 7.36

Transfer characteristic of the circuit in Figure 7.35

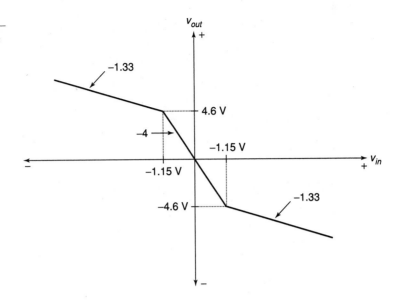

then the input break points are known, and the output points are found by multiplying by the slope. Both of these items are illustrated in the following example.

EXAMPLE 7-5

Draw the transfer curve for the circuit of Figure 7.37.

All diode pairs are identical, so symmetry is maintained. By inspection, the input break points are:

$$V_{break1-in} = 1.0 \text{ V} + .7 \text{ V} = 1.7 \text{ V}$$
$$V_{break2-in} = 2.2 \text{ V} + .7 \text{ V} = 2.9 \text{ V}$$

FIGURE 7.37

Multiple section circuit for Example 7-5

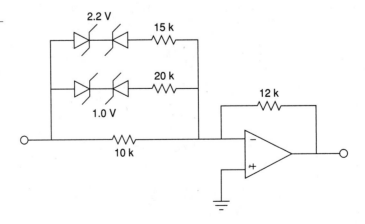

The base gain is

$$A_v = -\frac{R_f}{R_i}$$
$$A_v = -\frac{12\text{ k}}{10\text{ k}}$$
$$A_v = -1.2$$

The second level gain is

$$A_{v2} = -\frac{R_f}{R_i \| R_a}$$
$$A_{v2} = -\frac{12\text{ k}}{10\text{ k} \| 20\text{ k}}$$
$$A_{v2} = -1.8$$

The third level gain is

$$A_{v3} = -\frac{R_f}{R_i \| R_a \| R_b}$$
$$A_{v3} = -\frac{12\text{ k}}{10\text{ k} \| 20\text{ k} \| 15\text{ k}}$$
$$A_{v3} = -2.6$$

The corresponding output breaks are

$$V_{break1\text{-}out} = A_v V_{break1\text{-}in}$$
$$V_{break1\text{-}out} = -1.2 \times 1.7$$
$$V_{break1\text{-}out} = -2.04\text{ V}$$

$$V_{break2\text{-}out} = V_{break1\text{-}out} + A_{v2}(V_{break2\text{-}in} - V_{break1\text{-}in})$$
$$V_{break2\text{-}out} = -2.04\text{ V} + -1.8 \times (2.9\text{ V} - 1.7\text{ V})$$
$$V_{break2\text{-}out} = -4.2\text{ V}$$

The resulting curve is shown in Figure 7.38.

EXAMPLE 7-6

A temperature transducer's response characteristic is plotted in Figure 7.39. Unfortunately, the response is not consistent across a large temperature range. At very high or low temperatures, the device becomes more sensitive. Design a circuit that will compensate for this error, so that the output will remain at a sensitivity of 1 V/10 C° over a large temperature range.

First of all, in order to compensate for an increasing curve, a mirror image decreasing curve is called for. The circuit will only need a single diode section; however, it will not be symmetrical. The positive break occurs at 30 degrees (3 V), and the negative break occurs at −40 degrees (−4 V). In this range the gain will remain at unity. Outside of this range, the gain must fall. In order to bring a ratio of 1 V/8 C° back to 1 V/10 C°, it must be multiplied by its reciprocal—a gain of 8/10. The desired transfer curve is shown in Figure 7.40. If R_f and

250 NONLINEAR CIRCUITS

FIGURE 7.38

Characteristic of multiple-section circuit of Figure 7.37

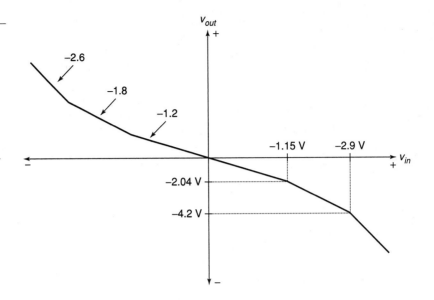

R_i are arbitrarily chosen to be 10 kΩ, the resulting combination of $R_a \| R_f$ must be 8 kΩ to achieve this gain. R_a can then be found:

$$R_a \| R_f = 8 \text{ k}$$
$$\frac{1}{R_a} + \frac{1}{10 \text{ k}} = \frac{1}{8 \text{ k}}$$
$$R_a = 40 \text{ k}$$

FIGURE 7.39

Temperature transducer's response for Example 7-6

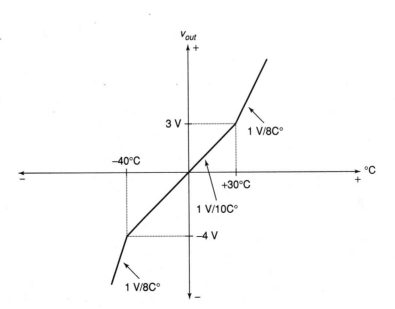

FIGURE 7.40

Desired transfer curve for Example 7-6

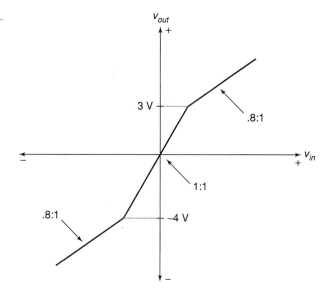

Because the break points have already been determined to be 3 and −4 V respectively, all that needs to be done is to compensate for the forward drop of the other diode.

$$V_{z+} = 3\text{ V} - .7\text{ V}$$
$$V_{z+} = 2.3\text{ V}$$

$$V_{z-} = -(4\text{ V} - .7\text{ V})$$
$$V_{z-} = -3.3\text{ V} \text{ (use absolute value)}$$

Because this circuit inverts the signal, it may be wise to include an inverting buffer to compensate. The resulting circuit is shown in Figure 7.41.

FIGURE 7.41

Correct circuit for the transducer in Example 7-6

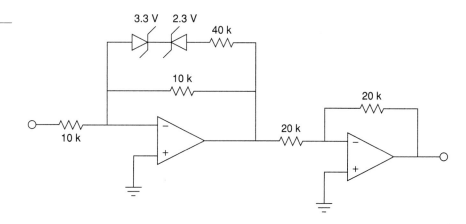

FIGURE 7.42

Biased diode function generator

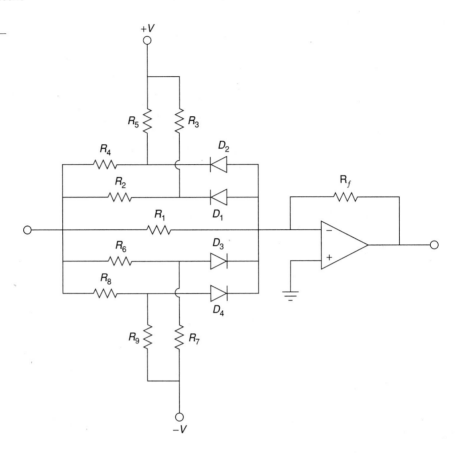

One of the problems with the Zener form is that the transfer curves change slope slowly. Because of this, fine control over the shape is not very easy to come by. Consequently, a good deal of trial and error is sometimes needed during the design sequence. Also, there is a limited range of Zener diode values available. A form that is somewhat more precise utilizes a biased-diode network.

An increasing gain-function circuit that utilizes a biased-diode approach is shown in Figure 7.42. Its corresponding transfer curve is shown in Figure 7.43. For each break point, one diode and a pair of resistors are required. Even if the break points are symmetrical for positive and negative inputs, one set of components will be required for each polarity. The basic concept is the same as that of the Zener approach; by turning on sections with higher and higher input voltages, resistors will be placed in parallel with the base feedback resistors, thus changing the gain. As with the Zener approach, a decreasing gain characteristic is formed by placing the network across R_f, instead of R_i. Let's take a look at how a single section reacts.

Each break point requires a blocking diode, a gain resistor and a bias resistor. Resistors R_2 and R_3, along with diode D_1 make up a single section for negative input voltages. Under low signal conditions, D_1 is off and blocks current flow. Effectively, the section is an open circuit. If the input potential goes negative enough, D_1 will turn

FIGURE 7.43

Transfer characteristic of biased diode function generator

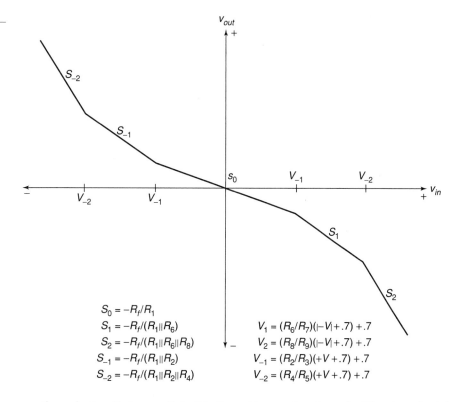

$S_0 = -R_f/R_1$
$S_1 = -R_f/(R_1 \| R_6)$
$S_2 = -R_f/(R_1 \| R_6 \| R_8)$
$S_{-1} = -R_f/(R_1 \| R_2)$
$S_{-2} = -R_f/(R_1 \| R_2 \| R_4)$

$V_1 = (R_6/R_7)(|-V|+.7) +.7$
$V_2 = (R_8/R_9)(|-V|+.7) +.7$
$V_{-1} = (R_2/R_3)(+V +.7) +.7$
$V_{-2} = (R_4/R_5)(+V +.7) +.7$

on, thus placing R_2 in parallel with R_1 and increasing the gain. The slope (gain) of the transfer curve is:

$$\text{Slope} = -\frac{R_f}{R_1 \| R_2}$$

The break point occurs when the cathode of D_1 goes to approximately $-.7$ V (assuming silicon). At this point, the drop across R_3 must be

$$V_{R_3} = V_{CC} + .7 \text{ V}$$

The drop across R_2 must be

$$V_{R_2} = V_{in} - .7 \text{ V}$$

Right before D_1 conducts, the current through R_2 must equal the current through R_3.

$$\frac{V_{R_3}}{R_3} = \frac{V_{R_2}}{R_2}$$

$$\frac{V_{CC} + .7 \text{ V}}{R_3} = \frac{V_{in} - .7 \text{ V}}{R_2}$$

At this instant, V_{in} equals the break point. Solving for V_{in} gives,

$$V_{in} = V_{breakpoint} = \frac{R_2}{R_3}(V_{CC} + .7\,\text{V}) + .7\,\text{V}$$

or to a rough approximation,

$$V_{breakpoint} \approx \frac{R_2}{R_3} V_{CC}$$

The approximation error may be minimized by using germanium diodes. Note that since the break points are set by a resistor ratio, very tight control of the curve is possible. When designing a function circuit, the required slopes dictate the values of the gain resistors. For ease of use the bias potentials are set to the positive and negative supply rails. Given these elements and the desired break point voltages, the necessary bias resistor values may be calculated.

multiSIM Computer Simulation

A simulation using the biased diode technique is presented in Figure 7.44 (pp. 255–256). Here the circuit has been designed to echo the response found in Figure 7.36 from Example 7-4. Note that the same values are used here for the feedback resistors as those found in the Zener version of the circuit. The only new calculation was for the biasing resistors (the 40.3 kΩ) to set the output break points. In order to see the gain effect clearly, a triangle wave is used as the input. Because the input waveform consists of nice, straight, diagonal lines, the gain response will be reflected in the output wave shape. This is obvious in the Transient Analysis graph. The signal inversion stands out, but the more interesting and important item is the change in slope that occurs when the output waveform reaches about 4.5 volts. This is approximately where the output break point was calculated to be, and the reduction in gain above this point is quite apparent.

7.5 Comparators

The simple open-loop op amp comparator was discussed in Chapter 2. Although this circuit is functional, it is not the final word on comparators. It suffers from two faults: (1) it is not particularly fast, and (2) it does not use *hysteresis*. Hysteresis provides a margin of safety and "cleans up" switching transitions. Providing a comparator with hysteresis means that its reference depends on its output state. As an example, for a positive going transition, the reference might be 2 V, but for a negative transition, the reference might be 1 V. This effect can dramatically improve performance when used with noisy inputs.

Take a look at the noisy signal in Figure 7.45. If this signal is fed into a simple comparator, the noise will produce a false turn-off spike. If this same signal is fed into

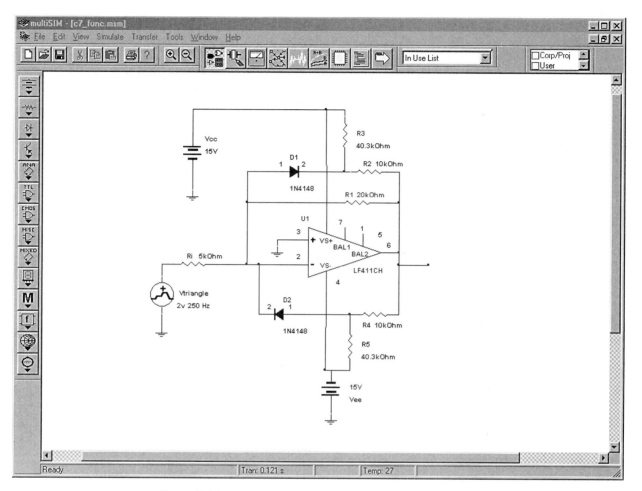

FIGURE 7.44a

Biased diode function synthesizer in MultiSIM

a comparator with hysteresis, as in Figure 7.46, a clean transition results. In order to change from low to high, the signal must exceed the upper reference. In order to change from high to low, the signal must drop below the lower reference. This is a very useful feature. A comparator with hysteresis is shown in Figure 7.47.

The output voltage of this circuit will be either $+V_{sat}$ or $-V_{sat}$. Let's assume that the device is at $-V_{sat}$. In order to change to $+V_{sat}$, the inverting input must go lower than the noninverting input. The noninverting input is derived from the $\frac{R_1}{R_2}$ voltage divider. The divider is driven by the output of the op amp, in this case, $-V_{sat}$. Therefore, to change state, V_{in} must be

$$V_{in} = V_{lowerthres} = -V_{sat}\frac{R_2}{R_1 + R_2} \tag{7.4}$$

FIGURE 7.44b

Function synthesizer waveforms

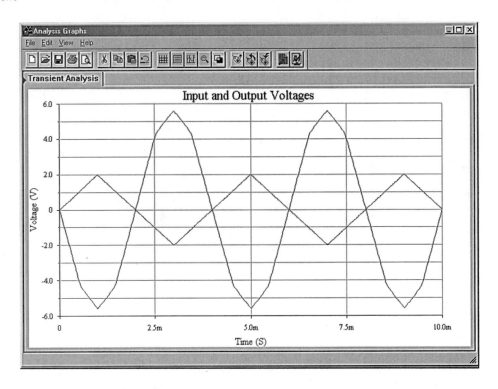

FIGURE 7.45

False turn-off spike

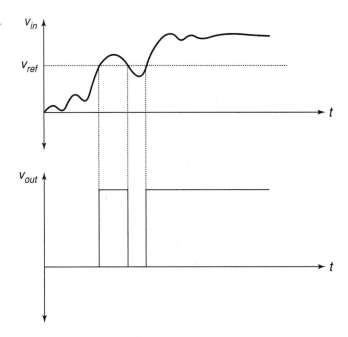

COMPARATORS

FIGURE 7.46
Clean transition using hysteresis

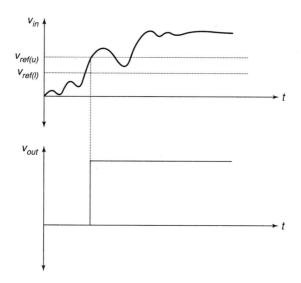

Similarly, to go from positive to negative,

$$V_{in} = V_{upperthres} = +V_{sat}\frac{R_2}{R_1 + R_2} \qquad (7.5)$$

The upper and lower trip points are commonly referred to as the upper and lower thresholds. The adjustment of R_1 and R_2 creates an "error band" around zero (ground). Note that when the trip point is reached, the comparator output state changes, thus changing the reference. This reinforces the initial change. In effect, the comparator is now using positive feedback (note how the feedback signal is tied to the noninverting terminal). This circuit is sometimes referred to as a *Schmitt Trigger*.

A noninverting version of the Schmitt Trigger is shown in Figure 7.48. Note that in order to change state, the noninverting terminal's voltage will be approximately zero. If the circuit is in the low state, the voltage across R_1 will equal $-V_{sat}$ at the time of transition. At this point, the voltage across R_2 will equal V_{in}. The current into the op amp is negligible, so the current through R_1 equals that through R_2.

$$\frac{V_{R_1}}{R_1} = \frac{V_{R_2}}{R_2}$$

$$\frac{-(-V_{sat})}{R_1} = \frac{V_{in}}{R_2}$$

Because V_{in} equals the threshold voltage at transition,

$$V_{in} = V_{upperthres} = V_{sat}\frac{R_2}{R_1} \qquad (7.6)$$

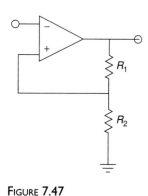

FIGURE 7.47
Comparator with hysteresis

FIGURE 7.48
Noninverting comparator with hysteresis

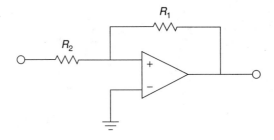

Similarly, for the opposite transition,

$$V_{in} = V_{lowerthres} = -V_{sat}\frac{R_2}{R_1} \qquad (7.7)$$

EXAMPLE 7-7

Sketch the output waveform for the circuit of Figure 7.49 if the input signal is a 5 V peak sine wave.

First, determine the upper and lower threshold voltages.

$$V_{upperthres} = V_{sat}\frac{R_2}{R_1}$$

$$V_{upperthres} = 13 \text{ V}\frac{2 \text{ k}}{20 \text{ k}}$$

$$V_{upperthres} = 1.3 \text{ V}$$

$$V_{lowerthres} = -V_{sat}\frac{R_2}{R_1}$$

$$V_{lowerthres} = -13 \text{ V}\frac{2 \text{ k}}{20 \text{ k}}$$

$$V_{lowerthres} = -1.3 \text{ V}$$

The output will go to $+13$ V when the input exceeds $+1.3$ V and will go to -13 V when the input drops to -1.3 V. The input/output waveform sketches are shown in Figure 7.50.

FIGURE 7.49
Comparator for Example 7-7

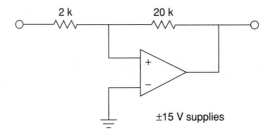

±15 V supplies

FIGURE 7.50

Comparator waveforms

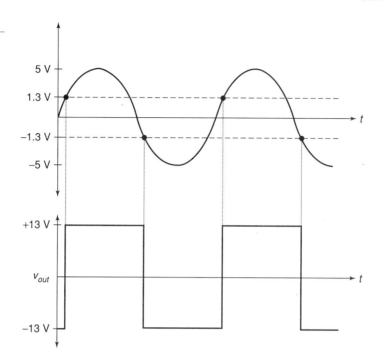

multiSIM COMPUTER SIMULATION

In Figure 7.51 (pp. 260–261), a simulation of the circuit and waveforms of Example 7-7 are shown using Electronics Workbench MultiSIM. A reasonably fast op amp, the LF411, was chosen for the simulation in order to minimize slewing effects. The input and output waveforms are superimposed in the Transient Analysis. This allows the switching levels to be determined precisely. From the graph, it is very clear that the high-to-low output transition occurs when the input drops below about −1.3 volts. Similarly, the low-to-high transition occurs when the input rises above about 1.3 volts. This is exactly as expected, and it reinforces the concept of hysteresis visually. The other effect that may be noted here is an effective delaying of the pulse signal relative to the timing at the input's zero crossings. This is an unfortunate side effect that is magnified by wide thresholds and slowing varying input signals.

Although the use of positive feedback and hysteresis is a step forward, switching speed is still dependent on the speed of the op amp. Also, the output levels are approximately equal to the power supply rails, so interfacing to other circuitry (such as TTL logic) requires extra circuitry. To cure these problems, specialized comparators circuits have evolved. Generally, comparator ICs can be broken down into a few major categories: general-purpose, high speed, and low power/low cost. A typical general-purpose device is the LM311, whereas the LM360 is a high-speed device with differential outputs. Examples of the low-power variety include the LM393

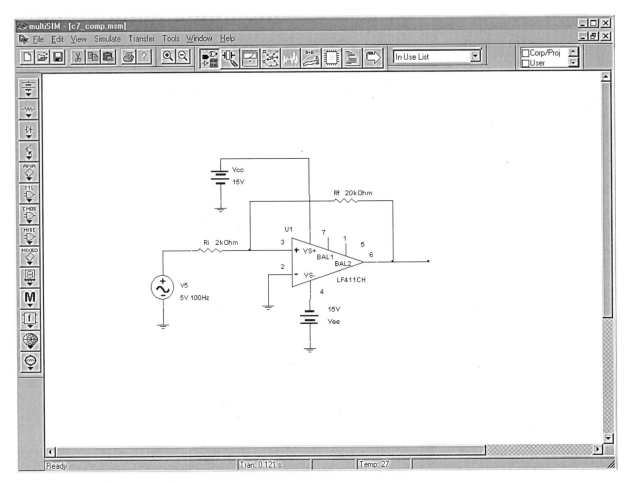

FIGURE 7.51a

Simulation of comparator

dual comparator and the LM339 quad comparator. As was the case with ordinary op amps, there is a definite trade-off between comparator speed and power consumption. As you might guess, the high-speed LM360 suffers from the highest power consumption, whereas the more miserly LM393 and LM339 exhibit considerably slower switching speeds. High-speed devices often exhibit high input bias and offset currents as well.

The general-purpose LM311 is one of the more popular comparators in use today. An FET input version, the LF311, is also available. Generally, all of the op amp-based comparator circuits already mentioned can be adapted for use with the LM311. The LM311 is far more flexible than the average op amp comparator, however. An outline and data sheet for the LM311 are shown in Figures 7.52a and 7.52b.

FIGURE 7.51b

Comparator waveforms

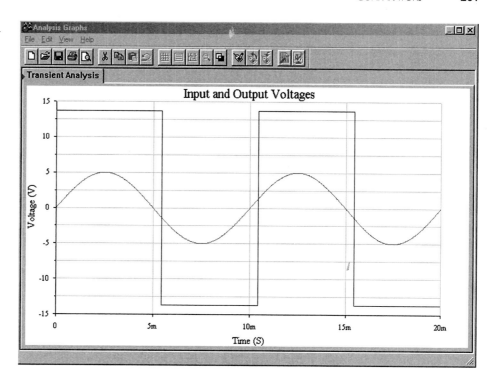

First of all, note that the LM311 is reasonably fast, producing a response time of approximately 200 nS. This places it squarely in the middle performance range, as it is 10 or so times faster than a low-power comparator, but at least 10 times slower than high-speed devices. The voltage gain is relatively high, at 200,000 typically. Input offset voltage is moderate at 2 mV typically, and 7.5 mV maximum, at room temperature. The input offset and bias currents are 50 nA and 250 nA worst case, respectively. Devices with improved offset and bias performance are available, but these values

FIGURE 7.52

Outline and data sheet for the LM311 comparator

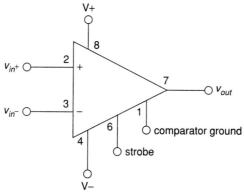

a. The LM311

Figure 7.52

(Continued)

Reprinted with permission of National Semiconductor Corporation

5.0 Absolute Maximum Ratings for the LM111/LM211 (Note 12)

If Military/Aerospace specified devices are required, please contact the National Semiconductor Sales Office/Distributors for availability and specifications.

Total Supply Voltage (V_{84})	36V
Output to Negative Supply Voltage (V_{74})	40V
Ground to Negative Supply Voltage (V_{14})	30V
Differential Input Voltage	±30V
Input Voltage (Note 13)	±15V
Power Dissipation (Note 14)	500 mW
ESD Rating (Note 19)	300V
Output Short Circuit Duration	10 sec
Operating Temperature Range	0° to 70°C
Storage Temperature Range	−65°C to 150°C
Lead Temperature (soldering, 10 sec)	260°C
Voltage at Strobe Pin	V^+−5V
Soldering Information	
Dual-In-Line Package	
Soldering (10 seconds)	260°C
Small Outline Package	
Vapor Phase (60 seconds)	215°C
Infrared (15 seconds)	220°C

See AN-450 "Surface Mounting Methods and Their Effect on Product Reliability" for other methods of soldering surface mount devices.

Electrical Characteristics (Note 15)

for the LM311

Parameter	Conditions	Min	Typ	Max	Units
Input Offset Voltage (Note 16)	T_A=25°C, R_S≤50k		2.0	7.5	mV
Input Offset Current (Note 16)	T_A=25°C		6.0	50	nA
Input Bias Current	T_A=25°C		100	250	nA
Voltage Gain	T_A=25°C	40	200		V/mV
Response Time (Note 17)	T_A=25°C		200		ns
Saturation Voltage	V_{IN}≤−10 mV, I_{OUT}=50 mA T_A=25°C		0.75	1.5	V
Strobe ON Current (Note 18)	T_A=25°C		2.0	5.0	mA
Output Leakage Current	V_{IN}≥10 mV, V_{OUT}=35V T_A=25°C, I_{STROBE}=3 mA V^- = Pin 1 = −5V		0.2	50	nA
Input Offset Voltage (Note 16)	R_S≤50K			10	mV
Input Offset Current (Note 16)				70	nA
Input Bias Current				300	nA
Input Voltage Range		−14.5	13.8, −14.7	13.0	V
Saturation Voltage	V^+≥4.5V, V^-=0 V_{IN}≤−10 mV, I_{OUT}≤8 mA		0.23	0.4	V
Positive Supply Current	T_A=25°C		5.1	7.5	mA
Negative Supply Current	T_A=25°C		4.1	5.0	mA

Note 12: "Absolute Maximum Ratings indicate limits beyond which damage to the device may occur. Operating Ratings indicate conditions for which the device is functional, but do not guarantee specific performance limits."

Note 13: This rating applies for ±15V supplies. The positive input voltage limit is 30V above the negative supply. The negative input voltage limit is equal to the negative supply voltage or 30V below the positive supply, whichever is less.

Note 14: The maximum junction temperature of the LM311 is 110°C. For operating at elevated temperature, devices in the H08 package must be derated based on a thermal resistance of 165°C/W, junction to ambient, or 20°C/W, junction to case. The thermal resistance of the dual-in-line package is 100°C/W, junction to ambient.

Note 15: These specifications apply for V_S=±15V and Pin 1 at ground, and 0°C < T_A < +70°C, unless otherwise specified. The offset voltage, offset current and bias current specifications apply for any supply voltage from a single 5V supply up to ±15V supplies.

Note 16: The offset voltages and offset currents given are the maximum values required to drive the output within a volt of either supply with 1 mA load. Thus, these parameters define an error band and take into account the worst-case effects of voltage gain and R_S.

Note 17: The response time specified (see definitions) is for a 100 mV input step with 5 mV overdrive.

Note 18: This specification gives the range of current which must be drawn from the strobe pin to ensure the output is properly disabled. Do not short the strobe pin to ground; it should be current driven at 3 to 5 mA.

Note 19: Human body model, 1.5 kΩ in series with 100 pF.

b. LM311 data sheet

are acceptable for most applications. (For example, the FET input LF311 shows input offset and bias currents of about 1000 times less, with only a slight increase in worst-case input offset voltage.) The saturation voltage indicates exactly how low a "low-state" potential really is. For a typical TTL-type load, the LM311 low output will be no more than .4 V. Finally, note the current draw from the power supplies. Worst-case values are 7.5 mA and 5.0 mA from the positive and negative supplies, respectively. By comparison, low-power comparators are normally in the 1 mA range, whereas high-speed devices can range up toward 20 to 30 mA.

The LM311 may be configured to drive TTL or MOS logic circuits and loads referenced to ground, the positive supply or the negative supply. Finally, it can directly drive relays or lamps with its 50 mA output current capability.

The operation of the LM311 is as follows:

1. If $V_{in_+} > V_{in_-}$ the output goes to an open collector condition. Therefore, a pull-up resistor is required to establish the high output potential. The pull-up does not have to go back to the same supply as the LM311. This aids in interfacing with various logic levels. For a TTL interface, the pull-up resistor will be tied back to the +5 V logic supply.
2. If $V_{in_-} > V_{in_+}$ the output will be shorted through to the "comparator ground" pin. Normally this pin goes to ground, indicating a low logic level of 0 V. It may be tied to other potentials if need be.
3. The strobe pin affects the overall operation of the device. Normally, it is left open. If it is connected to ground through a current-limiting resistor, the output will go to the open-collector state regardless of the input levels. Normally, an LM311 is strobed with a logic pulse that turns on a switching transistor.

An LM311 based comparator circuit is shown in Figure 7.53. It runs off of ±15 V supplies so that the inputs are compatible with general op amp circuits. The output uses a +5 V pull-up source so that the output logic is TTL compatible. A small-signal transistor such as a 2N2222 is used to strobe the LM311. A logic low on the base of the transistor turns the transistor off, thus leaving the LM311 in normal operation mode. A high on the base will turn on the transistor, thus placing the LM311's output at a logic high.

Our final type of comparator circuit is the *window comparator*. The window comparator is used to determine if a particular signal is within an allowable range of levels. This circuit features two different threshold inputs, an *upper threshold*, and a *lower threshold*. Do not confuse these items with the similarly named levels associated with Schmitt Triggers. A block diagram of a window comparator is shown in Figure 7.54. This circuit is comprised of two separate comparators, with common inputs and outputs. As long as the input signal is between the upper and lower thresholds, the output of both comparators will be high, thus producing a logic high at the circuit output. If the input signal is greater than the upper threshold or less than the lower threshold (i.e., outside of the allowed window), then one of the comparators will short its output to comparator ground, producing a logic low. The other comparator will go to an open collector condition. The net result is that the circuit output will be a logic low. Because the window comparator requires two separate comparators, a dual comparator package such as the LM319, can prove to be convenient.

Figure 7.53

LM311 with strobe

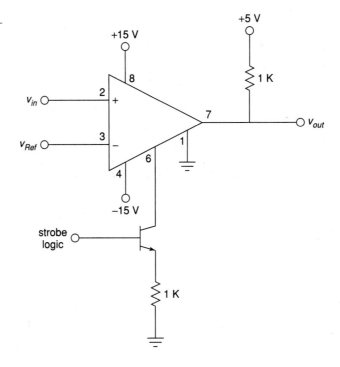

An interesting application that uses comparators and function generation is shown in Figure 7.55. This is a single neuron from a pulse-coded continuous-time recurrent neural network.[1] Neural networks are modeled after biological nervous systems and are used in a variety of applications such as robot motion control. Such networks may

Figure 7.54

Window comparator

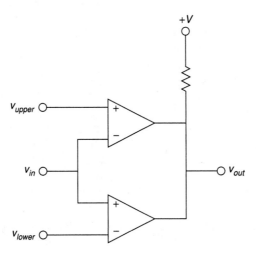

---

[1]For further details on pulse-coded continuous-time recurrent neural networks, see Alan Murray and Lionel Tarassenko, *Analogue Neural VLSI—A Pulse Stream Approach* (London: Chapman and Hall, 1994).

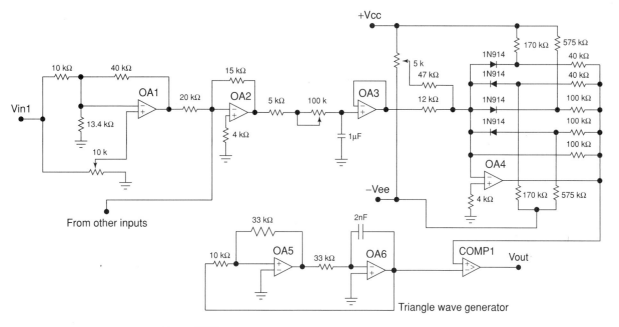

FIGURE 7.55

Pulse-coded continuous-time recurrent neural network (Single neuron shown)

be implemented using digital or analog techniques, with advantages and disadvantages in each approach. This particular system is largely analog, although its final output is a pulse-width modulated square wave. This avoids the signal attenuation problem that a pure analog approach might suffer from.

A complete system contains several neurons. Each neuron has a single output and several inputs. These inputs are fed from the outputs of the other neurons (including itself). The signals are weighted and then summed to create a composite excitation signal. The composite is summed with an offset or bias level, and then processed through a sigmoid function to produce the final output. In a pure analog scheme, the signals are just analog voltages. This scheme is a little different in that it uses pulse-width modulation to encode the signal levels. Low-level signals are represented as square waves with small-duty cycles. High-level signals are represented via large-duty cycles. One convenient aspect of this representation is that the signal is already in a form to directly control certain devices such as motors.

Let's follow the signal flow. The first stage is an adjustable gain inverting/non-inverting amplifier, as seen in Chapter 4. The gain of this stage corresponds to the neuron weighting. Because the weighting can be either positive or negative, a simple potentiometer by itself is insufficient. The second stage is a fairly stock summing amplifier (also from Chapter 4). Its job is to combine the signals from the various weighted inputs. Its output is a very complicated-looking waveform: It is a combination of pulse-width modulated square waves, each with a different duty cycle and amplitude. The average "area under the curve" represents the overall signal strength. This is obtained via the adjustable RC network. The time constant is much slower than the base square wave frequency, so an averaging of the signal takes place. The output of op amp 3 is a smooth, slowly varying signal. This signal is combined with an adjustable offset bias and fed into the function generation circuit built around op amp 4. The gains and breakpoints are designed to mimic the compressive nature of the sigmoid function $\frac{1}{1+\epsilon^{-x}}$. The resulting output level is then pulse coded. The pulse-width modulator is made from a simple triangle wave oscillator (covered in Chapter 9) and a comparator. The amplitude range of the triangle wave precisely matches the range of signals expected from the sigmoid circuit. The larger the sigmoid output is, the longer the comparator output will be high. In other words, the duty cycle will follow the sigmoid circuit's output level, creating a pulse-width modulated signal. This output signal will be fed to the inputs of the other neurons in the network, and may also be used as one of the final desired output signals. For example, this signal may be used to drive one of the leg motors of a walking robot.

7.6 Log and Antilog Amplifiers

It is possible to design circuits with logarithmic response. The uses for this are manyfold. Recalling your logarithm fundamentals, remember that processes such as multiplication and division turn into addition and subtraction for logs. Also, powers and roots turn into multiplication and division. Bearing this in mind, if an input signal is processed by a logarithmic circuit, multiplied by a gain, A, and then processed by an antilog circuit, the signal will have effectively been raised to the Ath power. This brings to light a great many possibilities. For example, in order to take the square root of a value, the signal would be passed through a log circuit, divided by two (perhaps with something as simple as a voltage divider), and then passed through an antilog circuit. One possible application is in true RMS detection circuits. Other applications for log/antilog amplifiers include signal compression and process control. Signals are often compressed in order to decrease their dynamic range (i.e., the difference between the highest and lowest level signals). In telecommunications systems, this may be required in order to achieve reasonable voice or data transmission with limited resources. Seeing their possible uses, our question then, is how do we design log/antilog circuits?

Figure 7.56 shows a basic log circuit. The input voltage is turned into an input current by R_i. This current feeds the transistor. Note that the output voltage appears across the base-emitter junction, V_{BE}. The transistor is being used as a current-to-voltage converter. The voltage/current characteristic of a transistor is logarithmic, thus the circuit produces a log response. In order to find an output equation, we start with the basic Shockley equation for PN junctions:

$$I_c = I_s \left(\epsilon^{\frac{qV_{BE}}{KT}} - 1 \right) \tag{7.8}$$

Where I_s is the reverse saturation current, ϵ is log base, q is the charge on one electron 1.6×10^{-19} Coulombs, K is Boltzmann's constant 1.38×10^{-23} Joules/Kelvin°, and T is the absolute temperature in Kelvin°.

FIGURE 7.56

Basic log circuit

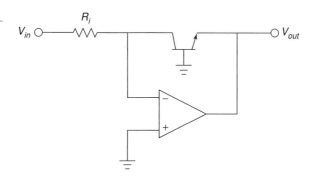

Using 300° K (approximately room temperature) and substituting these constants into Equation 7.8 produces

$$I_c = I_s(\epsilon^{38.6 V_{BE}} - 1) \tag{7.9}$$

Normally, the exponent term is much larger than one, so this may be approximated as

$$I_c = I_s \epsilon^{38.6 V_{BE}} \tag{7.10}$$

Using the inverse log relationship and solving for V_{BE}, this is reduced to

$$V_{BE} = .0259 \ln \frac{I_c}{I_s} \tag{7.11}$$

Earlier it was noted that the current I_c is a function of the input voltage and R_i. Also, note that $V_{out} = -V_{BE}$. Substituting these elements into Equation 7.11 yields

$$V_{out} = -.0259 \ln \frac{V_{in}}{R_i I_s} \tag{7.12}$$

We now have an amplifier that takes the log of the input voltage and also multiplies the result by a constant. It is very important that the antilog circuit multiply by the reciprocal of this constant, or errors will be introduced. If the input voltage (or current) is plotted against the output voltage, the result will be a straight line if plotted on a semilog graph, as shown in Figure 7.57. The rapid transition at approximately .6 V is due to the fast turn-on of the transistor's base-emitter junction. If output voltages greater than .6 V are required, amplifying stages will have to be added.

There are a couple of items to note about this circuit. First, the range of input signals is small. Large input currents will force the transistor into less-than-ideal logarithmic operation. Also, the input signal must be unipolar. Finally, the circuit is rather sensitive to temperature changes. Try using a slightly different value for T in the above derivation, and you shall see a sizable change in the resulting constant.

FIGURE 7.57

Input/output characteristic of log amplifier

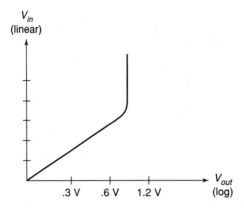

A basic antilog amplifier is shown in Figure 7.58. Note that the transistor is used to turn the input voltage into an input current, with a log function. This current then feeds R_f, which produces the output voltage. The derivation of the input/output equation is similar to the log circuit's:

$$V_{out} = -R_f I_c$$

Recalling Equation 7.10,

$$I_c = I_s \epsilon^{38.6 V_{BE}}$$
$$V_{out} = -R_f I_s \epsilon^{38.6 V_{BE}} \tag{7.13}$$

The same comments about stability and signal limits apply to both the antilog amplifier and the log amplifier. Also, there is one interesting observation worth remembering: in a general sense, the response of the op amp system echoes the characteristics of the elements used in the feedback network. When just resistors are used, which have a linear relation between voltage and current, the resulting amplifier exhibits a linear response. If a logarithmic PN junction is used, the result is an amplifier with a log or antilog response.

FIGURE 7.58

Basic antilog circuit

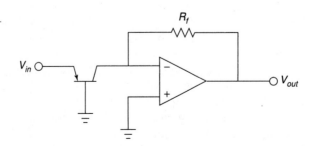

Example 7-8

Determine the output voltage for the circuit of Figure 7.56 if $V_{in} = 1$ V, $R_i = 50$ kΩ, and $I_s = 30$ nA. Assume $T = 300°$ Kelvin. Also determine the output for inputs of .5 V and 2 V.

For $V_{in} = 1$ V

$$V_{out} = -.0259 \ln \frac{V_{in}}{R_i I_s}$$
$$V_{out} = -.0259 \ln \frac{1 \text{ V}}{50 \text{ k } 30 \text{ nA}}$$
$$V_{out} = -.0259 \ln 666.6$$
$$V_{out} = -.1684 \text{ V}$$

For $V_{in} = .5$ V

$$V_{out} = -.0259 \ln \frac{.5 \text{ V}}{50 \text{ k } 30 \text{ nA}}$$
$$V_{out} = -.0259 \ln 333.3$$
$$V_{out} = -.1504 \text{ V}$$

For $V_{in} = 2$ V

$$V_{out} = -.0259 \ln \frac{2 \text{ V}}{50 \text{ k } 30 \text{ nA}}$$
$$V_{out} = -.0259 \ln 1333$$
$$V_{out} = -.1864 \text{ V}$$

Notice that for each doubling of the input signal, the output signal went up a constant 18 mV. Such is the nature of a log amplifier.

As noted, the basic log/antilog forms presented have their share of problems. It is possible to create more elaborate and accurate designs, but generally, circuit design of this type is not for the fainthearted, as the device-matching and temperature-tracking considerations are not minor.

Some manufacturers supply relatively simple-to-use log/antilog circuits in IC form. This takes much of the grind out of log circuit design. An example device is the 4127 from Burr-Brown. An equivalent circuit for the 4127 is shown in Figure 7.59. This device is a hybrid IC. The circuit uses matched transistors to increase the circuit accuracy. A recovery amplifier and a Zener regulated current source round out the package. The 4127 accepts bipolar input signals and has an input current dynamic range of $10^6 : 1$. The device can be configured to act as either a log or an antilog amplifier with the addition of just a few external resistors. Other circuits are also possible. Further details on the design and application of higher quality log amplifiers (such as the 4127) may be found in Section 7.7, Extended Topic.

Figure 7.59

4127 equivalent circuit

Reprinted, in whole or in part, with the permission of Burr-Brown Corporation

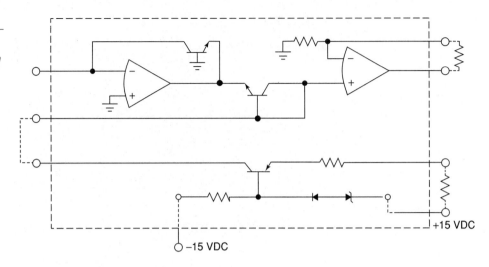

Four-Quadrant Multiplier

A *four-quadrant multiplier* is a device with two inputs and a single output. The output potential is the product of the two inputs along with a scaling factor, K.

$$V_{out} = K V_x V_y \qquad (7.14)$$

Typically, K is .1 in order to minimize the possibility of output overload. The schematic symbol for the multiplier is shown in Figure 7.60. It is called a four-quadrant device, as both inputs and the output may be positive or negative. An example device is the Motorola MC1494. The 1494 utilizes monolithic construction and can be powered by standard ±15 V power supplies. Although multipliers are not really "nonlinear" in and of themselves, they can be used in a variety of out-of-the-ordinary applications and in areas where log amps might be used.

Multipliers have many uses including squaring, dividing, balanced modulation/demodulation, frequency modulation, amplitude modulation, and automatic gain control. The most basic operation, multiplication, involves using one input as the signal input and the other input as the gain control potential. Unlike the simple VCA circuit discussed earlier, this gain control potential is allowed to swing both positive and negative. A negative polarity will produce an inverted output. This basic connection is shown in Figure 7.61. This same circuit can be used as a balanced modulator. This is very useful for creating dual sideband signals for communications work. Multiplier ICs may have external connections for scale factor and offset adjust potentiometers. Also, they may be modeled as current sources, so an external op amp connected as a current-to-voltage converter may be required.

Figure 7.60

Schematic symbol for multiplier

FIGURE 7.61

Multiplication circuit

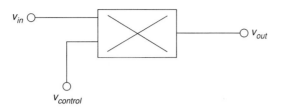

If the two inputs are tied together and fed from a single input, the result will be a squaring circuit. Squaring circuits can be very useful for RMS calculations and for frequency doubling. An example is shown in Figure 7.62.

The multiplier may also be used for division or square root functions. A divider circuit is shown in Figure 7.63. Here's how it works: First, note that the output of the multiplier, V_m, is a function of V_x and the output of the op amp, V_{out}.

$$V_m = K V_x V_{out} \tag{7.15}$$

V_m produces a current through R_2. Note that the bottom end of R_2 is at a virtual ground, so that all of V_m drops across R_2.

$$I_2 = \frac{V_m}{R_2}$$

As we have already seen,

$$I_1 = \frac{V_{in}}{R_1}$$

Now, because the current into the op amp is assumed to be zero, I_1 and I_2 must be equal and opposite. It is apparent that the arbitrary direction of I_2 is negative. Therefore, if R_1 is set equal to R_2,

$V_m = -V_{in}$ and using Equation 7.15,
$-V_{in} = K V_x V_{out}$
$$V_{out} = -\frac{V_{in}}{K V_x} \tag{7.16}$$

V_x then sets the magnitude of the division. There are definite size restrictions on V_x. If it is too small, output saturation will result. In a similar vein, if V_x is also tied back

FIGURE 7.62

Squaring circuit

FIGURE 7.63

Divider circuit

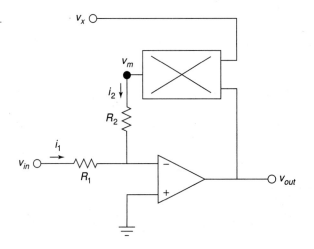

to the op amp's output, a square root function will result. Picking up the derivation at the next-to-last step,

$$-V_{in} = K V_{out} V_{out}$$

$$V_{out} = \sqrt{\frac{-V_{in}}{K}} \tag{7.17}$$

EXAMPLE 7-9

Determine the output voltage in Figure 7.62 if $K = .1$ and the input signal is $2 \sin 2\pi 60t$.

$$V_{out} = K V_x V_y$$

Because both inputs are tied together, this reduces to

$$V_{out} = K V_{in}^2$$
$$V_{out} = .1(2 \sin 2\pi 60t)^2 \tag{7.18}$$

A basic trig identity is

$$(\sin \omega)^2 = .5 - .5 \cos 2\omega$$

Substituting this into Equation 7.18,

$$V_{out} = .1(2 - 2 \cos 2\pi 120t)$$
$$V_{out} = .2 - .2 \cos 2\pi 120t$$

This means that the output signal is .2 V peak, and riding on a .2 V DC offset. It also indicates a phase shift of −90 degrees and a doubling of the input frequency. This last attribute makes this connection very useful. We have seen many approaches to increasing a signal's amplitude, but this circuit allows us to increase its frequency. The waveforms are shown in Figure 7.64.

FIGURE 7.64

Waveforms of the squaring circuit for Example 7-9

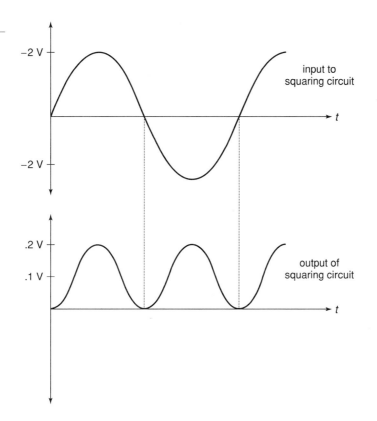

7.7 EXTENDED TOPIC: A PRECISION LOG AMP

The basic log amplifier discussed earlier suffers from two major problems: repeatability and temperature sensitivity. Let's take a closer look at Equation 7.12. First, if you work backward through the derivation, you will notice that the constant .0259 is actually $\frac{KT}{q}$ using a temperature of 300° K. We may rewrite this equation as

$$V_{out} = -\frac{KT}{q} \ln \frac{V_{in}}{R_i I_s} \tag{7.19}$$

Note that the output voltage is directly proportional to the circuit temperature, T. Normally, this is not desired. The second item of interest is I_s. This current can vary considerably between devices and is also temperature sensitive, approximately doubling for each 10 C° rise. For these reasons, commercial log amps such as the 4127 mentioned earlier are preferred. This does not mean that it is impossible to create stable log amps from the basic op amp building blocks. To the contrary, a closer look at practical circuit solutions will point out why ICs such as the 4127 are successful. We shall treat the two problems separately.

One way of removing the effect of I_s from our log amp is to subtract away an equal effect. The circuit of Figure 7.65 does exactly this. This circuit utilizes two log amps.

Figure 7.65

A high-quality log amplifier

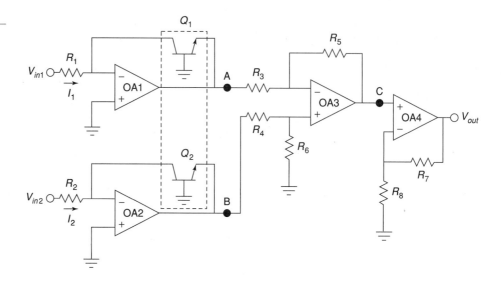

Although each is drawn with an input resistor and input voltage source, removal of R_1 and R_2 would allow current sensing inputs. Each log amp's transistor is part of a *matched pair*. These two transistors are fabricated on the same silicon wafer and exhibit nearly identical characteristics (in our case, the most notable being I_s). Based on Equation 7.19 we find

$$V_A = -\frac{KT}{q} \ln \frac{V_{in1}}{R_1 I_{s1}}$$

$$V_B = -\frac{KT}{q} \ln \frac{V_{in2}}{R_2 I_{s2}}$$

Where typically, $R_1 = R_2$. These two signals are fed into a differential amplifier comprised of op amp 3, and resistors R_3 through R_6. It is normally set for a gain of unity. The output of the differential amplifier (point C) is

$$V_C = V_B - V_A$$

$$V_C = -\frac{KT}{q} \ln \frac{V_{in2}}{R_2 I_{s2}} - \left(-\frac{KT}{q} \ln \frac{V_{in1}}{R_1 I_{s1}}\right)$$

$$V_C = \frac{KT}{q}\left(\ln \frac{V_{in1}}{R_1 I_{s1}} - \ln \frac{V_{in2}}{R_2 I_{s2}}\right) \qquad (7.20)$$

Using the basic identity that subtracting logs is the same as dividing their arguments, 7.20 becomes

$$V_C = \frac{KT}{q} \ln \frac{\frac{V_{in1}}{R_1 I_{s1}}}{\frac{V_{in2}}{R_2 I_{s2}}} \qquad (7.21)$$

Because R_1 is normally set equal to R_2, and I_{s1} and I_{s2} are identical due to the fact that Q_1 and Q_2 are matched devices, 7.21 simplifies to

EXTENDED TOPIC: A PRECISION LOG AMP

$$V_C = \frac{KT}{q} \ln \frac{V_{in1}}{V_{in2}} \tag{7.22}$$

As you can see, the effect of I_s has been removed. V_C is a function of the ratio of the two inputs. Therefore, this circuit is called a *log ratio amplifier*. The only remaining effect is that of temperature variation. Op amp 4 is used to compensate for this. This stage is little more than a standard SP noninverting amplifier. What makes it unique is that R_8 is a temperature-sensitive resistor. This component has a positive temperature coefficient of resistance, meaning that as temperature rises, so does its resistance. Because the gain of this stage is $1 + \frac{R_7}{R_8}$, a temperature rise causes a decrease in gain. Combining this with 7.22 produces

$$V_{out} = \left(1 + \frac{R_7}{R_8}\right) \frac{KT}{q} \ln \frac{V_{in1}}{V_{in2}} \tag{7.23}$$

If the temperature coefficient of R_8 is chosen correctly, the first two temperature dependent terms of 7.23 will cancel, leaving a temperature stable circuit. This coefficient is approximately $\frac{1}{300° K}$, or .33% per $C°$, in the vicinity of room temperature.

Our log ratio circuit is still not complete. Although the major stability problems have been eliminated, other problems do exist. It is important to note that the transistor used in the feedback loop loads the op amp, just as the ordinary feedback resistor R_f would. The difference lies in the fact that the effective resistance, which the op amp sees, is the dynamic base-emitter resistance, r'_e. This resistance varies with the current passing through the transistor and has been found to equal $\frac{26 \text{ mV}}{I_E}$ at room temperature. For higher input currents this resistance can be very small and might lead to an overload condition. A current of 1 mA, for example, would produce an effective load of only 26 Ω. This problem can be alleviated by inserting a large value resistance in the feedback loop. This resistor, labeled R_E, is shown in Figure 7.66. An appropriate value for R_E may be found by realizing that at saturation, virtually all of the output potential will be dropped across R_E, save V_{BE}. The current through R_E is the maximum expected input current plus load current. Using Ohm's law,

$$R_E = \frac{V_{sat} - V_{BE}}{I_{max}} \tag{7.24}$$

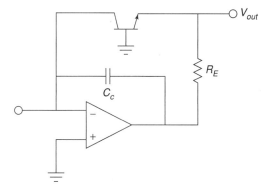

FIGURE 7.66

Compensation components

FIGURE 7.67

Light transmission measurement

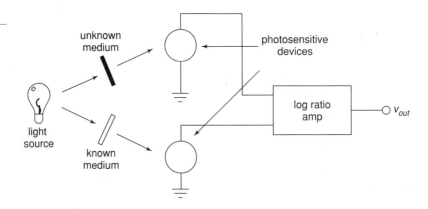

Figure 7.66 also shows a compensation capacitor, C_c. This capacitor is used to roll off high-frequency gain in order to suppress possible high-frequency oscillations. An optimum value for C_c is not easy to determine, as the feedback element's resistance changes with the input level. It may be found empirically in the laboratory. A typical value would be in the vicinity of 100 pF.[2]

One possible application for the log ratio circuit is found in Figure 7.67. This system is used to measure the light transmission of a given material. Because variations in the light source will affect a direct measurement, the measurement is instead made *relative to a known material*. In this example, light is passed through a known medium (such as a vacuum) while it is simultaneously passed through the medium under test. On the far side of both materials are light-sensitive devices such as photodioides, phototransistors, or photomultiplier tubes. These devices will produce a current proportional to the amount of light striking them. In this system, these currents are fed into the log ratio circuit, which will then produce an output voltage proportional to the light transmission abilities of the new material. This setup eliminates the problem of light source fluctuation, because each input will see an equal percentage change in light intensity. Effectively, this is a common-mode signal that is suppressed by the differential amplifier section. This system also eliminates the difficulty of generating calibrated light intensity readings for comparisons. By its very nature, this system performs relative readings.

SUMMARY

In this chapter, you have examined a variety of nonlinear op amp circuits. They are deemed nonlinear because their input/output characteristic is not a straight line.

Precision rectifiers use negative feedback to compensate for the forward drop of diodes. By doing so, very small signals may be successfully rectified. Both half-wave and full-wave versions are available. An extension of the precision rectifier is the precision peak detector. This circuit uses a capacitor to effectively lengthen the duration of peaks. It can also be used as an envelope detector.

Other active diode circuits include clampers and limiters. Clampers adjust the DC level of an input signal so that the output is either always positive or always negative. The furthest peaks will just hit 0 V. Clampers may also be used with a DC offset for greater control of the output signal. Limiters force their output to be no greater than a

---

[2] A detailed derivation for C_c may be found in Daniel H. Sheingold, ed., *Nonlinear Circuits Handbook*, 2nd ed., (Norwood, Mass.: Analog Devices, 1976), 174–178.

preset value. It may be thought of as a programmable signal clipper.

Function generation circuits are used to create piecewise linear approximations of transfer functions. These can be used to compensate for the nonlinearities of transducers and measurement devices. There are two basic approaches for realizing a design: (1) using Zener diodes, and (2) using a biased-signal diode network. The Zener scheme is a bit easier to set up, but does not offer the performance of the biased-diode form.

In order to improve the performance of comparators, positive feedback may be used. The resulting circuits are called Schmitt Triggers and have much greater noise immunity than simple open-loop op amp comparators do.

Specialized comparators are also available. These devices are generally faster than simple op amps, and offer more flexibility for logic family interfacing.

Log and antilog amplifiers utilize a transistor in their feedback loops. The result is a form of signal compression for the log amp, and an inverse signal expansion for the antilog amp. Specialized log/antilog ICs are available that take much of the tedium out of amplifier design and testing.

Finally, a four-quadrant multiplier produces an output that is proportional to the product of its two inputs. These devices may be used for a number of applications, including balanced modulation, frequency doubling, and more.

In short, although op amps are linear devices, they may be used with other components to form nonlinear circuits.

REVIEW QUESTIONS

1. What are the advantages of active rectifiers versus passive rectifiers?
2. What are the disadvantages of active rectifiers versus passive rectifiers?
3. What is a peak detector?
4. What is a limiter?
5. What is the function of a clamper?
6. What are the differences between active and passive clampers?
7. What is a transfer function generator circuit, and what is its use?
8. Explain how a function circuit might be used to linearize an input transducer.
9. What is a Schmitt trigger?
10. What is the advantage of a Schmitt-type comparator versus an ordinary open-loop op amp comparator?
11. What are the advantages of dedicated comparators such as the LM311 versus ordinary op amp comparators?
12. How are log and antilog amplifiers formed?
13. What is the effect of passing a signal through a log or antilog amplifier?
14. What is a four-quadrant multiplier?
15. How might a multiplier be used?
16. What is the difference between a multiplier and a VCA?

PROBLEMS

Analysis Problems

1. Sketch the output of Figure 7.2 if the input is a 100 Hz 2 V peak triangle wave.
2. Repeat Problem 1 for a square wave input.
3. Sketch the output of Figure 7.8 if $R_f = 20$ kΩ and $R_i = 5$ kΩ. Assume that V_{in} is a .5 V peak triangle wave.
4. Repeat Problem 3, but with the diode reversed.
5. Sketch the output of Figure 7.9 if $C = 50$ nF and $R = 5$ MΩ. Assume that the input is a 1 V peak pulse with 10% duty cycle. The input frequency is 1 kHz.
6. Repeat Problem 5 with the diode reversed.
7. Sketch the output of Figure 7.14 if $R = 25$ kΩ and $V_{in} = 2 \sin 2\pi 500t$.
8. Repeat Problem 7 with V_{in} equal to a 3 V peak square wave.
9. Sketch the output of Figure 7.15 if the diodes are reversed. Assume that V_{in} is a 100 mV peak-to-peak sine wave at 220 Hz.
10. In Figure 7.23, assume that $R = 10$ kΩ, $R_l = 1$ MΩ, $C = 100$ nF and $V_{offset} = 2$ V. Sketch the output for a 10 kHz 1 V peak sine wave input.
11. Sketch the output of Figure 7.27 if $R_i = 10$ kΩ, $R_f = 40$ kΩ and the Zener potential is 3.9 V. The input signal is a 2 V peak sine wave.
12. Sketch the transfer curve for the circuit of Problem 11.
13. In Figure 7.31, sketch the transfer curve if $R_i = 5$ kΩ, $R_f = 33$ kΩ, $R_a = 20$ kΩ and $V_{Zener} = 5.7$ V.
14. Sketch the output of the circuit in Problem 13 for an input signal equal to a 2 V peak triangle wave.

15. Repeat Problem 14 with a square wave input.
16. If $R_i = 4$ kΩ, $R_a = 5$ kΩ, $R_f = 10$ kΩ, and $V_z = 2.2$ V in Figure 7.33, determine the minimum and maximum input impedance.
17. Draw the transfer curve for the circuit of Problem 16.
18. Sketch the output voltage for the circuit of Problem 16 if the input signal is a 3 V peak triangle wave.
19. Sketch the transfer curve for the circuit of Figure 7.34 if $R_i = 1$ kΩ, $R_f = 10$ kΩ, $R_a = 20$ kΩ, $R_b = 18$ kΩ, $V_{Zener-a} = 3.9$ V, and $V_{Zener-b} = 5.7$ V.
20. Sketch the output of the circuit in Problem 19 if the input is a 1 V peak triangle wave.
21. If $R_1 = 10$ kΩ and $R_2 = 33$ kΩ in Figure 7.47, determine the upper and lower thresholds if the power supplies are ± 15 V.
22. Determine the upper and lower thresholds for Figure 7.48 if $R_1 = 4.7$ kΩ and $R_2 = 2.2$ kΩ, with ± 12 V power supplies.
23. Sketch the output of Figure 7.53 if $V_{in} = 2 \sin 2\pi 660 t$, $V_{ref} = 1$ V DC, and $V_{strobe} = 5$ V DC.
24. Repeat Problem 23 for $V_{strobe} = 0$ V DC.
25. Determine the output of Figure 7.56 if $V_{in} = .1$ V, $R_i = 100$ kΩ, and $I_s = 60$ nA.
26. Determine the output of Figure 7.58 if $V_{in} = 300$ mV, $R_f = 20$ kΩ and $I_s = 40$ nA.
27. Sketch the output signal of Figure 7.61 if $V_{in} = 2 \sin 2\pi 1000 t$, $K = .1$, and $V_{control} = 1$ V DC, -2 V DC, and 5 V DC.
28. Sketch the output of Figure 7.62 if $V_{in} = 1 \sin 2\pi 500 t$, and $K = .1$.
29. Sketch the output of Figure 7.63 if $V_{in} = 5 \sin 2\pi 2000 t$, $K = .1$, $V_x = 4$ V DC, and $R_1 = R_2 = 10$ kΩ.

Design Problems

30. Determine values of R and C in Figure 7.9 so that stage 1 slewing is at least 1 V/μS, along with a time constant of 1 mS. Assume that op amp 1 can produce at least 20 mA of current.
31. In Figure 7.21, assume that $C = 100$ nF. Determine an appropriate value for R_i if the input signal is at least 2 kHz. Use a time constant factor of 100 for your calculations.
32. Determine new values for R_a and R_f in Problem 13, so that the slopes are -5 and -3.
33. Determine new resistor values for the circuit of Figure 7.33 so that the slopes are -10 and -20. The input impedance should be at least 3 kΩ.
34. Determine the resistor values required in Figure 7.37 to produce slopes of -5, -8, and -12, if the input impedance must be at least 1 kΩ.
35. Determine the resistor values required in Figure 7.42 for $S_0 = -1$, $S_1 = S_{-1} = -5$, $S_2 = S_{-2} = -12$. Also, $V_1 = |V_{-1}| = 3$ V, and $V_2 = |V_{-2}| = 4$ V. Use $R_f = 20$ kΩ.
36. Sketch the transfer curve for Problem 35.
37. Determine a value for R_i in Figure 7.56 so that a 3 V input will produce an output of .5 V. Assume $I_s = 60$ nA.
38. Determine a value for R_f in Figure 7.58 such that a .5 V input will produce a 3 V output. Assume $I_s = 40$ nA.
39. If the multiplier of Figure 7.63 can produce a maximum current of 10 mA, what should the minimum sizes of R_1 and R_2 be? (Assume ± 15 V supplies.)

Challenge Problems

40. Assuming that 5% resistors are used in Figure 7.14, determine the worst-case mismatch between the two halves of the rectified signal.
41. Design a circuit that will light an LED if the input signal is beyond ± 5 V peak. Make sure that you include some form of pulse-stretching element so that the LED remains visible for short-duration peaks.
42. A pressure transducer produces an output of 500 mV per atmosphere up to 5 atmospheres. From 5 to 10 atmospheres, the output is 450 mV per atmosphere. Above 10 atmospheres, the output falls off to 400 mV per atmosphere. Design a circuit to linearize this response using the Zener form.
43. Repeat the problem above using the biased-diode form.
44. Design a circuit to produce the transfer characteristic shown in Figure 7.68.

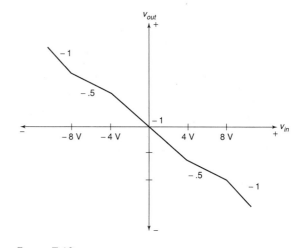

FIGURE 7.68

45. Sketch the output waveform of Figure 7.48 if $R_1 = 22$ kΩ, $R_2 = 4.7$ kΩ, and $V_{in} = 2 + 8\sin 2\pi 120t$. Assume that the power supplies are ± 15 V.
46. Determine the output of Figure 7.60 if $K = .1$, $V_x = 1\sin 2\pi 440t$, and $V_y = 3\sin 2\pi 200000t$.

Computer Simulation Problems

47. In preceding work it was noted that the use of an inappropriate device model can produce computer simulations that are way off mark. An easy way to see this is to simulate the circuit of Figure 7.2 using an accurate model (such as the 741 simulation presented in the chapter) and a simple model (such as the controlled-voltage source version presented in Chapter 2). Run two simulations of this circuit for each of these models. One simulation should use a lower-input frequency, such as 1 kHz, and the second simulation should use a higher frequency where the differences in the models is very apparent, such as 500 kHz.
48. Simulate the circuit designed in Problem 30 using a square wave input. Perform the simulation for several different input frequencies between 10 Hz and 10 kHz. Do the resulting waveforms exhibit the proper shapes?
49. Perform simulations for the circuit of Figure 7.14. Use $R = 12$ kΩ. For the input waveform, use both sine and square waves, each being 1 V peak at 200 Hz.
50. Component tolerances can directly affect the rectification accuracy of the full-wave circuit shown in Figure 7.14. This effect is easiest to see using a square wave input. If the positive and negative portions of the input signal see identical gains, the output of the circuit will be a DC level. Any inaccuracy or mismatch will produce a small square wave riding on this DC level. This effect can be simulated using the Monte Carlo analysis option. If Monte Carlo analysis is not available on your system, you can still see the effect by manually altering the resistor values within a preset tolerance band for each of a series of simulation runs.
51. Use a simulator to verify the results of the limiter examined in Example 7.3.
52. Use a simulator to verify the response of the circuit shown in Figure 7.37. For the stimulus, use a low-frequency triangle wave of 4 V peak amplitude. Does the resulting waveform exhibit the same hard "corners" shown on the transfer curve of Figure 7.37?

CHAPTER 8

Voltage Regulation

CHAPTER OBJECTIVES

After completing this chapter, you should be able to:

- Understand the need and usefulness of voltage regulators.
- Detail the difference between load and line regulation.
- Detail the need for a stable reference voltage.
- Describe the operation of a basic linear regulator.
- Describe the operation of a basic switching regulator.
- Outline the differences between linear and switching regulators, giving the advantages and disadvantages of each.
- Detail the function of a pass transistor in both linear and switching regulators.
- Utilize many of the popular linear IC regulators, such as the 78XX and 723.
- Analyze system heat sink requirements.

8.1 Introduction

In this chapter, the need for voltage regulation is examined. The different schemes for achieving voltage regulation and the typical ICs used are presented. By the end of this chapter, you should be able to use standard voltage regulator devices in your work and understand the advantages and disadvantages of the various types.

Generally, voltage regulators are used to keep power supply potentials constant in spite of changes in load current or source voltage. Without regulation, some circuits may be damaged by the fluctuations present in the power supply voltage. Even if devices are not damaged, the fluctuations may degrade circuit performance.

Two general forms are presented, the *linear regulator* and the *switching regulator*. Both forms have distinct advantages and drawbacks. The trick is to figure out which form works best in a given situation. Due to their wide use in modern power supply design, many regulators have achieved high levels of internal sophistication and robust performance. Often, the inclusion of voltage regulator ICs is a very straightforward—almost cookbook—affair.

The chapter concludes with a discussion of heat sink theory and application. As regulators are called upon to dissipate a bit of power, attention to thermal considerations is an important part of the design process.

8.2 THE NEED FOR REGULATION

Modern electronic circuits require stable power supply voltages. Without stable potentials, circuit performance may degrade, or if variations are large enough, the circuit may cease to function altogether, and various components may be destroyed. There are many reasons why a power supply may fluctuate. No matter what the cause, it is the job of the regulator to absorb the fluctuations, and thereby maintain constant operating potentials.

A basic power supply circuit is shown in Figure 8.1. First, a transformer is used to isolate the circuit from the AC power source. It is also used to step down (or step up) the AC source potential to a more manageable level. A rectifier then converts the scaled AC signal to pulsating DC, in the form of either half-wave or, more typically, full-wave rectification. The variations in the pulsating DC signal are then filtered out in order to produce a (hopefully) stable DC potential, which feeds the load. The filter may take the form of a simple capacitor or, possibly, more complex networks comprising both capacitors and inductors.

There are two main causes of power supply output variation. First, if the AC source signal changes, a proportional change will be seen at the output. If, for example, a *brownout*[1] occurs, the nominal 120 V AC source used in the US, may drop to, say, 100 V AC. This represents a decrease of one sixth, or about 16.7%. This same change will be reflected by the transformer, so the output of the transformer will be about 16.7% low as well. This reduction carries right through the rectifier and filter to the load. In some circuits, this will not present a major problem. For example, this may mean that an op amp will be running off a 12 V supply instead of a 15 V supply. On the other hand, a TTL logic circuit may not operate correctly if only 4 V instead of 5 V are applied. Of course, if an overvoltage occurred, both the op amp and the logic circuit could be damaged.

FIGURE 8.1

Basic power supply

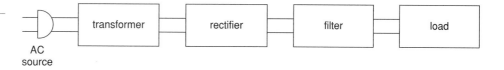

---

[1] A reduction in the standard supplied potential may be introduced by a power utility as a way of dealing with particularly heavy load conditions, such as on a very hot summer day when numerous air conditioners are running.

How well a circuit handles variations in the AC line signal is denoted by the parameter, *line regulation*.

$$\text{Line Regulation} = \frac{V_{max} - V_{min}}{V_{min}} \times 100\% \tag{8.1}$$

where V_{max} is the load voltage produced at the maximum AC line potential and V_{min} is the load voltage for the lowest AC line potential. Normally, this value is expressed as a percentage. Ideally, a power supply will always produce the same output potential, and therefore, the perfect line regulation figure would be 0%.

The other major source of supply variation is variation of load current. Indeed, load current variations can be far greater than the usual variations seen in the AC line. This normally has the effect of increasing or decreasing the amount of AC *ripple* seen in the output voltage. Ideally, there would be no ripple in the output. Ripple is caused by the fact that heavy load currents effectively reduce the discharge time constant of the filter. The result is that the filter gives up its stored energy faster and cannot successfully "fill in the gaps" of the pulsating DC signal fed to it. The ripple signal is effectively an AC signal that rides on the DC output. This rapid variation of supply potential can find its way into an audio or signal processing path and create a great deal of interference. Along with this variation, the effective DC value of the supply may drop as the load current is increased. The figure of merit for stability in spite of load changes is called *load regulation*.

$$\text{Load Regulation} = \frac{V_{max} - V_{min}}{V_{min}} \times 100\% \tag{8.2}$$

where V_{max} is the largest load voltage produced, and V_{min} is the minimum load voltage produced. These points usually occur at the minimum and maximum load currents, respectively. Again, this value is normally expressed as a percentage and would ideally be 0%.

In order to maintain a constant output voltage, the power supply regulator needs to sense its output and then compensate for any irregularities. This implies a number of things. First, some form of reference is needed for stable comparison to the output signal. Second, some form of comparator or amplifier is required in order to make use of this comparison. Finally, some form of control element is needed to absorb the difference between the input to the regulator circuit and the desired output. This control element may either appear in series or in parallel with the load, as seen in Figure 8.2. The first form is shown in Figure 8.2a and is normally referred to as a *series mode regulator*. The control element allows current to pass through to the load, but drops a specific amount of voltage. The voltage that appears across the control element is the difference between the filter's output and the desired load voltage.

Figure 8.2b shows a *shunt mode regulator*. Here, the control element is in parallel with the load and draws enough current to keep the output level constant. For many applications, shunt mode regulators are not as efficient as series mode regulators are. One example of a simple shunt mode regulator is the resistor/Zener diode arrangement shown in Figure 8.3. Note that under no-load conditions (i.e., when the load impedance is infinite) considerable current flows in the regulating circuit.

FIGURE 8.2

Two forms of regulator control elements: series and parallel

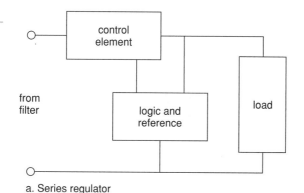

a. Series regulator

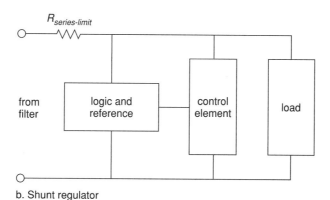

b. Shunt regulator

8.3 LINEAR REGULATORS

If a regulator's control element operates in its linear region, the regulator is said to be a *linear regulator*. Most linear regulator ICs are series mode types. The main advantages of linear regulators are their ease of use and accuracy of control. Their main disadvantage is low efficiency.

A basic linear regulator is shown in Figure 8.4. The series control element is the transistor Q_1. This component is most often referred to as a *pass transistor*, because it allows current to pass through to the load. The pass transistor is driven by the op amp. The pass transistor's job is to amplify the output current of the op amp. The op amp constantly monitors its two inputs. The noninverting input sees the Zener voltage of device D_1. R_1 is used to properly bias D_1. The op amp's inverting input sees the voltage drop across R_3. Note that R_2 and R_3 make up a voltage divider with V_{load} as the divider's source. Remembering the basics of negative feedback, recall that the op amp will produce enough current to keep its two inputs at approximately the same level, thus keeping V_{error} at zero. In other words, the voltage across R_3 should be equal to the Zener potential. As long as the op amp has a high enough output current capability, this equality will be maintained. Note that the V_{BE} drop produced by Q_1 is compensated for, as it is within the feedback loop. The difference between the filtered input signal and the final V_{load} is dropped across the collector/emitter of Q_1.

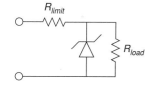

FIGURE 8.3

Simple zener shunt regulator

FIGURE 8.4

Basic op amp linear regulator

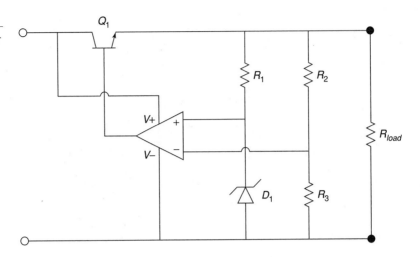

In essence, the circuit of Figure 8.4 is a noninverting amplifier. V_{zener} is the input potential, and R_2 and R_3 take the place of resistors R_f and R_i, respectively. V_{load} is found by using a variation of the basic gain formula:

$$V_{load} = V_{zener} \frac{R_2 + R_3}{R_3} \qquad (8.3)$$

By choosing the appropriate ratios of resistors along with a suitable Zener, a wide range of output potentials may be achieved. Note that the op amp is fed from the nonregulated input. Any ripple that exists on this line will hardly affect the op amp's function. The ripple will be reduced by the op amp's PSRR (power supply rejection ratio). The Zener bias resistor, R_1, is chosen so that it sets up a current that will guarantee Zener conduction. This is usually in the low milliamp range.

EXAMPLE 8-1

Determine the output of Figure 8.4 if $R_1 = 5$ kΩ, $R_2 = 20$ kΩ, $R_3 = 10$ kΩ and $V_{zener} = 3.9$ V. Assume that the filtered input is 20 V DC with no more than 3 V peak-to-peak ripple.

First, note that the input signal varies between a minimum of 18.5 V to a maximum of 21.5 V (20 V $\pm$1.5 V peak). As long as the circuit doesn't try to produce more than the minimum input voltage and as long as all current and power dissipation limits are obeyed, everything should function correctly.

$$V_{load} = V_{zener} \frac{R_2 + R_3}{R_3}$$
$$V_{load} = 3.9 \text{ V} \times \frac{20 \text{ k} + 10 \text{ k}}{10 \text{ k}}$$
$$V_{load} = 11.7 \text{ V}$$

Due to the V_{BE} drop, the op amp needs to produce about .7 V more than this. Also, note that the Zener current may now be found.

$$I_{zener} = \frac{V_{load} - V_{zener}}{R_1}$$
$$I_{zener} = \frac{11.7\text{ V} - 3.9\text{ V}}{5\text{ k}}$$
$$I_{zener} = 1.56\text{ mA}$$

This is a reasonable value.

EXAMPLE 8-2

Determine the power dissipation for Q_1 and the circuit efficiency for Example 8-1 if the effective load resistance is 20 Ω.

First, I_{load} must be determined.

$$I_{load} = \frac{V_{load}}{R_{load}}$$
$$I_{load} = \frac{11.7\text{ V}}{20}$$
$$I_{load} = .585\text{ A}$$

The dissipation of Q_1 is the product of the current passing through it and the voltage across it. The current through Q_1 is the load current. The voltage across Q_1 is the difference between the filter's output and the load voltage. Because the filter's output contains a relatively small AC signal riding on DC, the average will equal the DC value, in this case, 20 V.

$$V_{CE} = V_C - V_E$$
$$V_{CE} = 20\text{ V} - 11.7\text{ V}$$
$$V_{CE} = 8.3\text{ V}$$

$$P_D = I_C V_{CE}$$
$$P_D = .585\text{ A} \times 8.3\text{ V}$$
$$P_D = 4.86\text{ W}$$

So, the pass transistor must dissipate 4.86 W and tolerate a current of .585 amps. The maximum collector-emitter voltage will occur at the peak input of 21.5 V. Therefore, the maximum differential is 9.8 V.

Finally, note that a .585 amp output would require a minimum β of

$$\beta = \frac{I_C}{I_B}$$
$$\beta = \frac{585\text{ mA}}{20\text{ mA}}$$
$$\beta = 29.25$$

This assumes that the op amp can produce 20 mA and also ignores the small Zener and voltage divider currents at the output.

As far as efficiency is concerned, the input power and the load power need to be calculated. For load power,

$$P_{load} = I_{load} V_{load}$$
$$P_{load} = .585 \text{ A} \times 11.7 \text{ V}$$
$$P_{load} = 6.844 \text{ W}$$

Ignoring the current requirements of the op amp, Zener, and R_2/R_3 divider, the supplied current equals .585 A. The average input voltage is 20 V.

$$P_{in} = I_{in} V_{in}$$
$$P_{in} = .585 \text{ A} \times 20 \text{ V}$$
$$P_{in} = 11.7 \text{ W}$$

Efficiency, η, is defined as the ratio of useful output to required input, so,

$$\eta = \frac{P_{load}}{P_{in}}$$
$$\eta = \frac{6.844 \text{ W}}{11.7 \text{ W}}$$
$$\eta = .585 \text{ or } 58.5\%$$

Therefore, 41.5% of the input power is wasted. In order to minimize the waste and maximize the efficiency, the input/output differential voltage needs to be as small as possible. For proper operation of the op amp and pass transistor, this usually means a differential lower than 2 to 3 V is impossible. Consequently, this form of regulation is very inefficient when the power supply must produce low-output voltage levels.

multiSIM COMPUTER SIMULATION

An Electronics Workbench MultiSIM simulation of an op amp-based linear regulator, such as the one in Figure 8.4, is shown in Figure 8.5 (pp. 287–288). The input signal is 20 V average, with a 120 Hz 2 V peak sine wave riding on it representing ripple. A 741 op amp is used as the comparison element along with a generic transistor for the pass device. The reference is a 5.2 V Zener diode. Given the circuit values, a manual calculation shows

$$V_{out} = V_{zener}\left(1 + \frac{R_f}{R_i}\right)$$
$$V_{out} = 5.2 \text{ V} \times \left(1 + \frac{14 \text{ k}}{11 \text{ k}}\right)$$
$$V_{out} = 11.8 \text{ V}$$

The output plot indicates a constant output at approximately 12.6 V. The discrepancy can be attributed to the nonideal nature of the Zener diode (e.g., its internal resistance). Indeed, the initial solution indicates that the Zener potential (node 5) is approximately 5.5 V. A more accurate specification of the Zener diode (in particular, the BV and IBV parameters) would result in a much closer prediction. Another way of looking at this is to say that the 5.2 V Zener specification is too far below the actual Zener bias point for utmost accuracy (i.e., the manual calculation is somewhat sloppy and is, therefore, the one in error). This minor discrepancy aside, the time-domain plot does show the stability of the output signal in spite of the sizable input variations.

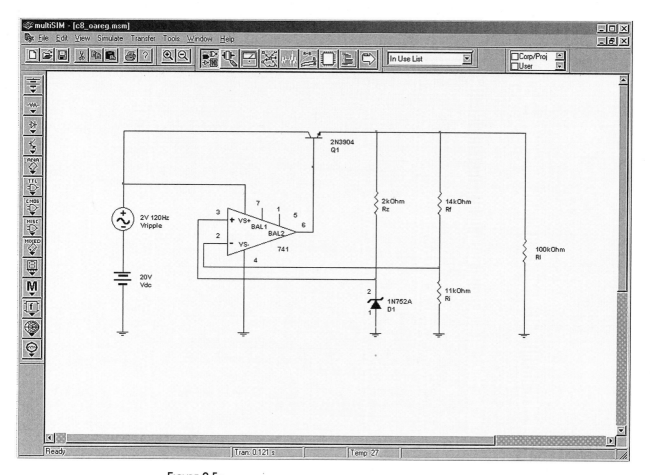

FIGURE 8.5a

Op amp regulator in MultiSIM

Three Terminal Devices

In an effort to make the designer's job ever easier, manufacturers provide circuits such as the one shown in Figure 8.4 in a single package. If you notice, the circuit really needs only three pins with which to connect to the outside world: input from the filter, ground, and output. These devices are commonly known as *three pin regulators*—hardly a creative name tag, but descriptive at least. Several different devices are available for varying current and power dissipation demands.

A typical "three pin" device family is the LM340-XX/LM78XX and LM360-XX/LM79XX. The LM340-XX/LM78XX series is for positive outputs, whereas the LM360-XX/LM79XX series produces negative outputs. The XX indicates the rated load voltage. For example, the LM340-05 is a +5 V regulator, and the LM7912 is a −12 V unit. The most popular sizes are the 5, 12, and 15 V units. For simplicity, the series will be referred to as the LM78XX from here on.

FIGURE 8.5b

Input and output waveforms from MultiSIM

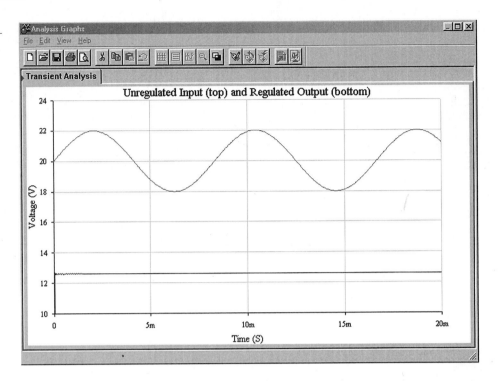

A data sheet for the LM78XX series is shown in Figure 8.6. This regulator comes in several variants including TO-3, TO-220, and surface-mount versions. The TO-3 case version offers somewhat higher power dissipation capability. Output currents in excess of 1 A are available. As a side note, for lighter loads with lower current demands, the LM78LXX may be used. This regulator offers a 100 mA output and comes in a variety of packages.

The practical power dissipation limit of the LM78XX series, like any power device, is highly dependent on the type of heat sink used. The device graphs show that it can be as high as 20 W with the TO-3 case. The typical output voltage is within approximately 5% of the nominal value. This indicates the inherent accuracy of the IC, and is not a measure of its regulation abilities. Figures for both load and line regulation are given in the data sheet. For a variance in the line voltage of over 2:1, we can see that the output voltage varies by no more than 1% of the nominal output. Worst-case load regulation is just as good, showing a 1% deviation for a load current change from 5 mA up to 1.5 A. The regulator can also be seen to draw very little standby current, only 8.5 mA over a wide temperature range.

The output of the LM78XX is fairly clean. The LM7815 shows an output noise voltage of 90 μV, typically, and an average ripple rejection of 70 dB. This means that ripple presented to the input of the regulator is reduced by 70 dB at the output. Because 70 dB translates to a factor of over 3000, this means that an input ripple signal of 1 V will be reduced to less than one-third of a millivolt at the output. It is worthwhile to note that this figure is frequency-dependent, as shown by the Ripple Rejection graph. Fortunately, the maximum value occurs in the desired range of 60 to 120 Hz.

FIGURE 8.6a

LM78XX data sheet

Absolute Maximum Ratings

* If Military/Aerospace specified devices are required, contact the National Semiconductor Sales Office/Distributors for availability and specifications.

Input Voltage (V_O = 5 V, 12 V and 15 V)	35 V
Internal Power Dissipation (Note 1)	Internally Limited
Operating Temperature Range (T_A)	0°C to +70°C
Maximum Junction Temperature	
(K Package)	150°C
(T Package)	150°C
Storage Temperature Range	−65°C to +150°C
Lead Temperature (Soldering, 10 sec.)	
TO-3 Package K	300°C
TO-220 Package T	230°C

Electrical Characteristics LM78XXC (Note 2) 0°C ≤ T_j ≤ 125°C unless otherwise noted.

			Output Voltage	5 V			12 V			15 V			
			Input Voltage (unless otherwise noted)	10 V			19 V			23 V			Units
Symbol	Parameter		Conditions	Min	Typ	Max	Min	Typ	Max	Min	Typ	Max	
V_O	Output Voltage		T_j = 25°C, 5 mA ≤ I_O ≤ 1A	4.8	5	5.2	11.5	12	12.5	14.4	15	15.6	V
			P_D ≤ 15 W, 5 mA ≤ I_O ≤ 1A	4.75		5.25	11.4		12.6	14.25		15.75	V
			V_{MIN} ≤ V_{IN} ≤ V_{MAX}	(7.5 ≤ V_{IN} ≤ 20)			(14.5 ≤ V_{IN} ≤ 27)			(17.5 ≤ V_{IN} ≤ 30)			V
ΔV_O	Line Regulation	I_O = 500 mA	T_j = 25°C		3	50		4	120		4	150	mV
			ΔV_{IN}	(7 ≤ V_{IN} ≤ 25)			(14.5 ≤ V_{IN} ≤ 30)			(17.5 ≤ V_{IN} ≤ 30)			V
			0°C ≤ T_j ≤ +125°C			50			120			150	mV
			ΔV_{IN}	(8 ≤ V_{IN} ≤ 20)			(15 ≤ V_{IN} ≤ 27)			(18.5 ≤ V_{IN} ≤ 30)			V
		I_O ≤ 1A	T_j = 25°C			50			120			150	mV
			ΔV_{IN}	(7.5 ≤ V_{IN} ≤ 20)			(14.6 ≤ V_{IN} ≤ 27)			(17.7 ≤ V_{IN} ≤ 30)			V
			0°C ≤ T_j ≤ +125°C			25			60			75	mV
			ΔV_{IN}	(8 ≤ V_{IN} ≤ 12)			(16 ≤ V_{IN} ≤ 22)			(20 ≤ V_{IN} ≤ 26)			V
ΔV_O	Load Regulation	T_j = 25°C	5 mA ≤ I_O ≤ 1.5 A		10	50		12	120		12	150	mV
			250 mA ≤ I_O ≤ 750 mA			25			60			75	mV
			5 mA ≤ I_O ≤ 1A, 0°C ≤ T_j ≤ +125°C			50			120			150	mV
I_Q	Quiescent Current	I_O ≤ 1A	T_j = 25°C			8			8			8	mA
			0°C ≤ T_j ≤ +125°C			8.5			8.5			8.5	mA
ΔI_Q	Quiescent Current Change	5 mA ≤ I_O ≤ 1A				0.5			0.5			0.5	mA
		T_j = 25°C, I_O ≤ 1A				1.0			1.0			1.0	mA
		V_{MIN} ≤ V_{IN} ≤ V_{MAX}		(7.5 ≤ V_{IN} ≤ 20)			(14.8 ≤ V_{IN} ≤ 27)			(17.9 ≤ V_{IN} ≤ 30)			V
		I_O ≤ 500 mA, 0°C ≤ T_j ≤ +125°C				1.0			1.0			1.0	mA
		V_{MIN} ≤ V_{IN} ≤ V_{MAX}		(7 ≤ V_{IN} ≤ 25)			(14.5 ≤ V_{IN} ≤ 30)			(17.5 ≤ V_{IN} ≤ 30)			V
V_N	Output Noise Voltage	T_A = 25°C, 10 Hz ≤ f ≤ 100 kHz			40			75			90		μV
$\Delta V_{IN} / \Delta V_{OUT}$	Ripple Rejection	f = 120 Hz { I_O ≤ 1A, T_j = 25°C or I_O ≤ 500 mA		62	80		55	72		54	70		dB
		0°C ≤ T_j ≤ +125°C		62			55			54			dB
		V_{MIN} ≤ V_{IN} ≤ V_{MAX}		(8 ≤ V_{IN} ≤ 18)			(15 ≤ V_{IN} ≤ 25)			(18.5 ≤ V_{IN} ≤ 28.5)			V
R_O	Dropout Voltage	T_j = 25°C, I_{OUT} = 1A			2.0			2.0			2.0		V
	Output Resistance	f = 1 kHz			8			18			19		mΩ
	Short-Circuit Current	T_j = 25°C			2.1			1.5			1.2		A
	Peak Output Current	T_j = 25°C			2.4			2.4			2.4		A
	Average TC of V_{OUT}	0°C ≤ T_j ≤ +125°C, I_O = 5 mA			0.6			1.5			1.8		mV/°C
V_{IN}	Input Voltage Required to Maintain Line Regulation	T_j = 25°C, I_O ≤ 1A			7.5			14.6			17.7		V

Note 1: Thermal resistance of the TO-3 package (K, KC) is typically 4°C/W junction to case and 35°C/W case to ambient. Thermal resistance of the TO-220 package (T) is typically 4°C/W junction to case and 50°C/W case to ambient.

Note 2: All characteristics are measured with capacitor across the input of 0.22 μF, and a capacitor across the output of 0.1 μF. All characteristics except noise voltage and ripple rejection ratio are measured using pulse techniques (t_w ≤ 10 ms, duty cycle ≤ 5%). Output voltage changes due to changes in internal temperature must be taken into account separately.

Figure 8.6b

(Continued)

Reprinted with permission of National Semiconductor Corporation

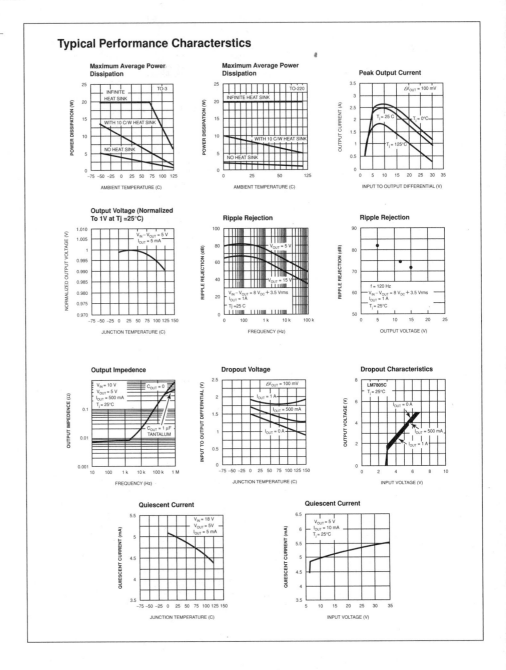

The last major item in the data sheet is an extremely important one: Input Voltage Required to Maintain Line Regulation. This value is approximately 2.5 to 2.7 V greater than the nominal rating of the regulator. Without this headroom, the regulator will cease to function properly. Although it is desirable to keep the input/output

FIGURE 8.7

Dual supply for op amp circuits

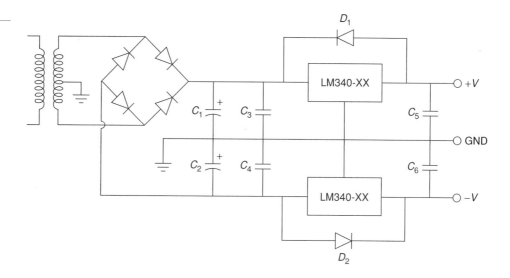

differential voltage low in order to decrease device power dissipation and maximize efficiency, too low of a value will cripple the regulator. As seen in the Peak Output Current graph, the highest load currents are achieved when the differential is in the 5 to 10 V range.

In all cases, the devices offer thermal shutdown and output short-circuit protection. The maximum input voltage is limited to 35 V. Besides the LM78XX series, special high-current types with outputs in the 3 A and higher range are available.

As an example, a bipolar power supply is shown in Figure 8.7. A center-tapped transformer is used to generate the required positive/negative polarities. C_1 and C_2 serve as the filter capacitors. The positive peak across each capacitor should not exceed 35 V or the regulators may be damaged. The minimum voltage across the capacitors should be at least 3 V greater than the desired output level. The exact size of the capacitor depends on the amount of load current expected. In a similar manner, the magnitude of the load current will determine which case style will be used. Capacitors C_3 and C_4 are .22 μF units and are only required if the regulators are located several inches or more from the filter capacitors. If their use is required, C_3 and C_4 must be located very close to the regulator. C_5 and C_6 are used to improve the transient response of the regulator. They are not used for filtering or for circuit stabilization. Through common misuse, C_5 and C_6 are often called "stability capacitors," although this is not their function. Finally, diodes D_1 and D_2 protect the regulators from output overvoltage conditions, such as those experienced with inductive loads.

As you can see, designing moderate fixed voltage regulated power supplies with this template can be a very straightforward exercise. Once the basic power supply is configured, all that needs to be added are the regulators and a few capacitors and diodes. Quite simply, the regulator is "tacked on" to a basic capacitor-filtered unregulated supply. For a simple single polarity supply, the additional components can be reduced to a single regulator IC, the extra capacitors and diode being ignored.

Figure 8.8

Pass transistors for increased I_{load}

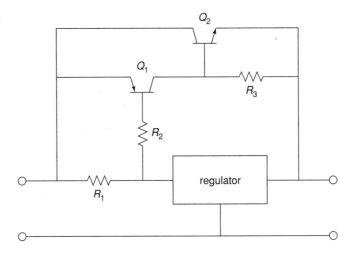

Current Boosting

For very high output currents, it is possible to utilize external pass transistors to augment the basic linear regulator ICs. An example is shown in Figure 8.8. When the input current rises to a certain level, the voltage drop across R_1 will be large enough to turn on transistor Q_1. When Q_1 starts to conduct, a current will flow through R_3. As this current rises, the potential across R_3 will increase to the point where the power transistor Q_2 will turn on. Q_2 will handle any further increases in load current. Normally, R_1 is set so that the regulator is running at better than one-half of its maximum rating before switchover occurs. A typical value would be 22 Ω. For example, if switchover occurs at 1 A, the regulator will supply all of the load's demand up to 1 A. If the load requires more than 1 A, the regulator will supply 1 A, and the power transistor will supply the rest. Circuits of this type often need a minimum load current to function correctly. This "bleed-off" can be achieved by adding a single resistor in parallel with the load.

Another possibility for increased output current is by paralleling devices and adding small ballast resistors. An example of this is shown in Figure 8.9. The ballast resistors are used to create local feedback. This reduces current hogging and forces the regulators to equally share the load current. (This is the same technique that is commonly used in high-power amplifiers so that multiple output transistors may be connected in parallel.) The size of the ballast resistors is rather low, usually .5 Ω or less.

Low Dropout Regulators

Low dropout regulators (usually shortened to LDO) are a special subclass of ordinary linear regulators. Generally, they operate the same way, but with one major exception. Unlike normal linear regulators, LDOs do not require a large input-output differential

FIGURE 8.9

Parallel devices for increased I_{load}

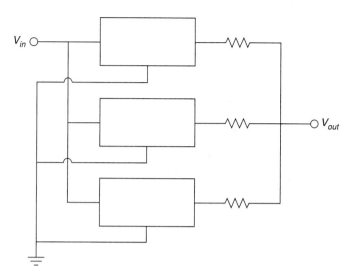

voltage of several volts. Instead, LDOs will regulate with a differential as little as a few tenths of a volt. This minimum input-output differential is known as the *dropout voltage*. With a lowered headroom requirement, the LDO is much more efficient than the standard regulator, especially with low voltage outputs. An example is the LM2940. Like the LM78XX, this regulator is available at many popular output potentials, including 5, 12, and 15 V. It is rated for a 1 A output current. At maximum current, the dropout voltage is typically .5 V. At an output of 100 mA, the dropout voltage is typically 110 mV.

Programmable and Tracking Regulators

Along with the simple three-pin fixed regulators, a number of adjustable or programmable devices are available. Some devices also include features such as programmable current limiting. Also, it is possible to configure multiple regulators so that they *track*, or follow, each other.

One popular adjustable regulator is the LM317. This device is functionally similar to the 340 series discussed in the previous section. In essence, it has an internal reference of 1.25 V. By using an external voltage divider, a wide range of output potentials is available. The LM317 will produce a maximum current of 1.5 A and a maximum voltage of 37 V. A basic connection diagram is shown in Figure 8.10. The R_1/R_2 divider sets the output voltage according to the formula

$$V_{out} = 1.25 \text{ V}\left(1 + \frac{R_2}{R_1}\right) + I_{adj}R_2$$

I_{adj} is the current flowing through the bottom *adjust pin*. I_{adj} is about 100 μA, and

Figure 8.10

Basic LM317 regulator

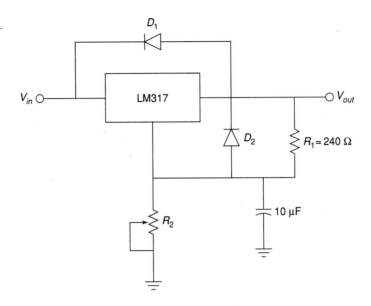

being this small, may be ignored to a first approximation. R_1 is set to 240 Ω. D_1 serves as a protection device for output overvoltages as seen in the preceding section. The 10 μF capacitor is used to increase the ripple rejection of the device. The addition of this capacitor will increase ripple rejection by at least 10 dB. Diode D_2 is used to prevent possible destructive discharges from the 10 μF capacitor. In practice, R_2 is set to a fixed size resistor for static supplies, or is a potentiometer for user-adjustable supplies.

Example 8-3

Determine a value for R_2 so that the output is adjustable from a minimum of 1.25 V to a maximum of 15 V.

For the minimum value, R_2 should be 0 Ω. Ignoring the effect of I_{adj}, the maximum value is found by

$$V_{out} = 1.25 \text{ V} \left(1 + \frac{R_2}{R_1}\right)$$

$$R_2 = R_1 \times \left(\frac{V_{out}}{1.25 \text{ V}} - 1\right)$$

$$R_2 = 240 \times \left(\frac{15 \text{ V}}{1.25 \text{ V}} - 1\right)$$

$$R_2 = 2.64 \text{ k}$$

Normally, such a value is not available for stock potentiometers. A 2.5 kΩ unit may be readily available, so R_1 may be reduced a bit in order to compensate.

multiSIM Computer Simulation

Figure 8.11 (pp. 295–296) shows a simulation of the regulator designed in Example 8-3. The LM117 model used is very similar to the LM317. The maximum potentiometer value of 2.5 kΩ is used here in order to see just how far off the design is from the 15-volt target. As in the earlier chapter simulation, a sine wave riding on a DC offset is used as the input to mimic the presence of ripple on the unregulated power source. The Transient Analysis of MultiSIM is used to plot the input and output waveforms. The circuit produces a very stable DC output as desired. Also, the level is only a few tenths of a volt shy of the desired 15-volt maximum, as expected. The simulation verifies the manual calculations quite well.

Figure 8.11a

LM317/LM117 simulation

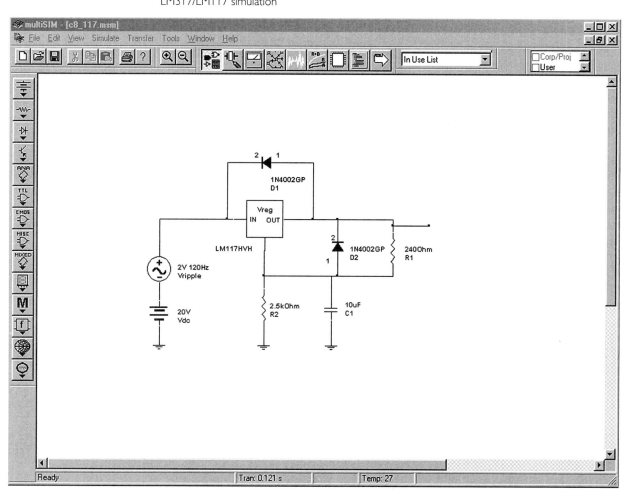

Figure 8.11b

Waveforms from MultiSIM

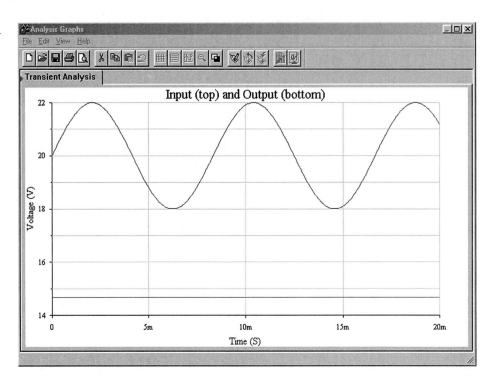

When using the LM317, if a minimum value greater than 1.25 V is needed, the pot may be placed in series with a fixed resistor. Alternately, precise preset values may be obtained through the use of a rotary switch and a bank of fixed resistors, as shown in Figure 8.12. As a matter of fact, these techniques may also be applied to the fixed three-pin devices presented in the previous section. For example, a LM7805 may be used as a 15 V regulator just by adding an external divider network as shown

Figure 8.12

Adjustable regulation

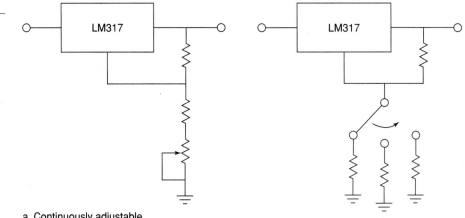

a. Continuously adjustable

b. Stepped

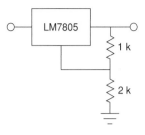

FIGURE 8.13

Varying V_{out} with a fixed regulator

in Figure 8.13. You can think of the LM7805 as an LM317 with a 5 V reference. The exact values are not critical; what is important is that the ratio of the two resistors is 2:1.

One of the major problems with adjustable linear regulators is a "bottom end" limitation. It is easier for a regulator of this type to generate a high-load current at a high-load voltage than it is to generate a high-load current at a low-load voltage. The reason for this is that at low-output voltages, the internal pass transistor sees a very high differential voltage. This results in very high power dissipation. If the device gets too hot, the thermal protection circuits may activate. From a user's standpoint, this means that the power supply may be able to produce a 15 V, 1 A output, but be unable to produce a 5 V, 1 A output.

Another popular regulator IC is the LM723. This device is of a more modular design and allows for preset current limiting. The equivalent circuit of 723 is shown in Figure 8.14. By itself, the LM723 is only capable of producing 150 mA. With the

FIGURE 8.14

LM723 equivalent circuit

Reprinted with permission of National Semiconductor Corporation

Connection Diagrams

Equivalent Circuit*

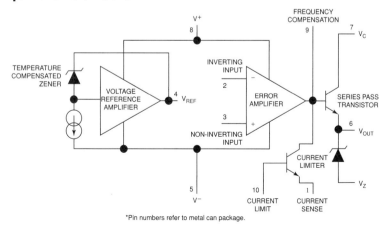

use of external pass transistors, an LM723-based regulator may produce several amps of load current. The internal reference is approximately 7.15 V. Because of this, two basic configurations exist: one for outputs below 7 V, and one for outputs above 7 V. These configurations and their output formulas are shown in Figure 8.15. By combining Figure 8.13 with Figure 8.14 and redrawing the results in Figures 8.16a and 8.16b, you can see a strong resemblance to the basic op amp regulator encountered earlier. Figure 8.16a is used for outputs greater than the 7.15 V reference. As such, resistors R_1 and R_2 are used to set the gain of the internal amplifier (i.e., the amount by which the reference will be multiplied). Resistor R_3 simply serves as DC bias compensation for R_1 and R_2. For outputs less than the reference, R_1 and R_2 serve as a voltage divider, which effectively reduces the reference, whereas the amplifier operates with a gain of unity. Again, R_3 serves as bias compensation. In short, the output voltage is once again determined by the reference voltage in conjunction with a pair of resistors. In both cases, the output current limit is set by

$$I_{limit} = \frac{V_{sense}}{R_{sc}}$$

Where V_{sense} is approximately .65 V at room temperature (for other temperatures, a V_{sense} graph is given in the LM723 data sheet). This equation may be found on the manufacturer's data sheet, but may be readily derived by inspecting Figure 8.16. Resistor R_{sc} is placed in series with the load, and thus, the current through it is the load current (ignoring the small current required by the adjustment resistors). The current limit transistor is connected so that the voltage across R_{sc} is applied to this transistor's base-emitter junction. The collector of the current limit transistor is connected to the base of the output pass transistor. If the output current rises to the point where the voltage across R_{sc} exceeds approximately .65 V, the current limit transistor will turn on, thus shunting output drive current away from the output pass transistor. As you can see, the formula for I_{limit} is little more than the direct application of Ohm's law. This scheme is similar to the method examined in Chapter 2 to safely limit an op amp's output current.

EXAMPLE 8-4

Design a +12 V regulator using the LM723, with a current limit of 50 mA.

The basic form for this is the version shown in 8.15.2. The appropriate output equation is

$$V_{out} = V_{ref} \frac{R_1 + R_2}{R_2}$$

Choosing an arbitrary value for R_2 of 10 kΩ, and then solving for R_1,

$$R_1 = \frac{V_{out}}{V_{ref}} R_2 - R_2$$

$$R_1 = \frac{12 \text{ V}}{7.15 \text{ V}} 10\text{k} - 10\text{k}$$

$$R_1 = 6.78 \text{ k}$$

LINEAR REGULATORS

TABLE I. Resistor Values (kΩ) for Standadrd Output Voltage

Positive Output Voltage	Applicable Figures (Note 4)	Fixed Output ±5%		Output Adjustable ±10% (Note 5)			Negative Output Voltage	Applicable Figures	Fixed Output ±5%		5% Output Adjustable ±10%		
		R1	R2	R1	P1	R2			R1	R2	R1	P1	R2
+3.0	1, 5, 6, 9, 12 (4)	4.12	3.01	1.8	0.5	1.2	+100	7	3.57	102	2.2	10	91
+3.6	1, 5, 6, 9, 12 (4)	3.57	3.65	1.5	0.5	1.5	+250	7	3.57	255	2.2	10	240
+5.0	1, 5, 6, 9, 12 (4)	2.15	4.99	0.75	0.5	2.2	−6(Note 6)	3, (10)	3.57	2.43	1.2	0.5	0.75
+6.0	1, 5, 6, 9, 12 (4)	1.15	6.04	0.5	0.5	2.7	−9	3, 10	3.48	5.36	1.2	0.5	2.0
+9.0	2, 4, (5, 6, 9, 12)	1.87	7.15	0.75	1.0	2.7	−12	3, 10	3.57	8.45	1.2	0.5	3.3
+12	2, 4, (5, 6, 9, 12)	4.87	7.15	2.0	1.0	3.0	−15	3, 10	3.65	11.5	1.2	0.5	4.3
+15	2, 4, (5, 6, 9, 12)	7.87	7.15	3.3	1.0	3.0	−28	3, 10	3.57	24.3	1.2	0.5	10
+28	2, 4, (5, 6, 9, 12)	21.0	7.15	5.6	1.0	2.0	−45	8	3.57	41.2	2.2	10	33
+45	7	3.57	48.7	2.2	10	39	−100	8	3.57	97.6	2.2	10	91
+75	7	3.57	78.7	2.2	10	68	−250	8	3.57	249	2.2	10	240

TABLE II. Formulae for Intermediate Output Voltages

Outputs from +2 to +7 volts (Figures 1, 5, 6, 9, 12 [4]) $V_{OUT} = \left(V_{REF} \times \frac{R2}{R1+R2} \right)$	Outputs from +4 to +250 volts (Figure 7) $V_{OUT} = \left(\frac{V_{REF}}{2} \times \frac{R2}{R1+R2} \right)$; R3 = R4	Current Limiting $I_{LIMIT} = \frac{V_{SENSE}}{R_{SC}}$
Outputs from +7 to +37 volts (Figures 2, 4, [5, 6, 9, 12]) $V_{OUT} = \left(V_{REF} \times \frac{R1+R2}{R2} \right)$	Outputs from −6 to −250 volts (Figures 3, 8, 10) $V_{OUT} = \left(\frac{V_{REF}}{2} \times \frac{R1+R2}{R1} \right)$; R3 = R4	Foldback Current Limiting $I_{KNEE} = \left(\frac{V_{OUT} R3}{R_{SC} R4} + \frac{V_{SENSE}(R3+R4)}{R_{SC} R4} \right)$ $I_{SHORT\ CKT} = \left(\frac{V_{SENSE}}{R_{SC}} \times \frac{R3+R4}{R4} \right)$

Typical Applications

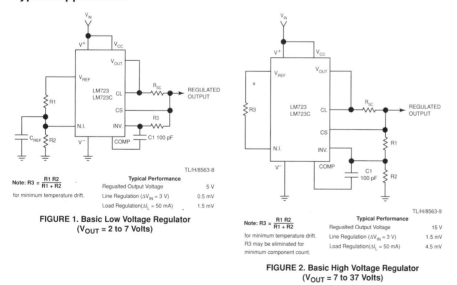

FIGURE 1. Basic Low Voltage Regulator (V_{OUT} = 2 to 7 Volts)

Note: $R3 = \frac{R1\ R2}{R1+R2}$ for minimum temperature drift.

Typical Performance
Regulated Output Voltage 5 V
Line Regulation (ΔV_IN = 3 V) 0.5 mV
Load Regulation (ΔI_L = 50 mA) 1.5 mV

FIGURE 2. Basic High Voltage Regulator (V_{OUT} = 7 to 37 Volts)

Note: $R3 = \frac{R1\ R2}{R1+R2}$ for minimum temperature drift. R3 may be eliminated for minimum component count.

Typical Performance
Regulated Output Voltage 15 V
Line Regulation (ΔV_IN = 3 V) 1.5 mV
Load Regulation (ΔI_L = 50 mA) 4.5 mV

FIGURE 8.15
LM723 "hook-ups"

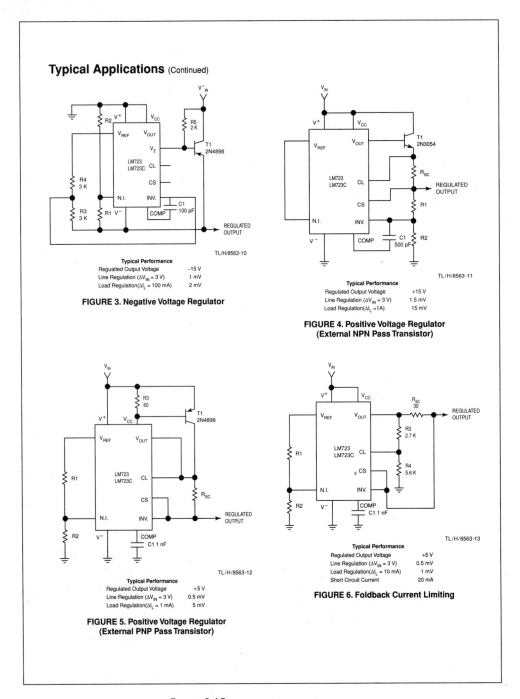

FIGURE 8.15

(Continued)

Reprinted with permission of National Semiconductor Corporation

FIGURE 8.16

Two basic configurations of an LM723 regulator

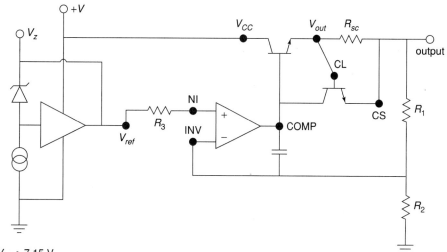

a. $V_{out} > 7.15$ V

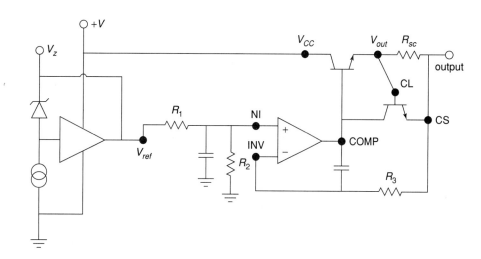

b. $V_{out} < 7.15$ V

For the current sense resistor,

$$I_{limit} = \frac{V_{sense}}{R_{sc}}$$

$$R_{sc} = \frac{V_{sense}}{I_{limit}}$$

$$R_{sc} = \frac{.65 \text{ V}}{50 \text{ mA}}$$

$$R_{sc} = 13 \text{ }\Omega$$

FIGURE 8.17

Completed 12 V circuit for Example 8-4

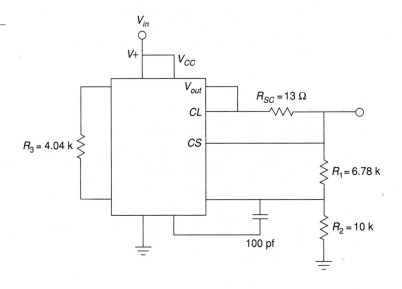

Finally, for minimum temperature drift, R_3 is included, and set to $R_1 \| R_2$

$$R_3 = R_1 \| R_2$$
$$R_3 = 10 \text{ k} \| 6.78 \text{ k}$$
$$R_3 = 4.04 \text{ k}$$

The completed circuit is shown in Figure 8.17.

EXAMPLE 8-5

Using the LM723, design a continuously adjustable 2 V to 5 V supply, with a current limit of 1.0 A.

The basic form for this is the version shown in Figure 8.15.1. The appropriate output equation is

$$V_{out} = V_{ref} \frac{R_2}{R_1 + R_2}$$

We need to make a few modifications to the basic form in order to accommodate the high output current and the output voltage adjustment. One possibility is shown in Figure 8.18. In order to produce a 1 A load current, an external pass transistor will be used. To obtain the desired voltage adjustment, resistor R_1 is replaced with a series potentiometer/resistor combination (R_{1a}, R_{1b}). In this fashion, the minimum value for R_1 will be R_{1b}, and the maximum value will be $R_{1a} + R_{1b}$. There are several ways in which we can approach the

FIGURE 8.18

Circuit for Example 8-5: 2 V–5 V regulator, 1 amp

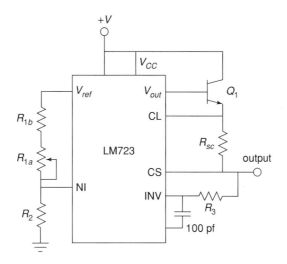

calculation of these three resistors. Perhaps the easiest is to pick a value for R_2 and then determine values for R_{1a} and R_{1b}. Though this is fairly straightforward, it is not very practical because you will most likely wind up with an odd size for the potentiometer. A better, though admittedly more involved, approach revolves around the selection of a reasonable pot value, such as 10 kΩ. Given the desired output potentials, the other two resistors may be found.

First of all, note that these three resistors are nothing more than a voltage divider. The output voltage equation may be rewritten as

$$\frac{R_1 + R_2}{R_2} = \frac{V_{ref}}{V_{out}}$$

For the maximum case, we have

$$\frac{R_{1b} + R_2}{R_2} = \frac{V_{ref}}{V_{out}}$$

$$\frac{R_{1b} + R_2}{R_2} = \frac{7.15 \text{ V}}{5 \text{ V}}$$

$$\frac{R_{1b} + R_2}{R_2} = 1.43$$

If we consider R_2 to be unity, we may say that the ratio of the two resistors to R_2 is 1.43:1, or, that the ratio of R_{1b} to R_2 is .43:1.

For the minimum case, we have

$$\frac{R_{1a} + R_{1b} + R_2}{R_2} = \frac{V_{ref}}{V_{out}}$$

$$\frac{R_{1a} + R_{1b} + R_2}{R_2} = \frac{7.15 \text{ V}}{2 \text{ V}}$$

$$\frac{R_{1a} + R_{1b} + R_2}{R_2} = 3.575$$

We may say that the ratio of the three resistors to R_2 is 3.575:1, or that the ratio of $R_{1a} + R_{1b}$ to R_2 is 2.575:1. Because we already know that the ratio of R_{1b} to R_2 is .43:1, the ratio of R_{1a} to R_2 must be the difference, or 2.145:1. Because we chose 10 kΩ for R_{1a},

$$R_2 = \frac{R_{1a}}{2.145}$$
$$R_2 = \frac{10\ k}{2.145}$$
$$R_2 = 4.66\ k$$

Similarly,

$$R_{1b} = R_2 .43$$
$$R_{1b} = 4.66\ k\ .43$$
$$R_{1b} = 2\ k$$

For the current sense resistor,

$$I_{limit} = \frac{V_{sense}}{R_{sc}}$$
$$R_{sc} = \frac{V_{sense}}{I_{limit}}$$
$$R_{sc} = \frac{.65\ V}{1.0\ A}$$
$$R_{sc} = .65\ \Omega$$

For minimum temperature drift, R_3 is included, and set to $R_1 \parallel R_2$. Because R_1 is adjustable, a midpoint value will be used.

$$R_3 = R_1 \parallel R_2$$
$$R_3 = 6\ k \parallel 4.66\ k$$
$$R_3 = 2.62\ k$$

The final calculation involves the external pass transistor. This design uses a 1 A output, so this device must be able to handle this current continuously. Also, a minimum β specification is required. As the LM723 will be driving the pass transistor, the LM723 only needs to produce base drive current. With a maximum output of 150 mA, this translates to a minimum β of

$$\beta_{min} = \frac{I_c}{I_b}$$
$$\beta_{min} = \frac{1\ A}{150\ mA}$$
$$\beta_{min} = 6.67$$

This value should pose no problem for a power transistor.

FIGURE 8.19

Current limiting

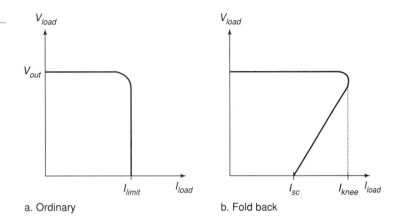

a. Ordinary

b. Fold back

Besides the applications we have just looked at, the LM723 can also be used to make negative regulators, switching regulators, and other types as well. One useful variation on the basic theme is the use of *foldback current limiting*, as seen in Figure 8.15.6. Unlike the ordinary form of current limiting, foldback limiting actually produces a decrease in output current once the limit point is reached. Figure 8.19a shows the effect of ordinary limiting. Once the limit point is reached, further demands by the load will be ignored. The problem with this arrangement is that under short-circuit conditions, the pass transistor will be under heavy stress. The load is shorted, so $V_{load} = 0$, and therefore, a large potential will drop across the pass transistor. This device is already handling the full current draw, thus the resulting power dissipation can be very high. Foldback limiting gets around this problem by lowering the output current as the pass transistor's voltage increases. Note how in Figure 8.19b, the current limit curve does not drop straight down to the limit point, but rather, as the load demand increases, the current drops back to I_{sc}. By limiting the current in this fashion, a much lower power dissipation is achieved.

Our last item of interest in this section is the LM325/LM326 dual tracking regulator. These devices are of particular interest to the op amp technician and designer. The LM325 is a ±15 V bipolar regulator with output current up to 100 mA. This is a tracking device, meaning that the positive and negative outputs are closely aligned. (The LM125 upgraded version exhibits outputs balanced to within 1%.) Like the LM723, the LM325 includes provisions for output current limiting as well as current boosting. The LM326 is similar to the LM325, except that its output is set up for ±12 V. A pin-out diagram and example circuit outlines are given in Figure 8.20. These devices are ideal for powering general-purpose op amp circuits.

The circuit labeled "Basic Regulator" in Figure 8.20a is a minimum parts-configuration dual regulator. All that is required ahead of this circuit is a standard transformer, rectifier, and filter capacitor arrangement, such as that found in Figure 8.7. The maximum output (unloaded) of the filter capacitors should be no more than 30 V in order to prevent damage to the LM325/LM326. This circuit is certainly simpler than the LM340-based regulator of Figure 8.7. The downside is that its power dissipation and maximum currents are considerably lower than its LM340 counterpart. Still, a 100 mA

FIGURE 8.20

LM325 pinout and examples
Reprinted with permission of National Semiconductor Corporation

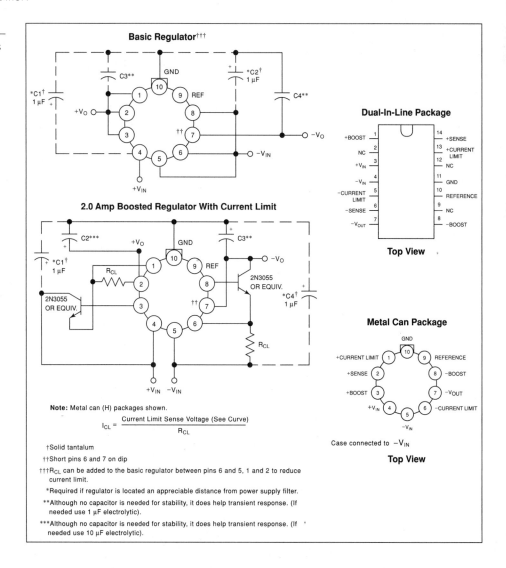

output capability is sufficient to drive a large number of op amps and other small signal devices. Current limiting is not shown on this circuit, but can be added through the inclusion of a pair of current-sensing resistors (using essentially the same scheme as the LM723), one each for the positive and negative outputs.

For designs demanding higher output current and power dissipation, the second circuit of Figure 8.20 may be considered. Note that this circuit is only marginally more complex than the basic circuit of Figure 8.20. To each output, a current sense resistor and external pass transistor have been added. The LM325/LM326 only needs to produce base drive current for the external pass transistor, so output currents of several amps are possible using this configuration. The actual rating depends on the current and power capabilities of the external pass transistors. In this case, the popular

2N3055 NPN power devices yield a 2 A load current. (Note that this indicates a transistor minimum β of $\frac{2 \text{ A}}{100 \text{ mA}}$, or 20. This is a reasonable value for a medium- to high-power transistor.)

There are many other linear regulator ICs available to the designer than have been presented here. Many of these devices are rather specialized, and all units seem to have their own special set of operational formulas and graphs. There are, however, a few common threads among them all. First, due to the relative internal complexity, manufacturers often give very specific application guidelines for their particular ICs. The resulting design sequence is rather like following a cookbook and makes the designer's life much easier. Second, as mentioned at the outset, all linear regulators tend to be rather inefficient. This inefficiency is inherent in the design and implementation of the linear regulation circuits, and cannot be avoided. At best, the inefficiency can be minimized for a given application. In order to achieve high efficiency, a different topology must be considered. One alternative is the switching regulator.

8.4 Switching Regulators

The major cause of the inefficiency of a linear regulator is that its pass transistor operates in the linear region. This means that it constantly sees both a high current and a high (or at least moderate) voltage. The result is a sizable power dissipation. In contrast, switching regulators rely on the efficiency of the transistor switch. When the transistor is off, no current flows, and thus, no power is dissipated. When the transistor is on (in saturation), a high current flows, but the voltage across the transistor ($V_{ce(sat)}$) is very small. This results in modest power dissipation. The only time that both the current and voltage are relatively large at the same time is during the switching interval. This time period is relatively short compared to cycle time, so again, the power dissipation is rather small.

Switching the pass transistor on and off can be very efficient, but unfortunately, the resulting pulses of current are not appropriate for most loads. Some way of smoothing out the pulses into a constant DC level is needed. One way to do this is through an inductor/capacitor arrangement. This concept is essential to the switching regulator.

Although the switching regulator does offer an increased efficiency over the linear regulator, it is not without its detractions. First of all, switching regulators tend to be more complex than linear types. This complexity may outstrip the efficiency advantage, particularly for low-power designs. Some of the newer single-chip switching regulators make switching power supply design almost as straightforward as linear design, so this problem is not quite as great as it once was. The other major problem is that the switching process can create a great deal of radiated electrical noise. Without proper shielding and similar precautions, the induced noise signals can produce grave interference in nearby analog or digital circuits. In spite of these obstacles, switching power supplies have seen great acceptance in a variety of areas, including powering personal computers.

A basic switching regulator outline is shown in Figure 8.21. Once again, a control device is used to adjust the incoming signal. Unlike the linear regulator, the signal out of the control element is a pulse train. In order to produce a smooth DC output, the

FIGURE 8.21

Basic switching regulator

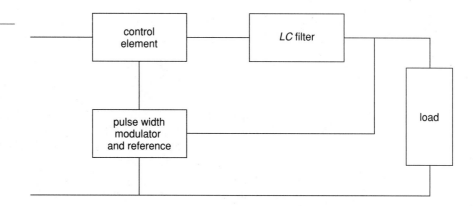

pulse train is passed through a capacitor/inductor network. The heart of the circuit is the pulse-width modulator. This circuit creates a variable duty cycle pulse waveform, which alternately turns the control element on and off. The duty cycle of the pulse is proportional to the load current demand. Low-load impedances require large currents and, thus, will create pulse trains with long "on" times. The peak value of the pulses will tend to be much greater than the average load current demand; however, if this current pulse is averaged over one cycle, the result will equal the load current demand. Effectively, the LC network serves to integrate the current pulses. This is shown graphically in Figure 8.22. Since the lower load impedance is receiving a proportionally larger current, the load voltage remains constant. This is shown graphically in Figure 8.23.

FIGURE 8.22

Current waveforms

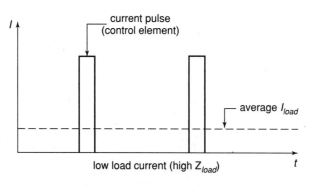

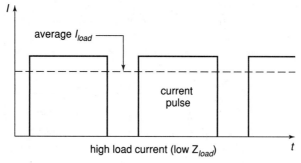

FIGURE 8.23

Voltage waveforms

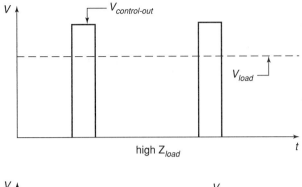

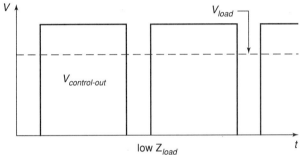

Due to the constant charging and discharging of the L and C components, switching regulators tend to respond to load changes somewhat slower than linear regulators do. Also, if the frequency of the switching pulse is high, the required sizes for L and C may be reduced, resulting in a smaller circuit and possible cost reductions. Many switching regulators run in the 20 kHz to 100 kHz region. The practical upper limit for switching speed is determined by the speed of the control element. Devices with fast switching times will prove to be more efficient. Power FETs are very attractive for this application because of their inherent speed and low drive requirements. Their negative temperature coefficient of transconductance also helps reduce thermal runaway problems. For the highest powers, bipolar devices are often the only choice. Finally, it is possible to configure many switchers for *step down*, *step up*, or *inverter* operation. As an example, we will investigate the step down, or *buck*, configuration.

A block diagram of a step down switching regulator is shown in Figure 8.24. Here is how it works: When the transistor, Q, is on, current flows through Q, L, and the load. The inductor current rises at a rate equal to the inductor voltage divided by the inductance. The inductor voltage is equal to the input voltage minus the load voltage and the transistor's saturation potential. While L is charging (i.e., $I_L < I_{load}$), C provides load current. When Q turns off, the inductor's magnetic field starts to collapse causing an effective polarity reversal. In other words, the inductor is now acting as a source for load current. At this point, diode D is forward biased, effectively removing the left portion of the circuit. L also supplies current to capacitor C during this time period. Eventually, the inductor current will drop below the value required by the load, and C will start to discharge, making up the difference. Before L completely discharges,

FIGURE 8.24

Basic stepdown switching regulator

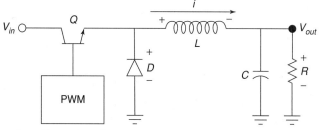

a. Q on: Source supplies current and charges inductor

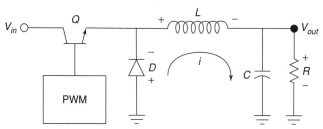

b. Q off: Inductor supplies current

the transistor switch will turn back on, repeating the cycle. Because the inductor is never fully discharged, this is called *continuous operation*. *Discontinuous operation* is also possible, although we will not pursue it here. One point worth noting is that for proper continuation of this cycle, some load current draw must always be present. This minimum level can be achieved through the use of a bleed resistor in parallel with the load.

The values for C and L are dependent on the input and output voltages, desired output current, the switching frequency, and the particulars of the switching circuit used. Manufacturers generally give look-up tables and charts for appropriate values and/or formulas. Earlier switching ICs contained the necessary base components and required only a moderate amount of external circuitry. The newest ICs may be configured with no more than three or four external passive parts.

One example of a switch mode regulator is the LM3578A. A functional diagram of the inner workings of this IC is shown in Figure 8.25. The AND gates, comparator, oscillator, and associated latches combine to form the pulse-width modulator.

Notice that this circuit includes an internal reference and medium-power switching transistor. Thermal shutdown and current limiting are available. The circuit will operate at inputs up to 40 V and can produce output currents up to .75 amp. The maximum switching frequency is 100 kHz. Step up, step down, and inverter configurations are all possible with this device. To make the design sequence as fast as possible, the manufacturer has included a design chart. This is shown in Figure 8.26. As with virtually all highly specialized ICs, specific design equations only apply to particular devices, and probably cannot be used with ICs that produce similar functions. Consequently, it is recommended that you do not memorize these formulas!

FIGURE 8.25

LM3578 equivalent circuit

Reprinted with permission of National Semiconductor Corporation

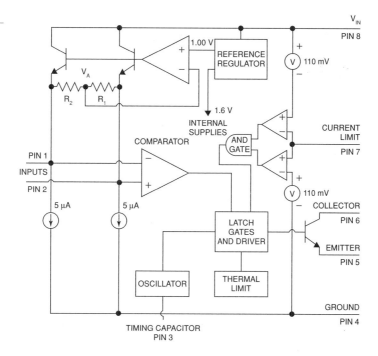

A step down regulator using the LM3578A is shown in Figure 8.27. C_1 is the frequency-selection capacitor and can be found from the manufacturer's chart. C_3 is necessary for continuous operation and is generally in the vicinity of 10 to 30 pF. D_1 should be a Schottky-type rectifier. R_1 and R_2 set the step down ratio (they are functionally the same as R_f and R_i in the previous work). R_3 sets the current limit. Finally, L_1 and C_2 are used for the final output filtering. The relevant equations from the manufacturer's data sheets are:

$$V_{out} = \frac{R_1}{R_2} + 1 \text{ (in volts)}$$

$$R_3 = \frac{.11 \text{ V}}{I_{sw(max)}}$$

$$C_2 >= V_{out} \frac{V_{in} - V_{out}}{8 f^2 V_{in} V_{ripple} L_1}$$

$$L_1 = V_{out} \frac{V_{in} - V_{out}}{\Delta I_{out} V_{in} f}$$

$$\Delta I_{out} = 2 I_{out} DiscontinuityFactor \text{ (typically .2)}$$

where $I_{sw(max)}$ is the maximum current through the switching element, and V_{ripple} is in peak-to-peak volts. Some values may be found using the look-up chart method shown in Figure 8.26.

312 VOLTAGE REGULATION

FIGURE 8.26

LM3578 design chart

Reprinted with permission of National Semiconductor Corporation

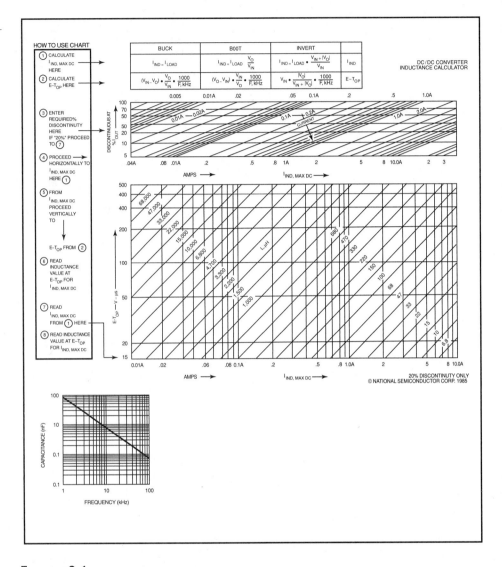

EXAMPLE 8-6

Using the LM3578A, design a step down regulator that delivers 12 V from a 20 V source, with 200 mA of load current. Use an oscillator frequency of 50 kHz, and a discontinuity factor of .2 (20%). Ripple should be no more than 40 mV.

First determine the R_1 and R_2 values. R_2 is arbitrarily chosen at 10 kΩ.

$$V_{out} = \frac{R_1}{R_2} + 1 \text{ (in volts)}$$
$$R_1 = R_2(V_{out} - 1\text{ V})$$
$$R_1 = 10\,\text{k}(12\,\text{V} - 1\,\text{V})$$
$$R_1 = 110\,\text{k}$$

Figure 8.27

Step-down regulator using an LM3578

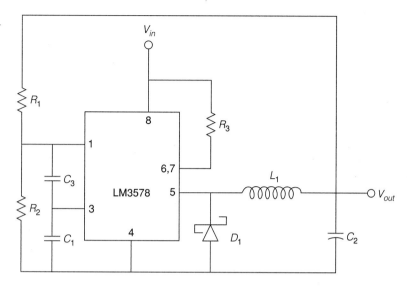

$$R_3 = \frac{.11\ V}{I_{sw(max)}}$$

$$R_3 = \frac{.11\ V}{.75\ A}$$

$$R_3 = .15\ \Omega$$

Note that R_3 will always be .15 Ω for this circuit form.

From the oscillator graph, C_1 is estimated at 1700 pF, and C_3 is set to 20 pF, as suggested by the manufacturer.

$$\Delta I_{out} = 2 I_{out} DiscontinuityFactor$$

$$\Delta I_{out} = 2 \times 200\ mA \times .2$$

$$\Delta I_{out} = 80\ mA$$

$$L_1 = V_{out} \frac{V_{in} - V_{out}}{\Delta I_{out} V_{in} F}$$

$$L_1 = 12\ V \times \frac{20\ V - 12\ V}{80\ mA \times 20\ V \times 50\ kHz}$$

$$L_1 = 1.2\ mH$$

$$C_2 >= V_{out} \frac{V_{in} - V_{out}}{8 f^2 V_{in} V_{ripple} L_1}$$

$$C_2 >= 12\ V \frac{20\ V - 12\ V}{8 \times 50\ kHz^2 \times 20\ V \times 40\ mV \times 1.2\ mH}$$

$$C_2 >= 5\ \mu F$$

To be on the conservative side, C_2 is usually set a bit higher, so a standard 33 μF or 47 μF might be used.

A basic inverting switcher is shown in Figure 8.28. This circuit produces a negative output potential from a positive input. It works as follows: When the transistor switch is closed, the inductor L is charged. Diode D is in reverse bias, as its anode is negative. When the switch opens, the inductor's collapsing field causes it to appear as a source of opposite polarity. The diode is forward-biased because its cathode is now forced to be lower than its anode. The inductor is now free to deliver current to the load. Eventually, the inductor will discharge to the point where the transistor switch will turn back on. While the inductor is charging, capacitor C will supply the load current. The process repeats, thus maintaining a constant output potential.

A basic step up switcher is shown in Figure 8.29. This variation is used when a potential greater than the input is desired, such as deriving a 15 V supply from an existing 5 V source. Here is how the circuit works: When the transistor switch is closed, the inductor L charges. During this time period, the capacitor C is supplying load current. Because the output potential will be much higher than the saturation voltage of the transistor switch, the diode D will be in reverse bias. When the transistor turns off, the magnetic field of the inductor collapses, causing the inductor to appear as a

FIGURE 8.28

Basic inverting switcher

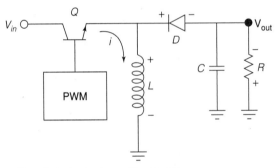

a. Q on: Inductor is charged by source

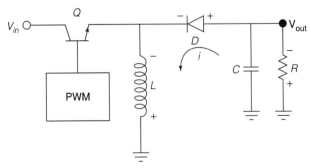

b. Q off: Inductor supplies current

FIGURE 8.29

Basic step-up switching regulator

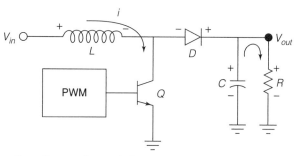

a. Q on: Source charges inductor

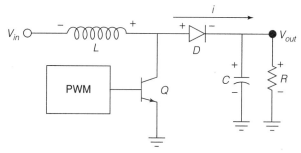

b. Q off: Inductor supplies current

source. This potential is added to the driving source potential, as these two elements are in series. This combined voltage is what the load sees; hence, the load voltage is greater than the driving source. Eventually, the inductor will discharge to the point where the transistor switches back on, thus reverse-biasing the diode and recharging L. The capacitor will continue to supply load current during this time. This process will continue in this fashion, producing the desired output voltage.

For the step up and inverter forms, other sets of equations are used for the LM3578A. Although the design sequence is certainly not quite as straightforward as in the linear regulator circuits, it is definitely not a major undertaking, either. If 750 mA is not sufficient, an external pass transistor may be added to the LM3578A. Other switching regulator ICs are available from different manufacturers. Each unit operates on the same basic principle, but the realization of the design may take considerably different routes. Specific device data sheets must be consulted for each model.

For specific applications, some manufacturers offer switching regulators that are almost drop-in replacements for basic three-pin linear regulators. These devices are not nearly as flexible as the generic switching regulator ICs, though. A good example is the LM2576 regulator. This is a 3 A output, step down regulator (i.e., buck-mode only). It is available at a variety of output potentials ranging from 3.3 V to 15 V. An adjustable version is also available. A block diagram and typical circuit are shown in Figure 8.30. As you can see, a typical fixed output design requires a minimum of external components: 2 capacitors, an inductor, and a diode. Other members of this family include the LM2574 .5 A step down regulator, and the LM2577 step up regulator.

FIGURE 8.30

LM2576 switching regulator

Reprinted with permission of National Semiconductor Corporation

LM2576/LM2576HV Series
SIMPLE SWITCHER® 3A Step-Down Voltage Regulator

General Description

The LM2576 series of regulators are monolithic integrated circuits that provide all the active functions for a step-down (buck) switching regulator, capable of driving 3A load with excellent line and load regulation. These devices are available in fixed output voltages of 3.3 V, 5 V, 12 V, 15 V, and an adjustable output version.

Requiring a minimum number of external components, these regulators are simple to use and include internal frequency compensation and a fixed-frequency oscillator.

The LM2576 series offers a high-efficiency replacement for popular three-terminal linear regulators. It substantially reduces the size of the heat sink, and in some cases no heat sink is required.

A standard series of inductors optimized for use with the LM2576 are available from several different manufacturers. This feature greatly simplifies the design of switch-mode power supplies.

Other features include a guaranteed ±4% tolerance on output voltage within specified input voltages and output load conditions, and ±10% on the oscillator frequency. External shutdown is included, featuring 50 µA (typical) standby current. The output switch includes cycle-by-cycle current limiting, as well as thermal shutdown for full protection under fault conditions.

Features

- 3.3 V, 5 V, 12 V, 15 V, and adjustable output versions
- Adjustable version output voltage range, 1.23 V to 37 V (57 V for HV version) ± 4% max over line and load conditions
- Guaranteed 3A output current
- Wide input voltage range, 40 V up to 60 V for HV version
- Requires only 4 external components
- 52 kHz fixed frequency internal oscillator
- TTL shutdown capability, low power standby mode
- High efficiency
- Uses readily available standard inductors
- Thermal shutdown and current limit protection
- P + Product Enhancement tested

Applications

- Simple high-efficiency step-down (buck) regulator
- Efficient pre-regulator for linear regulators
- On-card switching regulators
- Positive to negative converter (Buck-Boost)

Typical Application (Fixed Output Voltage Versions)

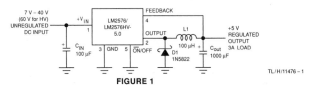

FIGURE 1

Block Diagram

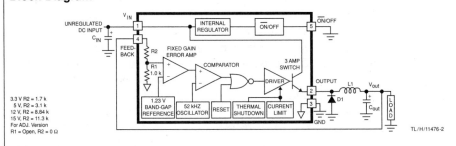

Patent Pending

8.5 Heat Sink Usage

Whenever appreciable amounts of power are dissipated by semiconductor devices, some form of cooling element needs to be considered. Power supply regulation circuits are no exception. The pass transistors used in both linear and switching regulators can be forced to dissipate large amounts of power. The result of this is the production of heat. Generally, the life span of semiconductors drops as the operation temperature rises. Most silicon-based devices exhibit maximum allowable junction temperatures in the 150°C range. Although power transistors utilize heavier metal cases, they are generally not suitable for high dissipation applications by themselves. In order to increase the thermal efficiency of the device, an external heat sink is used. Heat sinks are normally made of aluminum and appear as a series of fins. The fins produce a large surface, which enhances the process of heat convection. In other words, the heat sink can transfer heat to the surrounding atmosphere faster than the power transistor can. By bolting the transistor to the heat sink, the device will be able to dissipate more power at a given operating temperature. Some typical heat sinks are shown in Figure 8.31.

Physical Requirements

Heat sinks are designed to work with specific device-case styles. The most common case styles for regulators are the TO-220 "power tab," and the TO-3 "can." Heat sinks are available for these specific styles, including the requisite mounting hardware and insulation spacers. Some of the lower power regulators utilize TO-5 "mini can" or DIP-type cases. Heat sinks are available for these package types, too, but are not quite as common.

There are a couple of general rules that should be followed when using heat sinks:

1. Always use some form of heat sink grease or thermally conductive pad between the heat sink and the device. This will increase the thermal transfer between the two parts. Note that excessive quantities of heat sink grease will actually **decrease** performance.
2. Mount fins in the vertical plane for optimum natural convective cooling.
3. Do not overcrowd or obstruct devices that use heat sinks.
4. Do not block air flow around heat sinks—particularly directly above and below items that rely on natural convection.
5. If thermal demands are particularly high, consider using forced convection (e.g., fans).

Thermal Resistance

In order to specify a particular heat sink for a given application, a more technical explanation is in order. What we are going to do is create a *thermal circuit equivalent*. In this model, the concept of *thermal resistance* is used. Thermal resistance denotes how easy it is to transfer heat energy from one mechanical part to another. The symbol for

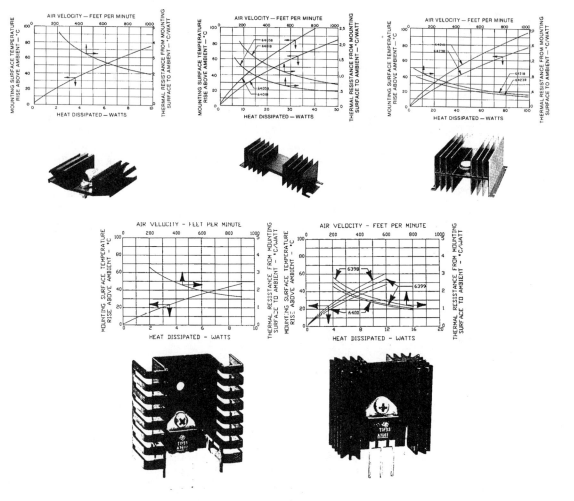

FIGURE 8.31

Typical heat sinks
Courtesy of Thermalloy, Inc.

thermal resistance is θ, and the units are Centigrade degrees per watt. In this model, temperature is analogous to voltage, and thermal power dissipation is analogous to current. A useful equation is

$$P_D = \frac{\Delta T}{\theta_{total}} \quad (8.4)$$

Where P_D is the power dissipated by the semiconductor device in watts, ΔT is the temperature differential, and θ_{total} is the sum of the thermal resistances. Basically, this is a thermal version of Ohm's law.

FIGURE 8.32

Device and heat sink

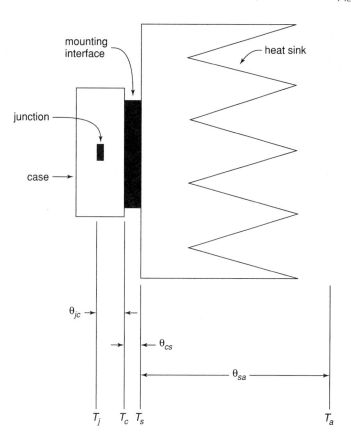

In order to construct our model, let's take a closer look at the power-device/heat-sink combination. This is shown in Figure 8.32. T_j is the semiconductor junction temperature. This heat energy source heats the device case to T_c. The thermal resistance between the two entities is θ_{jc}. The case, in turn, heats the heat sink via the interconnection. This thermal resistance is θ_{cs}, and the resulting temperature is T_s. Finally, the heat sink passes the thermal energy to the surrounding air, which is sitting at T_a. The thermal resistance of the heat sink is θ_{sa}. The equivalent thermal model is shown in Figure 8.33. (Although this does not have perfect correspondence with normal circuit analysis, it does illustrate the main points.)

In this model, ground represents a temperature of absolute zero. The circuit is sitting at an ambient temperature T_a, thus a voltage source of T_a is connected to ground and the heat sink. The three thermal resistances are in series and are driven by a current source that is set by the present power dissipation of the device. Note that if the power dissipation is high, the resulting "voltage drops" across the thermal resistances are high. Voltage is analogous to temperature in this model, so this indicates that a high temperature is created. Because there is a maximum limit to T_j, higher power dissipations require lower thermal resistances. As θ_{jc} is set by the device manufacturer, you have no control over that element. However, θ_{cs} is a function of the case style and

FIGURE 8.33

Equivalent thermal model of Figure 8.32

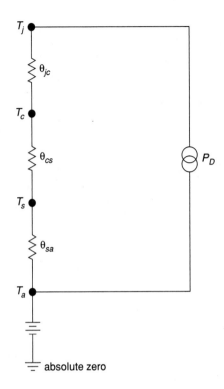

the insulation material used, so you do have some control (but not a lot) over that. On the other hand, as the person who specifies the heat sink, you have a great deal of control over θ_{sa}. Values for θ_{sa} are given by heat sink manufacturers. A useful variation of Equation 8.4 is

$$P_D = \frac{T_j - T_a}{\theta_{jc} + \theta_{cs} + \theta_{sa}} \quad (8.5)$$

Normally, power dissipation, junction and ambient temperatures, θ_{jc}, and θ_{cs} are known. The idea is to determine an appropriate heat sink. Both T_j and θ_{jc} are given by the semiconductor device manufacturer. The ambient temperature, T_a, may be determined experimentally. Due to localized warming, it tends to be higher than the actual "room temperature." Standard graphs, such as those found in Figure 8.34, may be used to determined θ_{cs}.

EXAMPLE 8-7

Determine the appropriate heat sink rating for a power device rated as follows: $T_{j(max)} = 150°C$, TO-220 case style, $\theta_{jc} = 3.0$ C°/W. The device will be dissipating a maximum of 6 W in an ambient temperature of 40°C. Assume that the heat sink will be mounted with heat sink grease and a .002 mica insulator.

First, find θ_{cs} from the TO-220 graph. Curve 3 is used. The approximate (conservative) value is 1.6 C°/W.

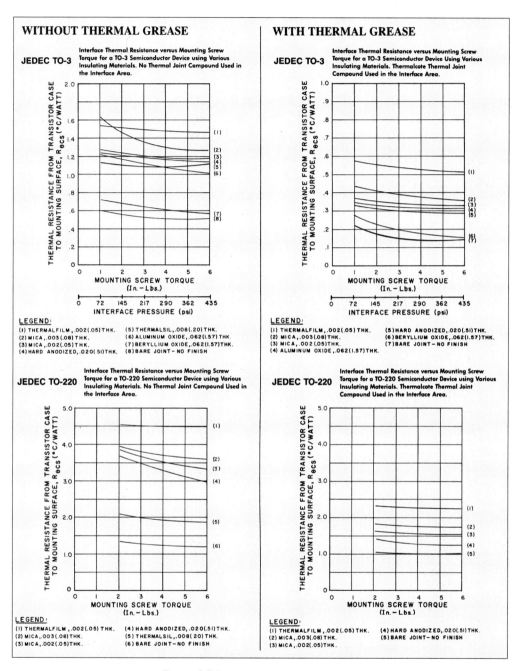

FIGURE 8.34

θ_{CS} for TO-3 and TO-220

Courtesy of Thermalloy, Inc.

$$P_D = \frac{T_j - T_a}{\theta_{jc} + \theta_{cs} + \theta_{sa}}$$

$$\theta_{sa} = \frac{T_j - T_a}{P_D} - \theta_{jc} - \theta_{cs}$$

$$\theta_{sa} = \frac{150°C - 40°C}{6\text{ W}} - 3.0 C°/W - 1.6 C°/W$$

$$\theta_{sa} = 13.73 C°/W$$

This is the maximum acceptable value for the heat sink's thermal resistance. Note that the use of heat sink grease gives us an extra 2 C°/W or so. Also, note the generally lower values of θ_{sa} for the TO-3 case relative to the TO-220. This is one reason why TO-3 cases are used for higher power devices. This case also makes it easier for the manufacturer to reduce θ_{jc}.

8.6 Extended Topic: Primary Switcher

The switching regulators examined earlier are referred to as *secondary switchers* because the switching elements are found on the secondary side of the power transformer. In contrast to this is the *primary* or *forward switcher*. The switching circuitry in these designs is placed prior to the primary of the power transformer. This positioning offers a distinct advantage over the secondary switcher. The power transformer and secondary rectifier will be handling much higher frequencies, thus they can be made much smaller. The result is a physically smaller and lighter design.

One possible configuration of a primary switcher is shown in Figure 8.35. This is known as a *push-pull* configuration. The two power transistors are alternately pulsed on

Figure 8.35

Push-pull primary switching regulator

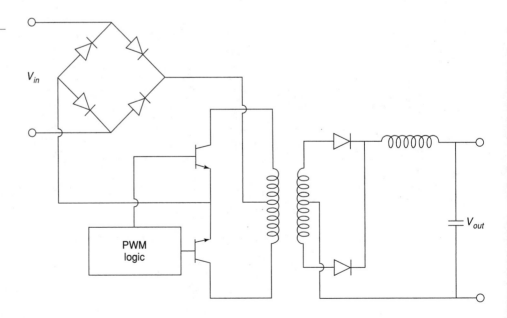

FIGURE 8.36

Full-bridge primary switching regulator

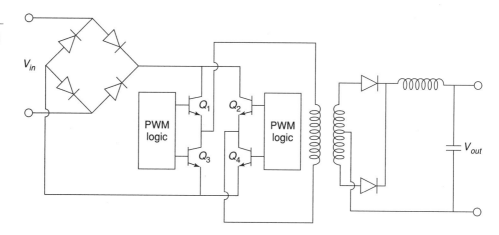

and off. That is, when one device is conducting, the other is off. By doing this, opposite polarity pulses are fed into the primary of the transformer, creating a high-frequency alternating current (as is the case with secondary switchers, primary switchers operate at frequencies well above the nominal 60 Hz power line). This high-frequency waveform is then stepped up or down to the secondary, where it is again rectified and then filtered, producing a DC output signal.

Although the transformer and secondary rectifier/filter may be reduced in size, it is important to note that the main input rectifier and switching transistors are not isolated from the AC power source, as is the case in other power supply designs. These devices must handle high input potentials. For an ordinary 120 V AC line, that translates to over 170 V peak. Also, the power transistors will see an off-state potential of twice V_{in}, or over 340 V in this case. Some form of in-rush current limiting will also need to be added.

A somewhat more sophisticated approach is taken in Figure 8.36. This circuit is known as a *full bridge* switcher. In this configuration, diagonal pairs of devices are simultaneously conducting (i.e., Q_1/Q_4 and Q_2/Q_3). By eliminating the center-tapped primary, each device sees a maximum potential of V_{in}, or one-half that of the push-pull switcher. The obvious disadvantage is the need for four power devices instead of just two.

Primary switchers do offer a size and weight advantage over secondary switchers and linear regulators. They also maintain the high efficiency characteristics of the secondary switcher. They do tend to be somewhat more complex, though, and their application is therefore best suited to cases where circuit size, weight, and efficiency are paramount.

SUMMARY

In this chapter you have examined the basic operation of voltage regulators. Their purpose is simple: to provide constant, nonvarying output voltages despite changes in either the AC source, or in the load current demand. Voltage regulation circuits are an integral part of just about every piece of modern electronic equipment. Due to their wide use, a number of specialized voltage regulator ICs are available from a variety of manufacturers.

Voltage regulation may be achieved through two main methods. These methods are linear regulation and switching regulation. In both cases, a portion of the output voltage is compared to a stable internal reference. The result of this comparison is used to drive a control element, usually a power transistor. If the output voltage is too low, the control element allows more current to flow to the load from the rectified AC source. Conversely, if the output is too high, the control element constricts the current flow. In the case of the linear regulator, the control element is always in the active, or linear, state. Because of this, the linear regulator tends to dissipate quite a bit of power and, as a result, is rather inefficient. On the plus side, the linear regulator is able to quickly react to load variations, and thus exhibits good transient response.

In contrast to the linear regulator, the control device in the switching regulator is either fully on or fully off. As a result, its power dissipation tends to be reduced. For best performance, fast control devices are needed. The control device is driven by a pulse-width modulator. The output of this modulator is a rectangular pulse whose duty cycle is proportional to the load-current demand. As the control device produces current pulses instead of a constant current, some means of smoothing the pulses is necessary. This function is performed by an LC filter. The main advantage of the switching regulator is its high efficiency. On the downside, switching regulators are somewhat more difficult to design, do not respond as fast to transient load conditions, and tend to radiate high-frequency interference.

No matter what type of regulator is used, power dissipation can be rather large in the control device, so heat sinking is generally advisable. Heat sinks allow for a more efficient transferal of heat energy to the surrounding atmosphere than the control device exhibits on its own.

Review Questions

1. What is the function of a voltage regulator?
2. What is the difference between load regulation and line regulation?
3. Why do regulators need a reference voltage?
4. What is the functional difference between a linear regulator and a switching regulator?
5. What are the main advantages of using linear regulators versus switching regulators?
6. What are the main advantages of using switching regulators versus linear regulators?
7. What is the function of a pass transistor?
8. Describe two ways in which to increase the output current of an IC-based regulator.
9. How can fixed "three-pin" regulators be used to regulate at other than their rated voltage?
10. What is the purpose of the output inductor and capacitor in the switching regulator?
11. Explain the correlation between the output current demand and the pulse-width modulator used in switching regulators.
12. What is the purpose of a heat sink?
13. What is meant by the term *thermal resistance*?
14. What are the thermal resistance elements that control heat flow in a typical power-device/heat-sink connection?
15. What are the general rules that should be considered when using heat sinks?

Problems

1. If the average input voltage to the circuit of Problem 9 is 22 V, determine the maximum device dissipation for a 900 mA output.
2. If the average input voltage is 25 V for the circuit of Problem 11, determine the maximum output current for each output voltage. Use the TO-220 case style ($P_D = 15$ W, $I_{limit} = 1.5$ A).
3. Draw a block diagram of a complete ±12 V regulated power supply using LM78XX and LM79XX series parts.
4. Determine the maximum allowable thermal resistance for a heat sink given the following: Ambient temperature = 50°C, maximum operating temperature = 150°C, TO-3 case style with thermal grease and Thermalfilm isolator, power dissipation is 30 W, and the device's thermal resistance is 1.1 C°/W, junction to case.
5. A pass transistor has the following specifications: maximum junction temperature = 125°C, TO-220 case, junction to case thermal resistance = 1.5 C°/W. Determine the maximum power dissipation allowed if this device is connected to a 20 C°/W heat sink with thermal grease, using a .003 mica insulator. The ambient temperature is 35°C.
6. The thermal resistance of the LM723 is 25 C°/W, junction to case. Its maximum operating temperature is 150°C. For a maximum dissipation of 500 mW and an ambient

temperature of 30°C, determine the maximum allowable thermal resistance for the heat-sink/insulator-interface combination.

Design Problems

7. Using Figure 8.4, design a 15 V regulator using a 3.3 V Zener. The Zener bias current should be 2 mA, the output should be capable of 500 mA.
8. Using Figure 8.4 as a guide, design a variable power supply regulator with a 5 to 15 V output range using a 3.9 V Zener. $I_{zener} = 3$ mA.
9. Design a +12 V regulator using the LM317. The output current capability should be at least 900 mA.
10. Design a +3 to +15 V regulator using the LM317. The output should be continuously variable.
11. Using the LM317, configure a regulator to produce either +5V, +12V, or +15V.
12. Design a +12 V regulator using the LM7805.
13. Design a +9 V regulator using the LM723. Use a current limit of 100 mA.
14. Design a +5 V regulator with 100 mA current limiting using the LM723.
15. Configure a ±12 V regulator with 70 mA current limiting. Use the LM326.
16. Reconfigure the circuit of Problem 15 for ±15 V.
17. Using the LM3578A, design a 5 V, 400 mA, regulator. The input voltage is 15 V. Use a discontinuity factor of .2, and an oscillator frequency of 75 kHz. No more than 10 mV of ripple is allowed.
18. Repeat Problem 17 for a 9 V output.

Challenge Problems

19. Based on the LM723 adjustable regulator example, design a regulator that will produce a continuously variable output from 5 V to 12 V.
20. The LM317 has a maximum operating temperature of 125°C. The TO-220 case version shows a thermal resistance of 4 C°/W, junction to case. It also shows 50 C°/W, junction to ambient (no heat sink used). Assuming an ambient temperature of 50°C, what is the maximum allowable power dissipation for each setup? Assume that the first version uses a 15 C°/W heat sink with a 2 C°/W case to heat sink interconnection.
21. Forced air cooling of a heat-sink/power-device can significantly aid in removing heat energy. As a rule of thumb, forced air cooling at a velocity of 1000 feet per minute will effectively increase the efficiency of a heat sink by a factor of 5. Assuming such a system is applied to the circuit of Problem 7, calculate the new power dissipation.
22. An LM317 (TO-3) is used for a 5 V, 1 A power supply. The average voltage into the regulator is 12 V. Assume a maximum operating temperature of 125°C, and an ambient temperature of 25°C. First, determine whether or not a heat sink is required. If it is, determine the maximum acceptable thermal resistance for the heat-sink/insulator combination. For the LM317, thermal resistance = 2.3 C°/W, junction to case, and 35 C°/W, junction to ambient.

Computer Simulation Problems

23. Using a simulator, plot the time-domain response of the circuit of Figure 8.13, assuming an input of 22 V with 3 V peak ripple. How does the simulation change if the ripple is increased to 8 V peak?
24. Verify the output waveform for the circuit of Figure 8.17 using a simulator. Use various loads in order to test the current limit operation. The source is 18 V DC, with 2 V peak ripple.
25. Verify the adjustment range for the regulator designed in Example 8-5 in the text using a simulator. Use a load of 200 Ω, and a source equal to 10 V, with 1 V peak ripple.
26. Use several different loads with a simulator in order to test the current limit portion of the regulator designed in Example 8-5 in the text.

CHAPTER 9

Oscillators and Frequency Generators

CHAPTER OBJECTIVES

After completing this chapter, you should be able to:

- Explain the differences between positive and negative feedback.
- Define the *Barkhausen criterion*, and relate it to individual circuits.
- Detail the need for level-detecting circuitry in a practical oscillator.
- Analyze the operation of Wien bridge and phase shift op amp oscillators.
- Analyze the operation of comparator-based op amp oscillators.
- Detail which factors contribute to the accuracy of a given oscillator.
- Explain the operation of a VCO.
- Explain the operation of a PLL, and define the terms *capture range* and *lock range*.
- Explain the operation of basic timers and waveform generators.

9.1 Introduction

Oscillators are signal sources. Many times, it is necessary to generate waveforms with a known wave shape, frequency, and amplitude. A laboratory signal generator perhaps first comes to mind, but there are many other applications. Signal sources are needed to create and receive radio and television signals, to time events, and to create electronic music, among other uses. Oscillators may produce very low frequencies (a fraction of a cycle per second) to very high frequencies (microwaves, >1 GHz). Oscillators employing op amps are generally used in the area below 1 MHz. Specialized linear circuits may be used at much higher frequencies. The output wave shape may be sinusoidal, triangular, pulse, or some other shape. Oscillators can generally be broken into two broad categories: fixed frequency or variable. For many fixed frequency oscillators, absolute accuracy and freedom from drift are of prime importance. For variable oscillators, ease of tuning and repeatability are usually important. Also, a variable frequency oscillator might not be directly controlled by human hands; rather, the oscillator may be tuned by another circuit. A VCO, or *voltage-controlled oscillator*, is one example of this. Depending on the application, other factors such as total harmonic distortion or risetime may be important.

No matter what the application or how the oscillator design is realized, the oscillator circuit will normally employ positive feedback. Unlike negative feedback, positive feedback is regenerative—it reinforces change. Generally speaking, without some form of positive feedback, oscillators could not be built. In this chapter we are going to look at positive feedback and the requirements for oscillation. A variety of small oscillators based on op amps will be examined. Finally, more powerful integrated circuits will be discussed.

9.2 Op Amp Oscillators

Positive Feedback and the Barkhausen Criterion

In earlier work, we examined the concept of negative feedback. Here, a portion of the output signal is sent back to the input and summed out of phase with the input signal. The difference between the two signals, then, is what is amplified. The result is stability in the circuit response because the large open-loop gain effectively forces the difference signal to be very small. Something quite different happens if the feedback signal is summed in phase with the input signal, as shown in Figure 9.1. In this case, the combined signal looks just like the output signal. As long as the open-loop gain of the amplifier is larger than the feedback factor, the signal can be constantly regenerated. This means that the signal source can be removed. In effect, the output of the circuit is used to create its own input. As long as power is maintained to the circuit, the output signal will continue practically forever. This self-perpetuating state is called *oscillation*. Oscillation will cease if the product of open-loop gain and feedback factor falls below unity or if the feedback signal is not returned perfectly in phase (0° or some integer multiple of 360°). This combination of factors is called the *Barkhausen Oscillation Criterion*. We may state this as follows:

In order to maintain self-oscillation, the closed-loop gain must be unity or greater, and the loop phase must be $N360°$, where $N = 0, 1, 2, 3\ldots$

Note that when we examined linear amplifiers, we looked at this from the opposite end. Normally, you don't want amplifiers to oscillate, and thus you try to guarantee that the Barkhausen Criterion is never met by setting appropriate gain and phase margins.

A good example of positive feedback is the "squeal" sometimes heard from improperly adjusted public address systems. Basically, the microphone is constantly picking up the ambient room noise, which is then amplified and fed to the loudspeakers. If the amplifier gain is high enough or if the acoustic loss is low enough (i.e., the loudspeaker is physically close to the microphone), the signal that the microphone picks up from the loudspeaker can be greater than the ambient noise. The result is that the signal constantly grows in proper phase to maintain oscillation. The result is the

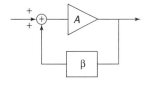

FIGURE 9.1

Positive feedback

FIGURE 9.2

Practical oscillator

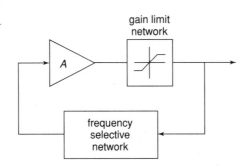

familiar squealing sound. In order to stop the squeal, either the gain or the phase must be disrupted. Moving the microphone may change the relative phase, but it is usually easier to just reduce the volume a little. It is particularly interesting to listen to a system that is on the verge of oscillation. Either the gain or the phase is just not quite perfect, and the result is a rather irritating ringing sound, as the oscillation dies out after each word or phrase.

There are a couple of practical considerations to be aware of when designing oscillators. First of all, it is not necessary to provide a "start-up" signal source as seen in Figure 9.1. Normally, there is enough energy in either the input noise level or possibly in a turn-on transient to get the oscillator started. Both the turn-on transient and the noise signal are broad spectrum signals, so the desired oscillation frequency is contained within either of them. The oscillation signal will start to increase as time progresses due to the closed-loop gain being greater than one. Eventually, the signal will reach a point where further level increases are impossible due to amplifier clipping. For a more controlled, low-distortion oscillator, it is desirable to have the gain start to roll off before clipping occurs. In other words, the closed-loop gain should fall back to exactly one. Finally, in order to minimize frequency drift over time, the feedback network should be selective. Frequencies either above or below the target frequency should see greater attenuation than the target frequency. Generally, the more selective (i.e., higher Q) this network is, the more stable and accurate the oscillation frequency will be. One simple solution is to use an RLC tank circuit in the feedback network. Another possibility is to use a piezo-electric crystal. A block diagram of a practical oscillator is shown in Figure 9.2.

A Basic Oscillator

A real-world circuit that embodies all of the elements is shown in Figure 9.3. This circuit is not particularly efficient or cost effective, but it does illustrate the important points. Remember, in order to maintain oscillation the closed-loop gain of the oscillator circuit must be greater than 1, and the loop phase must be a multiple of 360°. To provide gain, a pair of inverting amplifiers is used. Note op amp 2 serves to buffer the output signal. As each stage produces a 180° shift, the shift for the pair is 360°. The product of the gains has to be larger than the loss produced by the frequency selecting network. This network is made up of R_3, L, and C. Because the LC combination produces an impedance peak at the resonant frequency, f_o, a minimum loss will occur there. Also, at

FIGURE 9.3

A basic oscillator

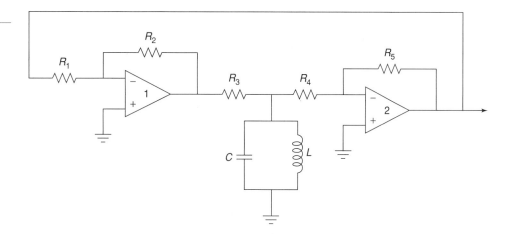

resonance, the circuit is basically resistive, so no phase change occurs. Consequently, this circuit should oscillate at the f_o set by L and C. This circuit can be easily tested in lab. For example, if you drop the gain of one of the op amp stages, there will not be enough system gain to overcome the tank circuit's loss, and thus, oscillation will cease. You can also verify the phase requirement by replacing one of the inverting amplifiers with a noninverting amplifier of equal gain. The resulting loop phase of 180° will halt oscillation. This circuit does not include any form of automatic gain adjustment, so the output signal may be clipped. If properly chosen, the slew rate of the op amp may be used as the limit factor. (A 741 will work acceptably for f_o in the low kHz range.) Although this circuit does work and points out the specifics, it is certainly not a top choice for an oscillator design based on op amps.

Wien Bridge Oscillator

A relatively straightforward design useful for general-purpose work is the *Wien bridge oscillator*. This oscillator is far simpler than the generalized design shown in Figure 9.3, and offers very good performance. The frequency selecting network is a simple lead/lag circuit, such as that shown in Figure 9.4. This circuit is a frequency-sensitive voltage divider. It combines the response of both the simple lead and lag networks. Normally, both resistors are set to the same value. The same may be said for the two capacitors. At very low frequencies, the capacitive reactance is essentially infinite, and thus, the upper series capacitor appears as an open. Because of this, the output voltage is zero. Likewise, at very high frequencies the capacitive reactance approaches zero, and the lower shunt capacitor effectively shorts the output to ground. Again, the output voltage is zero. At some middle frequency the output voltage will be at a peak. This will be the preferred, or selected, frequency and will become the oscillation frequency so long as the proper phase relation is held. We need to determine the phase change at this point as well as the voltage divider ratio. These items are needed in order to guarantee that the Barkhausen conditions are met.

Figure 9.4
Lead/lag network

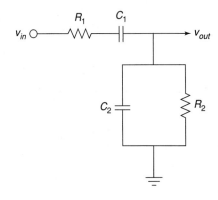

First, note that

$$\beta = \frac{Z_2}{Z_1 + Z_2}$$

where $Z_1 = R_1 - jX_{C_1}$, and $Z_2 = R_2 || - jX_{C_2}$.

$$Z_2 = \frac{-jX_{C_2} R_2}{-jX_{C_2} + R_2}$$

$$Z_2 = \frac{-jX_{C_2} R_2}{-jX_{C_2}\left(1 + \frac{R_2}{-jX_{C_2}}\right)}$$

$$Z_2 = \frac{R_2}{1 + \frac{R_2}{-jX_{C_2}}}$$

Recalling that $X_C = \frac{1}{\omega C}$, we find

$$Z_1 = R_1 - \frac{j}{\omega C_1}$$

$$Z_2 = \frac{R_2}{1 + j\omega R_2 C_2}$$

So

$$\beta = \frac{\frac{R_2}{1+j\omega R_2 C_2}}{\frac{R_2}{1+j\omega R_2 C_2} + R_1 - \frac{j}{\omega C_1}}$$

$$\beta = \frac{R_2}{R_2 + R_1 - \frac{j}{\omega C_1} + j\omega R_1 R_2 C_2 + \frac{R_2 C_2}{C_1}}$$

$$\beta = \frac{R_2}{R_2\left(1 + \frac{C_2}{C_1}\right) + R_1 + j\left(\omega R_1 R_2 C_2 - \frac{1}{\omega C_1}\right)} \quad (9.1)$$

We can determine the desired frequency from the imaginary portion of Equation 9.1.

$$\omega R_1 R_2 C_2 = \frac{1}{\omega C_1}$$

$$\omega^2 = \frac{1}{R_1 R_2 C_1 C_2}$$

$$\omega = \frac{1}{\sqrt{R_1 R_2 C_1 C_2}} \tag{9.2}$$

Normally, $C_1 = C_2$ and $R_1 = R_2$, so Equation 9.2 reduces to

$$\omega = \frac{1}{RC}$$

$$\text{or, } f_o = \frac{1}{2\pi RC} \tag{9.3}$$

To find the magnitude of the feedback factor, and thus the required forward gain of the op amp, we need to examine the real portion of Equation 9.1.

$$\beta = \frac{R_2}{R_2\left(1 + \frac{C_2}{C_1}\right) + R_1}$$

Assuming that equal components are used, this reduces to

$$\beta = \frac{R}{3R}$$

$$\text{or simply, } \beta = \frac{1}{3}$$

The end result is that the forward gain of the amplifier must have a gain of slightly over 3 and a phase of 0° in order to maintain oscillation. It also means that the frequency of oscillation is fairly easy to set and can even be adjusted if potentiometers are used to replace the two resistors. The final circuit is shown in Figure 9.5. This circuit uses a combination of negative feedback and positive feedback to achieve oscillation. The positive feedback loop utilizes R_t and C. The negative feedback loop utilizes R_a and R_b. R_b must be approximately twice the size of R_a. If it is smaller, the $A\beta$ product will be less than unity and oscillation cannot be maintained. If the gain is significantly larger, excessive distortion may result. Indeed, some form of gain reduction at higher output voltages is desired for this circuit. One possibility is to replace R_a with a lamp. As the signal amplitude increases across the lamp, its resistance increases, thus decreasing gain. At a certain point the lamp's resistance will be just enough to produce an $A\beta$ product of exactly 1. Another technique is shown in Figure 9.6. Here an opposite approach is taken. Resistor R_b is first broken into two parts, the smaller part, R_{b2}, is shunted by a pair of signal diodes. For lower amplitudes, the diodes are off and do not affect the circuit operation. At higher amplitudes, the diodes start to turn

FIGURE 9.5

Wien bridge oscillator

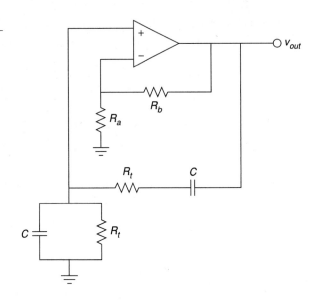

on, and thus start to short R_{b2}. If correctly implemented, this action is not instantaneous and does not produce clipping. It simply serves to reduce the gain at higher amplitudes.

Another way of drawing the Wien bridge oscillator is shown in Figure 9.7. This form clearly shows the Wien bridge configuration. Note that the output of the bridge is the differential input voltage (i.e., error voltage). In operation, the bridge is balanced, and thus, the error voltage is zero.

FIGURE 9.6

Wien bridge oscillator with gain adjustment

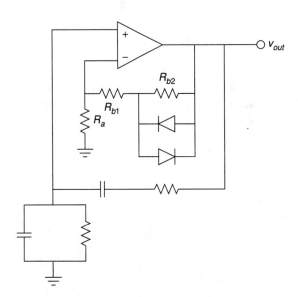

FIGURE 9.7

Wien bridge oscillator, redrawn

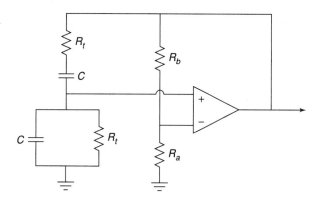

EXAMPLE 9-1

Determine the frequency of oscillation for the circuit of Figure 9.8.

$$f_o = \frac{1}{2\pi RC}$$

$$f_o = \frac{1}{2\pi \times 50\text{ k} \times .01\ \mu\text{F}}$$

$$f_o = 318\text{ Hz}$$

For other frequencies, either R or C may be altered as required. Also, note that the forward gain works out to exactly 3, thus perfectly compensating the positive feedback factor of $\frac{1}{3}$. In reality, component tolerances make this circuit impractical. To overcome this difficulty, a small resistor/diode combination may be placed in series with the 20 kΩ, as shown in Figure 9.6. A typical resistor value would be about one-fourth to one-half the value of R_f, or about 5 kΩ to 10 kΩ in this example. R_f would be decreased slightly as well (or, R_i might be increased).

FIGURE 9.8

Oscillator for Example 9-1

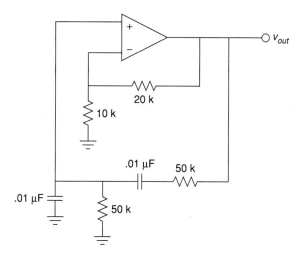

The ultimate accuracy of f_o depends on the tolerances of R and C. If 10% parts are used in production, a variance of about 20% is possible. Also, at higher frequencies, the op amp circuit will produce a moderate phase shift of its own. Thus, the assumption of a perfect noninverting amplifier is no longer valid, and some error in the output frequency will result. With extreme values in the positive feedback network, it is also possible that some shift of the output frequency may occur due to the capacitive and resistive loading effects of the op amp. Normally, this type of loading is not a problem, as the op amp's input resistance is very high, and its input capacitance is quite low.

EXAMPLE 9-2

Figure 9.9 shows an adjustable oscillator. Three sets of capacitors are used to change the frequency range, whereas a dual-gang potentiometer is used to adjust the frequency within a given range. Determine the maximum and minimum frequency of oscillation within each range.

First, note that the capacitors are spaced by decades. This means that the resulting frequency ranges will also change by factors of 10. The .1 μF capacitor will produce the lowest range, the .01 μF will produce a range 10 times higher, and the .001 μF range will be 10 times higher still. Thus, we only need to calculate the range produced by the .1 μF.

FIGURE 9.9

Adjustable oscillator

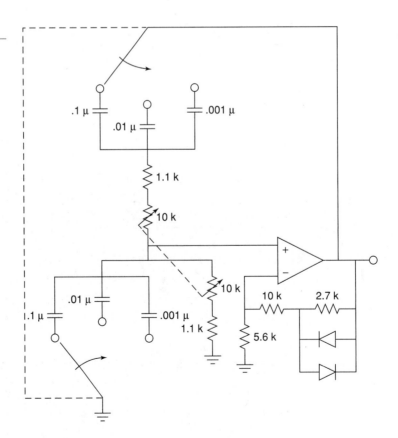

The maximum frequency of oscillation within a given range will occur with the lowest possible resistance. The minimum resistance is seen when the 10 kΩ pot is fully shorted, the result being 1.1 kΩ. Conversely, the minimum frequency will occur with the largest resistance. When the pot is fully in the circuit, the resulting sum is 11.1 kΩ. Note that a dual-gang potentiometer means that both units are connected to a common shaft; thus, both pots track in tandem.

For $f_{minimum}$ with .1 μF:

$$f_o = \frac{1}{2\pi RC}$$
$$f_o = \frac{1}{2\pi \times 11.1\,k \times .1\,\mu F}$$
$$f_o = 143.4\text{ Hz}$$

For $f_{maximum}$ with .1 μF:

$$f_o = \frac{1}{2\pi RC}$$
$$f_o = \frac{1}{2\pi \times 1.1\,k \times .1\,\mu F}$$
$$f_o = 1.447\text{ kHz}$$

For the .01 μF, the ranges would be 1.434 kHz to 14.47 kHz, and for the .001 μF, the ranges would be 14.34 kHz to 144.7 kHz. Note that each range picks up where the previous one left off. In this way, there are no "gaps," or unobtainable frequencies. For stable oscillation, this circuit must have a gain of 3. For low-level outputs, the diodes will not be active, and the forward gain will be

$$A_v = 1 + \frac{R_f}{R_i}$$
$$A_v = 1 + \frac{10\,k + 2.7\,k}{5.6\,k}$$
$$A_v = 3.27$$

As the signal rises, the diodes begin to turn on, thus shunting the 2.2 kΩ resistor and dropping the gain back to exactly 3.

Phase Shift Oscillator

Considering the Barkhausen Criterion, it should be possible to create an oscillator by using a simple phase shift network in the feedback path. For example, if the circuit uses an inverting amplifier (−180° shift), a feedback network with an additional 180° shift should create oscillation. The only other requirement is that the gain of the inverting amplifier be greater than the loss produced by the feedback network. This is illustrated in Figure 9.10. The feedback network can be as simple as three cascaded lead networks. The lead networks will produce a combined phase shift of 180° at only one frequency. This will become the frequency of oscillation. In general, the feedback network will look

Figure 9.10

Block diagram of phase shift oscillator

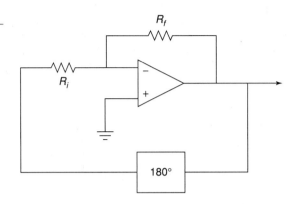

something like the circuit of Figure 9.11. This form of *RC* layout is usually referred to as a *ladder network*.

There are many ways in which R and C may be set in order to create the desired 180° shift. Perhaps the most obvious scheme is to set each stage for a 60° shift. The components are determined by finding a combination that produces a 60° shift at the desired frequency. In order to avoid loading effects, each stage must be set to higher and higher impedances. For example, R_2 might be set to 10 times the value of R_1, and R_3 set to 10 times the value of R_2. The capacitors would see a corresponding decrease. Because the tangent of the phase shift yields the ratio of X_C to R, at our desired 60° we find

$$\tan 60 = \frac{X_C}{R}$$
$$1.732 = \frac{X_C}{R}$$
$$X_C = 1.732R$$

Using this in the general reactance formula produces

$$f_o = \frac{1}{2\pi \, 1.732 \, RC} \tag{9.4}$$

Figure 9.11

Phase shift network

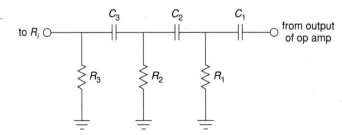

Likewise, a look at the magnitude shows the approximate loss per stage of

$$\beta = \frac{R}{\sqrt{R^2 + X_C^2}}$$

$$\beta = \frac{R}{\sqrt{R^2 + (1.732R)^2}}$$

$$\beta = .5$$

Because there are three stages, the total loss for the feedback network would be .125. Therefore, the inverting amplifier needs a gain of 8 in order to set the $A\beta$ product to unity. Remember, these results are approximate and depend on minimum interstage loading. A more exacting analysis will follow shortly.

EXAMPLE 9-3

Determine the frequency of oscillation in Figure 9.12.

$$f_o = \frac{1}{2\pi \, 1.732 \, RC}$$

$$f_o = \frac{1}{2\pi \times 1.732 \times 1\,k \times .1\,\mu F}$$

$$f_o = 919 \text{ Hz}$$

Figure 9.12 graphically points up the major problem with the "60° per stage" concept. In order to prevent loading, the final resistors must be very high. In this case a feedback resistor of 8 MΩ is required. It is possible to simplify the circuit somewhat

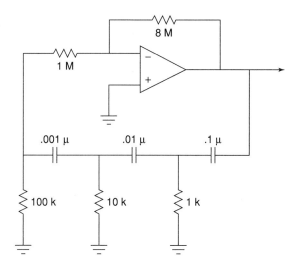

FIGURE 9.12

Phase shift oscillator (minimum loading form)

FIGURE 9.13

Improved phase shift oscillator

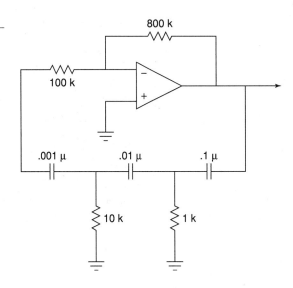

by omitting the 1 MΩ and connecting the 100 kΩ directly to the op amp as shown in Figure 9.13. This saves one part and does allow the feedback resistor to be dropped in value, but the resulting component spread is still not ideal. If we accept the loading effect, we can simplify things a bit by making each resistor and capacitor the same size, as shown in Figure 9.14. With these values, we can be assured that the resulting frequency will no longer be 919 Hz, as Equation 9.4 is no longer valid. Also, it is quite likely that the loss produced by the network will no longer be .125. We need to determine the general input/output ($\frac{V_0}{V_3}$) ratio of the ladder network, and from this, find the gain and frequency relations for a net phase shift of $-180°$. One technique involves the use of simultaneous loop equations. Because all resistors and capacitors are equal in this variation, we will be able to simplify our equations readily. By inspection, the three loop equations are (from left to right):

$$V_0 = (R + X_C)I_1 - RI_2 \tag{9.5}$$
$$0 = -RI_1 + (2R + X_C)I_2 - RI_3 \tag{9.6}$$
$$0 = -RI_2 + (2R + X_C)I_3 \tag{9.7}$$

FIGURE 9.14

Phase shift analysis

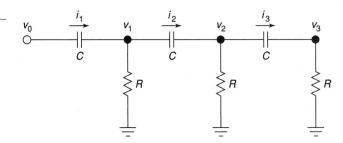

Also, note that

$$V_3 = I_3 R \qquad (9.8)$$

We now have expressions for V_0 and V_3, however, V_0 is in terms of I_1 and I_2, and V_3 is in terms of I_3. Write I_1 and I_2 in terms of I_3 so that we can substitute these back into Equation 9.5. Rewriting Equation 9.7 yields an expression for I_2

$$I_2 = I_3\left(2 + \frac{X_C}{R}\right) \qquad (9.9)$$

For I_1, rewrite Equation 9.6

$$I_1 = \left(2 + \frac{X_C}{R}\right)I_2 - I_3 \qquad (9.10)$$

Substituting Equation 9.9 into Equation 9.10 yields

$$I_1 = I_3\left(\left(2 + \frac{X_C}{R}\right)^2 - 1\right) \qquad (9.11)$$

Thus, V_0 may be rewritten as

$$V_0 = I_3(R + X_C)\left(\left(2 + \frac{X_C}{R}\right)^2 - 1\right) - I_3 R\left(2 + \frac{X_C}{R}\right) \qquad (9.12)$$

Equation 9.12 may be simplified to

$$V_0 = I_3\left(R\left(2 + \frac{X_C}{R}\right)^2 + X_C\left(2 + \frac{X_C}{R}\right)^2 - 3R - 2X_C\right) \qquad (9.13)$$

The input/output expression is simplified as follows.

$$\frac{V_0}{V_3} = \left(R\left(2 + \frac{X_C}{R}\right)^2 + X_C\left(2 + \frac{X_C}{R}\right)^2 - 3R - 2X_C\right)\left(\frac{1}{R}\right) \qquad (9.14)$$

$$\frac{V_0}{V_3} = \left(2 + \frac{X_C}{R}\right)^2 + \frac{X_C}{R}\left(2 + \frac{X_C}{R}\right)^2 - 3 - \frac{2X_C}{R}$$

$$\frac{V_0}{V_3} = 1 + \frac{6X_C}{R} + \frac{5X_C^2}{R^2} + \frac{X_C^3}{R^3} \qquad (9.15)$$

At this point, we are nearly done with the general equation. All that is left is to substitute $\frac{1}{j\omega C}$ in place of X_C. Remember, $j^2 = -1$.

$$\frac{V_0}{V_3} = 1 + \frac{6}{j\omega C R} - \frac{5}{\omega^2 C^2 R^2} - \frac{1}{j\omega^3 C^3 R^3} \qquad (9.16)$$

This equation contains both real and imaginary terms. For this equation to be satisfied, the imaginary components ($\frac{6}{j\omega CR}$ and $\frac{1}{j\omega^3 C^3 R^3}$) must sum to zero, and the real components must similarly sum to zero. (As there are only two terms, their magnitudes must be equal.) We can use these facts to find both the gain and the frequency.

$$\frac{6}{j\omega CR} = \frac{1}{j\omega^3 C^3 R^3}$$

$$\frac{1}{j\omega CR} = \frac{1}{6j\omega^3 C^3 R^3}$$

$$1 = \frac{1}{6\omega^2 C^2 R^2}$$

$$\omega^2 = \frac{1}{6C^2 R^2} \tag{9.17}$$

$$\omega = \frac{1}{\sqrt{6}CR} \tag{9.18}$$

$$\text{or, } f_o = \frac{1}{2\pi\sqrt{6}CR} \tag{9.19}$$

For the gain, we solve Equation 9.16 in terms of the voltage ratio and zero the imaginary terms, because the result must be real.

$$\frac{V_0}{V_3} = 1 - \frac{5}{\omega^2 C^2 R^2} \tag{9.20}$$

Substituting Equation 9.17 into Equation 9.20 yields

$$\frac{V_0}{V_3} = 1 - \frac{5}{\frac{1}{6C^2 R^2} C^2 R^2} \tag{9.21}$$

$$\frac{V_0}{V_3} = 1 - 5 \times 6$$

$$\frac{V_0}{V_3} = -29 \tag{9.22}$$

The gain of the ladder network is $\frac{V_3}{V_0}$, or the reciprocal of Equation 9.22, or

$$\beta = \frac{1}{-29} \tag{9.23}$$

The loss produced will be 1/29. This has the disadvantage of requiring a forward gain of 29 instead of 8 (as in the previous form). This disadvantage is minor compared to the advantage of reasonable component values.

FIGURE 9.15

Equal component phase shift oscillator

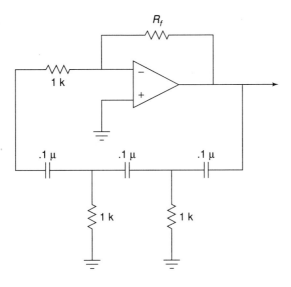

EXAMPLE 9-4

Determine a value for R_f in Figure 9.15 in order to maintain oscillation. Also determine the oscillation frequency.

Equation 9.23 shows that the inverting amplifier must have a gain of 29.

$$A_v = -\frac{R_f}{R_i}$$
$$R_f = -R_i A_v$$
$$R_f = -1\,k \times -29$$
$$R_f = 29\,k$$

Of course, the higher standard value will be used. Also, in order to control the gain at higher levels, a diode/resistor combination (as used in the Wien bridge circuits) should be placed in series with R_f. Without a gain limiting circuit, excessive distortion may occur.

$$f_o = \frac{1}{2\pi \sqrt{6}\, RC}$$
$$f_o = \frac{1}{2\pi \sqrt{6} \times 1\,k \times .1\,\mu F}$$
$$f_o = 650\,Hz$$

multiSIM COMPUTER SIMULATION

In order to verify Equations 9.3 and 9.19, the gain and phase responses of the feedback networks of Figures 9.13 and 9.15 are found in Figure 9.16 (pp. 342–345). These graphs were obtained via the standard Electronics Workbench MultiSIM route. Note that the phase for both circuits hits $-180°$, very close to the predicted frequencies. The equal value network prediction is very accurate, whereas the staggered network prediction is off by only a few percent. Once the phase exceeds $-180°$, the MultiSIM grapher wraps it back to $+180°$, so it is very easy to see this frequency. Likewise, the gain response agrees with the derivations for attenuation at the oscillation frequency. It can be very instructive to analyze these circuits for the gain and phase response at each stage as well.

FIGURE 9.16a

Equal value network in MultiSIM

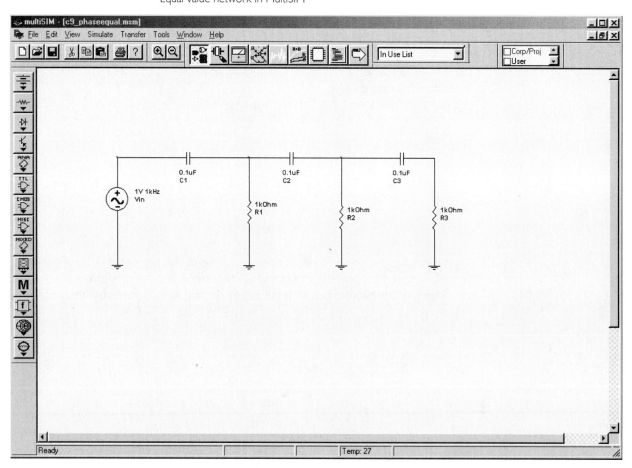

FIGURE 9.16b

Response of equal value network

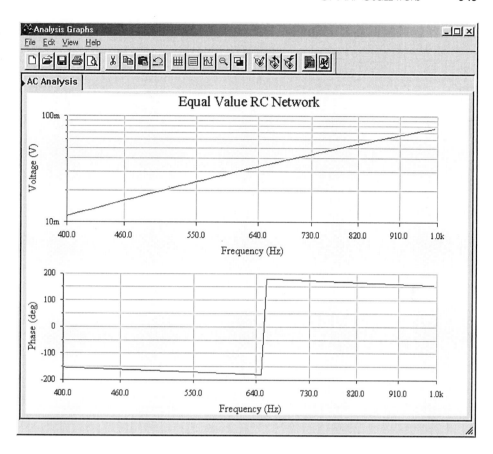

Square/Triangle Function Generator

Besides generating sine waves, op amp circuits may be employed to generate other wave shapes such as ramps, triangle waves, or pulses. Generally speaking, square- and pulse-type waveforms may be derived from other sources through the use of a comparator. For example, a square wave may be derived from a sine wave by passing it through a comparator, such as those seen in Chapter 7. Linear waveforms such as triangles and ramps may be derived from the charge/discharge action of a capacitor. As you may recall from basic circuit theory, the voltage across a capacitor will rise linearly if it is driven by a constant current source. One way of achieving this linear rise is with the circuit of Figure 9.17. In essence, this circuit is an inverting amplifier with a capacitor taking the place of R_f. The input resistor, R, turns the applied input voltage into a current. Because the current into the op amp itself is negligible, this current flows directly into capacitor C. As in a normal inverting amplifier, the output voltage is equal to the voltage across the feedback element, though inverted. The relationship between the capacitor current and voltage is

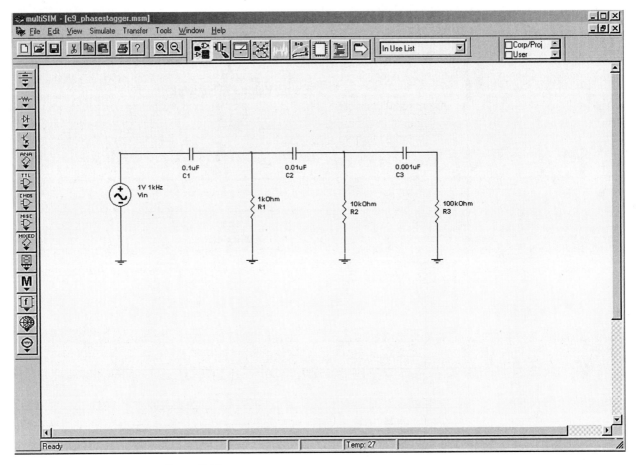

Figure 9.16c

Staggered value network in MultiSIM

$$\frac{dv}{dt} = \frac{i}{C} \quad (9.24)$$

$$V(t) = \frac{1}{C} \int i\, dt$$

$$V_{out} = -\frac{1}{C} \int i\, dt \quad (9.25)$$

As expected, a fast rise can be created by either a small capacitor or a large current. (As a side note, this circuit is called an *integrator* and will be examined in greater detail in the next chapter.)

By choosing appropriate values for R and C, the V_{out} ramp may be set at a desired rate. The polarity of the ramp's slope is determined by the direction of the input current; a positive source will produce a negative going ramp and vice versa. If the polarity of the

FIGURE 9.16d

Response of staggered value network

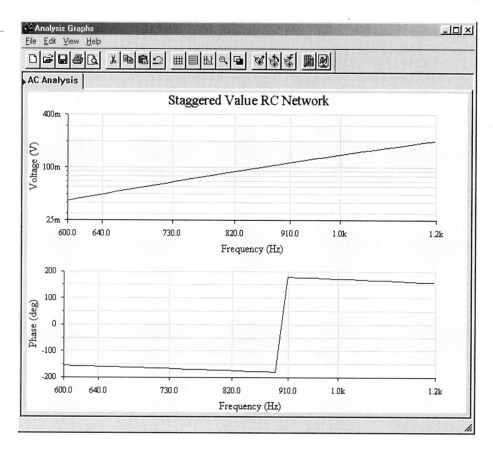

input changes at a certain rate, the output ramp will change direction in tandem. The net effect is a triangle wave. A simple way to generate the alternating input polarity is to drive R with a square wave. As the square wave changes from plus to minus, the ramp changes direction. This is shown in Figure 9.18. So, we are now able to generate a triangle wave. The only problem is that a square wave source is needed. How do we produce the square source? As mentioned earlier, a square wave may be derived by passing an AC signal through a comparator. Logically then, we should be able to pass the output

FIGURE 9.17

Ramp generator

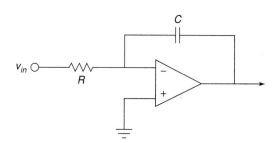

FIGURE 9.18

Ramp generator waveforms

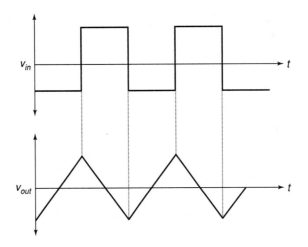

triangle wave into a comparator in order to create the needed square wave. The resulting circuit is shown in Figure 9.19. A comparator with hysteresis is used to turn the triangle into a square wave. The square then drives the ramp circuit. The circuit produces two simultaneous outputs: a square wave that swings to ± saturation and a triangle wave that swings to the upper and lower comparator thresholds. This is shown in Figure 9.20. The thresholds may be determined from the equations presented in Chapter 7. In order to determine the output frequency, the V/S rate of the ramp is determined from Equation 9.24. Knowing the peak-to-peak swing of the triangle is $V_{upperthres} - V_{lowerthres}$, the period of the wave may be found. The output frequency is the reciprocal of the period.

EXAMPLE 9-5

Determine the output frequency and amplitudes for the circuit of Figure 9.21. Use $V_{sat} = \pm 13$ V.

First, note that the comparator always swings between $+V_{sat}$ and $-V_{sat}$. Now, determine the upper and lower thresholds for the comparator.

FIGURE 9.19

Triangle/square generator

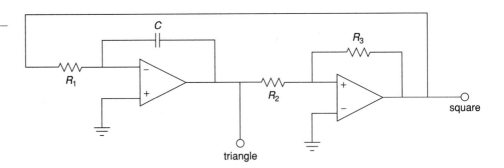

Figure 9.20
Output waveforms of triangle/square generator

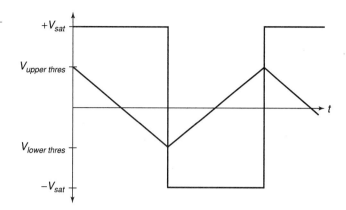

$$V_{upperthres} = V_{sat}\frac{R_2}{R_3}$$

$$V_{upperthres} = 13\text{ V}\frac{10\text{ k}}{20\text{ k}}$$

$$V_{upperthres} = 6.5\text{ V}$$

The lower threshold will be -6.5 V. We now know that the triangle wave output will be 13 V peak-to-peak. From this we may determine the output period. Because the ramp generator is driven by a square wave with an amplitude of V_{sat}, Equation 9.24 may be rewritten as

$$\frac{dv}{dt} = \frac{V_{sat}}{RC}$$

$$\frac{dv}{dt} = \frac{13\text{ V}}{33\text{ k} \times .01\text{ }\mu\text{F}}$$

$$\frac{dv}{dt} = 39,394\text{ V/S}$$

The time required to produce the 13 V peak-to-peak swing is

$$T = \frac{13\text{ V}}{39,394\text{ V/S}}$$

$$T = .33\text{ mS}$$

Figure 9.21
Signal generator for Example 9-5

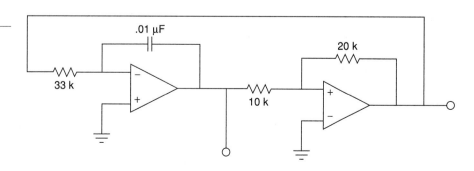

This represents one half-cycle of the output wave. To go from +6.5 V to −6.5 V and back will require .66 mS. Therefore, the output frequency is

$$f = \frac{1}{T}$$

$$f = \frac{1}{.66 \text{ mS}}$$

$$f = 1.52 \text{ kHz}$$

This frequency in Example 9.5 may be adjusted by changing either the 33 kΩ resistor or the .01 μF capacitor. Changing the comparator's resistors can alter the thresholds, and thus alter the frequency, but this is generally not recommended, as a change in output amplitude will occur as well. By combining steps, the above process may be reduced to a single equation:

$$f = \frac{1}{\frac{2V_{pp}}{V_{sat}} RC} \tag{9.26}$$

where V_{pp} is the difference between $V_{upperthres}$ and $V_{lowerthres}$. Note that if R_3 is 4 times larger than R_2 in the comparator, Equation 9.26 reduces to

$$f = \frac{1}{RC}$$

and the peak triangle wave amplitude is one-fourth of V_{sat}.

Generally, circuits such as this are used for lower frequency work. For clean square waves, very fast op amps are required. Finally, for lower impedance loads, the outputs should be buffered with voltage followers.

multiSIM Computer Simulation

The Electronics Workbench MultiSIM simulation for the signal generator of Example 9-5 is shown in Figure 9.22 (pp. 349–350). The square and triangle outputs are plotted together so that the switching action can be seen. Note how each wave is derived from the other. The output plot is delayed 5 milliseconds in order to guarantee a plot of the steady state output. Failure to delay the plotting times will result in a graph of the initial turn-on transients. It may take many milliseconds before the waveforms finally stabilize, depending on the desired frequency of oscillation and the initial circuit conditions. Finally, note the sharp rising and falling edges of the square wave. This is due to the moderately fast slew rate of the LF411 op amp chosen. Had a slower device such as the 741 been used, the quality of the output waveforms would have suffered.

If an accurate triangle wave is not needed, and only a square wave is required, the circuit of Figure 9.19 may be reduced to a single op amp stage. This is shown in

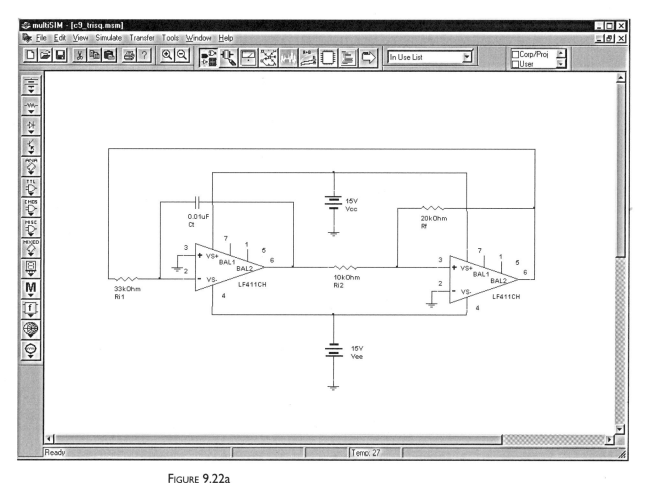

FIGURE 9.22a

Triangle/square generator in MultiSIM

Figure 9.23. This circuit is, in essence, a comparator. Resistors R_1 and R_2 form the positive feedback portion and set the effective comparator trip point, or threshold. The measurement signal is the voltage across the capacitor. The potentials of interest are shown in Figure 9.24. If the output is at positive saturation, the noninverting input will see a percentage of this, depending on the voltage divider produced by R_1 and R_2. This potential is $V_{upperthres}$. Because the output is at positive saturation, the capacitor C will be charging towards it. Because it is charging through resistor R, the waveform is an exponential type. Once the capacitor voltage reaches $V_{upperthres}$, the noninverting input will no longer be greater than the inverting input, and the device will change to the negative state. At this point, C will reverse its course and move towards negative saturation. At the lower threshold, the op amp will again change state, and the process repeats. In order to determine the frequency of oscillation, we need to find how long it takes the capacitor to charge between the two threshold points. Normally the

FIGURE 9.22b

Output waveforms from simulator

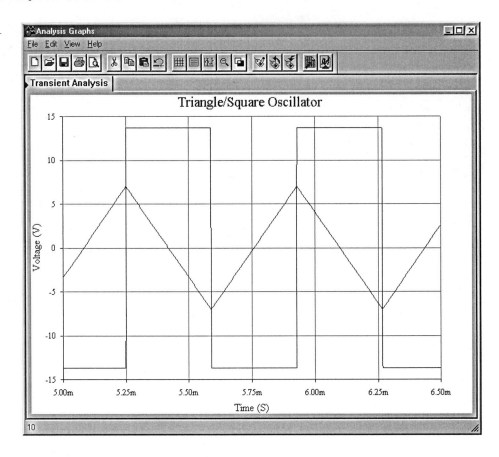

FIGURE 9.23

Simple square wave generator

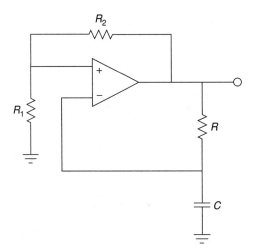

Op Amp Oscillators   351

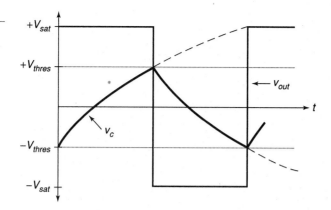

Figure 9.24

Waveforms of a simple square wave generator

circuit will be powered from equal magnitude supplies, and therefore $+V_{sat} = -V_{sat}$ and $V_{upperthres} = V_{lowerthres}$. By inspection,

$$V_{thres} = V_{sat} \frac{R_1}{R_1 + R_2} \qquad (9.27)$$

The capacitor voltage is

$$V_C(t) = V_k \left(1 - \epsilon^{\frac{-t}{RC}}\right) \qquad (9.28)$$

where V_k is the total potential applied to the capacitor. Because the capacitor will start at one threshold and attempt to charge to the opposite saturation limit, this is

$$V_k = V_{sat} + V_{thres} \qquad (9.29)$$

Combining Equation 9.27, Equation 9.28, and Equation 9.29 yields

$$V_C(t) = (V_{sat} + V_{thres})\left(1 - \epsilon^{\frac{-t}{RC}}\right) \qquad (9.30)$$

At the point where the comparator changes state,

$$V_C = 2V_{thres} \qquad (9.31)$$

Combining Equation 9.30 and Equation 9.31 produces

$$2V_{thres} = (V_{sat} + V_{thres})\left(1 - \epsilon^{\frac{-t}{RC}}\right)$$

$$\frac{2V_{thres}}{V_{sat} + V_{thres}} = 1 - \epsilon^{\frac{-t}{RC}}$$

$$\frac{2V_{sat}\frac{R_1}{R_1+R_2}}{V_{sat}\left(1 + \frac{R_1}{R_1+R_2}\right)} = 1 - \epsilon^{\frac{-t}{RC}}$$

$$\frac{2R_1}{2R_1 + R_2} = 1 - \epsilon^{\frac{-t}{RC}}$$

$$\frac{R_2}{2R_1 + R_2} = \epsilon^{\frac{-t}{RC}}$$

$$\ln\left(\frac{R_2}{2R_1 + R_2}\right) = \frac{-t}{RC}$$

$$t = RC \ln\left(\frac{2R_1 + R_2}{R_2}\right)$$

This represents the charge time of the capacitor. One period requires two such traverses, so we may say

$$T = 2RC \ln\left(\frac{2R_1 + R_2}{R_2}\right)$$

$$\text{or, } f_o = \frac{1}{2RC \ln\left(\frac{2R_1+R_2}{R_2}\right)} \tag{9.32}$$

We can transform Equation 9.32 into "nicer" forms by choosing values for R_1 and R_2 such that the log term turns into a convenient number, such as 1 or .5. If we set $R_1 = .859 R_2$, the log term is unity, and Equation 9.32 becomes

$$f_o = \frac{1}{2RC}$$

EXAMPLE 9-6

Design a 2 kHz square wave generator using the circuit of Figure 9.23.
For convenience, set $R_1 = .859 R_2$. If R_1 is arbitrarily set to 10 kΩ, then

$$R_1 = .859 R_2$$
$$R_2 = \frac{R_1}{.859}$$
$$R_2 = \frac{10\text{ k}}{.859}$$
$$R_2 = 11.64\text{ k}$$

In order to set the oscillation frequency, R is arbitrarily set to 10 kΩ, and C is then determined.

$$f_o = \frac{1}{2RC}$$
$$C = \frac{1}{2R f_o}$$
$$C = \frac{1}{2 \times 10\text{ k} \times 2\text{ kHz}}$$
$$C = 25\text{ nF}$$

multiSIM COMPUTER SIMULATION

A simulation of the square wave generator of Example 9-6 is shown in Figure 9.25 (pp. 353–355). In order to graphically illustrate the importance of the op amp having sufficient bandwidth and slew rate, the simulation is run twice, once using the moderately fast LF411 and a second time using the much slower 741. Both the output and capacitor voltages are plotted from the Transient Analysis. Using the LF411, the output waveform is very crisp with sharp rising and falling edges. The capacitor voltage appears exactly as it should. The resulting frequency is just a little lower than the 2 kHz target. In contrast, the 741 plots show some problems. First, the square wave has noticeable slew rate limiting on the transitions. Second, due to the slewing problems, the capacitor voltage waveshape appears distorted (note the excessive rounding of the peaks). These effects combine to produce a frequency about 15% lower than the target, or about 1.7 kHz. The end result is a lackluster output waveform.

FIGURE 9.25a

Square wave generator in MultiSIM

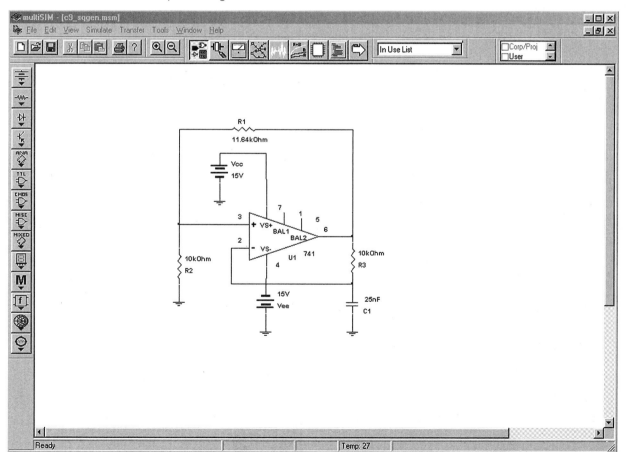

Figure 9.25b

Waveforms using LF411

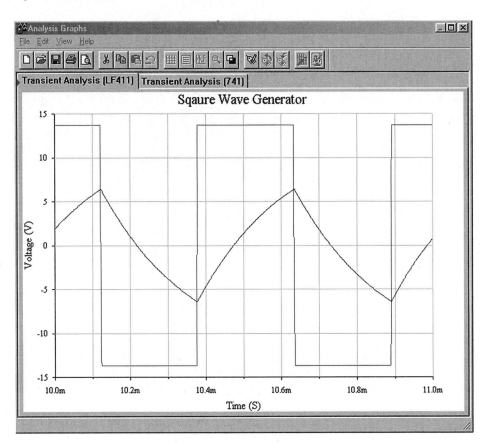

9.3 Single Chip Oscillators and Frequency Generators

The generation of signals is a basic requirement for a wide variety of applications, thus a number of manufacturers produce a selection of single IC oscillators and frequency generators. The majority of these tend to work in the range below 1 MHz and usually require some form of external resistor/capacitor network to set the operating frequency. Highly specialized circuits for targeted applications are also available. In this section we shall examine four popular ICs that lend themselves to a wide range of applications: the NE566 voltage-controlled oscillator, the NE565 *phase locked loop*, the NE555 timer, and the ICL8038 waveform generator.

Voltage-Controlled Oscillator

A voltage-controlled oscillator (usually abbreviated as VCO) does not produce a fixed output frequency. As its name suggests, the output frequency of a VCO is dependent on a control voltage. There is a fixed relationship between the control voltage and the output frequency. Theoretically, just about any oscillator can be turned into a VCO. For example, if a resistor is used as part of the tuning circuit, it could be replaced with some

FIGURE 9.25c

Waveforms using 741

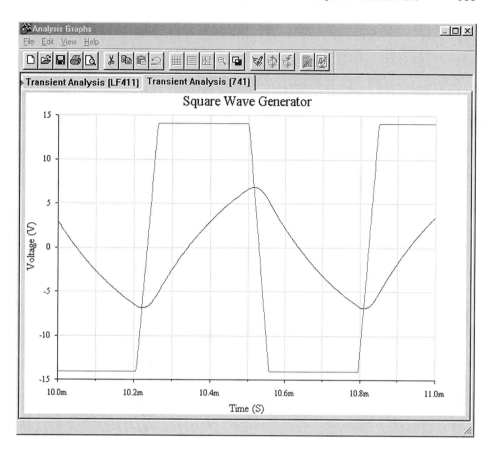

form of voltage-controlled resistor, such as an FET or a light-dependent resistor/lamp combination. By doing this, an external potential can be used to set the frequency of oscillation. This is very useful if the frequency needs to be changed quickly or accurately swept through some range.

A classic example of the usefulness of a VCO is shown in Figure 9.26. This is a simplified schematic of a monophonic musical keyboard synthesizer. The keys on the synthesizer are little more than switches. These switches tap potentials off a voltage divider. As the musician plays up the keyboard, the switches engage higher and higher potentials. These levels are used to control a very accurate VCO. The higher the control voltage, the higher the output frequency or pitch will be. VCOs can be used for a number of other applications, including swept frequency spectrum analyzers, frequency modulation and demodulation, and control systems. It is also an integral part of the phase-locked loop.

The NE566 is one popular IC VCO. Its block diagram is shown in Figure 9.27. In general, it is not too much different than the square/triangle generator examined in the last section. The major functional difference is the inclusion of the modulation, or control voltage, input. This serves to adjust the capacitor's charge/discharge current. By altering this rate, the frequency may be changed. As with most specialty ICs, the

FIGURE 9.26

Simplified music synthesizer using VCO

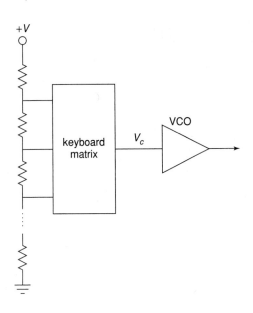

manufacturer gives specific design guidelines. For the Signetics NE566,

1. f_o should be less than 1 MHz.
2. V_+ should be between 6 and 12 V.
3. R_1 should be between 2 kΩ and 20 kΩ.
4. V_c (the control voltage) should be between V_+ and $3/4\ V_+$.

FIGURE 9.27

NE566 VCO block diagram

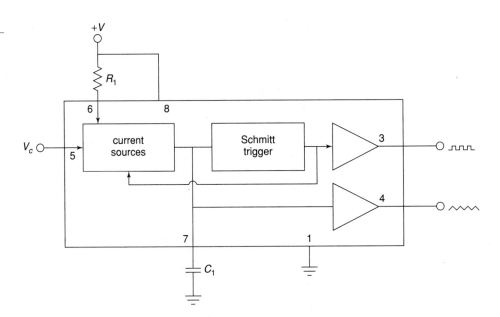

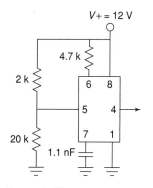

FIGURE 9.28

NE566 VCO fixed frequency oscillator for Example 9-7

The free-running, or center, frequency is found from

$$f_o = 2\frac{V_+ - V_c}{V_+ R_1 C_1} \quad (9.33)$$

For a 12 V supply, the typical triangle wave output will be 1 V peak riding on a 5 V DC offset. The typical square wave output will be 3 V peak riding on an 8.5 V DC offset.

EXAMPLE 9-7

The circuit of Figure 9.28 is connected as a fixed-frequency oscillator. Determine the output frequency.

The control voltage is set by the 2 kΩ, 20 kΩ voltage divider.

$$V_c = 12\text{ V}\,\frac{20\text{ k}}{2\text{ k} + 20\text{ k}}$$

$$V_c = 10.91\text{ V}$$

Note that this conforms to design guideline number 4.

$$f_o = 2\frac{V_+ - V_c}{V_+ R_1 C_1}$$

$$f_o = 2\frac{12\text{ V} - 10.91\text{ V}}{12\text{ V} \times 4.7\text{ k} \times 1.1\text{ nF}}$$

$$f_o = 35.14\text{ kHz}$$

EXAMPLE 9-8

If the circuit of Figure 9.29 is used to couple a control signal to the NE566, determine the maximum and minimum output frequencies if the control signal is 1 V peak-to-peak.

Because the DC bias on the control voltage input has not changed, the center frequency will remain at 35.14 kHz. Equation 9.33 indicates that the maximum output frequency will occur when V_c is at a minimum. This signal will be the addition of the 10.91 V DC bias and the negative .5 V peak from the control source.

$$V_{c,min} = 10.91\text{ V} + (-.5\text{ V})$$

$$V_{c,min} = 10.41\text{ V}$$

Note that this conforms to design guideline number 4.

$$f_o = 2\frac{V_+ - V_c}{V_+ R_1 C_1}$$

$$f_o = 2\frac{12\text{ V} - 10.41\text{ V}}{12\text{ V} \times 4.7\text{ k} \times 1.1\text{ nF}}$$

$$f_o = 51.26\text{ kHz}$$

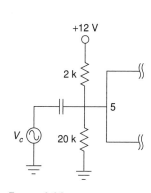

FIGURE 9.29

Circuit for Example 9-8: applying a control voltage to the NE566

For the minimum frequency, use the maximum V_c,

$$V_{c.max} = 10.91 \text{ V} + .5 \text{ V}$$
$$V_{c.max} = 11.41 \text{ V}$$

Note that this conforms to design guideline number 4.

$$f_o = 2\frac{V_+ - V_c}{V_+ R_1 C_1}$$
$$f_o = 2\frac{12 \text{ V} - 11.41 \text{ V}}{12 \text{ V} \times 4.7 \text{ k} \times 1.1 \text{ nF}}$$
$$f_o = 19 \text{ kHz}$$

Thus, a range of 51.26:19, or 2.7:1, is achieved. Another way of looking at this is to note that a 1 V change in control voltage produced a 32.26 kHz change in frequency. The sensitivity of this circuit may be stated as 32.26 kHz per volt. If values are properly chosen, the NE566 is capable of producing a 10:1 frequency sweep. Control signals close to the limits set in design guideline number 4 will produce some nonlinearity. In other words, the relationship between the control voltage and the resulting frequency will no longer be a straight line. In this example, the sensitivity will deviate from the 32.26 kHz per volt ideal.

In closing, note that the way in which the frequency sweeps depends on the wave shape of V_c. If a sinusoid is used, the output frequency will vary smoothly between the stated limits. On the other hand, if the wave shape for V_c is a ramp, the output frequency will start at one extreme and then move smoothly to the other limit as the V_c ramp continues. When the ramp resets itself, the output frequency will jump back to its starting point. An example of this is shown in Figure 9.30. Finally, if the control

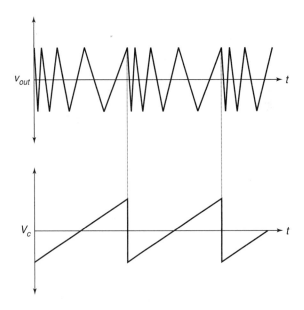

FIGURE 9.30

VCO frequency sweep using a ramp

FIGURE 9.31

VCO 2-tone output using a square wave

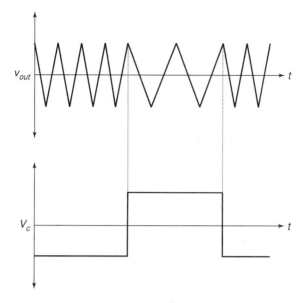

wave shape is a square, the output frequency will abruptly jump from the minimum to the maximum frequency and back. This effect is shown in Figure 9.31 and can be used to generate FSK (frequency shift key) signals. FSK is used in the communications industry to transmit binary information.

Phase-Locked Loop

One step up from the VCO is the **Phase-Locked Loop**, or PLL. The PLL is a self-correcting circuit; it can lock onto an input frequency and adjust to track changes in the input. PLLs are used in modems, for FSK systems, frequency synthesis, tone decoders, FM signal demodulation, and other applications. A block diagram of a basic PLL is shown in Figure 9.32.

In essence, the PLL uses feedback in order to lock an oscillator to the phase and frequency of an incoming signal. It consists of three major parts; a *phase comparator*, a *loop filter* (typically, a lag network of some form), and a VCO. An amplifier may also exist within the loop. The phase comparator is driven by the input signal and the output of the VCO. It produces an error signal that is proportional to the phase difference between its inputs. This error signal is then filtered in order to remove spurious high-frequency

FIGURE 9.32

Phase-locked loop

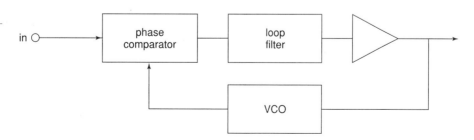

FIGURE 9.33

Operating ranges for phase-locked loop

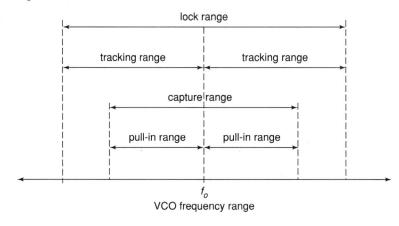

signals and noise. The resulting error signal is used as the control voltage for the VCO, and as such, sets the VCO's output frequency. As long as the error signal is not too great, the loop will be self-stabilizing. In other words, the error signal will eventually drive the VCO to be in perfect frequency and phase synchronization with the input signal. When this happens, the PLL is said to be in *lock* with the input. The range of frequencies over which the PLL can stay in lock as the input signal changes is called the *lock range*. Normally, the lock range is symmetrical about the VCO's *free-running*, or center, frequency. The deviation from the center frequency out to the edge of the lock range is called the *tracking range* and is therefore one-half of the lock range. This is illustrated in Figure 9.33.

Although a PLL may be able to track changes throughout the lock range, it may not be able to initially acquire sync with frequencies at the range limits. A somewhat narrower band of frequencies, called the *capture range*, indicates frequencies that the PLL will always be able to lock on to. Again, the capture range is usually symmetrical about f_o. The deviation on either side of f_o is referred to as the *pull-in range*. For a PLL to function properly, the input frequency must first be within the capture range. Once the PLL has locked onto the signal, the input frequency may vary throughout the larger lock range. The VCO center frequency is usually set by an external resistor or capacitor. The loop filter may also require external components. Depending on the application, the desired output signal from the PLL may be either the VCO's output, or the control voltage for the VCO.

The block diagram for one popular PLL, the NE565, is shown in Figure 9.34. Not all of the items are internally connected as you might expect. By bringing specific signals out to IC pins (such as the VCO output), the circuit is more flexible and can be used in a wider variety of applications. The free-running frequency of the VCO is set by R_1 and C_1. The loop filter is controlled by C_2 and the internal 3.6 kΩ resistor. The following design equations are given by the manufacturer:

VCO center frequency:

$$f_o = \frac{.3}{R_1 C_1} \tag{9.34}$$

(Note: R_1 should be between 2 kΩ and 20 kΩ, with an optimal value of 4 kΩ.)

FIGURE 9.34

Block diagram of the NE565

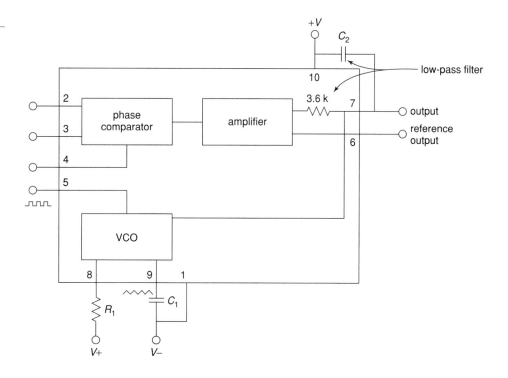

Lock range:

$$f_l = \pm \frac{8 f_o}{V_{cc}} \text{ (in Hertz)} \tag{9.35}$$

Capture range:

$$f_c = \pm \frac{\sqrt{\frac{2\pi f_l}{3600 C_2}}}{2\pi} \tag{9.36}$$

For best lock and capture range, the input signal should be at least 100 mV. At this point, we'll take a look at a couple of practical uses for the NE565.

One way to transmit binary signals is via FSK. This may be used to allow two computers to exchange data over telephone lines. Due to limited bandwidth, it is not practical to directly transmit the digital information in its normal pulse-type form. Instead, logic high and low can be represented by distinct frequencies. A square wave, for example, would be represented as an alternating set of two tones. FSK is very easy to generate. All you need to do is drive a VCO with the desired logic signal. To recover the data, the reception circuit needs to create a high or low level, depending on which tone is received. A PLL may be used for this purpose. The output signal will be the error signal that drives the VCO. The logic behind the circuit operation is deceptively simple. If the PLL is in lock, the output frequency of its VCO must

Figure 9.35

FSK circuit for 300 bits per second transmission

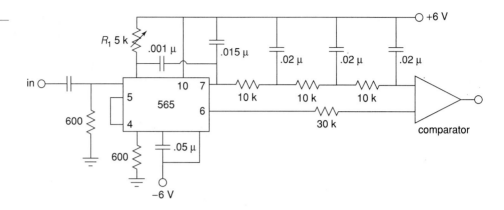

be the same as the input signal. Remembering that the incoming FSK signal is itself derived from a VCO, for the VCOs to be in lock, they must be driven with identical control signals. Therefore, the control signal that drives the PLL's internal VCO must be the same as the control signal that originally generated the FSK signal. The PLL control signal can then be fed to a comparator in order to properly match the signal to the following logic circuitry. Figure 9.35 shows a circuit that works with the standard FSK tones of 1270 Hz and 1070 Hz. This circuit works with data rates up to 300 bits per second. The loop filter needs to be a bit more stringent for this application, and as such, a multistage filter is employed. Under free-running conditions, R_1 is adjusted so that the DC voltage at the output (pin 7) is the same as the reference (pin 6). By doing so, detection of the 1070 Hz tone will produce a positive differential, thus producing a logic high out of the comparator. A 1270 Hz tone will produce a lower potential at pin 7, thus causing the comparator to produce a logic low.

Along the same lines as the FSK demodulator is the standard FM signal demodulator. Again, the operational logic is the same. In order for the PLL to remain in lock, its VCO control signal must be the same as the original modulating signal. In the case of typical radio broadcasts, the modulating signal is either voice or music. The output signal will need to be AC coupled and amplified further. The PLL serves as the intermediate frequency amplifier, limiter, and demodulator. The result is a very cost-effective system.

Another usage for the PLL is in frequency synthesis. From a single, accurate signal reference, a PLL may be used to derive a number of new frequencies. A block diagram is shown in Figure 9.36. The major change is in the addition of a programmable divider between the VCO and the phase comparator. The PLL can only remain in lock with the reference oscillator by producing the same frequency out of the divider. This means that the VCO must generate a frequency N times higher than the reference oscillator. We can use the VCO output as desired. In order to change the output frequency, all that needs to be changed is the divider ratio. Normally, a highly accurate and stable reference, such as a quartz crystal oscillator, is used. In this way, the newly synthesized frequencies will also be very stable and accurate.

FIGURE 9.36

PLL frequency synthesizer

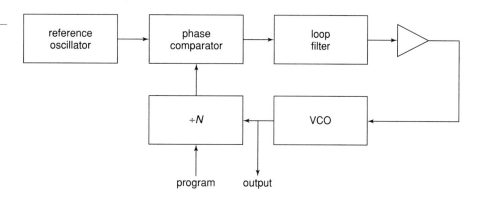

EXAMPLE 9-9

For the circuit of Figure 9.37, determine the free-running frequency, the capture range, and the lock range.

$$f_o = \frac{.3}{R_1 C_1}$$

$$f_o = \frac{.3}{4\text{ k} \times 330\text{ pF}}$$

$$f_o = 227.3 \text{ kHz}$$

FIGURE 9.37

Phase-locked loop for Example 9-9

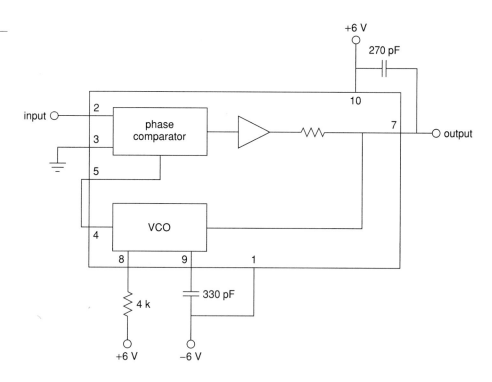

$$f_l = \pm \frac{8 f_o}{V_+}$$

$$f_l = \pm \frac{8 \times 227.3 \text{ kHz}}{6 \text{ V}}$$

$$f_l = \pm 303 \text{ kHz}$$

$$f_c = \pm \frac{\sqrt{\frac{2\pi f_l}{3600 C_2}}}{2\pi}$$

$$f_c = \pm \frac{\sqrt{\frac{2\pi \times 303 \text{ kHz}}{3600 \times 270 \text{ pF}}}}{2\pi}$$

$$f_c = \pm 222.7 \text{ kHz}$$

Under free-running conditions, the DC level at pin 7 should be around 4.25 V, according to the manufacturer's data sheets. As the input frequency rises, the voltage at this pin drops. Conversely, as the frequency falls, the output voltage increases.

555 Timer

The 555 timer is a versatile integrated circuit first introduced in the early 1970s by Signetics. It is a popular building block in a variety of applications ranging from simple square wave oscillators to burglar alarms to pulse-width modulators and beyond. In its most basic forms, the one-shot, or monostable, and the astable oscillator, the 555 requires only a handful of external components. Usually, only two capacitors and two resistors are needed for basic functions. The 555 is made by different manufacturers and in a few forms. The 556, for example, is a dual 555. The 555 can produce frequencies up to approximately 500 kHz. The output current is specified as 200 mA, although this entails fairly high internal voltage drops. A more reasonable expectation would be below 50 mA. The circuit may be powered from supplies as low as 5 volts and as high as 18 volts. This makes the 555 suitable for both TTL digital logic and typical op amp systems. Rise and fall times for the output square wave are typically 100 nS.

A block diagram of the 555 is shown in Figure 9.38. It is comprised of a pair of comparators tied to a string of three equal-valued resistors. Note that the upper, or Threshold, comparator sees approximately 2/3 of V_{cc} at its inverting input, assuming no external circuitry is tied to the Control pin. (If the Control pin is unused, a .01 μF capacitor should be placed between the pin and ground.) The lower, or Trigger, comparator sees approximately 1/3 of V_{cc} at its noninverting input. These two comparators feed a flip-flop, which in turn feeds the output circuitry and Discharge and Reset transistors. If the flip-flop output is low, the Discharge transistor will be off. Note that the output stage is inverting, so that when the flip-flop output is low, the circuit output is high. In contrast, if the input to the Reset transistor is low, this will inhibit the output signal. If Reset capabilities are not needed, the Reset pin should be tied to V_{cc}.

Returning to the comparators, if the noninverting input of the Threshold comparator were to rise above 2/3 V_{cc}, the comparator's output would change state, triggering

FIGURE 9.38

Block diagram of 555 Timer

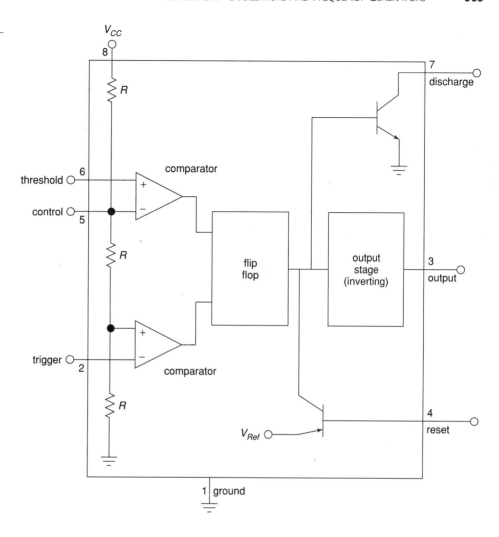

the flip-flop and producing a low out of the 555. Similarly, if the input to the inverting input of the Trigger comparator were to drop below $1/3\ V_{cc}$, the comparator's output would change, and ultimately, the 555 output would go high.

555 Monostable Operation

The basic monostable circuit is shown in Figure 9.39. In this form, the 555 will produce a single pulse of predetermined width when a negative going pulse is applied to the trigger input. Note that the three input components, R_{in}, C_{in}, and D serve to limit and differentiate the applied pulse. In this way, a very narrow pulse will result which reduces the possibility of false triggers. To see how the circuit works, refer to the waveforms presented in Figure 9.40. Assume that the output of the 555 is initially low. This implies that the Discharge transistor is on, shorting the timing capacitor C. A narrow low pulse

FIGURE 9.39

555 monostable connection

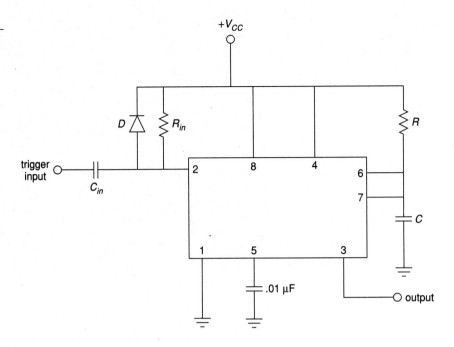

FIGURE 9.40

555 monostable waveforms

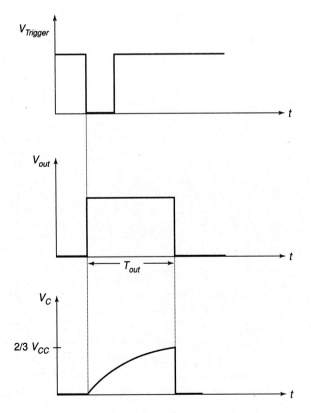

is applied to the input of the circuit. This will cause the Trigger comparator to change state, firing the flip-flop, which in turn will cause the output to go high and also turn off the Discharge transistor. At this point, C begins to charge toward V_{cc} through R. When the capacitor voltage reaches 2/3 V_{cc}, the Threshold comparator fires, setting the output low and turning on the Discharge transistor. This drains the timing capacitor, and the circuit is ready for the application of a new input pulse. Note that without the input waveshaping network, the trigger pulse must be narrower than the desired output pulse. The equation for the output pulse width is

$$T_{out} = 1.1RC$$

An interesting item to note is that the value of V_{cc} does not enter into the equation. This is because the comparators are always comparing the input signals to specific percentages of V_{cc} rather than to specific voltages.

EXAMPLE 9-10

Determine values for the timing resistor and capacitor to produce a 100 μS output pulse from the 555.

A reasonable choice for R would be 10 kΩ.

$$T_{out} = 1.1RC$$
$$C = \frac{T_{out}}{1.1R}$$
$$C = \frac{100\,\mu S}{1.1 \times 10\,k}$$
$$C = 9.09\text{ nF}$$

The nearest standard value would be 10 nF, so a better choice for R might be 9.1 kΩ (also a standard value). This pair would yield the desired pulse width quite accurately.

555 Astable Operation

Figure 9.41 shows the basic astable, or free-running form, for a square wave generator. Note the similarities to the monostable circuit. The obvious difference is that the former trigger input is now tied into the resistor-capacitor timing network. In effect, the circuit will trigger itself continually. To see how the circuit works, refer to Figure 9.42 for the waveforms of interest.

Assume initially that the 555 output is in the high state. At this point, the Discharge transistor is off and capacitor C is charging toward V_{cc} through R_A and R_B. Eventually, the capacitor voltage will exceed 2/3 V_{cc} causing the Threshold comparator to trigger the flip-flop. This will turn on the Discharge transistor and make the 555 output go low. The Discharge transistor effectively places the upper end of R_B at ground, removing R_A and V_{cc} from consideration. C now discharges through R_B toward 0. Eventually, the capacitor voltage will drop below 1/3 V_{cc}. This will fire the Trigger comparator, which

FIGURE 9.41

555 astable connection

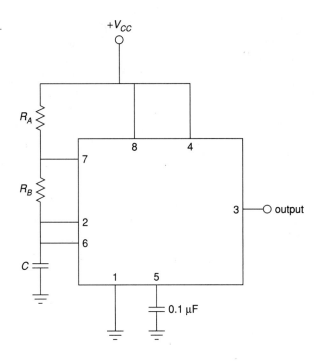

FIGURE 9.42

555 astable waveforms

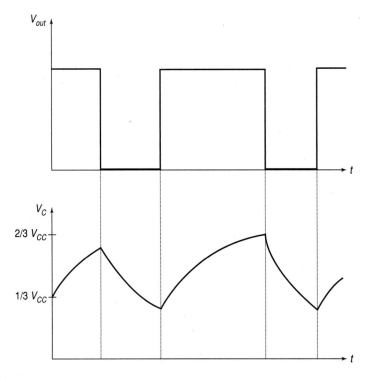

will in turn place the circuit back to its initial state, and the cycle will repeat. The frequency of oscillation clearly depends only on C, R_A, and R_B. The time periods are

$$T_{high} = 0.69(R_A + R_B)C$$
$$T_{low} = 0.69 R_B C$$

This results in a frequency of

$$f = \frac{1.44}{R_A + 2R_B}$$

The duty cycle is normally defined as the high time divided by the period. The 555 documentation often reverses this definition, but we will stick with the industry norm.

$$\text{Duty Cycle} = \frac{R_A + R_B}{R_A + 2R_B}$$

A quick examination of the duty cycle equation shows that there is no reasonable combination of resistors that will yield 50% duty cycle, let alone anything smaller. There is a simple trick to solve this problem, though. All you need to do is place a diode in parallel with R_B as illustrated in Figure 9.43. The diode will be forward-biased during

FIGURE 9.43

555 with shunting diode for duty cycles ≤50%

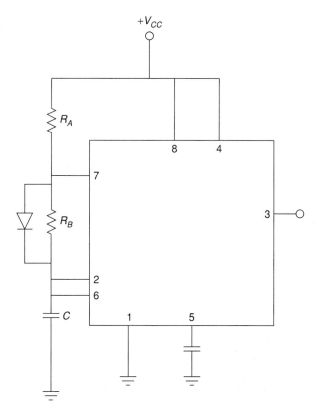

the high time period and will effectively short out R_B. During the low time period the diode will be reverse-biased, and R_B will still be available for the discharge phase. If R_A and R_B are set to the same value, the end result will be 50% duty cycle. Of course, due to the nonideal nature of the diode, this will not be perfect, so some adjustment of the resistor values may be in order. Further, note that if R_A is also replaced with a potentiometer (and perhaps a series limiting resistor) a tunable square wave generator will result.

EXAMPLE 9-11

Determine component values for a 2 kHz square wave generator with an 80% duty cycle.
    First, note the period is the reciprocal of the desired frequency, or 500 μS. For an 80% duty cycle, that yields

$$T_{high} = \text{Duty Cycle} \times T$$
$$T_{high} = 0.8 \times 500\ \mu S$$
$$T_{high} = 400\ \mu S$$

$$T_{low} = T - T_{high}$$
$$T_{low} = 500\ \mu S - 400\ \mu S$$
$$T_{low} = 100\ \mu S$$

Choosing $R_B = 10\ k\Omega$,

$$T_{low} = 0.69\ R_B C$$
$$C = \frac{T_{low}}{0.69\ R_B}$$
$$C = \frac{100\mu}{0.69\ 10\ k}$$
$$C = 14.5\ nF$$

$$T_{high} = 0.69(R_A + R_B)C$$
$$R_A = \frac{T_{high}}{0.69 C} - R_B$$
$$R_A = \frac{400\mu}{0.69\ 14.5\ nF} - 10\ k$$
$$R_A = 30\ k$$

Waveform Generator

Given the usefulness of general-purpose waveform generators, it is no surprise that semiconductor manufacturers have created entire systems on a single monolithic IC. One good example is the ICL8038 waveform generator IC. This device is capable of

FIGURE 9.44

ICL8038 block diagram
Copyright by Intersil Corporation and used with permission

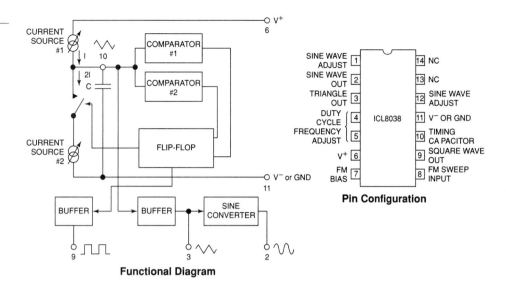

producing simultaneous sine, square, and triangle waves. Output frequencies may range from as low as .001 Hz to as high as 300 kHz. Sine wave distortion may be trimmed to under 1% THD. The unit also has provision for variable duty cycle and swept frequency operation. A block diagram of the ICL8038 is shown in Figure 9.44.

The operation of the generator revolves around two current sources and an associated comparator/flip-flop section. A single external capacitor is used to establish the base operating frequency. The technique used here is similar to that established earlier with the triangle/square wave generator. In this circuit, current source 1 charges the timing capacitor, producing a positive ramp. Eventually, this ramp will trigger the positive comparator, which in turn toggles the flip-flop. The flip-flop now connects the capacitor to current source 2. Current source 2 actually *sinks* current, thus forcing the capacitor to discharge. Eventually, this negative-going ramp will trip the negative comparator, which will again toggle the flip-flop, causing current source 1 to charge the capacitor. This process repeats indefinitely, producing a triangle wave across the capacitor and a square wave from the flip-flop. These signals are buffered and then sent to the appropriate output pins. The triangle wave is also passed through a *sine shaper*. The sine shaper is a function synthesizer, as studied in Chapter 7. This produces a decreasing gain function that "rounds off" the positive and negative extremes of the triangle wave, via a piecewise linear sine approximation. The resulting sine wave is of medium quality. It is not suitable for critical applications such as a distortion test source, but is adequate for a number of general-purpose uses. Although not perfect, the function synthesis approach has the advantage of reliable operation over a very wide frequency range. It is worth noting though, that sine wave distortion begins to increase rapidly above about 20 kHz.

Oscillation frequency is set by the external capacitor and the two associated current sources. These current sources are independent and programmed with separate resistors. It is quite possible to set the charge and discharge currents to different values

Figure 9.45

Effect of changing duty cycle

Copyright by Intersil Corporation and used with permission

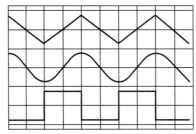
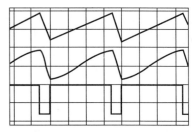

Square-Wave Duty Cycle—50%   **Square-Wave Duty Cycle—80%**

Phase Relationship of Waveforms

by using unequal resistors. This has the effect of altering the duty cycle of the square wave and changing the symmetry of the triangle and sine waves. This effect may be seen in Figure 9.45.

Output connections are fairly straightforward with the ICL8038. If high impedance loads are utilized, they may be connected directly to the sine and triangle outputs. Lower impedance loads should be buffered, as the output impedance of the sine wave section is on the order of 1 kΩ. The square wave section is an open-collector type, and thus requires a pull-up resistor for proper operation. This has the advantage of allowing you to set the upper limit of the voltage swing by tying the pull-up resistor to a desired potential. This makes logic circuit interfacing much easier. The upper limit of this potential is set by the internal breakdown limit of the ICL8038, which is approximately 30 V. If a split polarity supply is used, the output waveforms will be centered around ground. For a single-polarity supply (i.e., V_- is 0 V), the signals will be centered around one-half of the supply potential. The amplitudes of each section are not the same. The square wave will swing from just below the pull-up potential down to just above the negative supply potential. The triangle wave's peak-to-peak output will be approximately 30% of the total supplied potential, while the sine wave's peak-to-peak swing will be about 20% of the total supplied potential.

Typical circuit connections are shown in Figure 9.46. Figure 9.46a shows the most basic connection. Resistors R_A and R_B individually set the charge and discharge currents. Capacitor C is the external timing capacitor, R_L is the pull-up resistor for the square wave section, and the 82 kΩ is part of the sine wave converter. The disadvantage of this configuration is that frequency selection and duty cycle adjustment are interdependent. A better scheme is shown in Figure 9.46b. Here, fixed resistors are used for R_A and R_B, and a small potentiometer is tied between them. This potentiometer may be adjusted for the desired duty cycle. The effective value of R_A or R_B is the fixed value plus the portion of the pot between the resistor and the pot's wiper. If a 1 kΩ pot is used and is set at its midpoint, then 500 Ω is effectively added to both R_A and R_B. Another item worth noting is the replacement of the 82 kΩ resistor with a 100 kΩ potentiometer. This allows you to adjust the circuit for minimum sine wave distortion. Another variation, Figure 9.46c, shows R_A and R_B combined into a single potentiometer. This is useful if duty cycle adjustment is not required. Due to internal offsets, however, it is quite likely that this arrangement will not result in a perfect 50% duty cycle wave. The remaining pins 7 and 8 are used for *frequency modulation*. This allows you to turn the ICL8038 into a VCO.

Single Chip Oscillators and Frequency Generators   373

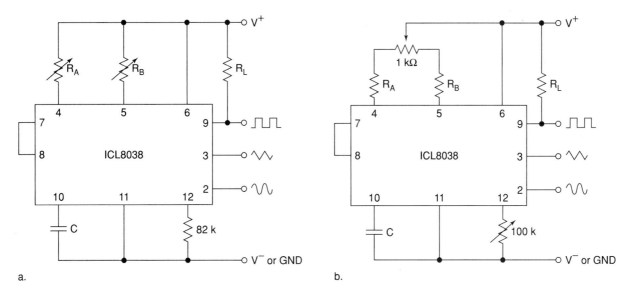

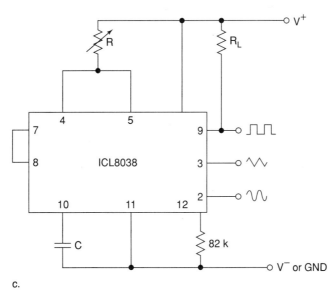

Figure 9.46

Typical connections

Copyright by Intersil Corporation and used with permission

Due to this circuit's rather specialized nature, the manufacturer gives the following set of equations:

The general oscillation frequency is

$$f_o = \frac{1}{\frac{5}{3} R_A C \left(1 + \frac{R_B}{2R_A - R_B}\right)}$$

If $R_A = R_B$, this reduces to

$$f_o = \frac{.3}{RC}$$

If a single timing resistor is used, as in Figure 9.46c, this reduces to

$$f_o = \frac{.15}{RC}$$

When setting values for the timing resistors, the charge and discharge currents should be kept within 10 µA to 1 mA. The currents are found by

$$I_A = \frac{V_+}{5R_A}$$

and

$$I_B = \frac{V_+}{5R_B}$$

Finally, when using the FM capability, the voltage at pin 8 must always be kept within these limits:

$$V_+ > V_{pin8} > \frac{2}{3}V_{total} + (V_-) + 2 \text{ V}$$

where V_{total} is the total supplied potential from the positive to the negative supply, and V_- is the negative supply potential *including sign, not just magnitude*. (Remember, it is possible to set V_- to 0 for single-supply applications.)

EXAMPLE 9-12

A waveform generator with adjustable frequency, amplitude, and wave shape is shown in Figure 9.47. Determine the range of output frequencies.

This circuit operates within one of three ranges, which are set by the three timing capacitors. The largest capacitor will produce the lowest frequency. The single-resistor approach is used here in place of separate values for R_A and R_B. The minimum value is 10 kΩ (pot shorted) and the maximum value is 110 kΩ (pot open). The maximum resistance will give the lowest frequency.

For the low range with lowest frequency:

$$f_{o.min} = \frac{.15}{RC}$$

$$f_{o.min} = \frac{.15}{110\text{k} \times .1\,\mu\text{F}}$$

$$f_{o.min} = 13.6 \text{ Hz}$$

FIGURE 9.47

Waveform generator for Example 9-10

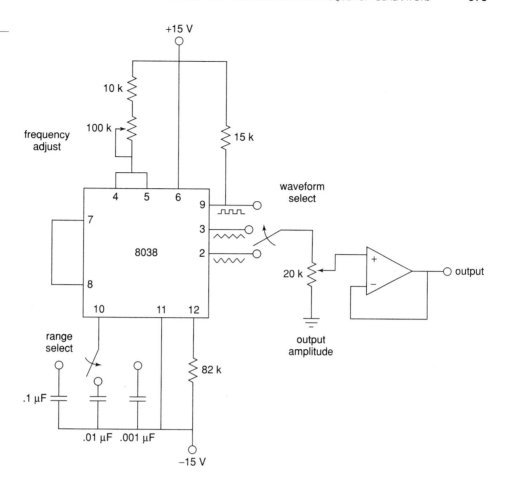

For the low range with highest frequency:

$$f_{o.max} = \frac{.15}{RC}$$

$$f_{o.max} = \frac{.15}{10\,k \times .1\,\mu F}$$

$$f_{o.max} = 150\ Hz$$

In like manner, the middle range, using $C = .01\ \mu F$ produces

$$f_{o.min} = 136\ Hz$$
$$f_{o.max} = 1500\ Hz$$

The high range, using $C = .001\ \mu F$, produces

$$f_{o.min} = 1360\ Hz$$
$$f_{o.max} = 15\ kHz$$

The output buffer serves to isolate the ICL8038 from low impedance loads. The 20 kΩ pot acts as an output level control by tapping off a portion of the output signal. For reasonable square wave outputs, a high slew rate op amp is desired. By itself, the square wave section exhibits rise and fall times of 180 nanoseconds and 40 nanoseconds, respectively.

Summary

Oscillators and frequency generators find use in a wide variety of applications. They may be realized from simple single op amp topologies or use more elaborate special-purpose integrated circuits. Basic op amp oscillators are usually constrained to the frequency range below 1 MHz. Two sine wave oscillators that are based on op amps are the Wien bridge and phase shift types. Both oscillators rely on positive feedback in order to create their outputs. To maintain oscillation, the amplifier/feedback loop must conform to the Barkhausen criterion. This states that in order to maintain oscillation the loop phase must be 0°, or an integer multiple of 360°. Also, the product of the positive feedback loss and the forward gain must be greater than unity, to start oscillations and revert to unity to maintain oscillation. In order to make the gain fall back to unity, some form of gain limiting device, such as a diode or lamp, is included in the amplifier's negative feedback loop. The oscillation frequency is usually set by a simple resistor/capacitor network. As such, the circuits are relatively easy to tune. Their ultimate accuracy will depend on the tolerance of the tuning components and, to a lesser degree, on the characteristics of the op amp used.

Besides sine waves, other shapes such as triangles and squares may be produced. A simultaneous square/triangle generator may be formed from a ramp generator/comparator combination. Only two op amps are required to realize this design.

The voltage-controlled oscillator, or VCO, produces an output frequency that is dependent on an external control voltage. The free-running, or center, frequency is normally set via a resistor/capacitor combination. The control voltage may then be used to increase or decrease the frequency about the center point. The VCO is an integral part of the phase-locked loop, or PLL. The PLL has the ability to lock onto an incoming frequency. That is, its internal VCO frequency will match the incoming frequency. Should the incoming frequency change, the internal VCO frequency will change along with it. Two important parameters of the PLL are the capture range and the lock range. Capture range is the range of frequencies over which the PLL can acquire lock. Once lock is achieved, the PLL can maintain lock over a somewhat wider range of frequencies called the lock range. The PLL is in wide use in the electronics industry and is found in such applications as FM demodulation, FSK-based communication systems, and frequency synthesis.

Timers can be used to generate rectangular waves of various duty cycles as well as single-shot pulses. Waveform generators are useful for generating continuous sines, triangles, and rectangular waveforms across a range of frequencies.

Review Questions

1. How does positive feedback differ from negative feedback?
2. Define the Barkhausen criterion.
3. Explain the operation of the Wien bridge op amp oscillator.
4. Detail the operation of the phase shift op amp oscillator.
5. How might a square wave be generated from a sinusoidal or triangular source?
6. Give two ways to make the output frequency of a Wien bridge oscillator user-adjustable.
7. What factors contribute to the accuracy of a Wien bridge oscillator's output frequency?
8. What is a VCO, and how does it differ from a fixed-frequency oscillator?
9. Draw a block diagram of a PLL and explain its basic operation.
10. What is the difference between capture range and lock range for a PLL?
11. Give at least two applications for a fixed-frequency oscillator or VCO.
12. Give at least two applications for the PLL.
13. Explain the difference between astable and monostable operation of a timer.
14. What is the purpose of a sine shaper in a waveform generator?

Problems

Analysis Problems

Unless otherwise specified, all circuits use ±15 V power supplies.

1. Given the circuit of Figure 9.48, determine the frequency of oscillation if $R_1 = 1.5$ kΩ, $R_2 = R_3 = R_4 = 3$ kΩ, and $C_1 = C_2 = .022$ μF.

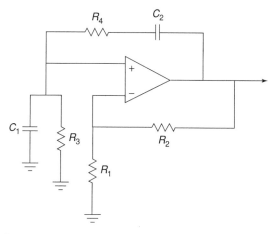

Figure 9.48

2. Given the circuit of Figure 9.48, determine the frequency of oscillation if $R_2 = 22$ kΩ, $R_1 = R_3 = R_4 = 11$ kΩ, and $C_1 = C_2 = 33$ nF.
3. Given the circuit of Figure 9.49, determine the maximum and minimum f_o if $R_1 = 5.6$ kΩ, $R_2 = 12$ kΩ, $R_3 = R_4 = 1$ kΩ, $P_1 = P_2 = 10$ kΩ, $C_1 = C_2 = 39$ nF.

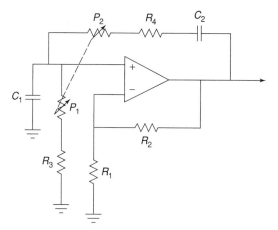

Figure 9.49

4. Given the circuit of Figure 9.50, determine f_o if $R_4 = 2$ kΩ, $R_3 = 20$ kΩ, $R_2 = 200$ kΩ, $R_1 = 1.6$ MΩ, $C_1 = 30$ nF, $C_2 = 3$ nF, $C_3 = .3$ nF.

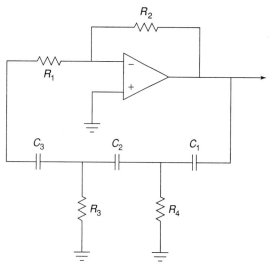

Figure 9.50

5. Given the circuit of Figure 9.50, determine f_o if $R_1 = R_3 = R_4 = 3.3$ kΩ, $R_2 = 100$ kΩ, $C_1 = C_2 = C_3 = 86$ nF.
6. Given the circuit of Figure 9.51, determine f_o if $R_1 = R_2 = 22$ kΩ, $R_3 = 33$ kΩ, $C = 3.3$ nF.

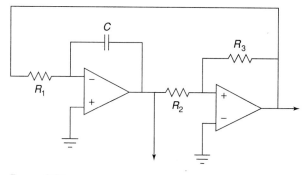

Figure 9.51

7. Using the circuit of Figure 9.52, find f_o if $R = 3.3$ kΩ, $R_2 = 30$ kΩ, $R_3 = 10$ kΩ, and $C = 2$ nF.

378 Oscillators and Frequency Generators

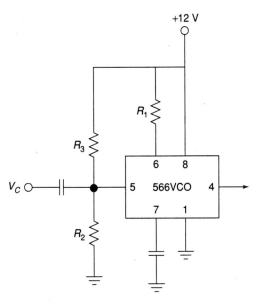

Figure 9.52

8. Give two different ways of moving the f_o (i.e., the free-running, or center, frequency) of circuit 9.45 to 30 kHz.
9. If a control voltage of $.4 \sin 2\pi 60t$ is used for the circuit of Figure 9.52, find the resulting maximum and minimum output frequencies.
10. Sketch the output waveform for Problem 9.
11. If $R = 8$ kΩ and $C = 800$ pF in Figure 9.53, determine the output frequency if $V_C = -50$ mV DC.

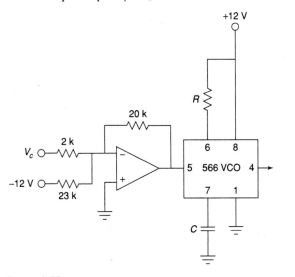

Figure 9.53

12. For Problem 11, determine the output frequencies if V_C is a 1 kHz square wave, .1 V peak.
13. Sketch the output waveform for Problem 12.
14. Determine f_o, f_l, and f_c for the NE565 PLL of Figure 9.54. Use $C_1 = 390$ pF, $C_2 = 680$ pF, and $R_1 = 10$ kΩ.

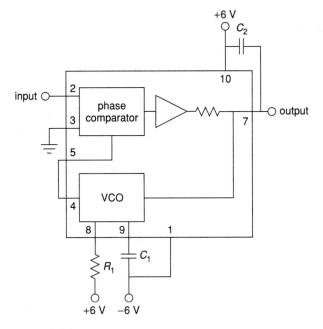

Figure 9.54

15. Given the circuit of Figure 9.51, determine f_o if $R_1 = R_2 = 22$ kΩ, $R_3 = 33$ kΩ, and $C = 3.3$ nF.
16. Determine f_o, f_l, and f_c for the PLL of Figure 9.54. Use $C_1 = 410$ pF, $C_2 = 750$ pF, and $R_1 = 5.6$ kΩ.
17. Determine the output frequency range of the circuit of Figure 9.47 if the potentiometer is lowered to 50 kΩ and the series resistor is increased from 10 kΩ to 60 kΩ.

Design Problems

18. For the circuit of Figure 9.48, determine values for C_1 and C_2 if $R_1 = 6.8$ kΩ, $R_2 = R_3 = R_4 = 15$ kΩ, and $f_o = 30$ kHz.
19. For the circuit of Figure 9.48, determine values for R_3 and R_4 if $R_1 = 2.2$ kΩ, $R_2 = 4.7$ kΩ, $C_1 = C_2 = .047$ μF, and $f_o = 400$ Hz.
20. For the circuit of Figure 9.48, determine values for R_2, C_1, and C_2 if $R_1 = 7.2$ kΩ, $R_3 = R_4 = 3.9$ kΩ, and $f_o = 19$ kHz.

21. Determine the values required for R_3, R_4, P_1, and P_2 in Figure 9.49 if $C_1 = C_2 = 98$ nF, $R_1 = 5.6$ kΩ, $R_2 = 12$ kΩ, $f_{o.min} = 2$ kHz, and $f_{o.max} = 20$ kHz.
22. Repeat Problem 21 for $f_{o.min} = 10$ kHz and $f_{o.max} = 30$ kHz.
23. Redesign the circuit of Problem 1 so that exact gain resistors are not needed. Use Figure 9.6 as a model.
24. Redesign the circuit of Problem 3 so that clipping does not occur. Use Figure 9.6 as a model.
25. Determine new values for the capacitors of Problem 4 if f_o is changed to 10 kHz.
26. Given the circuit of Figure 9.50, determine values for the capacitors if $R_1 = R_3 = R_4 = 3.3$ kΩ, $R_2 = 100$ kΩ, and $f_o = 7.6$ kHz.
27. Given the circuit of Figure 9.50, determine values for the resistors if the capacitors all equal 1100 pF and $f_o = 15$ kHz.
28. Determine the capacitor and resistor values for the circuit of Figure 9.50 if $R_2 = 56$ kΩ and $f_o = 1$ kHz.
29. Find C in Figure 9.51 if $f_o = 5$ kHz, $R_1 = R_3 = 39$ kΩ, $R_2 = 18$ kΩ.
30. Determine the resistor values in Figure 9.51 if $f_o = 20$ kHz and $C = 22$ nF. Set $R_1 = R_2$ and $R_2 = \frac{R_3}{2}$. Sketch the output waveforms as well.
31. Determine the required ratio for R_2/R_3 to set the triangle wave output to 5 V peak in Figure 9.51.
32. Using the circuit of Figure 9.52, find f_o if $R = 3.3$ kΩ, $R_2 = 30$ kΩ, $R_3 = 10$ kΩ, and $C = 2$ nF.
33. Determine values for R and C in Figure 9.53 to set the center frequency to 50 kHz with $V_C = 0$.
34. Design a square wave generator that is adjustable from 5 kHz to 20 kHz.
35. For the circuit of Figure 9.54, determine the capacitor values if R_1 is set to its optimal value (4 kΩ), and the free-running frequency is 65 kHz. Try to make the capture range at least half as large as the lock range.
36. Redesign the circuit of Figure 9.47 for a frequency range spanning 50 Hz to 50 kHz.

Challenge Problems

37. Using Figure 9.9 as a guide, design a sine wave oscillator that will operate from 2 Hz to 20 kHz, in decade ranges.
38. Design a 10 kHz TTL-compatable square wave oscillator using a Wien bridge oscillator and a 311 comparator.
39. Using a triangle or sine wave oscillator and a comparator, design a variable duty cycle pulse generator. *Hint*: Consider varying the comparator reference.
40. Using a function synthesizer (Chapter 7) and the oscillator of Figure 9.51, outline a simple laboratory frequency generator with sine, triangle, and square wave outputs.
41. For the preceding problem, outline how amplitude and DC offset controls could be implemented as well.
42. Assuming an accurate 50 kHz oscillator as a source, generate a stable 400 kHz square wave using a PLL.
43. Using Figures 9.46b and 9.47 as a guide, extend the ICL8038 circuit to include a span from 1 Hz to 100 kHz (in decade ranges), duty cycle adjustment from at least 60% to 40%, DC offset, and equal output amplitude when switching between the three waveforms.

Computer Simulation Problems

44. Perform a simulation of the circuit of Problem 1. Perform a frequency-domain analysis of the positive feedback loop's gain and phase, and verify that the Barkhausen criterion is met.
45. Perform a simulation for the circuit of Problem 4. Perform a frequency-domain analysis of the positive feedback loop's gain and phase, and verify that the Barkhausen criterion is met.
46. Perform a time-domain simulation analysis for the circuit of Problem 6. Make sure that you check both outputs (a simultaneous plot would be best).

CHAPTER 10

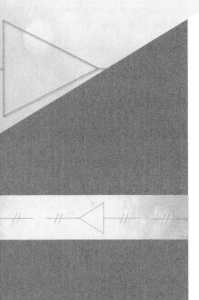

INTEGRATORS AND DIFFERENTIATORS

CHAPTER OBJECTIVES

After completing this chapter, you should be able to:

- Describe the fundamental usefulness and operation of an integrator.
- Describe the fundamental usefulness and operation of a differentiator.
- Detail the modifications required in order to make a practical op amp integrator or differentiator.
- Plot the useful frequency range of a given integrator or differentiator.
- Analyze the operation of integrator circuits using both time-continuous and time-discrete methods.
- Analyze the operation of differentiator circuits using both time-continuous and time-discrete methods.
- Explain the operation of an *analog computer*.

10.1 INTRODUCTION

Up to this point, we have examined a number of different op amp circuits and applications. Viewed from a purely mathematical perspective, the circuits perform basic functions. Amplifiers multiply an input quantity by a constant. Voltage-controlled or transconductance amplifiers can be used to multiply a quantity by a variable. Absolute value and logarithmic functions can also be produced. There is no reason to stop with just these tools; higher mathematical functions may be enlisted. In this chapter we shall examine circuits that perform integration and differentiation. Although these circuits may appear to be somewhat esoteric at first glance, they can prove to be quite useful. *Integrators* perform the function of summation over time. They may be used whenever an integration function is required, for example, to solve differential equations. The *differentiator* is the mirror of the integrator and may be used to find rates of change. One possible application is finding acceleration if the input voltage represents a velocity.

Integrators and differentiators may be combined with summing amplifiers and simple gain blocks to form *analog computers*, which can be used to model physical systems. This can be a valuable aid in the initial design and testing of such things as mechanical suspension systems or loudspeakers. Unlike their digital counterparts, analog computers do not require the use of a programming language, per se. Also, they can respond in real time to an input stimulus.

Besides the obvious use as a direct mathematical tool, integrators and differentiators can be used for wave-shaping purposes. As an example, the square/triangle generator of Chapter 9 used an integrator. With the exception of simple sinusoids, both the integrator and differentiator will change the fundamental shape of a waveform. This might be contrasted with a simple amplifier that only makes a wave larger and does not alter the basic shape.

For practical work, certain alterations will be required to the ideal circuits. The effects of these changes, both good and bad, will also be studied.

10.2 INTEGRATORS

The basic operation of an integrator is shown in Figure 10.1. The output voltage is the result of the definite integral of V_{in} from time $= 0$ to some arbitrary time t. Added to this will be a constant that represents the output of the network at $t = 0$. Remember, integration is basically the process of summation. You can also think of this as finding the "area under the curve." The output of this circuit always represents the sum total of the input values up to that precise instant in time. Consequently, if a static value (nonzero) is used as an input, the output will continually grow over time. If this growth continues unchecked, output saturation will occur. If the input quantity changes polarity, the output may also change in polarity. Having established the basic operation of the circuit, we are left with the design realization and limitations to be worked out.

As noted in earlier work, the response of an op amp circuit with feedback will reflect the characteristics of the feedback elements. If linear elements are used, the resulting response will be linear. If a logarithmic device is used in the feedback loop, the resulting response will have a log or antilog character. In order to achieve integration, then, the feedback network requires the use of an element that exhibits this characteristic. In other words, the current through the device must be proportional to either the integral or differential of the voltage across it. Inductors and capacitors answer these

FIGURE 10.1

A basic integrator

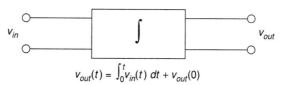

$v_{out}(t) = \int_0^t v_{in}(t)\, dt + v_{out}(0)$

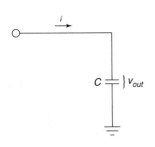

FIGURE 10.2

Capacitor as integration element

requirements, respectively. It should be possible, then, to create an integrator with either an inductor or a capacitor. Capacitors tend to behave in a more ideal fashion than do inductors. Capacitors exhibit virtually no problems with stray magnetic field interference, do not have the saturation limitations of inductors, are relatively inexpensive to manufacture, and operate reliably over a wide frequency range. Because of these factors, the vast majority of integrators are made using capacitors rather than inductors. The characteristic equation for a capacitor is

$$i(t) = C \frac{dv(t)}{dt} \tag{10.1}$$

This tells us that the current charging the capacitor is proportional to the differential of the input voltage. By integrating Equation 10.1, it can be seen that the integral of the capacitor current is proportional to the capacitor voltage.

$$v(t) = \frac{1}{C} \int_0^t i(t)\, dt \tag{10.2}$$

Assume for a moment that the capacitor voltage is the desired output voltage, as in Figure 10.2. If the capacitor current can be derived from the input voltage, the output voltage will be proportional to the integral of the input. The circuit of Figure 10.3 will satisfy our requirements. Note that it is based on the parallel-parallel inverting amplifier studied earlier. The derivation of its characteristic equation follows.

First, note that the voltage across the capacitor is equal to the output voltage. This is due to the virtual ground at the inverting op amp terminal. Given the polarities marked, the capacitor voltage is of opposite polarity.

$$V_{out}(t) = -V_c(t) \tag{10.3}$$

Due to the virtual ground at the op amp's inverting input, all of V_{in} drops across R_i. This creates the input current, i.

$$i(t) = \frac{V_{in}(t)}{R_i} \tag{10.4}$$

Because the current drawn by the op amp's inputs is negligible, all of the input current flows into the capacitor, C. Combining Equations 10.2 and 10.4 yields

FIGURE 10.3

A simple op amp integrator

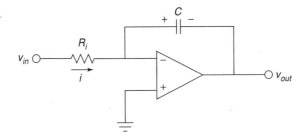

$$V_c(t) = \frac{1}{C} \int \frac{V_{in}(t)}{R_i} dt$$

$$V_c(t) = \frac{1}{R_i C} \int V_{in}(t) dt \tag{10.5}$$

Combining Equation 10.5 with Equation 10.3 produces the characteristic equation for the circuit.

$$V_{out}(t) = -\frac{1}{R_i C} \int V_{in}(t) dt \tag{10.6}$$

Note that this equation assumes that the circuit is initially relaxed (i.e., there is no charge on the capacitor at time = 0). The only difference between Equation 10.6 and the general equation as presented in Figure 10.1 is the preceding constant $-\frac{1}{R_i C}$. If the inversion or magnitude of this constant creates design problems, it can usually be corrected by gain/attenuation networks and/or inverting buffers. Finally, it is worth noting that due to the virtual ground at the inverting input, the input impedance of this circuit is approximately equal to R_i, as is the case with the ordinary parallel-parallel inverting amplifier.

Accuracy and Usefulness of Integration

As long as the circuit operation follows the model presented above, the accuracy of Equation 10.6 is very high. Small errors occur due to the approximations made. One possible source is the op amp's input bias current. If the input signal current is relatively low, the idealization that all of the input current will bypass the op amp and flow directly into the capacitor is no longer realistic. One possible solution is to use an FET input op amp. The limited frequency response and noise characteristics of the op amp will also play a role in the ultimate circuit accuracy. Generally, slew rate performance is not paramount, as integrators tend to "smooth out" signal variations. As long as the input signal stays within the op amp's frequency limits, well above the noise floor, and the output does not saturate, accurate integration results.

Circuits of this type can be used to model any number of physical processes. For example, the heat stress on a given transducer might be found by integrating the signal across it. The resulting signal could then be compared to a predetermined maximum. If the temperature proved to be too high, the circuit could be powered down as a safety measure. Although the initial reaction might be to directly measure the temperature with some form of thermal sensor, this is not always practical. In such a case, simulation of the physical process is the only reasonable route.

Optimizing the Integrator

For best performance, high-quality parts are a must. Low offset and drift op amps are needed, as the small DC values that they produce at the input will eventually force the circuit into saturation. Although they are not shown in Figure 10.3, bias and offset compensation additions are usually required. Very low offset devices such as chopper-stabilized (or commutating auto-zero) op amps may be used for lower frequency applications. For maximum accuracy, a small input bias current is desired, and thus, FET input op amps should be considered. Also, the capacitor should be a stable, low-leakage type, such as polypropylene film. Electrolytic capacitors are generally a poor choice for these circuits.

Even with high-quality parts, the basic integrator can still prove susceptible to errors caused by small DC offsets. The input offset will cause the output to gradually move toward either positive or negative saturation. When this happens, the output signal is distorted, and therefore, useless. To prevent this it is possible to place a shorting switch across the capacitor. This can be used to reinitialize the circuit from time to time. Although this does work, it is hardly convenient. Before we can come up with a more effective cure, we must first examine the cause of the problem.

We can consider the circuit of Figure 10.3 to be a simple inverting amplifier, where R_f is replaced by the reactance of the capacitor, X_C. In this case, we can see that the magnitude of the voltage gain is

$$A_v = -\frac{X_C}{R_i} = -\frac{1}{2\pi f C R_i}$$

Because X_C is frequency-dependent, it follows that the voltage gain must be frequency-dependent. Because X_C increases as the frequency drops, the voltage gain must increase as the frequency decreases. To be more specific, the gain curve will follow a −6 dB per octave slope. Eventually, the amplifier's response reaches the open-loop response. This is shown in Figure 10.4. The DC gain of the system is at the open loop-level, thus it is obvious that even very small DC inputs can play havoc with the circuit response. If the low-frequency gain is limited to a more modest level, saturation problems can be reduced, if not completely eliminated. Gain limiting may be produced by shunting the integration capacitor with a resistor, as shown in Figure 10.5. This resistor sets the upper limit voltage gain to

$$A_{max} = -\frac{R_f}{R_i}$$

As always, there will be a trade-off with modification. By definition, integration occurs where the amplitude response rolls off at −6 dB per octave, as this is the

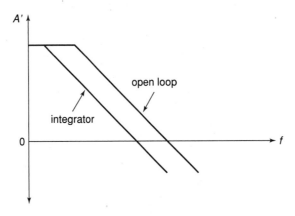

Figure 10.4

Response of a simple integrator

FIGURE 10.5

A practical integrator

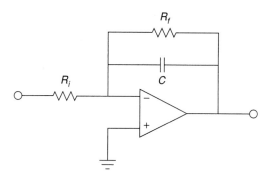

response of our idealized capacitor model. Any alteration to the response curve will impact the ultimate accuracy of the integration. By limiting the low-frequency gain, the Bode amplitude response will no longer maintain a constant −6 dB per octave rolloff. Instead, the response will flatten out below the critical frequency set by C and R_f, as shown in Figure 10.6. This critical frequency, f_{low}, is found in the standard manner,

$$f_{low} = \frac{1}{2\pi R_f C} \tag{10.7}$$

This is a very important frequency. It tells us where the useful integration range starts. If the input frequency is less than f_{low}, the circuit acts like a simple inverting amplifier and no integration results. The input frequency must be higher than f_{low} for useful integration to occur. The question then, is how much higher? A few general rules of thumb are useful. First, if the input frequency equals f_{low}, the resulting calculation will only be about 50% accurate.[1] This means that significant signal amplitude and phase changes will exist relative to the ideal. Second, at about 10 times f_{low}, the accuracy is about 99%. At this point, the agreement between the calculated and actual measured values will be quite solid.

FIGURE 10.6

Response of a practical integrator

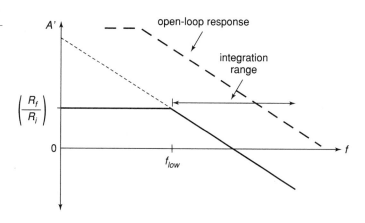

---

[1] The term *50% accurate* is chosen for convenience. The actual values are 50% of the calculated for power and phase angle and 70.7% of calculated for voltage.

FIGURE 10.7

Integrator for Example 10-1

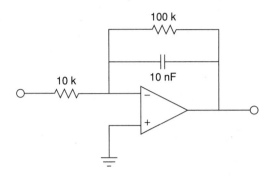

EXAMPLE 10-1

Determine the equation for V_{out}, and the lower frequency limit of integration for the circuit of Figure 10.7.

The general form of the output equation is given by Equation 10.6.

$$V_{out}(t) = -\frac{1}{R_i C} \int V_{in}(t)\, dt$$

$$V_{out}(t) = -\frac{1}{10\text{ k} \times 10\text{ nF}} \int V_{in}(t)\, dt$$

$$V_{out}(t) = -10^4 \int V_{in}(t)\, dt$$

The lower limit of integration is set by f_{low}.

$$f_{low} = \frac{1}{2\pi R_f C}$$

$$f_{low} = \frac{1}{2\pi\, 100\text{ k} \times 10\text{ nF}}$$

$$f_{low} = 159\text{ Hz}$$

This represents our 50% accuracy point. For 99% accuracy, the input frequency should be **at least** one decade above f_{low}, or 1.59 kHz. Accurate integration will continue to higher and higher frequencies.

The upper limit to useful integration is set by two factors: the frequency response of the op amp, and the signal amplitude versus noise. Obviously, at much higher frequencies, the basic assumptions of the circuit's operation will no longer be valid. The use of wide-bandwidth op amps reduces this limitation but cannot eliminate it. All amplifiers will eventually reach an upper frequency limit. Perhaps not so obvious is the limitation caused by signal strength. As noted earlier, the response of the integrator progresses at -6 dB per octave, which is equivalent to -20 dB per decade. In other words, a tenfold increase in input frequency results in a tenfold reduction in output amplitude. For higher frequencies, the net attenuation can be very great. At this point, the output signal runs the risk of being lost in the output noise. The only way around this is to ensure that the op amp is a low-noise type, particularly if a wide range of input frequencies will be used. For integration over a narrow band of frequencies, the RC integration constant may be optimized to prevent excessive loss.

Now that we have examined the basic structure of the practical integrator, it is time to analyze its response to different input waveforms. There are two basic ways of calculating the output: time-continuous or time-discrete. The time-continuous method involves the use of the indefinite integral and is well suited for simple sinusoidal inputs. Waveforms that require a more complex time domain representation, such as a square wave, may be analyzed with the time-discrete method. This corresponds to the definite integral.

Analyzing Integrators with the Time-Continuous Method

The time-continuous analysis approach involves finding a time-domain representation of the input waveform and then inserting it into Equation 10.6. The indefinite integral is taken, and the result is the time-domain representation of the output waveform. The constant term produced by the indefinite integral is ignored. As long as the input waveform can be written in a time-domain form, this method may be used. Although any waveform may be expressed in this way, it is not always practical or the most expedient route. Indeed, relatively common waveforms such as square waves and triangle waves require an infinite series time-domain representation. We shall not deal with these waveforms in this manner. The time-continuous integration of these functions is left as an exercise in the Challenge Problems at the end of this chapter.

Example 10-2

Using the circuit of Figure 10.7, determine the output if the input is a 1 V peak sine wave at 5 kHz.

First, write the input signal as a function time.

$$V_{in}(t) = 1 \sin(2\pi 5000 t)$$

Substitute this input into Equation 10.6:

$$V_{out}(t) = -\frac{1}{R_i C} \int V_{in}(t)\, dt$$

$$V_{out}(t) = -\frac{1}{10\text{k} \times 10\text{nF}} \int 1 \sin(2\pi 5000 t)\, dt$$

$$V_{out}(t) = -10^4 \int 1 \sin(2\pi 5000 t)\, dt$$

$$V_{out}(t) = -10^4 \frac{1}{2\pi 5000} \int (2\pi 5000) \sin(2\pi 5000 t)\, dt$$

$$V_{out}(t) = -.318(-\cos(2\pi 5000 t))$$

$$V_{out}(t) = .318 \cos(2\pi 5000 t)$$

Thus, the output is a sinusoidal wave that is leading the input by 90° and is .318 V peak. Again, note that the constant produced by the integration is ignored. In Example 1 it was noted that f_{low} for this circuit is 159 Hz. Because the input frequency is over 10 times f_{low}, the accuracy of the result should be better than 99%. Note that if the input frequency is changed, both the output frequency and amplitude change. For example, if the input were raised by a factor of 10 to 50 kHz, the output would also be at 50 kHz, and the output amplitude would be reduced by a factor of 10, to 31.8 mV. The output would still be a cosine, though.

---

multiSIM COMPUTER SIMULATION

Figure 10.8 (pp. 389–390) shows the Electronics Workbench MultiSIM simulation of the circuit of Example 10-2. The steady-state response is graphed. Note that in order to achieve steady-state results, the output is plotted only after several cycles of the input have passed through. Even after some 20 cycles, the output is not perfectly symmetrical. In spite of this, the accuracy of both the amplitude and phase is quite good when compared to the calculated results. This further reinforces the fact that the circuit is operating within its useful range. It can be instructive to rerun the simulation to investigate the initial response as well.

Analyzing Integrators with the Time-Discrete Method

Unlike the time-continuous approach, this method uses the definite integral and is used to find an output level at specific instances in time. This is useful if the continuous time-domain representation is somewhat complex, and yet the wave shape is relatively simple, as with a square wave. Often, a little logic may be used to ascertain the shape of the resulting wave. The integral is then used to determine the exact amplitude. The time-discrete method also proves useful for more complex waves when modeled within a computer program. Here, several calculations may be performed per cycle, with the results joined together graphically to form the output signal.

The basic technique revolves around finding simple time-domain representations of the input for specific periods of time. A given wave might be modeled as two or more sections. The definite integral is then applied over each section, and the results joined.

Example 10-3

Sketch the output of the circuit shown in Figure 10.7 if the input signal is a 10 kHz, 2 V peak square wave with no DC component.

The first step is to break down the input waveform into simple-to-integrate components and ascertain the basic shape of the result. The input is sketched in Figure 10.9. The input may be broken into two parts: a nonchanging 2 V potential from 0 to 50 μSec, and a nonchanging -2 V potential from 50 μSec to 100 μSec. The sequence repeats after this.

$$V_{in}(t) = 2 \quad \text{from} \quad t = 0, \quad \text{to} \quad t = 50 \ \mu\text{Sec}$$
$$V_{in}(t) = -2 \quad \text{from} \quad t = 50 \ \mu\text{Sec}, \quad \text{to} \quad t = 100 \ \mu\text{Sec}$$

In essence, we are saying that the input may be treated as either ± 2 V DC for short periods of time. Because integration is a summing operation, as time progresses, the area swept out from underneath the input signal increases, and then decreases due to the polarity changes. It follows that the output will grow (or shrink) in a linear fashion, as the input is constant. In other words, straight ramps will be produced during these time periods. The only

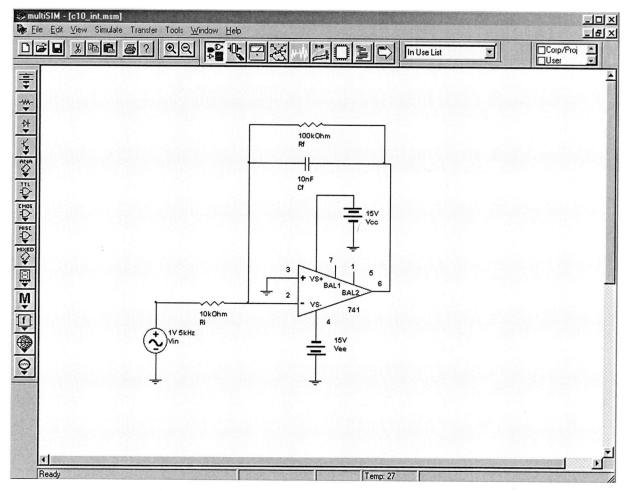

FIGURE 10.8a

Integrator in MultiSIM

difference between them will be the polarity of the slope. From this we find that the expected output waveform is a triangle wave. All we need to do now is determine the peak value of the output. To do this, perform the definite integral using the first portion of the input wave.

$$V_{out}(t) = -\frac{1}{R_i C} \int V_{in}(t)\, dt$$

$$V_{out}(t) = -\frac{1}{10\text{ k} \times 10\text{ nF}} \int_0^{50\mu} 2\, dt$$

$$V_{out}(t) = -10^4 \times 2 \times t \big|_{t=0}^{t=50\mu}$$

$$V_{out} = -20000 \times 50\,\mu$$

$$V_{out} = -1\text{ V}$$

FIGURE 10.8b

Integrator input and output waveforms

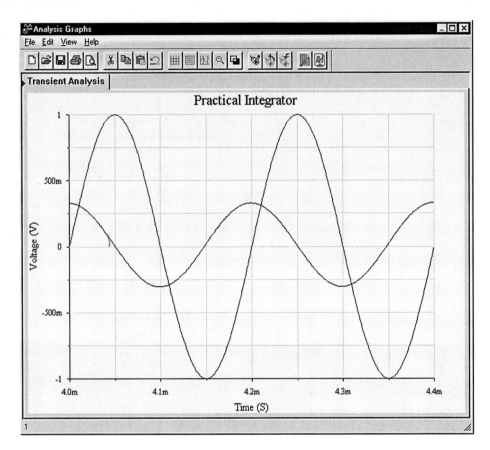

This represents the total change over the 50 μSec half-cycle interval. This is a peak-to-peak change, saying that the output is 1 V negative relative to its value at the end of the preceding half-cycle. Integration for the second half-cycle is similar, and produces a positive change. The resulting waveform is shown in Figure 10.10. Note that the wave is effectively inverted. For positive inputs, the output slope is negative. Because the input frequency is

FIGURE 10.9

Input waveform for Example 10-3

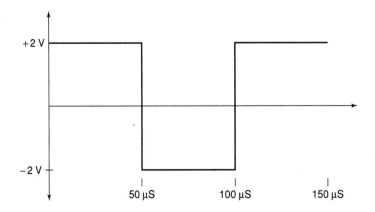

FIGURE 10.10

Output waveform for Example 10-3

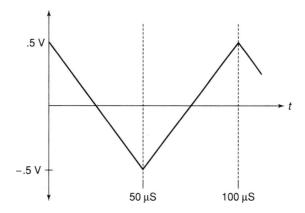

much higher than f_{low}, we can once again expect high accuracy. Although the solution seems to imply that the output voltage should swing between 0 and -1 V, a real-world integrator does indeed produce the indicated $\pm .5$ V swing. This is due to the fact that the integration capacitor in this circuit will not be able to maintain the required DC offset indefinitely.

EXAMPLE 10-4

Figure 10.11a shows an integrator connected to an *accelerometer*. This device produces a voltage that is proportional to the acceleration it experiences.[2] Accelerometers may be fastened to a variety of physical devices in order to determine how the devices respond to various mechanical inputs. This would be useful, for example, in experimentally determining the mechanical resonance characteristics of a surface. Another possibility is the determination of the lateral acceleration of an automobile as it travels through a corner. By integrating this signal, it is possible to determine velocity, and a further integration will produce position. If the accelerometer produces the voltage shown in Figure 10.11b, determine the shape of the velocity curve (assume a nonrepetitive input wave).

First, check the frequency limit of the integrator in order to see if high accuracy may be maintained. If the input waveform was repetitive, it would be approximately 200 Hz ($\frac{1}{5\ mSec}$).

$$f_{low} = \frac{1}{2\pi R_f C}$$
$$f_{low} = \frac{1}{2\pi\ 400\ k \times 20\ nF}$$
$$f_{low} = 19.9\ Hz$$

The input signal is well above the lower limit.

Note that the input waveshape may be analyzed in piecewise fashion, as if it were a square wave. The positive and negative portions are both 1 mS in duration, and .6 V peak. The only difference is the polarity. As these are square pulses, we expect ramp sections for

---

[2]A mechanical accelerometer consists of a small mass, associated restoring springs, and some form of transducer that is capable of reading the mass's motion. More recent designs may be fashioned using micromachined chips, where the deflection of a mass-supporting beam is measured indirectly.

FIGURE 10.11
Accelerometer with integrator

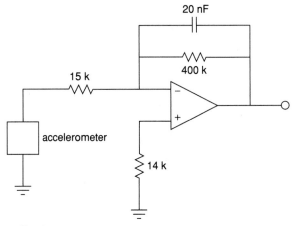

a. Circuit

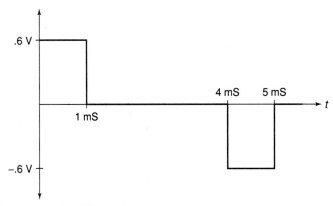

b. Signal produced by accelerometer

the output. For the positive pulse ($t = 0$ to $t = 1$ mS),

$$V_{out}(t) = -\frac{1}{R_i C} \int V_{in}(t)\, dt$$

$$V_{out}(t) = -\frac{1}{15\text{ k} \times 20\text{ nF}} \int_0^{10^{-3}} .6\, dt$$

$$V_{out}(t) = -3333 \times .6 \times t \Big|_{t=0}^{t=10^{-3}}$$

$$V_{out} = -2000 \times 10^{-3}$$

$$V_{out} = -2\text{ V}$$

This tells us that we will see a negative going ramp with a -2 V change. For the time period between 1 mS and 4 mS, the input signal is zero, and thus the change in output potential will be zero. This means that the output will remain at -2 V until $t = 4$ mS.

For the period between 4 mS and 5 mS, a positive going ramp will be produced. Because only the polarity is changed relative to the first pulse, we can quickly find that the change

FIGURE 10.12

Integrator output

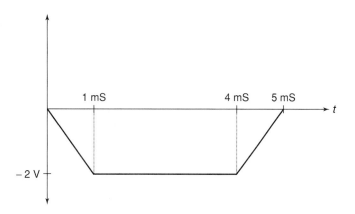

is +2 V. The input waveform is nonrepetitive, so the output waveform appears as shown in Figure 10.12. (A repetitive input would naturally cause the integrator to "settle" around ground over time, producing the same basic wave shape, but shifted positively.) This waveform tells us that the velocity of the device under test increases linearly up to a point. After this point, the velocity remains constant until a linear decrease in velocity occurs. (Remember, the integrator inverts the signal, so the output curve is effectively upside down. If the input wave indicates an initial positive acceleration, the actual output is an initial positive velocity.) On a different time scale, these are the sort of waveforms that might be produced by an accelerometer mounted on an automobile that starts at rest, smoothly climbs to a fixed speed, and then smoothly brakes to a stop.

10.3 DIFFERENTIATORS

Differentiators perform the complementary function to the integrator. The base form of the differentiator is shown in Figure 10.13. The output voltage is the differential of the input voltage. This is very useful for finding the rate at which a signal varies over time. For example, it is possible to find velocity given distance and acceleration given velocity. This can be very useful in process control work.

Essentially, the differentiator tends to reinforce fast signal transitions. If the input waveform is nonchanging, (i.e., DC), the slope is zero, and thus the output of the differentiator is zero. On the other hand, an abrupt signal change such as the rising edge of a square wave produces a very large slope, and thus the output of the differentiator will be large. In order to create the differentiation, an appropriate device needs to be associated with the op amp circuit. This was the approach taken with the integrator, and it remains valid here. In fact, we are left with the same two options: using either

FIGURE 10.13

A basic differentiator

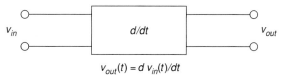

$v_{out}(t) = d\, v_{in}(t)/dt$

Figure 10.14

A simple op amp differentiator

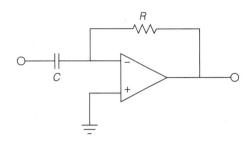

an inductor, or a capacitor. Again, capacitors tend to be somewhat easier to work with than inductors and are preferred. The only difference between the integrator and the differentiator is the position of the capacitor. Instead of placing it in the R_f position, the capacitor will be placed in the R_i position. The resulting circuit is shown in Figure 10.14. The analysis starts with the basic capacitor equation (Equation 10.1):

$$i(t) = C\frac{dv(t)}{dt}$$

We already know from previous work that the output voltage appears across R_f, though inverted.

$$V_{out} = -V_{R_f}$$

Also, by Ohm's law,

$$V_{R_f} = iR_f$$

By using the approximation that all input current flows through R_f (as the op amp's input current is zero), and then substituting Equation 10.1 for the current, we find

$$V_{out}(t) = -R_f C\frac{dv(t)}{dt}$$

A quick inspection of the circuit shows that all of the input voltage drops across the capacitor, because the op amp's inverting input is a virtual ground. Bearing this in mind, we arrive at the final output voltage equation,

$$V_{out}(t) = -R_f C\frac{dV_{in}(t)}{dt} \tag{10.8}$$

As with the integrator, a leading constant is added to the fundamental form. Again, it is possible to scale the output as required through the use of gain or attenuation networks.

Accuracy and Usefulness of Differentiation

Equation 10.8 is an accurate reflection of the circuit response so long as the base assumptions remain valid. As with the integrator, practical considerations tend to force limits on the circuit's operating range. If the circuit is analyzed at discrete points in the

FIGURE 10.15

Response of a simple differentiator

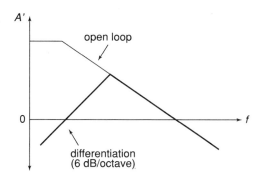

frequency domain, it can be modeled as an inverting amplifier with the following gain equation:

$$A_v = -\frac{R_f}{X_C} = -2\pi f C R_f$$

Note that as the frequency decreases, X_C grows, thus reducing the gain. Conversely, as the input frequency is raised, X_C falls in value, causing the gain to rise. This rise will continue until it intersects the open-loop response of the op amp. The resulting amplitude response is shown in Figure 10.15. This response is the mirror image of the basic integrator response and exhibits a slope of 6 dB per octave. Note that DC gain is zero, and therefore the problems created by input bias and offset currents are not nearly as troublesome as in the integrator. Because of this, there is no limit as to how low the input frequency may be, excluding the effects of signal-to-noise ratio. Things are considerably different on the high end though. Once the circuit response breaks away from the ideal 6 dB per octave slope, differentiation no longer takes place.

Optimizing the Differentiator

There are a couple of problems with the general differentiator of Figure 10.14. First of all, it is quite possible that the circuit may become unstable at higher frequencies. Also, the basic shape of the amplitude response suggests that high frequencies are accentuated, thus increasing the relative noise level. Both of these problems may be reduced by providing an artificial upper-limit frequency, f_{high}. This tailoring may be achieved by shunting R_f with a small capacitor. This reduces the high-frequency gain, and thus reduces the noise. The resulting response is shown in Figure 10.16.

We find f_{high} in the standard manner:

$$f_{high(fdbk)} = \frac{1}{2\pi R_f C_f} \qquad (10.9)$$

f_{high} represents the highest frequency for differentiation. It is the 50% accuracy point. For higher accuracy, the input frequency must be kept well below f_{high}. At about .1 f_{high}, the accuracy of Equation 10.8 is about 99%. Generally, you have to be somewhat more conservative in the estimation of accuracy than with the integrator. This

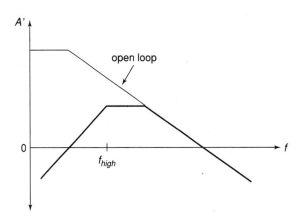

FIGURE 10.16

Response of a partially optimized differentiator

is because complex waves contain harmonics that are higher than the fundamental. Even though the fundamental may be well within the high accuracy range, the upper harmonics may not be.

The other major problem of the basic circuit is that the input impedance is inversely proportional to the input frequency. This is because X_C is the sole input impedance factor. This may present a problem at higher frequencies because the impedance will approach zero. To circumvent this problem, a resistor may be placed in series with the input capacitor in order to establish a minimum impedance value. Unfortunately, this will also create an upper break frequency, f_{high}.

$$f_{high(in)} = \frac{1}{2\pi R_i C} \qquad (10.10)$$

The resulting response is shown in Figure 10.17. The effective f_{high} for the system will be the lower of Equations 10.9 and 10.10. The completed, practical differentiator is shown in Figure 10.18. Note that a bias compensation resistor may be required at the noninverting input, although it is not shown.

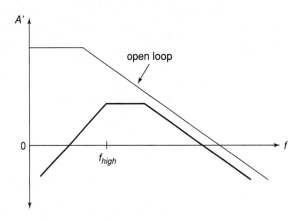

FIGURE 10.17

Response of a practical differentiator

FIGURE 10.18

A practical differentiator

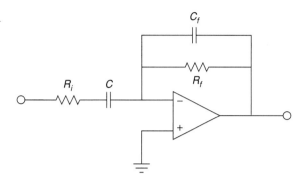

Analyzing Differentiators with the Time-Continuous Method

The time-continuous method will be used when the input signal may be easily written in the time domain (e.g., sine waves). For more complex waveforms, such as a triangle wave, a discrete time method will be used. The continuous method will lead directly to a time-domain representation of the output waveform. Specific voltage/time coordinates will not be evaluated.

EXAMPLE 10-5

Determine the useful range for differentiation in the circuit of Figure 10.19. Also determine the output voltage if the input signal is a 2 V peak sine wave at 3 kHz.

The upper limit of the useful frequency range will be determined by the lower of the two RC networks.

$$f_{high(fdbk)} = \frac{1}{2\pi R_f C_f}$$

$$f_{high(fdbk)} = \frac{1}{2\pi \times 5\,k \times 100\,pF}$$

$$f_{high(fdbk)} = 318.3\,kHz$$

FIGURE 10.19

Differentiator for Example 10-5

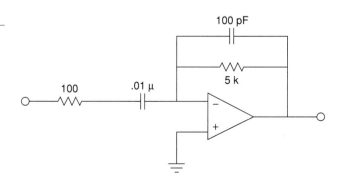

$$f_{high(in)} = \frac{1}{2\pi R_i C}$$

$$f_{high(in)} = \frac{1}{2\pi \times 100 \times .01\,\mu F}$$

$$f_{high(in)} = 159.2 \text{ kHz}$$

Therefore, the upper limit is 159.2 kHz. Remember, the accuracy at this limit is relatively low, and normal operation will typically be several octaves lower than this limit. Note that the input frequency is 3 kHz, so high accuracy should result. First, write V_{in} as a time-domain expression:

$$V_{in}(t) = 2 \sin 2\pi 3000 t$$

$$V_{out}(t) = -R_f C \frac{dV_{in}(t)}{dt}$$

$$V_{out}(t) = -5\,\text{k} \times .01\,\mu F \frac{d\,2\sin 2\pi 3000\,t}{dt}$$

$$V_{out}(t) = -10^{-4} \frac{d\,\sin 2\pi 3000\,t}{dt}$$

$$V_{out}(t) = -1.885 \cos 2\pi 3000\,t$$

This tells us that the output waveform is also sinusoidal, but it lags the input by 90°. Note that the input frequency has not changed, but the amplitude has. The differentiator operates with a 6 dB per octave slope, thus it can be seen that the output amplitude is directly proportional to the input frequency. If this example is rerun with a frequency of 6 kHz, the output amplitude will be double the present value.

multiSIM Computer Simulation

The Electronics Workbench MultiSIM simulation for the circuit of Example 10-5 is shown in Figure 10.20 (pp. 399–400). Note the excellent correlation for both the phase and amplitude of the output. As was the case with the integrator simulation, the Transient Analysis output plot is started after the initial conditions have settled.

Analyzing Differentiators with the Time-Discrete Method

For more complex waveforms, it is sometimes expedient to break the waveform into discrete chunks, differentiate each portion, and then combine the results. The idea is to break the waveform into equivalent straight-line segments. Differentiation of a straight-line segment will result in a constant (i.e., the slope, which does not change over that time). The process is repeated until one cycle of the input waveform is completed. The resultant levels are then joined together graphically to produce the output waveform.

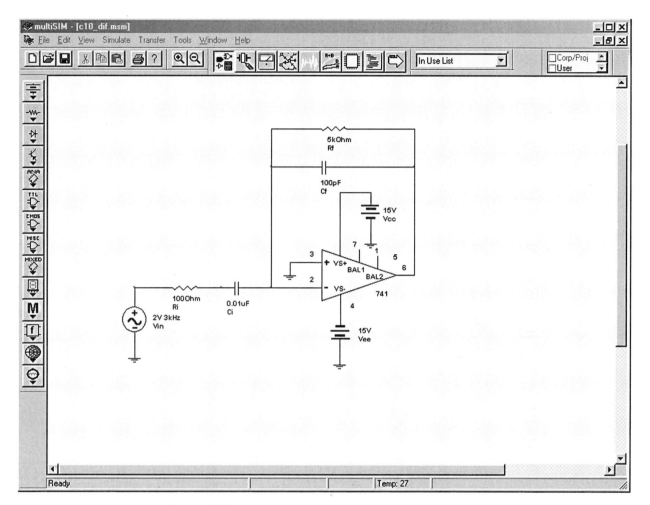

FIGURE 10.20a

Differentiator in MultiSIM

Often, waveforms are symmetrical, and only part of the calculation need be performed: a sign change is all that will be needed for the mirror image portions. As an example, a triangle wave may be broken into a positive-going line segment and a negative going-line segment. The slopes should be equal, only the direction (i.e., sign) has changed. A square wave may be broken into four parts: a positive-going edge, a static positive value, a negative-going edge, and a static negative value. The "flat" portions have a slope of zero, so only one calculation must be performed, and that's the positive-going edge. We shall take a look at both of these waveforms in the next two examples.

EXAMPLE 10-6

Sketch the output waveform for the circuit of Figure 10.19 if the input is a 3 V peak triangle wave at 4 kHz.

Figure 10.20b

Differentiator input and output waveforms

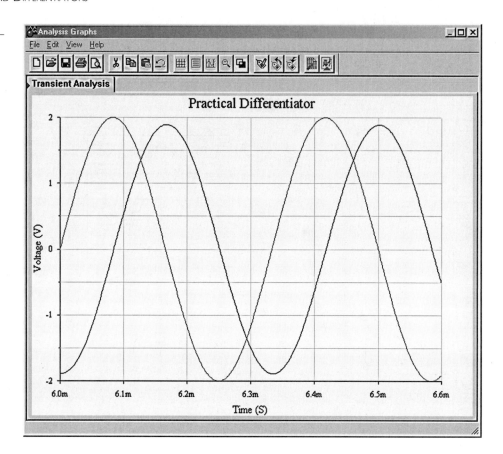

First, note that the input frequency is well within the useful range of this circuit, as calculated in Example 10-5. (Note that the highest harmonics will still be out of range, but the error introduced will be minor.)

The triangle wave may be broken into a positive-going portion and a negative-going portion. In either case, the total voltage change will be 6 V in one half-cycle. The period of the waveform is

$$T = \frac{1}{4 \text{ kHz}}$$
$$T = 250 \ \mu S$$

Therefore, for the positive-going portion, a 6 V change will be seen in 125 μSec (-6 V in 125 μS for the negative-going portion). The slope is

$$\text{Slope} = \frac{6 \text{ V}}{125 \ \mu S}$$
$$\text{Slope} = 48000 \text{ V/S}$$

Which, as a time-domain expression, is

$$V_{in}(t) = 48000t$$

FIGURE 10.21

Input/output waveforms

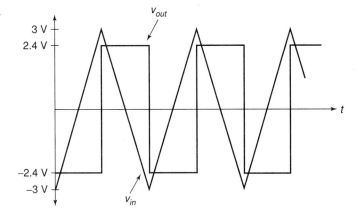

Substituting this equation into Equation 10.8 yields

$$V_{out}(t) = -R_f C \frac{dV_{in}(t)}{dt}$$

$$V_{out}(t) = -5\text{k} \times .01\ \mu\text{F}\ \frac{d48000t}{dt}$$

$$V_{out}(t) = -2.4\text{ V}$$

During $t = 0$ through $t = 125\ \mu$Sec, the output is -2.4 V. Differentiation of the second half of the wave is similar, but produces a positive output, $+2.4$ V. The result is a 4 kHz square wave that is 2.4 V peak. The resulting waveform is shown in Figure 10.21.

multiSIM COMPUTER SIMULATION

Some circuits are more sensitive to the choice of op amp than others are, and the effects of an improper choice may not always be immediately apparent without first building or simulating the circuit. A good example of this is shown in Figure 10.22 (pp. 402–404). MultiSIM was used to create the Transient Analysis for the circuit of Example 10-6 with two different op amps. Accurate differentiation requires excellent high-frequency response from the op amp. In the first simulation, a 741 op amp is used. This device is not particularly fast. As a result, the output waveform suffers from excessive overshoot and ringing. Also, slew rate limiting is fairly obvious, slowing the transitions of the output waveform. In contrast, using an LF411 in the same circuit yields far superior response. Some overshoot still exists, but its magnitude has been restrained, as has the ringing. Also, slew rate limiting is reduced by a wide margin. Clearly, the second result is much closer to the ideal calculation than the first run.

EXAMPLE 10-7

Repeat Example 10-6 with a 3 V peak, 4 kHz square as the input. Assume that the rising and falling edges of the square wave have been slew limited to 5 V/μS.

During the time periods that the input is at ± 3 V, the output will be zero. This is because the input slope is, by definition, zero when the signal is "flat." An output is only noted during the

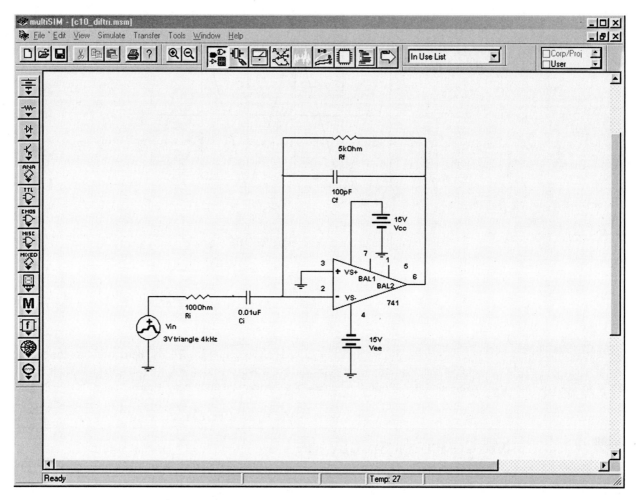

Figure 10.22a

MultiSIM schematic of differentiator

transitions between the ±3 V levels. Therefore, we need to find the slope of the transitions. It was stated that due to slew rate limiting (perhaps from some previous amplifier stage) the transitions run at 5 V/μS. As a time-domain expression, this is

$$V_{in}(t) = 5 \times 10^6 t$$

Substituting this equation into Equation 10.8 yields

$$V_{out}(t) = -R_f C \frac{dV_{in}(t)}{dt}$$

$$V_{out}(t) = -5\,k \cdot .01\,\mu F \frac{d5 \times 10^6 t}{dt}$$

$$V_{out}(t) = -250\,V$$

FIGURE 10.22b

Input and output waveforms using 741

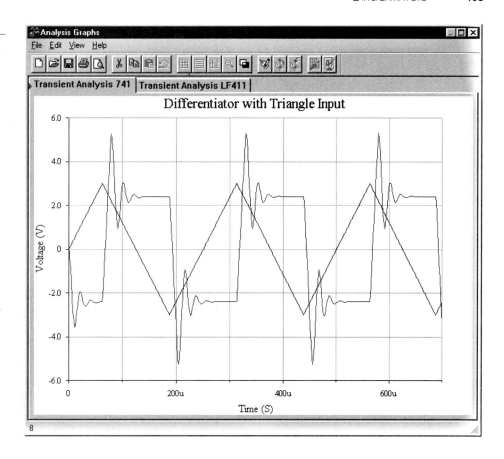

Obviously, when using a standard op amp and ±15 V power supply, clipping will occur in the vicinity of −13.5 V. For the negative-going edge, a similar result will be seen (+250 V calculated, with clipping at +13.5 V). The resulting waveform is shown in Figure 10.23. Note that the output waveform spikes will also be limited by the slew rate of the differentiator's op amp.

EXAMPLE 10-8

Figure 10.24a shows a differentiator receiving a signal from an LVDT, or *linear variable differential transformer*.[3] An LVDT can be used to accurately measure the position of objects with displacements of less than one-thousandth of an inch. This could be useful in a computer-aided

---

[3] An LVDT is a transformer with dual secondary windings and a movable core. The core is connected to a shaft, which is in turn actuated by some external object. The movement of the core alters the mutual inductance between primary and secondary. A carrier signal is fed into the primary, and the changing mutual inductance alters the strength of the signal induced into the secondaries. This signal change is turned into a simple DC voltage by a demodulator. The resulting DC potential is proportional to the position of the core, and thus, proportional to the position of the object under measurement.

FIGURE 10.22c

Input and output waveforms using LF411

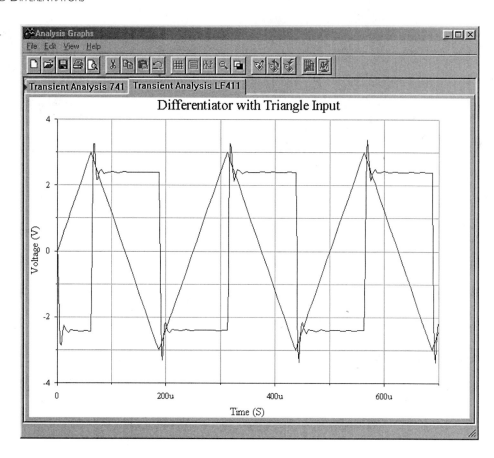

FIGURE 10.23

Differentiated square wave (Note output clipping.)

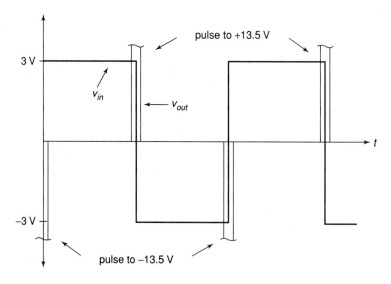

FIGURE 10.24

Differentiator with LVDT

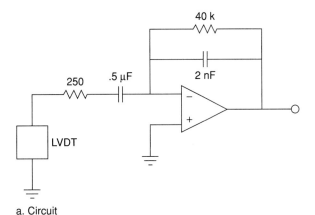

a. Circuit

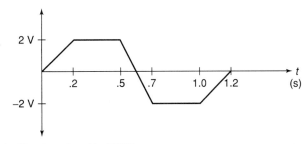

b. Signal produced by LVDT

manufacturing system. By differentiating this position signal, a velocity signal may be derived. A second differentiation will produce acceleration. If the LVDT produces the wave shown in Figure 10.24b, determine the velocity/time curve for the object being tracked.

First, check the upper frequency limit for the circuit.

$$f_{high(fdbk)} = \frac{1}{2\pi R_f C_f}$$

$$f_{high(fdbk)} = \frac{1}{2\pi \times 40 \text{ k} \times 2 \text{ nF}}$$

$$f_{high(fdbk)} = 1.99 \text{ kHz}$$

$$f_{high(in)} = \frac{1}{2\pi R_i C}$$

$$f_{high(in)} = \frac{1}{2\pi \times 250 \times .5 \text{ }\mu\text{F}}$$

$$f_{high(in)} = 1.273 \text{ kHz}$$

The limit will be the lower of the two, or 1.273 kHz. This is well above the slowly changing input signal, and therefore, high accuracy should be possible.

This wave can be analyzed in piece-wise fashion. The ramp portions will produce constant output levels and the flat portions will produce an output of 0 V (i.e., the rate of change is zero). For the first section,

$$\text{Slope} = \frac{2\text{ V}}{.2\text{ Sec}}$$

$$\text{Slope} = 10\text{ V/S}$$

Which, as a time-domain expression, is

$$V_{in}(t) = 10t$$

$$V_{out}(t) = -R_f C \frac{dV_{in}(t)}{dt}$$

$$V_{out}(t) = -40\text{ k} \times .5\ \mu\text{F}\ \frac{d10t}{dt}$$

$$V_{out}(t) = -.2\text{ V}$$

So, during $t = 0$ through $t = .2$ S, the output is $-.2$ V. The time period between .2 S and .5 S will produce an output of 0 V. For the negative-going portion,

$$\text{Slope} = \frac{-4\text{ V}}{.2\text{ S}}$$

$$\text{Slope} = -20\text{ V/S}$$

thus, $V_{in}(t) = -20t$

$$V_{out}(t) = -R_f C \frac{dV_{in}(t)}{dt}$$

$$V_{out}(t) = -40\text{ k}\ .5\ \mu\text{F}\ \frac{d20t}{dt}$$

$$V_{out}(t) = .4\text{ V}$$

The output is .4 V between .5 S and .7 S. The third section has the same slope as the first section and will also produce a $-.2$ V level. The output waveform is drawn in Figure 10.25.

FIGURE 10.25

Differentiator output

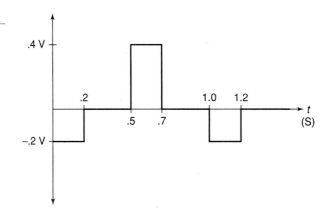

10.4 Analog Computer

Analog computers are used to simulate physical systems. These systems may be electrical, mechanical, acoustical, or what have you. An analog computer is basically a collection of integrators, differentiators, summers, and amplifiers. Due to their relative stability, integrators are favored over differentiators. It is not uncommon for analog computers to be made without any differentiators. Because physical systems may be described in terms of differential equations, analog computers may be used to solve these equations, thus producing as output some system parameter.

The basic advantage of simulation is that several variations of a given system may be examined in real time without actually constructing the system. For a large project this is particularly cost efficient. The process starts by writing a differential equation (first-, second-, or third-order) that describes the system in question. The equation is then solved for its highest-order element, and the result used to create a circuit.

Example 10-9

Let's investigate the system shown in Figure 10.26. This is a simple mechanical system that might represent (to a rough approximation) a variety of physical entities, including the suspension of an automobile. This system is comprised of a body with mass M, which is suspended from a spring. The spring has a spring constant, K. The mass is also connected to a shock absorber which produces damping, R. If an external force, F, excites the mass, it will move, producing some displacement, X. This displacement depends on the mass, force, spring constant, and damping. Essentially, the spring and shock absorber will create reactionary forces. From basic physics, $F = MA$, where A is the acceleration of the body. If X is the position of the body,

Figure 10.26

Mechanical system

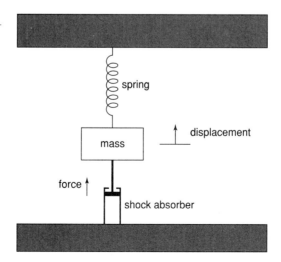

then $\frac{dX}{dt}$ is its velocity, and $\frac{d^2X}{dt^2}$ is its acceleration. Therefore, we can say

$$F = M\frac{d^2X}{dt^2}$$

In this system, the total force is comprised of the excitation force, F, and the forces produced by the spring and shock absorber.

$$F - F_{spring} - F_{shock} = M\frac{d^2X}{dt^2}$$

The spring's force is equal to the displacement times the spring constant:

$$F_{spring} = KX$$

The shock absorber's force is equal to the damping constant times the velocity of the body:

$$F_{shock} = R\frac{dX}{dt}$$

By substituting and rearranging the above elements we find that

$$F = M\frac{d^2X}{dt^2} + R\frac{dX}{dt} + KX$$

Here F is seen as the input signal, and X as the output signal. A somewhat less busy notation form is the dot convention. A single dot represents the first derivative with respect to time, two dots represent the second derivative, and so on. The above equation may be rewritten as

$$F = M\ddot{X} + R\dot{X} + KX$$

This is the final differential equation. Note how it contains only derivatives and no integrals. The last step is to solve the equation for the highest-order differential. By setting it up in this form, the simulation circuit may be realized without using differentiators. This will indicate how many integrators will be required.

$$\ddot{X} = \frac{F}{M} - \frac{R}{M}\dot{X} - \frac{K}{M}X$$

This says that the second differential of X is the sum of three components. To realize the circuit, start with a summing amplifier with the three desired signals as inputs. This is shown in block form in Figure 10.27. Note that two of the inputs use X and the first derivative of X. These elements may be produced by integrating the output of Figure 10.27. Appropriate

FIGURE 10.27

Circuit realization (block form)

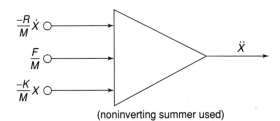

(noninverting summer used)

FIGURE 10.28

Circuit realization using function blocks

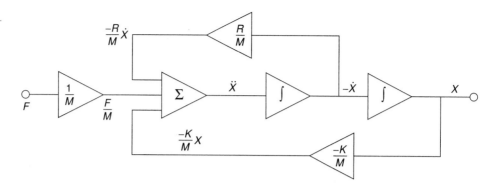

constants may be used to achieve the desired signal levels. This is shown in Figure 10.28. Certain elements may be combined; for example, a weighted summing amplifier may be used to eliminate unneeded amplifiers.

In use, the constants R, K, and M are set by potentiometers (they are essentially nothing more than scaled gain factors). A voltage representing the excitation force is applied to the circuit, and the desired output quantity is recorded. Note that the output of interest could be the acceleration, velocity, or displacement of the body. In order to test the system with a new spring or damping constant, all that is needed is to adjust the appropriate potentiometer. In this manner, a large number of combinations may be tried quickly. The most successful combinations may then be built and tested for the final design. An analog computer such as this would be very useful in testing such items as the suspension of an automobile or a loudspeaker system. To ease the design of the simulation circuit, commercial analog computers are available. Construction (or programming) of the circuit involves wiring integrator, amplifier, and summer blocks, together with the appropriate potentiometers. In this way, the details of designing and optimizing individual integrators or amplifiers is bypassed.

10.5 ALTERNATIVES TO INTEGRATORS AND DIFFERENTIATORS

There are alternatives to using op amp based integrators and differentiators. As long as systems can be described by a reasonable set of equations, simulation using digital computers is possible. The primary advantage of a digital computer-based simulation scheme is that it is very flexible. It can also be very accurate. Digital simulation does have limits, though. The first limitation is the speed of response. Analog computers can be configured as real-time or faster-than-real-time devices: they respond at the same speed or faster than the system that is being simulated. The typically heavy computation load of the digital computer requires prodigious calculation speed to keep up with reasonably fast processes. Given the increases in desktop computing power in the past decade, this problem is not nearly as large as it once was. The other advantage of the analog computer is its immediacy. It is by nature, interactive. This means that an operator can change simulation parameters and immediately see the result. If a simulation does not require real-time performance with interactive adjustment, a digital-based

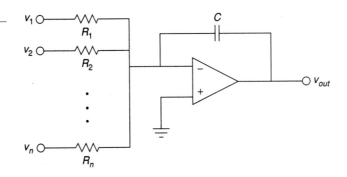

FIGURE 10.29

Summing integrator

simulation will most likely be preferred. The idea is to calculate the response for several closely spaced points in time. The results may be used to create graphs that can then be studied at leisure. Fundamentally, this process is the same as that used by circuit simulation programs. Indeed, it is quite possible to design an op amp based simulator, and then simulate its response by using a circuit simulator! As you might guess, this double simulation is not particularly efficient, especially with large systems.

10.6 Extended Topic: Other Integrator and Differentiator Circuits

The basic integrator and differentiator circuits examined earlier may be extended into other forms. Perhaps the most obvious extension is to add multiple inputs, as in an ordinary summing amplifier. In complex systems, this concept may save the use of several op amps. A summing integrator is shown in Figure 10.29. Note its similarity to a normal summing amplifier. In this circuit, the input currents are summed at the inverting input of the op amp. If the resistors are all set to the same value, we can quickly derive the output equation by following the original derivation. The result is

$$V_{out} = -\frac{1}{RC} \int V_1 + V_2 + \cdots + V_n \, dt$$

The output is the negative integral of the sum of the inputs.

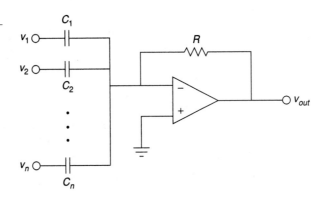

FIGURE 10.30

Summing differentiator

FIGURE 10.31

Augmenting integrator

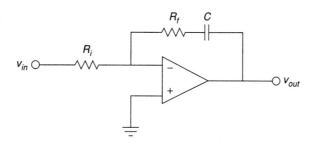

In a similar vein, a summing differentiator may be formed. This is shown in Figure 10.30. Again, the proof of its output equation follows the original differentiator derivation.

$$V_{out} = -RC_1 \frac{dV_1}{dt} - RC_2 \frac{dV_2}{dt} - \cdots - RC_n \frac{dV_n}{dt}$$

Another interesting adaptation of the integrator is the augmenting integrator. This circuit adds a constant gain portion to the output equation. An augmenting integrator is shown in Figure 10.31. The addition of the feedback resistor R_f provides the augmenting action. As you might surmise, the gain portion is directly related to R_f and R_i.

$$V_{out} = -V_{in} \frac{R_f}{R_i} - \frac{1}{RC} \int V_{in}\, dt$$

The augmenting integrator can also be turned into a summing/augmenting integrator by adding extra input resistors as in Figure 10.29. Note that the gain portion will be the same for all inputs if the input summing resistors are of equal value.

The final variant that we shall note is the double integrator. This design requires two reactive portions in order to achieve double integration. One possibility is shown in Figure 10.32. In this circuit, a pair of RC "Tee" networks is used. The output equation is

$$V_{out} = -\frac{4}{(RC)^2} \int\int V_{in}\, dt$$

FIGURE 10.32

Double integrator

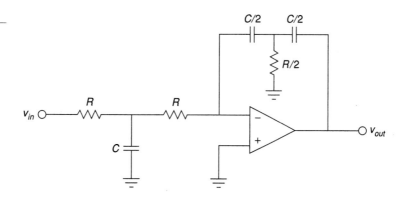

When properly used, the double integrator can cut down the parts requirement of larger circuit designs.

Summary

In this chapter we have examined the structure and use of integrator and differentiator circuits. Integrators produce a summing action whereas differentiators find the slope of the input. Both types are based on the general parallel-parallel inverting voltage feedback model. In order to achieve integration and differentiation, a capacitor is used in the feedback network in place of the standard resistor. Because the capacitor current is proportional to the rate of change of the capacitor's voltage, a differential or integral response is possible. For the integrator, the capacitor is placed in the normal R_f position, and for the differentiator, the capacitor is placed in the normal R_i position. As a result, the integrator exhibits a -6 dB per octave slope through the useful integration range. The differentiator exhibits the mirror image, or $+6$ dB per octave slope, throughout its useful range.

In both circuits, practical limitations require the use of additional components. In the case of the integrator, small DC offsets at the input can force the output into saturation. To avoid this, a resistor is placed in parallel with the integration capacitor in order to limit the low frequency gain. This has the unfortunate side effect of limiting the useful integration range to higher frequencies. In the case of the differentiator, noise, stability, and input impedance limits can pose problems. In order to minimize noise and aid in stability, a small capacitor may be placed in parallel with R_f. This reduces the high-frequency gain. In order to place a lower limit on the input impedance, a resistor may be placed in series with the differentiation capacitor. The addition of either component will limit the upper range of differentiation.

We examined two general techniques for determining the output signal. The first form is referred to as the time-continuous method and although it may be used with virtually any waveform, our use was with simple sine waves only. In the case of the integrator, it corresponds to the indefinite integral—a time-domain equation for the output is the result. The second form is the time-discrete method and is useful for waves that may be easily broken into segments. Each segment is analyzed, and the results joined graphically. This has the advantage of an immediate graphical result, whereas the shape of the time-continuous result may not be immediately apparent. The time-discrete method is also useful for producing output tables and graphs with a digital computer. In the case of the integrator, the time-discrete method corresponds to the definite integral.

Integrators and differentiators may be used in combination with summers and amplifiers to form analog computers. Analog computers may be used to model a variety of physical systems in real time. Unlike their digital counterparts, programming an analog computer only requires the proper interconnection of the various building blocks and appropriate settings for the required physical constants.

Finally, it is noted that integrators and differentiators may be used as wave-shaping circuits. Integrators may be used to turn square waves into triangle waves. Differentiators can be used to turn triangle waves into square waves. Indeed, as differentiators tend to "seek" rapid changes in the input signal, they can be quite useful as edge detectors.

Review Questions

1. What is the basic function of an integrator?
2. What is the basic function of a differentiator?
3. What is the function of the capacitor in the basic integrator and differentiator?
4. Why are capacitors used in favor of inductors?
5. What practical modifications need to be done to the basic integrator, and why?
6. What practical modifications need to be done to the basic differentiator, and why?
7. What are the negative side effects of the practical versus basic integrator and differentiator?
8. What is an analog computer, and what is it used for?
9. What are some of the advantages and disadvantages of the analog computer versus the digital computer?
10. How might integrators and differentiators be used as wave-shaping circuits?

Problems

Analysis Problems

1. Sketch the output waveform for the circuit of Figure 10.33 if the input is a 4 V peak square wave at 1 kHz.

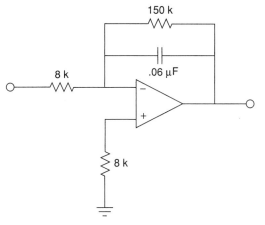

FIGURE 10.33

2. Repeat Problem 1 if $V_{in}(t) = 5 \sin 2\pi 318 t$.
3. Repeat Problem 1 if $V_{in}(t) = 20 \cos 2\pi 10000 t$.
4. Using the circuit of Problem 1, if the input is a ramp with a slope of 10 V/S, find the output after 1 mS, 10 mS, and 4 S. Sketch the resulting wave.
5. Determine the low-frequency gain for the circuit of Figure 10.34.

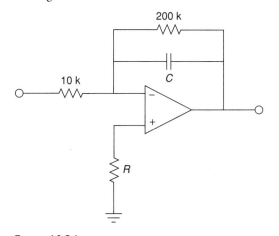

FIGURE 10.34

6. If $C = .033 \ \mu F$ in Figure 10.34, determine V_{out} if V_{in} is a 200 mV peak sine wave at 50 kHz.
7. Repeat Problem 6 using a 200 mV peak square wave at 50 kHz.
8. Sketch the output of the circuit of Figure 10.7 with the input signal given in Figure 10.35.

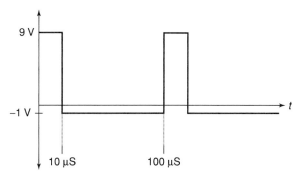

FIGURE 10.35

9. Assume that the input to the circuit of Figure 10.7 is 100 mV DC. Sketch the output waveform, and determine if it goes to saturation.
10. Sketch the output waveform for Figure 10.7, given an input of: $V_{in}(t) = .5 \cos 2\pi 9000 t$.
11. Given the circuit of Figure 10.36, sketch the output waveform if the input is a 100 Hz, 1 V peak triangle wave.

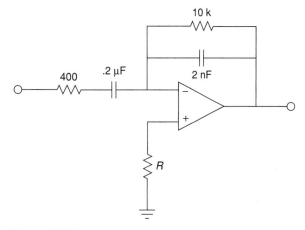

FIGURE 10.36

12. Repeat Problem 11 if the input is a 2 V peak square wave at 500 Hz. Assume that the rise and fall times are 1 µS and are linear (vs exponential).
13. Repeat Problem 11 for the following input: $V_{in}(t) = 3\cos 2\pi 60t$.
14. Repeat Problem 11 for the following input: $V_{in}(t) = .5\sin 2\pi 1000t$.
15. Repeat Problem 11 for the following input: $V_{in}(t) = 10t^2$.
16. Given the input shown in Figure 10.37, sketch the output of the circuit of Figure 10.19.

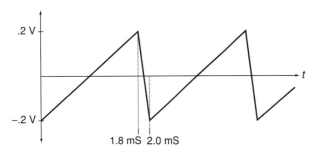

FIGURE 10.37

17. Given the input shown in Figure 10.38, sketch the output of the circuit of Figure 10.19.

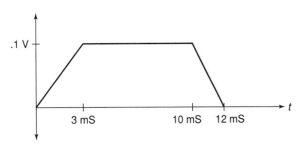

FIGURE 10.38

18. Sketch the output waveform for Figure 10.19, given an input of $V_{in}(t) = .5\cos 2\pi 4000t$.
19. Sketch the output waveform for Figure 10.19, if V_{in} is a 300 mV peak triangle wave at 2500 Hz.

Design Problems

20. Given the circuit of Figure 10.34,
    a. Determine the value of C required in order to yield an integration constant of −2000.
    b. Determine the required value of R for minimum integration offset error.
    c. Determine f_{low}.
    d. Determine the 99% accuracy point.
21. Given the differentiator of Figure 10.39, determine the value of R that will set the differentiation constant to -10^3.

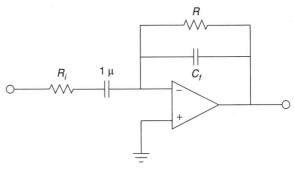

FIGURE 10.39

22. Determine the value of R_i such that the maximum gain is 20 for the circuit of Figure 10.39. (Use the values of Problem 21.)
23. Determine the value of C_f in Problem 22 such that noise above 5 kHz is attenuated.
24. Design an integrator to meet the following specifications: integration constant of −4500, f_{low} no greater than 300 Hz, Z_{in} at least 6 kΩ, and DC gain no more than 32 dB.
25. Design a differentiator to meet the following specifications: differentiation constant of -1.2×10^{-4}, f_{high} at least 100 kHz, and a minimum Z_{in} of 50 Ω.

Challenge Problems

26. Sketch the output waveform for the circuit of Figure 10.7 if the input is the following damped sinusoid: $V_{in}(t) = 3\epsilon^{-200t}\sin 2\pi 1000t$.
27. A 1.57 V peak, 500 Hz square wave may be written as the following infinite series:

$$V_{in}(t) = 2\sum_{n=1}^{\infty}\frac{1}{2n-1}\sin 2\pi 500(2n-1)t$$

Using this as the input signal, determine the infinite series output equation for the circuit of Figure 10.7.
28. Sketch the output for the preceding problem. Do this by graphically adding the first few terms of the output.
29. Remembering that the voltage across an inductor is equal to the inductance times the rate of change of current, determine the output equation for the circuit of Figure 10.40.

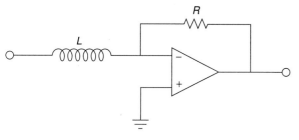

FIGURE 10.40

30. Repeat the preceding Problem for the circuit of Figure 10.41.

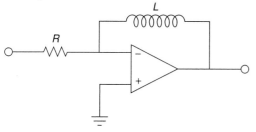

FIGURE 10.41

31. A 2.47 V peak, 500 Hz triangle wave may be written as the following infinite series:

$$V_{in}(t) = 2 \sum_{n=1}^{\infty} \frac{1}{(2n-1)^2} \cos 2\pi 500(2n-1)t$$

Using this as the input signal, determine the infinite series output equation for the circuit of Figure 10.19.

32. Sketch the output for the preceding problem. Do this by graphically adding the first few terms of the output.

Computer Simulation Problems

33. Simulate Problem 1 and determine the steady-state response.
34. Model Problem 8 using a simulator. Determine both the steady-state and initial outputs.
35. Simulate Problem 14. Determine the steady-state output.
36. Model Problem 22 using a simulator and determine the output signal.
37. Compare the resulting waveform produced by the circuit of Problem 8 using both the 741 op amp and the medium-speed LF411. Does the choice of op amp make a discernable difference in this application?

CHAPTER 11

ACTIVE FILTERS

CHAPTER OBJECTIVES

After completing this chapter, you should be able to:

- Describe the four main types of filters.
- Detail the advantages and disadvantages of active versus passive filters.
- Describe the importance of filter *order* on the shape of a filter's response.
- Describe the importance of filter *alignment* on the shape of a filter's response.
- Compare the general gain and phase response characteristics of the popular filter alignments.
- Analyze high- and low-pass Sallen and Key filters.
- Analyze both low and high Q band-pass filters.
- Analyze state-variable filters.
- Analyze notch filters.
- Analyze cascaded filters of high order.
- Detail the operation of basic bass and treble audio equalizers.
- Explain the operation of switched capacitor filters, and list their relative advantages and disadvantages compared to ordinary op amp filters.

11.1 INTRODUCTION

Generally speaking, a filter is a circuit that inhibits the transfer of a specific range of frequencies. Conversely, you can think of a filter as a circuit that allows only certain frequencies to pass through. Filters are used to remove undesirable frequency components from a complex input signal. The uses for this operation are many, including the suppression of power-line hum, reduction of very low or high-frequency interference and noise, and specialized spectral shaping. One very common use for filters is bandwidth limiting, which, as you'll see in Chapter 12, is an integral part of any analog-to-digital conversion system.

There are numerous variations on the design and implementation of filters. Indeed, an in-depth discussion of filters could easily fill more than one textbook. Our discussion must by necessity be of a limited and introductory nature. This chapter deals with the implementation of a number of popular op amp filter types. Due to the finite space available, every mathematical proof for the design sequences will not be detailed here, but may be found in the references listed at the end of the chapter.

Filter implementations may be classified into two very broad, yet distinct, camps: *digital filters* and *analog filters*. Digital filters work entirely in the digital domain, using numeric data as the input signal. The design of digital filters is an advanced topic and will not be examined here. The second category, analog filters, utilizes standard linear circuit techniques for their construction. Analog filter implementations can be broken into two subcategories: *passive* and *active*. Passive filters utilize only resistors, inductors, and capacitors, whereas active filters make use of active devices (i.e., discrete transistors or op amps) as well. Although we will be examining only one subcategory in the world of filters, it is important to note that many of the circuits that we will design can be realized through passive analog filters or digital filters.

11.2 Filter Types

No matter how a filter is realized, it will usually conform to one of four basic response types. These types are: *high-pass, low-pass, band-pass,* and *band-reject* (also known as a *notch* type). A high-pass filter allows only frequencies above a certain break point to pass through. In other words, it attenuates low-frequency components. Its general amplitude frequency response is shown in Figure 11.1. A low-pass filter is the logical mirror of the high-pass, that is, it allows only low-frequency signals to pass through, while suppressing high-frequency components. Its response is shown in Figure 11.2. The band-pass filter can be thought of as a combination of high- and low-pass filters. It only allows frequencies within a specified range to pass through. Its logical inverse is the band-reject filter that allows everything to pass through, with the exception of a specific range of frequencies. The amplitude frequency response plots for these last two types are shown in Figures 11.3 and 11.4, respectively. In complex systems, it is possible to combine several different types together in order to achieve a desired overall response characteristic.

In each plot, there are three basic regions. The flat area where the input signal is allowed to pass through is known as the *pass band*. The edge of the pass band is denoted by the *break frequency*. The break frequency is usually defined as the point at which the response has fallen 3 dB from its pass-band value. Consequently, it is also referred to as the *3 dB down frequency*. It is important to note that the break frequency is **not** always equal to the natural critical frequency, f_c. The area where the input signal is fully suppressed is called the *stop band*. The section between the pass band and the stop band is referred to as the *transition band*. This is shown for the low-pass filter

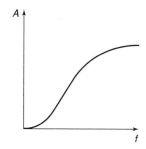

FIGURE 11.1

High-pass response

418 Active Filters

Figure 11.2
Low-pass response

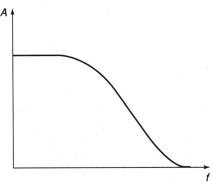

Figure 11.3
Band-pass response

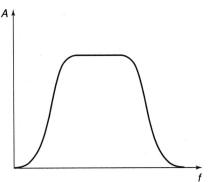

Figure 11.4
Notch response

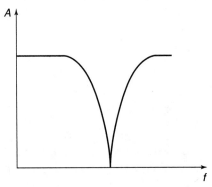

Figure 11.5
Filter regions

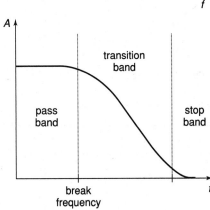

FIGURE 11.6

Passive low-pass filter

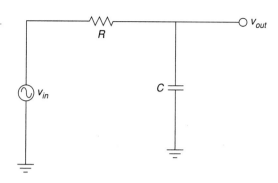

in Figure 11.5. Note that plots 11-1 through 11-4 only define the general response of the filters. The actual shape of the transition band for a given design will be refined in following sections.

These response plots may be achieved through the use of simple RLC circuits. For example, the simple lag network shown in Figure 11.6 may be classified as a low-pass filter. You know from earlier work that lag networks attenuate high frequencies. In a similar vein, a band-pass filter may be constructed as shown in Figure 11.7. Note that at resonance, the tank circuit exhibits an impedance peak, whereas the series combination exhibits its minimum impedance. Therefore, a large portion of V_{in} will appear at V_{out}. At very high or low frequencies the situation is reversed (the parallel combination shows a low impedance, and the series combination shows a high impedance), and very little of V_{in} appears at the output. Note that no active components are used in this circuit. This is a passive filter.

11.3 THE USE AND ADVANTAGES OF ACTIVE FILTERS

If filters may be made with only resistors, capacitors, and inductors, you might ask why anyone would want to design a variation that required the use of an op amp. This is a good question. Obviously, there must be certain shortcomings or difficulties associated with passive filter designs, or active filters would not exist. Active filters

FIGURE 11.7

Passive band-pass filter

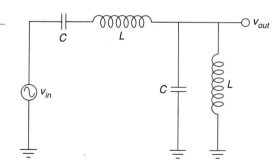

offer many advantages over passive implementations. First of all, active filters do not exhibit *insertion loss*. This means that the pass-band gain will equal 0 dB. Passive filters always show some signal loss in the pass band. Active filters may be made with pass-band gain, if desired. Active filters also allow for interstage isolation and control of input and output impedances. This alleviates problems with interstage loading and simplifies complex designs. It also produces modest component sizes (e.g., capacitors tend to be smaller for a given response). Another advantage of the active approach is that complex filters may be realized without using inductors. This is desirable, as practical inductors tend to be far less ideal than typical resistors and capacitors and are generally more expensive. The bottom line is that the active approach allows for the rapid design of stable, economical filters in a variety of applications.

Active filters are not perfect. First, by their very nature, active filters require a DC power supply whereas passives do not. This is usually not a problem, as the remainder of the circuit will probably require a DC supply anyway. Active filters are also limited in their frequency range. An op amp has a finite gain-bandwidth product, and the active filter produced can certainly not be expected to perform beyond it. For example, it would be impossible to design a filter that only passes frequencies above 10 MHz when using a standard μA741. Passive circuits do not have this limitation and can work well into the hundreds of MHz. Finally, active filters are not designed to handle large amounts of power. They are low signal-level circuits. With appropriate component ratings, passive filters may handle hundreds of watts of input power. A classic example of this is the crossover network found in most home loudspeaker systems. The crossover network splits the music signal into two or more bands and routes the results to individual transducers that are optimized to work within a given frequency range. Because the input to the loudspeaker may be as high as a few hundred watts, a passive design is in order.[1] Consequently, we can say that active filters are appropriate for designs at low to moderate frequencies (generally no more than 1 MHz with typical devices) that do not have to handle large amounts of power. As you might guess, that specification covers a great deal of territory, and therefore, active filters based on op amps have become rather popular.

11.4 Filter Order and Poles

The rate at which a filter's response falls in the transition band is determined by the filter's *order*. The higher the order of a filter, the faster its rolloff rate is. The order of a filter is given as an integer value and is derived from the filter's transfer function. As an example, all other factors being equal, a fourth-order filter will roll off twice as fast as a second-order filter, and four times faster than a first-order unit. The order of a filter also indicates the minimum number of reactive components that the filter will require. For example, a third-order filter requires at least three reactive components: one capacitor and two inductors, two capacitors and one inductor, or in the case of an

---

[1] In more advanced playback systems, active filters can be used. Examples include recording studio monitors and public-address systems. We'll take a look at just how this is done in one of the upcoming examples.

active filter, three capacitors. Related to this is the number of *poles* that a filter utilizes. It is common to hear descriptions such as "a four-pole filter." For most general-purpose high- or low-pass filters, the terms *pole* and *order* may be used interchangeably and completely describe the rolloff rate. For more complex filters this isn't quite the case, and you may also hear descriptions such as "a six-pole, two-zero filter." Because this chapter is an introduction to filters, we will not detail the operation of these more esoteric types. Suffice it to say that when a circuit is described as an Nth-order filter, you may assume that it is an N-pole filter, as well.

A general observation can be given that the rolloff rate of a filter will eventually approach 6 dB per octave per pole (20 dB per decade per pole). Therefore, a third-order filter (i.e., three-pole) eventually rolls off at a rate of 18 dB per octave (60 dB per decade). We say "eventually" because the response around the break frequency may be somewhat faster or slower than this value. Figure 11.8 compares the effect of order on four otherwise identical low-pass filters. Note that the higher-order filters offer greater attenuation at any frequency beyond the break point. As with most response plots, Figure 11.8 utilizes decibel instead of ordinary gain. Also, these filters are shown with unity gain in the pass band, although this doesn't have to be the case. High-order filters are used when the transition band needs to be as narrow as possible. It is not

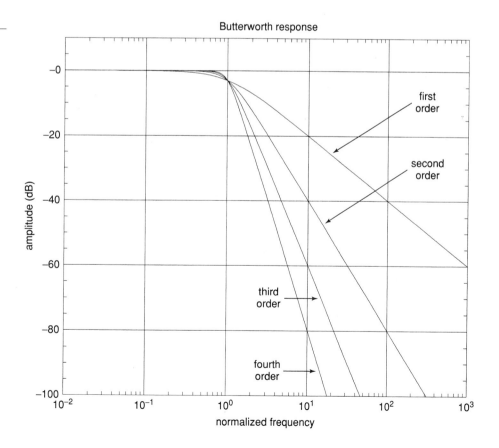

FIGURE 11.8

Effect of order on low-pass filters

uncommon to see twelfth-order and higher filters used in special applications. As you might guess, higher-order filters are more complex and costly to design and build. For many typical applications, orders in the range of two to six are common.

11.5 FILTER CLASS OR ALIGNMENT

Besides order, the shape of the transition band is determined by a filter's *class*, or *alignment*. These terms are synonymous and reflect the filter's *damping factor*. Damping factor is the reciprocal of Q, the quality factor. You should be familiar with Q from earlier work with inductors and resonant circuits. The symbol for damping factor is alpha, α. Alignment plays a key role in determining the shape of the transition region and, in some cases, the pass-band or stop-band shape also. There are a great number of possible filter alignments. We shall look at a few of the more popular types. A graph comparing the relative responses of the major types is shown in Figure 11.9. For simplicity, only second-order types are shown. Figure 11.9a shows filters with the same critical frequency and identical DC gains. In Figure 11.9b the responses have been adjusted for a peak gain of 0 dB and identical break frequencies (f_{3dB}).

FIGURE 11.9a

A comparison of the major filter alignments

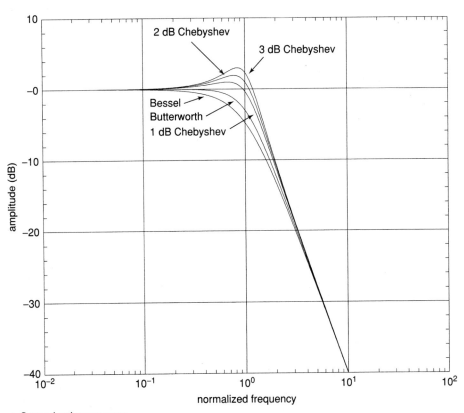

a. Second order response

FIGURE 11.9b

(Continued)

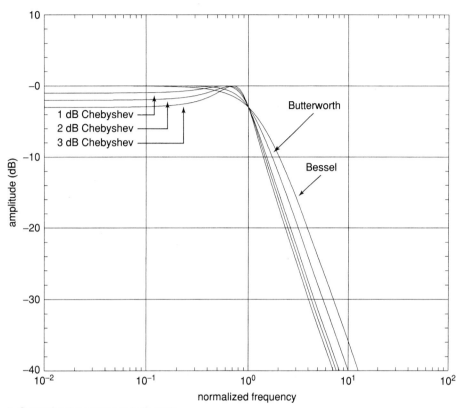

b. Second order response (adjusted)

Butterworth

Perhaps the most popular alignment type is the Butterworth. The Butterworth is characterized by its moderate amplitude and phase response. It exhibits the fastest rolloff of any monotonic (i.e., single slope or smooth) filter. In the time-domain, moderate ringing on pulses may be observed. This is also the **only** filter whose 3 dB down frequency equals its critical frequency ($f_{3\,dB} = f_c$). The Butterworth makes an excellent general-purpose filter and is widely used.

Bessel

Like the Butterworth, the Bessel is also monotonic, so it shows a smooth pass-band response. The transition region is somewhat elongated though, and the initial rolloff is less than 6 dB per octave per pole. The Bessel does exhibit a linear phase response that produces little ringing in the time domain. It is therefore a good choice for filtering pulses when the overall shape of the pulse must remain coherent (i.e., smooth and undistorted in time).

Chebyshev

The Chebyshev is actually a class of filters all its own. They are based on Chebyshev polynomials. There are many possible variations on this theme. In general, the Chebyshev exhibits initial rolloff rates in excess of 6 dB per octave per pole. This extra-fast transition is paid for in two ways: first, the phase response tends to be rather poor, resulting in a great deal of ringing when filtering pulses or other fast transients. The second effect is that the Chebyshev is nonmonotonic. The pass-band response is not smooth; instead, ripples may be noticed. In fact, the height of the ripples defines a particular Chebyshev response. It is possible to design an infinite number of variations from less than .1 dB ripple to more than 3 dB ripple. Generally, the more ripple you can tolerate, the greater the rolloff will be, and the worse the phase response will be. The choice is obviously one of compromise. The basic differences between the various Chebyshev types are characterized in Figure 11.10. Figure 11.10a compares two different low-pass Chebyshev filters of the same order. Note that only the height of the ripples is different. Figure 11.10b compares equal ripple Chebyshevs, but of different orders. Note that the higher-order filter exhibits a greater number of ripples. Also, note that even-order Chebyshevs exhibit a dip at DC whereas odd-order units show a crest

FIGURE 11.10a

Variation of Chebyshev parameter

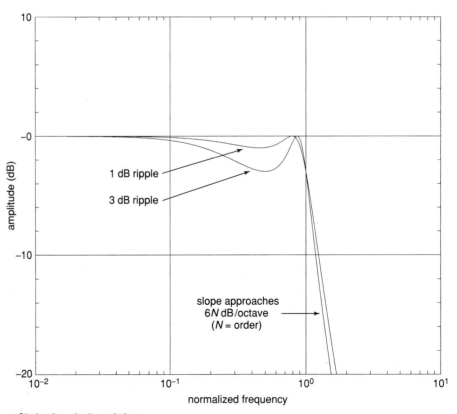

a. Chebyshev ripple variation

FIGURE 11.10b

(Continued)

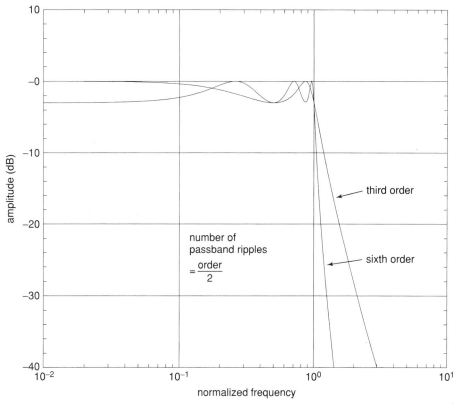

b. Chebyshev order variation

at DC. The number of ripples in the pass band is equal to the order of the filter divided by 2. When compared to the Butterworth and Bessel alignments in Figure 11.9, it is apparent that heavy damping produces the smoothest curves. The final note on the Chebyshev concerns, of all things, its spelling. You will often see "Chebyshev" spelled in a variety of ways, including "Chebycheff" and "Tschebycheff." It merely depends on how the name of the Russian mathematician is transliterated from Russian Cyrillic into English. The spellings all refer to the same filter.

Elliptic

The elliptic is also known as the *Cauer* alignment. It is a somewhat more advanced filter. It achieves very fast initial rolloff rates. Unlike the other alignments noted above, the elliptic does not "roll off forever." After its initial transition, the response rises back up, exhibiting ripples in the stop band. A typical response is shown in Figure 11.11. The design of elliptics is an advanced topic and will not be considered further. It should be noted, though, that they are popular in analog-to-digital conversion systems that require very narrow transition bands.

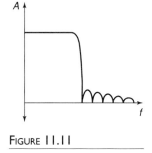

FIGURE 11.11

Elliptic response

Other Possibilities

In order to optimize the time-domain and frequency-domain characteristics for specific applications, a number of other alignments may be used. These include such alignments as *Paynter* and *Linkwitz-Reilly* and are often treated as being midway between Bessel and Butterworth, or Butterworth and Chebyshev.

11.6 Realizing Practical Filters

Usually, filter design starts with a few basic, desired parameters. This usually includes the break frequency, the amount of ripple that may be tolerated in the pass band (if any), and desired attenuation levels at specific points in the transition and stop bands. Phase response and associated time delays may also be specified. If phase response is paramount, the Bessel is normally chosen. Likewise, if pass-band ripple cannot be allowed, Chebyshevs are not considered. With the use of comparative curves such as those found in Figures 11.8 through 11.10, the filter order and alignment may be determined from the required attenuation values. At this point, the filter performance will be fully specified, for example, a 1 kHz, low-pass, third-order Butterworth. There are many ways in which this specification may be realized.

Sallen and Key VCVS Filters

There are many possible ways to create an active filter. Perhaps the most popular forms for realizing active high- and low-pass filters are the *Sallen and Key Voltage-Controlled Voltage Source* models. As the name implies, the Sallen and Key forms are based on a VCVS; in other words, they use series-parallel negative feedback. A general circuit for these models is shown in Figure 11.12. This circuit is a two-pole (second-order) section and can be configured for either high- or low-pass filtering. For our purposes,

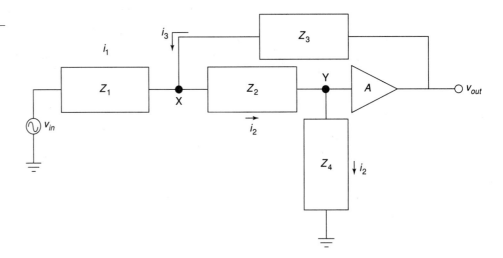

FIGURE 11.12

General VCVS 2-pole filter section

the amplifier block will utilize an op amp, although a discrete amplifier is possible. Besides the amplifier, there are four general impedances in the circuit. Usually, each element is a single resistor or capacitor. As we shall see, the selection of the component type will determine the type of filter.

At this point, we need to derive the general transfer equation for the circuit. Once the general equation is established, we will be able to refine the circuit for special cases. Being a higher-order active circuit, this procedure will naturally be somewhat more involved than the derivations found in Chapter 1 for the simple first-order lead and lag networks. The concepts are consistent though. This derivation utilizes a nodal analysis. In order to make the analysis a bit more convenient, it will help if we declare the voltage V_y as 1 V. This will save us from carrying an input voltage factor through our calculations, which would need to be factored out of the general equation anyway. By inspection we note the following:

$$V_{out} = AV_y = A \tag{11.1}$$

$$i_1 = \frac{V_{in} - V_x}{Z_1}$$

$$i_2 = \frac{1}{Z_4} \tag{11.2}$$

$$i_3 = \frac{V_{out} - V_x}{Z_3} = \frac{A - V_x}{Z_3}$$

$$V_x = i_2 Z_2 + V_y = i_2 Z_2 + 1 \tag{11.3}$$

By substituting Equation 11.2 into Equation 11.3, V_x may be expressed as

$$V_x = \frac{Z_2}{Z_4} + 1$$

We now sum the currents according to the figure,

$$i_2 = i_1 + i_3$$

substitute our current equivalences,

$$\frac{1}{Z_4} = \frac{V_{in} - V_x}{Z_1} + \frac{A - V_x}{Z_3}$$

and solve the equation in terms of V_{in}:

$$\frac{V_{in} - V_x}{Z_1} = \frac{1}{Z_4} - \frac{A - V_x}{Z_3}$$

$$V_{in} - V_x = \frac{Z_1}{Z_4} - \frac{Z_1(A - V_x)}{Z_3}$$

$$V_{in} = \frac{Z_1}{Z_4} - \frac{Z_1(A - V_x)}{Z_3} + V_x$$

$$V_{in} = \frac{Z_1}{Z_4} - \frac{Z_1(A-V_x)}{Z_3} + \frac{Z_2}{Z_4} + 1$$

$$V_{in} = \frac{Z_1}{Z_4} - \frac{Z_1}{Z_3}\left(A - \frac{Z_2}{Z_4} - 1\right) + \frac{Z_2}{Z_4} + 1$$

$$V_{in} = \frac{Z_1}{Z_4} - \frac{Z_1}{Z_3}A + \frac{Z_1 Z_2}{Z_3 Z_4} + \frac{Z_1}{Z_3} + \frac{Z_2}{Z_4} + 1$$

$$V_{in} = \frac{Z_1}{Z_4} + \frac{Z_1}{Z_3}(1-A) + \frac{Z_1 Z_2}{Z_3 Z_4} + \frac{Z_2}{Z_4} + 1 \qquad (11.4)$$

We can now write our general transfer equation using Equations 11.1 and 11.4.

$$\frac{V_{out}}{V_{in}} = \frac{A}{\frac{Z_1}{Z_4} + \frac{Z_1}{Z_3}(1-A) + \frac{Z_1 Z_2}{Z_3 Z_4} + \frac{Z_2}{Z_4} + 1} \qquad (11.5)$$

Equation 11.5 is applicable to any variation on Figure 11.12 that we wish to make. All we need to do is substitute the appropriate circuit elements for Z_1 through Z_4. As you might guess, direct substitution of resistance and reactance values would make for considerable work. In order to alleviate this difficulty, designers rely on the *Laplace transform* technique. This is also known as the *s domain* technique. A detailed analysis of the Laplace transform is beyond the scope of this text, but some familiarity will prove helpful. The basic idea is to replace complex terms with simpler ones. This is performed by setting the variable s equal to $j\omega$. As you will see, this substitution leads to far simpler and more generalized circuit derivations and equations. A capacitive reactance may be reduced as follows.

$$\text{Capacitive Reactance} = -jX_C$$

$$\text{Capacitive Reactance} = \frac{-j}{\omega C}$$

$$\text{Capacitive Reactance} = \frac{1}{j\omega C}$$

$$\text{Capacitive Reactance} = \frac{1}{sC}$$

Therefore, whenever a capacitive reactance is needed, the expression $\frac{1}{sC}$ is used. Equations can then be manipulated using basic algebra.

Sallen and Key Low-Pass Filters

Let's derive a general expression based on Equation 11.5 for low-pass filters. A low-pass filter is a lag network, so to echo this, we will use resistors for the first two elements and capacitors for the third and forth. Using the s operator, we find $Z_1 = R_1$, $Z_2 = R_2$, $Z_3 = \frac{1}{sC_1}$, and $Z_4 = \frac{1}{sC_2}$.

$$\frac{V_{out}}{V_{in}} = \frac{A}{\frac{Z_1}{Z_4} + \frac{Z_1}{Z_3}(1-A) + \frac{Z_1 Z_2}{Z_3 Z_4} + \frac{Z_2}{Z_4} + 1}$$

$$\frac{V_{out}}{V_{in}} = \frac{A}{sR_1C_2 + sR_1C_1(1-A) + s^2R_1R_2C_1C_2 + sR_2C_2 + 1}$$

$$\frac{V_{out}}{V_{in}} = \frac{A}{s^2R_1R_2C_1C_2 + s(R_1C_2 + R_2C_2 + R_1C_1(1-A)) + 1}$$

The highest power of s in the denominator determines the number of poles in the filter. Because this is a 2 here, the filter must be a 2-pole type (second-order), as expected. Usually, it is most convenient if the denominator coefficient for s^2 is unity. This makes the equation easier to factor.

$$\frac{V_{out}}{V_{in}} = \frac{A/R_1R_2C_1C_2}{s^2 + s\left(\frac{1}{R_2C_1} + \frac{1}{R_1C_1} + \frac{1}{R_2C_2}(1-A)\right) + \frac{1}{R_1R_2C_1C_2}} \quad (11.6)$$

Second-order systems appear in a variety of areas including mechanical, acoustical, hydraulic, and electrical. They have been widely studied, and a generalized form of one group of second-order responses is given by

$$G = \frac{A\omega^2}{s^2 + \alpha\omega s + \omega^2} \quad (11.7)$$

where A is the gain of the system, ω is the resonant frequency in radians, and α is the damping factor.

By comparing the general form of Equation 11.7 to the low-pass filter Equation 11.6, we find that

$$\omega^2 = \frac{1}{R_1R_2C_1C_2} \quad (11.8)$$

$$\alpha\omega = \frac{1}{R_2C_1} + \frac{1}{R_1C_1} + \frac{1}{R_2C_2}(1-A) \quad (11.9)$$

Based on the general expression of Equation 11.7, we can derive equations for the gain magnitude versus frequency and phase versus frequency, as we did in Chapter 1 for the first-order systems. For most filter work, it is convenient to work with *normalized frequency* instead of a true frequency. This means that the critical frequency will be set to 1 radian per second, and a generalized equation developed. For circuits using other critical frequencies, the general equation is used and its results are simply scaled by a factor equal to this new frequency. The normalized version of Equation 11.7 is

$$G = \frac{A}{s^2 + \alpha s + 1} \quad (11.10)$$

In order to determine the gain and phase expressions, Equation 11.10 must be split into its real and imaginary components. The first step is to replace s with its equivalent, $j\omega$, and then group the real and imaginary components.

$$G = \frac{A}{(j\omega)^2 + j\alpha\omega + 1}$$

$$G = \frac{A}{-\omega^2 + j\alpha\omega + 1}$$

$$G = \frac{A}{(1 - \omega^2) + j\alpha\omega}$$

To split this into separate real and imaginary components, we must multiply the numerator and denominator by the complex conjugate of the denominator, $(1 - \omega^2) - j\alpha\omega$.

$$G = \frac{A}{(1 - \omega^2) + j\alpha\omega} \frac{(1 - \omega^2) - j\alpha\omega}{(1 - \omega^2) - j\alpha\omega}$$

$$G = \frac{A((1 - \omega^2) - j\alpha\omega)}{(1 - \omega^2)^2 + \alpha^2\omega^2}$$

$$G = \frac{A(1 - \omega^2)}{(1 - \omega^2)^2 + \alpha^2\omega^2} - j\frac{A\alpha\omega}{(1 - \omega^2)^2 + \alpha^2\omega^2} \quad (11.11)$$

For the gain magnitude, recall that $\text{Mag} = \sqrt{\text{real}^2 + \text{imaginary}^2}$. Applying Equation 11.11 to this relation yields

$$\text{Mag} = \sqrt{\left(\frac{A(1 - \omega^2)}{(1 - \omega^2)^2 + \alpha^2\omega^2}\right)^2 + \left(-\frac{A\alpha\omega}{(1 - \omega^2)^2 + \alpha^2\omega^2}\right)^2}$$

$$\text{Mag} = \sqrt{\frac{A^2(1 - \omega^2)^2}{((1 - \omega^2)^2 + \alpha^2\omega^2)^2} + \frac{A^2\alpha^2\omega^2}{((1 - \omega^2)^2 + \alpha^2\omega^2)^2}}$$

$$\text{Mag} = \sqrt{\frac{A^2((1 - \omega^2)^2 + \alpha^2\omega^2)}{((1 - \omega^2)^2 + \alpha^2\omega^2)^2}}$$

$$\text{Mag} = \frac{A\sqrt{(1 - \omega^2)^2 + \alpha^2\omega^2}}{(1 - \omega^2)^2 + \alpha^2\omega^2}$$

$$\text{Mag} = \frac{A}{\sqrt{(1 - \omega^2)^2 + \alpha^2\omega^2}}$$

$$\text{Mag} = \frac{A}{\sqrt{1 + \omega^2(\alpha^2 - 2) + \omega^4}} \quad (11.12)$$

For the phase response, recall that $\theta = \arctan \frac{\text{imaginary}}{\text{real}}$. Applying Equation 11.11 to this relation yields

$$\theta = \arctan \frac{-\frac{A\alpha\omega}{(1-\omega^2)^2+\alpha^2\omega^2}}{\frac{A(1-\omega^2)}{(1-\omega^2)^2+\alpha^2\omega^2}}$$

$$\theta = \arctan \frac{-A\alpha\omega}{A(1 - \omega^2)}$$

$$\theta = -\arctan \frac{\alpha\omega}{1 - \omega^2} \quad (11.13)$$

To use Equations 11.12 and 11.13 with a particular alignment, substitute the appropriate damping value and simplify. Derivation of specific damping factors is beyond the scope of this chapter, but can be found in texts specializing on filter design. For our purposes, tables of damping factors for specific alignments will be presented. As one possible example, the damping factor for a second-order Butterworth alignment is $\sqrt{2}$. Substituting this into Equation 11.12 produces

$$\text{Mag} = \frac{A}{\sqrt{1 + \omega^2(\alpha^2 - 2) + \omega^4}}$$

$$\text{Mag} = \frac{A}{\sqrt{1 + \omega^2((\sqrt{2})^2 - 2) + \omega^4}}$$

$$\text{Mag} = \frac{A}{\sqrt{1 + \omega^4}}$$

There are a large number of ways of configuring a low-pass filter given the above equations. So that we might put some consistency into what appears to be a chaotic mess, we'll look at two distinct and useful variations. They are the *equal component* realization and the *unity-gain* realization.

The equal-component version Here, we will set $R_1 = R_2$ and $C_1 = C_2$. To keep the resulting equation generic, we shall use a *normalized frequency* of 1 radian per second. For other tuning frequencies, we will just scale our results to the desired value. From Equation 11.8, if $\omega = 1$, then $R_1 R_2 C_1 C_2 = 1$, and the most straightforward solution would be to set $R_1 = R_2 = C_1 = C_2 = 1$ (units of Ohms and Farads). Equation 11.6 simplifies to

$$\frac{V_{out}}{V_{in}} = \frac{A}{s^2 + s(1 + 1 + 1(1 - A)) + 1}$$

Note the damping factor is now given by

$$\alpha = 1 + 1 + 1(1 - A)$$
$$\alpha = 3 - A$$

We see that the gain and damping of the filter are linked together. Indeed, for a certain damping factor, only one specific gain will work properly:

$$A = 3 - \alpha$$

As the gain of a noninverting amplifier is

$$A = 1 + \frac{R_f}{R_i}$$

Figure 11.13

Low-pass equal-component VCVS

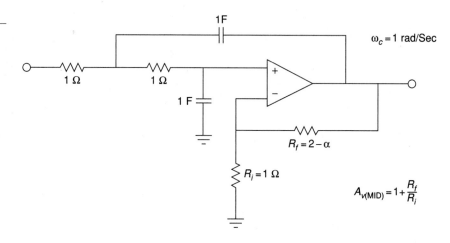

we may find the required value for R_f by combining these two equations:

$$R_f = 2 - \alpha$$

The finished prototype is shown in Figure 11.13.

The unity-gain version Here we shall set $A = 1$, and $R_1 = R_2$. From Equation 11.8, if $\omega = 1$, then $R_1 R_2 C_1 C_2 = 1$, and therefore $C_1 = \frac{1}{C_2}$. In effect, the ratio of the capacitors will set the damping factor for the system. Equation 11.6 may be simplified to

$$\frac{V_{out}}{V_{in}} = \frac{A}{s^2 + s(\frac{1}{C_1} + \frac{1}{C_1}) + 1}$$

The damping factor is now given by

$$\alpha = \frac{1}{C_1} + \frac{1}{C_1}$$

$$\alpha = \frac{2}{C_1}$$

$$\text{or, } C_1 = \frac{2}{\alpha}$$

Because $C_1 = \frac{1}{C_2}$, we find

$$C_2 = \frac{\alpha}{2}$$

The finished prototype is shown in Figure 11.14.

As you can see, there is quite a bit of similarity between the two versions. It is important to note that the inputs to these circuits must return to ground via a

FIGURE 11.14

Low-pass unity-gain VCVS

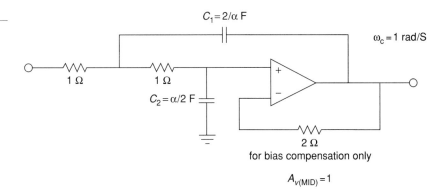

low-impedance DC path. If the signal source is capacitively coupled, the op amp's input bias current cannot be set up properly, and thus, some form of DC return resistor must be used at the source. Also, you can see that the damping factor (i.e., alignment) of the filter plays a role in setting component values. If the values shown are taken as having units of Ohms and Farads, the critical frequency will be 1 radian per second. It is an accepted practice to normalize the basic forms of circuits such as these, so that the critical frequency works out to this convenient value. This makes it very easy to scale the component values to fit your desired critical frequency. Because the critical frequency is inversely proportional to the tuning resistor and capacitor values, you only need to shrink R or C in order to increase f_c. (Remember, $f_c = \frac{1}{2\pi RC}$). A second scaling step normally follows this, in order to create practical values for R and C. This procedure is best shown with an example.

EXAMPLE 11-1

Design a 1 kHz low-pass, second-order Butterworth filter. Examine both the equal-component and the unity-gain forms as drawn in Figures 11.13 and 11.14, respectively. The required damping factor is 1.414.

Let's start with the equal-component version. First, find the required value for R_f from the damping factor, as given on the diagram.

$$R_f = 2 - \alpha$$
$$R_f = 2 - 1.414$$
$$R_f = .586 \, \Omega$$

Note that this will produce a pass-band gain of

$$A_v = 1 + \frac{R_f}{R_i}$$
$$A_v = 1 + \frac{.586}{1}$$
$$A_v = 1.586$$

Figure 11.15

Initial damping calculation for Example 11-1 (equal-component version)

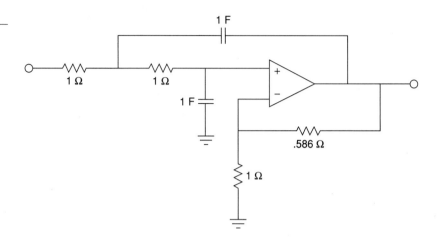

Figure 11.15 shows the second-order Butterworth low-pass filter. Its critical frequency is 1 radian per second. We need to scale this to 1 kHz.

$$\omega_c = 2\pi f_c$$
$$\omega_c = 2\pi 1 \text{ kHz}$$
$$\omega_c = 6283 \text{ radians per second}$$

Our desired critical frequency is 6283 times higher than the normalized base. As $\omega_c = \frac{1}{RC}$, to translate the frequency up, all we need to do is divide R or C by 6283. It doesn't really matter which one you choose, although it is generally easier to find "odd" sizes for resistors than capacitors, so we'll use R.

$$R = \frac{1}{6283}$$
$$R = 1.59 \times 10^{-4} \, \Omega$$

Figure 11.16 shows our 1 kHz, low-pass, second-order Butterworth filter. As you can see, though, the component values are not very practical. It is therefore necessary to perform the final scaling operations. First, consider multiplying R_f and R_i by 10 k. Note that this will have no effect on the damping, as it is the **ratio** of these two elements that determines damping. This scaling will not affect the critical frequency either, as f_c is set by the tuning resistors and capacitors. Second, we need to increase R to a reasonable value. A factor of 10^7 will place it at 1.59 kΩ. In order to compensate, the tuning capacitors must be dropped by an equal amount, which brings them to 100 nF. The completed design is shown in Figure 11.17. Other scaling factors could also be used. Also, if bias compensation is important, the R_i and R_f values will need to be scaled further, in order to balance the resistance seen at the noninverting input.

FIGURE 11.16

The filter for Example 11-1 after frequency scaling (equal-component version)

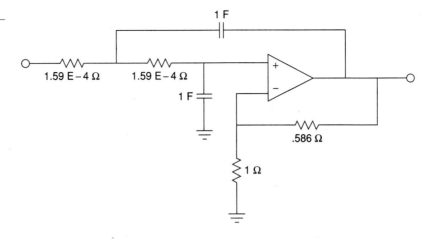

The approach for the unity-gain version is similar. First, adjust the capacitor values in order to achieve the desired damping, as specified in Figure 11.14.

$$C_1 = \frac{2}{\alpha}$$

$$C_1 = \frac{2}{1.414}$$

$$C_1 = 1.414 \text{ F}$$

$$C_2 = \frac{\alpha}{2}$$

$$C_2 = \frac{1.414}{2}$$

$$C_2 = .707 \text{ F}$$

The resulting circuit is shown in Figure 11.18. The circuit must be scaled to the desired f_c. The factor is 6283 once again, and the result is shown in Figure 11.19. The final component scaling is seen in Figure 11.20.

FIGURE 11.17

Final impedance scaling for Example 11-1 (equal-component version)

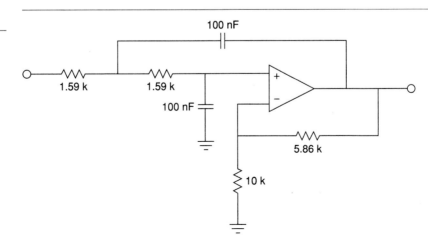

Figure 11.18

Initial damping calculation for the filter of Example 11-1 (unity-gain version)

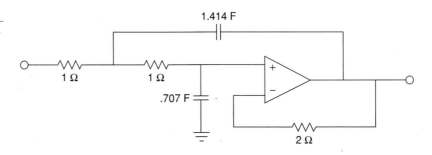

multiSIM COMPUTER SIMULATION

An Electronics Workbench MultiSIM simulation of the filter design of Example 11-1 is shown in Figure 11.21 (pp. 437–439). The analysis shows the Bode plot, ranging from 50 Hz to better than 20 kHz. This yields over one decade on either side of the 1 kHz critical frequency. The graph clearly shows the −3 dB point at approximately 1 kHz, with an attenuation slope of −12 dB per octave. As this is the unity-gain version, the low-frequency gain is set at 0 dB. Also, note that no peaking is evident in the response curve, as is expected for a Butterworth alignment. The phase response is also shown. Some graphing tools continue the phase shift below −180 degrees, whereas others will flip it back to +180, as MultiSIM does. Note that if the frequency plot range is extended, the phase shift starts to increase at the highest frequencies instead of leveling off. This is due to the extra phase shift produced by the op amp as the operating frequency approaches f_{unity}. A simpler op amp model would not create this real-world effect. Finally, a Monte Carlo analysis is used to mimic the effects seen due to component tolerance in a production environment (in this example, approximately 10% variation of nominal for each component). A total of 10 runs were generated. Although the overall shape of the curve remains consistent, there is some variation in the corner frequency, certainly more than the 10% offered by any single component. As you might guess, a Monte Carlo analysis is very tedious to do by hand, but quite straightforward to set up on a simulator.

As you can see, the realization process is little more than a scaling sequence. This makes filter design very rapid. The operation for high-pass filters is essentially the same.

Figure 11.19

The filter for Example 11-1 after frequency scaling (unity-gain version)

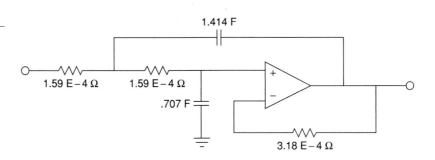

REALIZING PRACTICAL FILTERS   437

FIGURE 11.20

Final impedance scaling for the filter of Example 11-1 (unity-gain version)

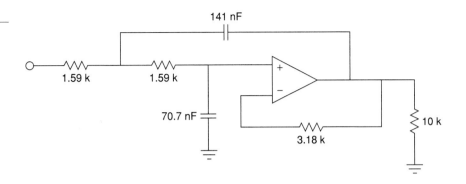

FIGURE 11.21a

VCVS low-pass filter in MultiSIM

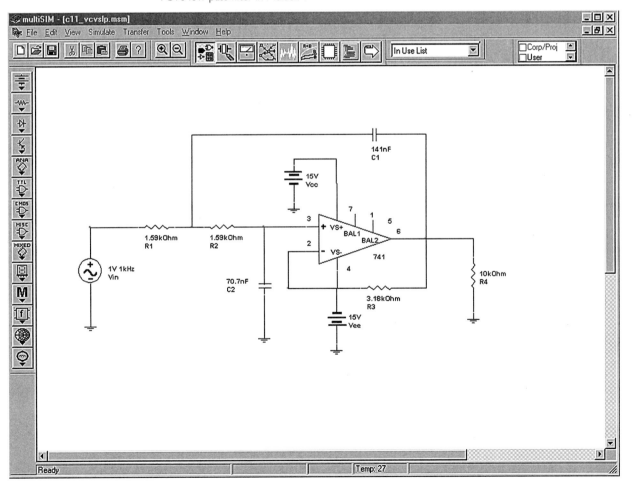

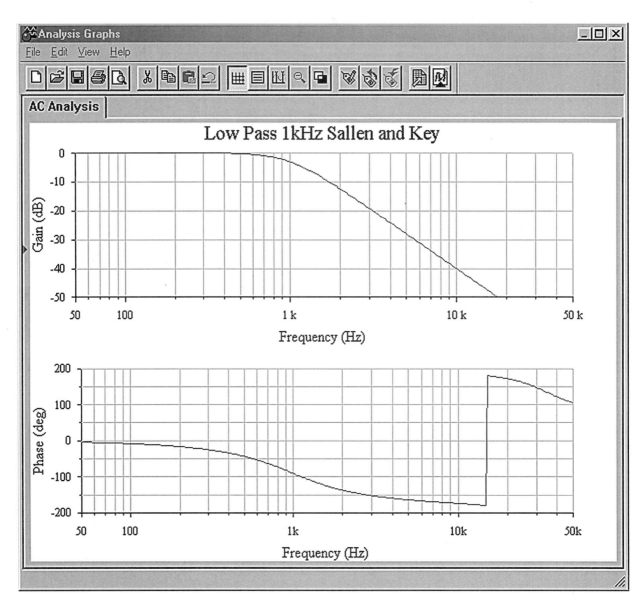

FIGURE 11.21b

Gain and phase plots for VCVS low-pass filter

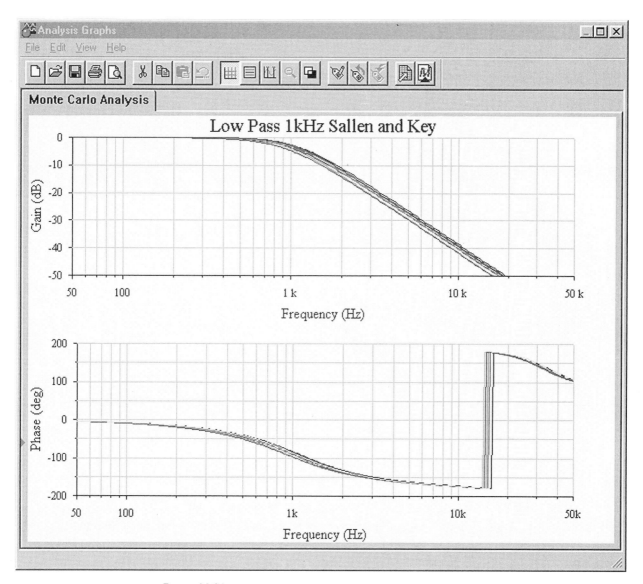

FIGURE 11.21c

Monte Carlo analysis for VCVS low-pass filter

Sallen and Key High-Pass Filters

We can derive a general expression for high-pass filters, based on Equation 11.5. A high-pass filter is a lead network, so to echo this, we will use capacitors for the first two elements and resistors for the third and forth. Using the s operator, we find, $Z_1 = \frac{1}{sC_1}$, $Z_2 = \frac{1}{sC_2}$, $Z_3 = R_1$, and $Z_4 = R_2$.

$$\frac{V_{out}}{V_{in}} = \frac{A}{\frac{Z_1}{Z_4} + \frac{Z_1}{Z_3}(1-A) + \frac{Z_1 Z_2}{Z_3 Z_4} + \frac{Z_2}{Z_4} + 1}$$

$$\frac{V_{out}}{V_{in}} = \frac{A}{\frac{1}{sR_2 C_1} + \frac{1}{sR_1 C_1}(1-A) + \frac{1}{s^2 R_1 R_2 C_1 C_2} + \frac{1}{sR_2 C_2} + 1}$$

$$\frac{V_{out}}{V_{in}} = \frac{As^2}{s^2 + s\left(\frac{1}{R_2 C_1} + \frac{1}{R_2 C_2} + \frac{1}{R_1 C_1}(1-A)\right) + \frac{1}{R_1 R_2 C_1 C_2}} \quad (11.14)$$

As with the low-pass filters we have two basic realizations: equal-component and unity-gain. In both cases, we start with Equation 11.14 and use normalized frequency (1 radian per second). The derivations are very similar to the low-pass case, and the results are summarized below.

The equal-component version

$$\alpha = 3 - A$$

The unity-gain version

$$R_2 = \frac{2}{\alpha}$$
$$R_1 = \frac{\alpha}{2}$$

These forms are shown in Figures 11.22 and 11.23. You may at this point ask two questions: One, how do you find the damping factor, and two, what about higher-order filters?

In order to find the damping factor needed, a chart such as Figure 11.24 may be consulted. This chart also introduces a new item, and that is the *frequency factor*, k_f. Normally, the critical frequency and 3 dB down frequency (break frequency) of a filter are not the same value. They are identical only for the Butterworth alignment. For any other alignment, the desired break frequency must first be translated to the appropriate critical frequency before scaling is performed. This is illustrated in the following example.

FIGURE 11.22

High-pass unity-gain VCVS

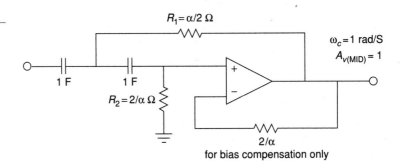

FIGURE 11.23

High-pass equal-component VCVS

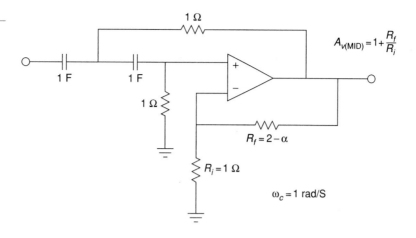

EXAMPLE 11-2

Design a second-order, high-pass Bessel filter, with a break frequency ($f_{3\,dB}$) of 5 kHz.

For this example, let's use the unity-gain form shown in Figure 11.22. First, obtain the damping and frequency factors from Figure 11.24.

$k_f = 1.274$, damping $= 1.732$.

Using the damping factor, the two tuning resistors may be found:

$$R_1 = \frac{\alpha}{2}$$
$$R_1 = \frac{1.732}{2}$$
$$R_1 = .866$$

$$R_2 = \frac{2}{\alpha}$$
$$R_2 = \frac{2}{1.732}$$
$$R_2 = 1.155$$

FIGURE 11.24

Second-order filter parameters

From Lancaster, Don. *Active Filter Cookbook.* Second edition, Newnes 1996. Reprinted with permission

Type	Damping	f_c Factor (k_f)
Bessel	1.732	1.274
Butterworth	1.414	1.0
1 dB Chebyshev	1.045	.863
2 dB Chebyshev	.895	.852
3 dB Chebyshev	.767	.841

$f_c = f_{3\,dB}/f_c$ Factor for high-pass
$f_c = f_{3\,dB} \times f_c$ Factor for low-pass

FIGURE 11.25

Initial damping calculation for Example 11-2

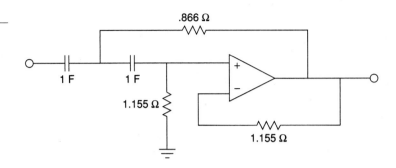

The intermediate result is shown in Figure 11.25. In order to do the frequency scaling, the desired break frequency of 5 kHz must first be translated into the required critical frequency. Because this is a high-pass filter,

$$f_c = \frac{f_{3\,dB}}{k_f}$$

$$f_c = \frac{5\text{ kHz}}{1.274}$$

$$f_c = 3925\text{ Hz}$$

$$\omega_c = 2\pi f_c$$

$$\omega_c = 2\pi\,3925$$

$$\omega_c = 24.66\text{ k radians per second}$$

Either the tuning resistors or capacitors may now be scaled.

$$R_1 = \frac{.866}{24.66\text{ k}}$$

$$R_1 = 3.51 \times 10^{-5}$$

$$R_2 = \frac{1.155}{24.66\text{ k}}$$

$$R_2 = 4.68 \times 10^{-5}$$

We now have a second-order, high-pass, 5 kHz Bessel filter. This is shown in Figure 11.26. A final scaling of 10^8 will give us reasonable values, and is shown in Figure 11.27.

Filters of Higher Order

There is a common misconception among novice filter designers that higher-order filters may be produced by cascading a number of lower-order filters of the same type. This is not true. For example, cascading three second-order 10 kHz Butterworth filters will **not** produce a sixth-order 10 kHz Butterworth filter. A quick inspection reveals why this is

FIGURE 11.26

Frequency scaling for Example 11-2

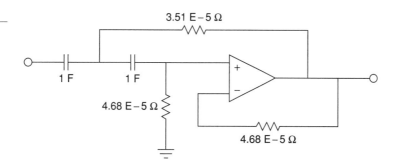

not the case: A single filter of any order will show a 3 dB loss at its break frequency by definition (in this case, 10 kHz). If three filters of the same type are cascaded, each filter will produce a 3 dB loss at the break frequency, which means an overall loss of 9 dB occurs. This much is true: a higher-order filter will require a number of individual sections, each with specific damping and frequency factors. Each section will be based on the second-order forms already examined.[2] In order to make odd-ordered filters, we will introduce a simple single-pole filter. The high- and low-pass versions of this unit are shown in Figure 11.28. The damping factor for this circuit is always unity. When working with it, you need only worry about the frequency factor.

Designing higher-order filters is, conceptually, no different from designing second-order filters. The reality is that new charts are needed for the required damping and frequency factors. A set of compatible charts is shown in Figure 11.29 for orders 3 through 6. To show the design sequence flows, let's look at an example.

EXAMPLE 11-3

We wish to design a filter suitable for removing subsonic tones from a stereo system. This could be used to reduce turntable rumble in a vintage Hi-Fi or DJ system, or to reduce stage

FIGURE 11.27

Impedance scaling for Example 11-2

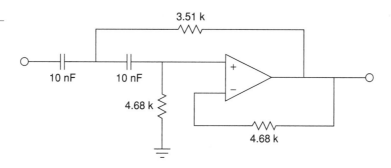

---

[2] The transfer function of a higher-order filter contains a high-order polynomial in the denominator of the form $s^n + b_{n-1}s^{n-1} + \cdots + b_1 s + b_0$. The n indicates the order of the filter and the b coefficients determine the alignment. This polynomial is factored into a product of second-order expressions (with a possible first-order unit for odd-ordered systems). Each of these expressions corresponds to a single section in the larger filter.

Figure 11.28

Single-pole sections

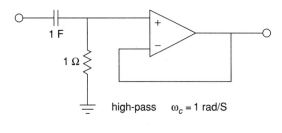

high-pass $\omega_c = 1$ rad/S

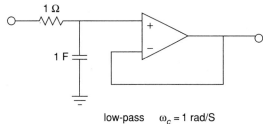

low-pass $\omega_c = 1$ rad/S

vibration in a public address system. The filter should attenuate frequencies below the lower limit of human hearing (about 20 Hz), while allowing all higher frequencies to pass. Transient response may be important here, so we'll choose a Bessel alignment. We will also specify a fifth-order system. This will create an attenuation of about 15 dB, one octave below the break frequency.

First, note that the specification requires the use of a high-pass filter. This filter may be realized with either the equal-component or the unity-gain forms. As this design will require multiple sections, excessive gain may result from the equal-component version. Our fifth-order system will be comprised of two second-order sections and a first-order section. An overview of the design is shown in Figure 11.30, with the appropriate damping and frequency factors taken from Figure 11.29. We'll break the analysis down stage by stage.

First, find the desired break frequency in radians.

$$\omega_{3\,dB} = 2\pi f_{3\,dB}$$
$$\omega_{3\,dB} = 2\pi 20 \text{ Hz}$$
$$\omega_{3\,dB} = 125.7 \text{ radians per second}$$

Stage 1

The break frequency must be translated to the required critical frequency. Because this is a high-pass filter, we need to divide by the frequency factor.

$$\omega_c = \frac{\omega_{3\,dB}}{k_f}$$
$$\omega_c = \frac{125.7}{1.557}$$
$$\omega_c = 80.7 \text{ radians per second}$$

FIGURE 11.29a

Filter design tools

From Lancaster, Don. Active Filter Cookbook. Second edition, Newnes 1996. Reprinted with permission

---

THIRD-ORDER FILTER PARAMETERS

Type	Second-order section Damping	f_c Factor	First-order section f_c Factor
Bessel	1.447	1.454	1.328
Butterworth	1.0	1.0	1.0
1 dB Chebyshev	.496	.911	.452
2 dB Chebyshev	.402	.913	.322
3 dB Chebyshev	.326	.916	.299

FOURTH-ORDER FILTER PARAMETERS

Type	Second-order section Damping	f_c Factor	Second-order section Damping	f_c Factor
Bessel	1.916	1.436	1.241	1.610
Butterworth	1.848	1.0	.765	1.0
1 dB Chebyshev	1.275	.502	.281	.943
2 dB Chebyshev	1.088	.466	.224	.946
3 dB Chebyshev	.929	.443	.179	.950

FIFTH-ORDER FILTER PARAMETERS

Type	Second-order section Damping	f_c Factor	Second-order section Damping	f_c Factor	First-order section f_c Factor
Bessel	1.775	1.613	1.091	1.819	1.557
Butterworth	1.618	1.0	.618	1.0	1.0
1 dB Chebyshev	.714	.634	.180	.961	.280
2 dB Chebyshev	.578	.624	.142	.964	.223
3 dB Chebyshev	.468	.614	.113	.967	.178

SIXTH-ORDER FILTER PARAMETERS

Type	Second-order section Damping	f_c Factor	Second-order section Damping	f_c Factor	Second-order section Damping	f_c Factor
Bessel	1.959	1.609	1.636	1.694	.977	1.910
Butterworth	1.932	1.0	1.414	1.0	.518	1.0
1 dB Chebyshev	1.314	.347	.455	.733	.125	.977
2 dB Chebyshev	1.121	.321	.363	.727	.0989	.976
3 dB Chebyshev	.958	.298	.289	.722	.0782	.975

$f_c = f_{3\,dB} / f_c$ Factor for high-pass

$f_c = f_{3\,dB} \times f_c$ Factor for low-pass

a. Parameters for third- through sixth-order filters

We will scale the resistor by 80.7 to achieve this tuning frequency.

$$R = \frac{1}{80.7}$$
$$R = .0124$$

The R and C values must now be scaled for practical values. A factor of 10^6 would be reasonable. The final result is

$$R = 12.4\,k$$
$$C = 1\,\mu F$$

FIGURE 11.29b

(Continued)

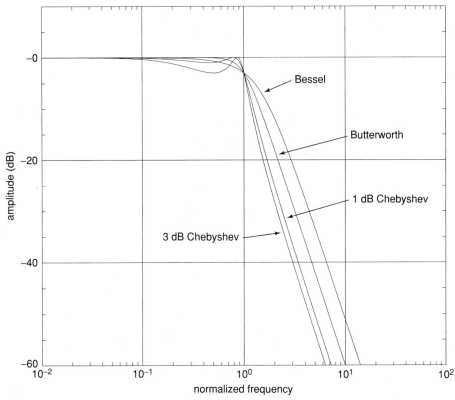

b. Third-order response (adjusted)

Stage 2

First, determine the values for the two resistors from the given damping factor.

$$R_1 = \frac{\alpha}{2}$$
$$R_1 = \frac{1.775}{2}$$
$$R_1 = .8875$$

$$R_2 = \frac{2}{\alpha}$$
$$R_2 = \frac{2}{1.775}$$
$$R_2 = 1.127$$

Now, the break frequency must be translated to the required critical frequency.

$$\omega_c = \frac{\omega_{3\,dB}}{k_f}$$

FIGURE 11.29c
(Continued)

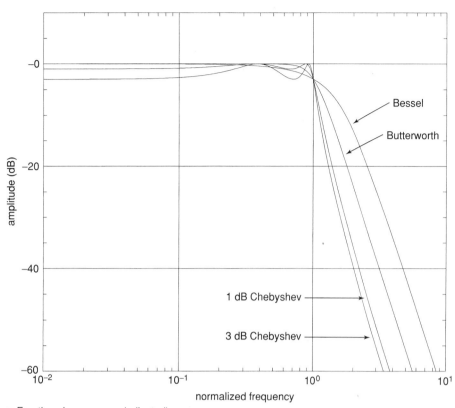

c. Fourth-order response (adjusted)

$$\omega_c = \frac{125.7}{1.613}$$
$$\omega_c = 77.9 \text{ radians per second}$$

We will scale the resistors by 77.9 to achieve this tuning frequency.

$$R_1 = \frac{.8875}{77.9}$$
$$R_1 = .0114$$

$$R_2 = \frac{1.127}{77.9}$$
$$R_2 = .0145$$

Again, R and C must be scaled for practical values. A factor of 10^6 would be reasonable. The final result is

$$R_1 = 11.4 \text{ k}$$
$$R_2 = 14.5 \text{ k}$$
$$C = 1 \text{ } \mu\text{F}$$

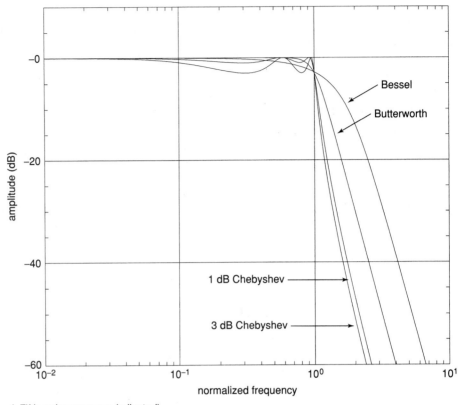

d. Fifth-order response (adjusted)

Stage 3
First, determine the values for the two resistors from the given damping factor.

$$R_1 = \frac{\alpha}{2}$$
$$R_1 = \frac{1.091}{2}$$
$$R_1 = .5455$$

$$R_2 = \frac{2}{\alpha}$$
$$R_2 = \frac{2}{1.091}$$
$$R_2 = 1.833$$

Now, the break frequency must be translated to the required critical frequency.

$$\omega_c = \frac{\omega_{3\,dB}}{k_f}$$

FIGURE 11.29e
(Continued)

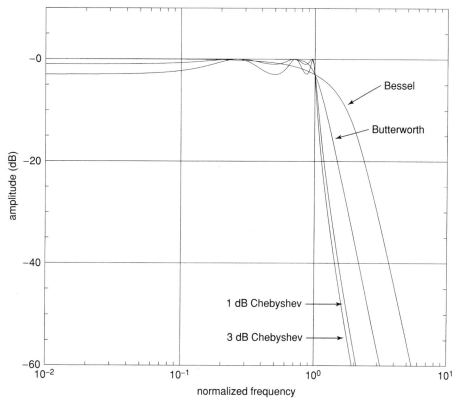

e. Sixth-order response (adjusted)

$$\omega_c = \frac{125.7}{1.819}$$
$$\omega_c = 69.1 \text{ radians per second}$$

We will scale the resistors by 69.1 to achieve this tuning frequency.

$$R_1 = \frac{.5455}{69.1}$$
$$R_1 = 7.89 \times 10^{-3}$$

FIGURE 11.30
Circuit outline (Note: all values in ohms and farads)

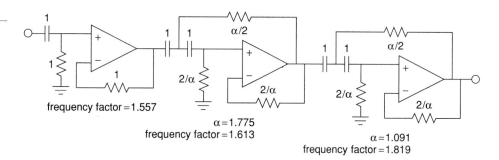

FIGURE 11.31

Completed filter for Example 11-3

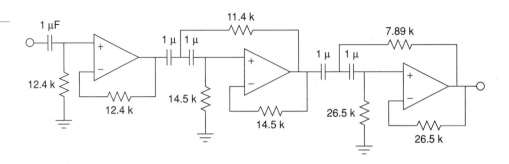

$$R_2 = \frac{1.833}{69.1}$$
$$R_2 = .0265$$

Again, R and C must be scaled for practical values. A factor of 10^6 would be reasonable. The final result is

$$R_1 = 7.89 \text{ k}$$
$$R_2 = 26.5 \text{ k}$$
$$C = 1 \ \mu\text{F}$$

The complete design is shown in Figure 11.31. Note that all of the capacitors are set at 1 μF. This certainly helps to cut inventory and parts placement costs.

We will complete our discussion of high- and low-pass VCVS filters with the following example.

EXAMPLE 11-4

As mentioned earlier, it is common for loudspeaker systems to rely on passive filters to create their crossover networks. More demanding applications such as recording studio monitoring or large public-address systems (i.e., concert systems) cannot afford the losses associated with passive crossovers. Instead, these applications utilize active crossovers composed of active filters, such as the one shown in Figure 11.32. Before the audio signal is fed to a

FIGURE 11.32

Commercial electronic crossover
Courtesy of Furman Sound, Inc.

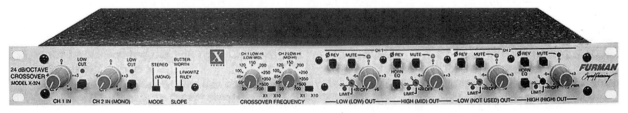

FIGURE 11.33

Electronic crossover system

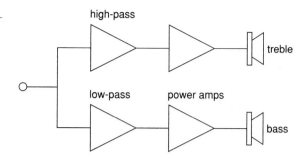

power amplifier, it is split into two or more frequency bands. The resulting signals each feed their own power amplifier/loudspeaker section. A block diagram of this approach is shown in Figure 11.33. This one is a two-way system. Large concert sound reinforcement systems may break the audio spectrum into four or five segments. The resulting system will be undoubtedly expensive, but will show lower distortion and higher output levels than a passively crossed system. A typical two-way system might be crossed at 800 Hz. In other words, frequencies above 800 Hz will be sent to a specialized high-frequency transducer, whereas frequencies below 800 Hz will be sent to a specialized low-frequency transducer. In essence, the crossover network is a combination of an 800 Hz low-pass filter, and an 800 Hz high-pass filter. The filter order and alignment vary considerably depending on the application. Let's design an 800 Hz crossover with second-order Butterworth filters.

The basic circuit layout is shown in Figure 11.34. We're using the equal-component value version here. Exact gain is normally not a problem in this case, as some form of volume control needs to be added anyway, in order to compensate between the sensitivity of the low- and high-frequency transducers. (This is most easily produced by adding a simple voltage divider/potentiometer at the output of the filters.)

For second-order Butterworth filters, the damping factor is found to be 1.414, and the frequency factor is unity (indicating that f_c and $f_{3\,dB}$ are the same). Note that the design for both halves is almost the same. Both sections show an f_c of 800 Hz, and a damping factor of 1.414. With identical characteristics, it follows that the component values will be the same in both circuits.

The required value for R_f is

$$R_f = 2 - \alpha$$
$$R_f = 2 - 1.414$$
$$R_f = .586$$

The critical frequency in radians is

$$\omega_c = 2\pi f_c$$
$$\omega_c = 2\pi\, 800 \text{ Hz}$$
$$\omega_c = 5027 \text{ radians per second}$$

Again, we shall scale the tuning resistors to yield

$$R = \frac{1}{5027}$$

FIGURE 11.34

Basic filter sections for crossover of Example 11-4

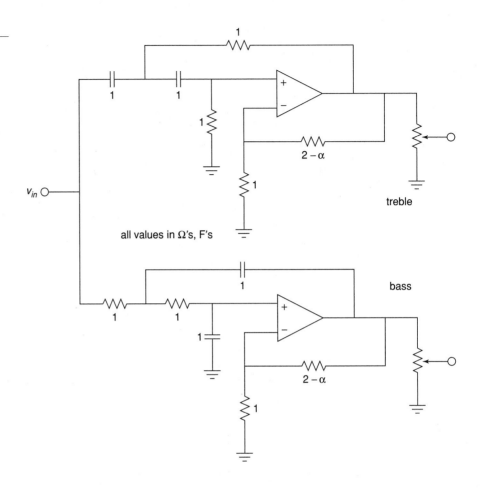

$$R = 1.989 \times 10^{-4}$$

A final RC scaling by 10^8 produces

$$R = 19.9 \text{ k}$$
$$C = 10 \text{ nF}$$

R_i and R_f are scaled by 10 k, and 10 kΩ log taper potentiometers may be used for the output volume trimmers. The resulting circuit is shown in Figure 11.35.

11.7 BAND-PASS FILTER REALIZATIONS

There are many ways to form a band-pass filter. Before we introduce a few of the possibilities, we must define a number of important parameters. As in the case of the high- and low-pass filters, the concept of damping is important. For historical reasons, band-pass filters are normally specified with the parameter Q, the quality factor, which

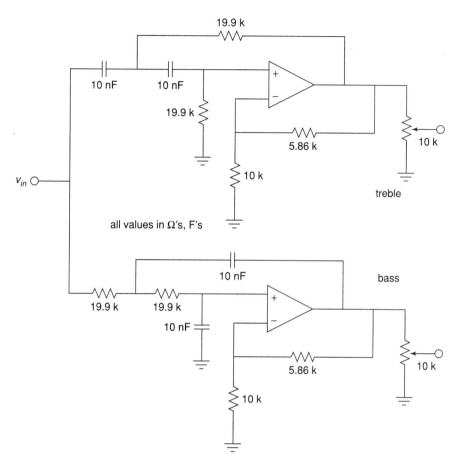

FIGURE 11.35

Completed crossover design for Example 11-4

is the reciprocal of the damping factor. Comparable to the break frequency is the center, or peak, frequency of the filter. This is the point of maximum gain. In *RLC* circuits, it is usually referred to as the resonance frequency. The symbol for center frequency is f_o. Because a band-pass filter produces attenuation on either side of the center frequency, there are two "3 dB down" frequencies. The lower frequency is normally given the name f_1, and the upper is given f_2. The difference between f_2 and f_1 is called the bandwidth of the filter and is abbreviated as *BW*. The ratio of center frequency to bandwidth is equal to the filter's *Q*.

$$BW = f_2 - f_1 \tag{11.15}$$

$$Q = \frac{f_o}{BW} \tag{11.16}$$

It is important to note that the center frequency is not equal to the arithmetic average of f_1 and f_2. Instead, it is equal to the geometric average of f_1 and f_2.

$$f_o = \sqrt{f_1 f_2} \tag{11.17}$$

FIGURE 11.36

Band-pass response

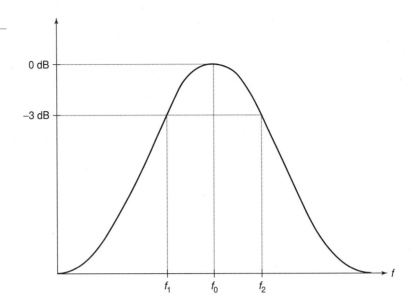

These parameters are shown graphically in Figure 11.36. If a filter requires a fairly low Q, say unity or less, the filter is best realized as a cascade of separate low- and high-pass filters. For higher Qs, we will examine two possible realizations. Multiple-feedback filters will be used for Qs up to about 10. For Qs above 10, the state-variable filter is presented.

Multiple-Feedback Filters

The basic multiple-feedback filter is a second-order type. It contains two reactive elements as shown in Figure 11.37. One pair of elements creates the low-pass response ($R_1 C_1$), and the other pair creates the high-pass response ($R_2 C_2$). Because of this, the ultimate attenuation slopes are ±6 dB. As with the VCVS high- and low-pass designs, the circuit of Figure 11.37 is normalized to a 1 radian per second center frequency.

FIGURE 11.37

Multiple feedback band-pass filter

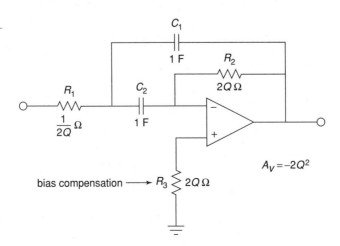

Extrapolation to new center frequencies is performed in the same manner as shown earlier. The peak gain for this circuit is

$$A_v = -2Q^2 \qquad (11.18)$$

You can see from Equation 11.18 that higher Qs will produce higher gains. For a Q of 10, the voltage gain will be 200. For this circuit to function properly, the open-loop gain of the op amp used must be greater than 200 at the chosen center frequency. Usually, a safety factor of 10 is included in order to keep stability high and distortion low. By combining these factors, we may determine the minimum acceptable f_{unity} for the op amp.

$$f_{unity} \geq 10 f_o A_v \qquad (11.19a)$$

or more directly,

$$f_{unity} \geq 20 f_o Q^2 \qquad (11.19b)$$

For a Q of 10 and a center frequency of 2 kHz, the op amp will need an f_{unity} of at least 4 MHz. It is not possible to use this type of filter for high-frequency, high-Q work, as standard op amps soon "run out of steam." This difficulty aside, the high gains produced by even moderate values for Q may well be impractical. For many applications, a unity gain version would be preferred. This is not particularly difficult to achieve. All that we need to do is attenuate the input signal by a factor equal to the voltage gain of the filter. Because the gain magnitude of the filter is $2Q^2$ the attenuation should be

$$\text{Attenuation} = \frac{1}{2Q^2} \qquad (11.20)$$

Although it is possible to place a pair of resistors in front of the filter to create a voltage divider, there is a more efficient way. We can split R_1 into two components, as shown in Figure 11.38. As long as the Thevenin equivalent of R_{1a} and R_{1b} as seen from the op amp equals the value of R_1, the tuning frequency of the filter will

FIGURE 11.38

Multiple feedback filter with unity-gain variation

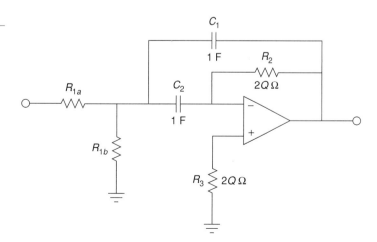

not be changed. Also required is that the voltage divider ratio produced by R_{1a} and R_{1b} satisfies Equation 11.20. First, let's determine the ratio of the two resistors. We can start by setting R_{1b} to the arbitrary value K. Using the voltage divider rule and Equation 11.20, R_{1a} is found:

$$\text{Attenuation} = \frac{R_{1b}}{R_{1a} + R_{1b}}$$

$$\frac{1}{2Q^2} = \frac{K}{R_{1a} + K}$$

$$R_{1a} + K = K2Q^2$$

$$R_{1a} = K(2Q^2 - 1) \tag{11.21}$$

So, we see that R_{1a} must be $2Q^2 - 1$ times larger than R_{1b}. Now we must determine the value of K, which will set the parallel combination of R_{1a} and R_{1b} to the required value of $\frac{1}{2Q}$, as based on Figure 11.37.

$$R_{Thevenin} = R_{1a} \| R_{1b}$$

$$R_{Thevenin} = \frac{R_{1a} R_{1b}}{R_{1a} + R_{1b}}$$

$$R_{Thevenin} = \frac{K^2(2Q^2 - 1)}{K(2Q^2 - 1) + K}$$

$$R_{Thevenin} = \frac{K^2(2Q^2 - 1)}{K2Q^2}$$

$$R_{Thevenin} = \frac{K(2Q^2 - 1)}{2Q^2}$$

Because $R_{Thevenin} = \frac{1}{2Q}$,

$$\frac{1}{2Q} = \frac{K(2Q^2 - 1)}{2Q^2}$$

$$1 = \frac{K(2Q^2 - 1)}{Q}$$

$$K = \frac{Q}{2Q^2 - 1} \tag{11.22}$$

Because R_{1b} was set to K,

$$R_{1b} = \frac{Q}{2Q^2 - 1} \Omega \tag{11.23}$$

Substituting 11.22 into 11.21 yields

$$R_{1a} = Q\Omega \tag{11.24}$$

By using these values for R_{1a} and R_{1b}, the filter will have a peak gain of unity. Note that as this scheme only attenuates the signal prior to gain, the f_{unity} requirement set in Equation 11.19 still holds true.

BAND-PASS FILTER REALIZATIONS

EXAMPLE 11-5

Design a filter that will only pass frequencies from 800 Hz to 1200 Hz. Make sure that this is a unity-gain realization.

First, we must determine the center frequency, bandwidth, and Q.

$$BW = f_2 - f_1$$
$$BW = 1200 \text{ Hz} - 800 \text{ Hz}$$
$$BW = 400 \text{ Hz}$$

$$f_o = \sqrt{f_1 f_2}$$
$$f_o = \sqrt{800 \text{ Hz} \times 1200 \text{ Hz}}$$
$$f_o = 980 \text{ Hz}$$

$$Q = \frac{f_o}{BW}$$
$$Q = \frac{980 \text{ Hz}}{400 \text{ Hz}}$$
$$Q = 2.45$$

The Q is too high to use separate high- and low-pass filters, but sufficiently low so that a multiple feedback type may be used. Before proceeding, we should check to make sure that the required f_{unity} for the op amp is reasonable.

$$A_v = -2Q^2$$
$$A_v = -2 \times 2.45^2$$
$$A_v = -12$$

$$f_{unity} >= 10 A_v f_o$$
$$f_{unity} >= 10 \times 12 \times 980 \text{ Hz}$$
$$f_{unity} >= 117.6 \text{ kHz}$$

Just about any modern op amp will exceed the f_{unity} specification. As this circuit shows a gain of 12, the unity gain variation shown in Figure 11.38 will be used. The calculations for the normalized components follow.

$$R_2 = 2Q$$
$$R_2 = 2 \times 2.45$$
$$R_2 = 4.9 \, \Omega$$

$$R_{1b} = \frac{Q}{2Q^2 - 1}$$
$$R_{1b} = \frac{2.45}{2 \times 2.45^2 - 1}$$
$$R_{1b} = .2226 \, \Omega$$

FIGURE 11.39

Initial damping calculation for Example 11-5

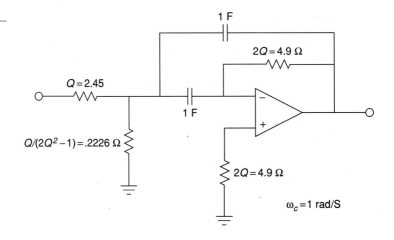

The resulting normalized circuit is shown in Figure 11.39. We must now find the frequency scaling factor.

$$\omega_o = 2\pi f_o$$
$$\omega_o = 2\pi 980 \text{ Hz}$$
$$\omega_o = 6158 \text{ radians per second}$$

In order to translate our circuit to this frequency, we must divide either the resistors or the capacitors by 6158. In this example, let's use the capacitors.

$$C = \frac{1}{6158}$$
$$C = 162.4 \ \mu F$$

A further impedance scaling is needed for practical component values. A factor of a few thousand or so would be appropriate here. To keep the calculations simple, we'll choose 10 k. Each resistor will be increased by 10 k, and each capacitor will be reduced by 10 k. The final scaled filter is shown in Figure 11.40.

FIGURE 11.40

Final impedance and frequency scaling for Example 11-5

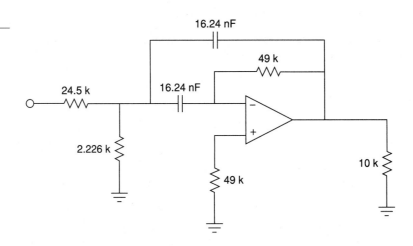

multiSIM COMPUTER SIMULATION

The Electronics Workbench MultiSIM simulation of the circuit of Example 11-5 is shown in Figure 11.41 (pp. 460–462). Note that the gain is 0 dB at the approximate center frequency (about 1 kHz). Also, the −3 dB breakpoints of 800 Hz and 1200 Hz are clearly seen. The phase response of this filter is also plotted. Note the very fast phase transition in the area around f_o. If the Q of this circuit was increased, this transition would be faster still.

In simulations such as this, it is very important that realistic op amp models be employed. If an overidealized version is used, nonideal behavior due to a reduction of loop gain will go unnoticed. This error is most likely to occur in circuits with high center frequencies and/or high Qs. You can verify this by translating the filter to a higher frequency and rerunning the simulation. For example, if C_1 and C_2 are decreased by a factor of 1000, the center frequency should move up to about 1 MHz. If the simulation is run again with an appropriate range of test frequencies, you will see that the limited bandwidth of the μA741 op amp prematurely cuts off the filter response. The result is a peaking frequency more than one octave below target, a maximum amplitude several dB below 0, and an asymmetrical response curve. This response graph is shown in Figure 11.41c. The accompanying phase plot also shows a great deviation from the ideal filter. An excessive phase shift at the middle and higher frequencies is clearly evident.

State-Variable Filter

As noted earlier, the multiple-feedback filter is not suited to high-frequency or high-Q work. For applications requiring Qs of about 10 or more, the state-variable filter is the form of choice. The state-variable is often referred to as the universal filter, as band-pass, high-pass, and low-pass outputs are all available. With additional components, a band-reject output may be formed as well. Unlike the earlier filter forms examined, the basic state-variable filter requires three op amps. Also, it is a second-order type, although higher-order types are possible. This form gets its name from state-variable analysis. One of the earliest uses for op amps was in the construction of analog computers (see Chapter 10). Interconnections of differentiators, amplifiers, summers, and integrators were used to electronically solve differential equations that described physical systems. State-variable analysis provides a technique for solving involved differential equations. The equations may, in fact, describe a required filter's characteristics. Although we will not examine state-variable analysis, this does not preclude a study of the state-variable filter. Designing with state-variable filters is really no more complex than our previous work.

Besides its ability to provide stable filters with relatively high Qs, the state-variable has other unique characteristics:

1. It is relatively easy to tune electronically over a broad frequency range.
2. It is possible to independently adjust the Q and tuning frequency.
3. It offers the ability to create other, more complex, filters as it has multiple outputs.

The state-variable filter is based on integrators. The general form utilizes a summing amplifier and two integrators, as shown in Figure 11.42. To understand how this circuit works on an intuitive level, recall that integrators are basically first-order,

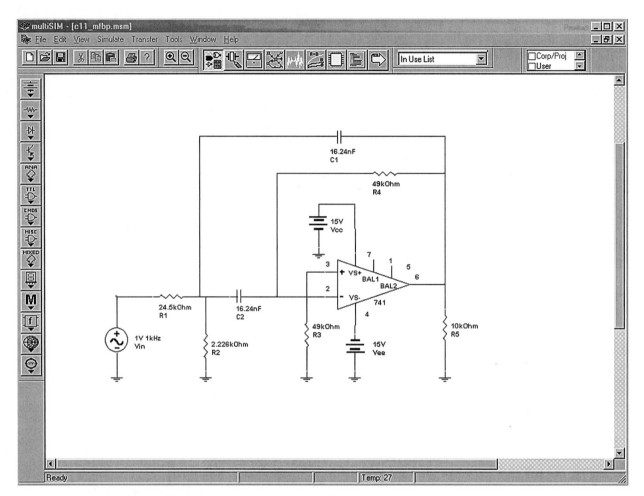

FIGURE 11.41a

Band-pass filter in MultiSIM

low-pass filters. As you can see, the extreme right-side output has passed through the integrators and produces a low-pass response. If the low-pass output is summed out of phase with the input signal, the low-frequency information will cancel, leaving just the high-frequency components. Therefore, the output of the summer is the high-pass output of the filter. If the high-pass signal is integrated (using the same critical frequency), the result will be a band-pass response. This is seen at the output of the first integrator. The band-pass signal is also routed back to the input summing amplifier. By changing the amount of the signal that is fed back, the response near the critical frequency may be altered, effectively setting the filter Q. Finally, the loop is completed by integrating the band-pass response, which yields the low-pass output. In effect, the second integrator's −6 dB per octave rolloff perfectly compensates for the rising band-pass response below f_o. This produces flat response below f_o. Above f_o, the combination

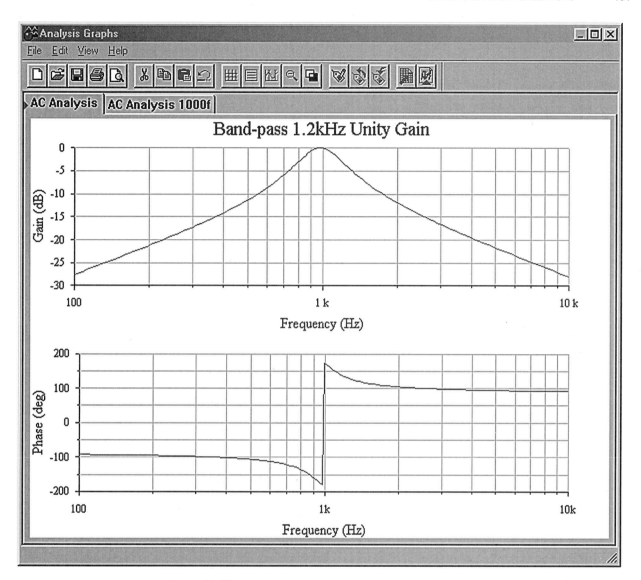

FIGURE 11.41b

Gain and phase plots for band-pass filter

of the two falling response curves produces the expected second-order, low-pass response.

Two popular ways of configuring the state-variable filter are the fixed-gain and adjustable-gain forms. The fixed-gain form is shown in Figure 11.43. This circuit uses a total of three op amps. The Q of the circuit is set by a single resistor, R_Q. Qs up to 100 are possible with state-variable filters. For the high- and low-pass outputs, the gain of this circuit is unity. For the band-pass output, the gain is equal to Q. Figure 11.44 shows

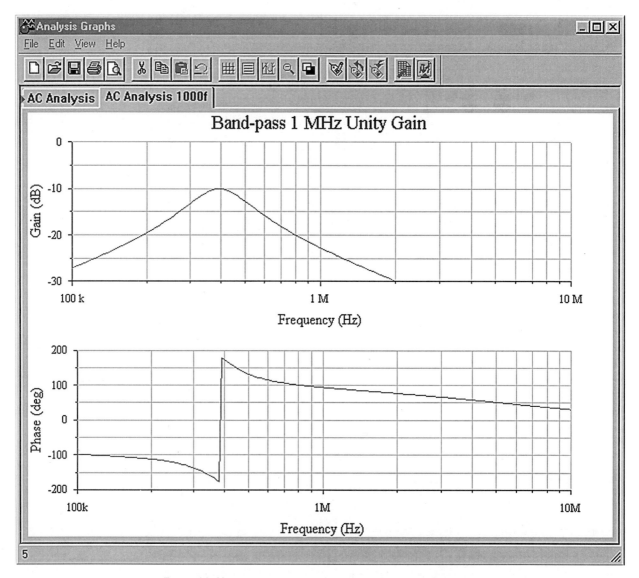

FIGURE 11.41c

Gain and phase plots for 1000 times frequency shift

an adjustable-gain version. For high- or low-pass use, the gain is equal to the arbitrary value K, whereas for band-pass use, the gain is equal to KQ. This variation requires a fourth op amp in order to isolate the Q and gain settings. Although four op amps may sound like a large number of devices, remember that a variety of quad op amp packages exist, indicating that the actual physical layout may be quite small. Also, even though three different outputs are available, it is not possible to individually optimize each one for simultaneous use. Consequently, the state-variable is most often used as a stable

FIGURE 11.42

Block diagram of state-variable filter

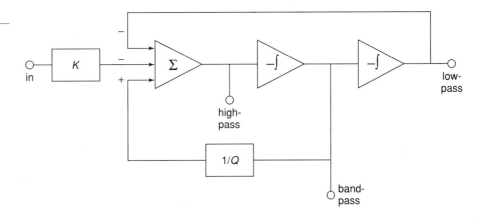

and switchable high/low-pass filter, or as a high-Q band-pass filter. Finally, in keeping with our previous work, the circuits are shown normalized to a critical frequency of one radian per second. Although we shall concentrate on band-pass design in this section, it is possible to use these circuits to realize various high- and low-pass filters, such as those generated with the Sallen and Key forms. The procedure is nearly identical and uses the same frequency and damping factors (Figures 11.24 and 11.29).

EXAMPLE 11-6

Design a band-pass filter with a center frequency of 4.3 kHz and a Q of 25. Use the fixed-gain form.

First, determine the damping resistor value. Then, scale the components for the desired center frequency. Note that a Q of 25 produces a bandwidth of only 172 Hertz for this filter $(\frac{4.3 \text{ kHz}}{25})$.

FIGURE 11.43

Fixed-gain version of state-variable filter

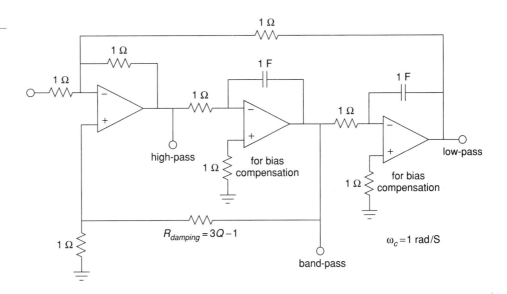

FIGURE 11.44

Variable-gain version of state-variable filter

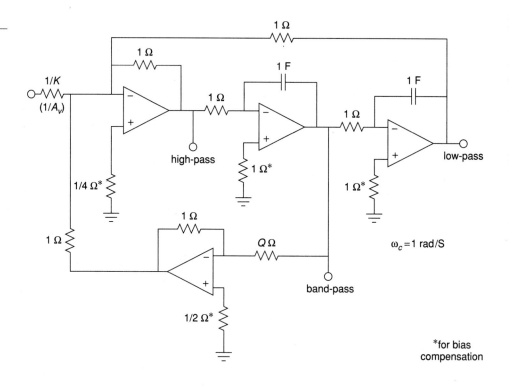

$$R_{damping} = 3Q - 1$$
$$R_{damping} = 3 \times 25 - 1$$
$$R_{damping} = 74 \, \Omega$$

$$\omega_o = 2\pi 4.3 \text{ kHz}$$
$$\omega_o = 27.02 \text{ k radians per second}$$

In order to translate the filter to our desired center frequency, we need to divide either the resistors or the capacitors by 27,020. For this example, we'll use the capacitors.

$$C = \frac{1}{27.02 \text{ k}}$$
$$C = 37 \, \mu F$$

A final impedance scaling is required to achieve reasonable component values. A reasonable value might be a factor of 5000.

$$C = \frac{37 \, \mu F}{5000}$$
$$C = 7.4 \text{ nF}$$

$$R_{damping} = 74 \times 5000$$
$$R_{damping} = 370 \text{ k}$$

FIGURE 11.45

Completed design of band-pass filter for Example 11-6

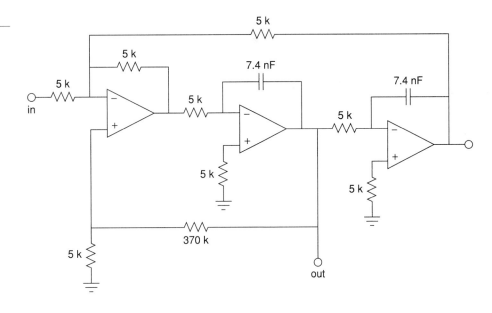

All remaining resistors will equal 5 kΩ.
Because this is a band-pass filter,

$$A_v = Q$$
$$A_v = 25$$

The completed filter is shown in Figure 11.45. The value for $R_{damping}$ is considerably larger than the other resistors. This effect gets worse as the required Q is increased. If this value becomes too large for practical components, it may be reduced to a more reasonable value as long as the associated divider resistor (from the noninverting input to ground) is reduced by the same amount. The ratio of these two resistors is what sets the filter Q, not their absolute values. Lowering these values will upset the ideal input bias current compensation, but this effect can be ignored in many cases, or reduced through the use of FET input op amps.

Altering this circuit for a variable-gain configuration requires the addition of a fourth amplifier as shown in Figure 11.44. The calculation for the damping resistor is altered, and a value for the input gain determining resistor is needed. The remaining component calculations are unchanged from the example above. Note that by setting the gain constant K to $\frac{1}{Q}$, the final filter gain may be set to unity.

11.8 NOTCH FILTER (BAND-REJECT) REALIZATIONS

By summing the high-pass and low-pass outputs of the state-variable filter, a notch, or band-reject, filter may be formed. Filters of this type are commonly used to remove interference signals. The summation is easily performed with a simple parallel-parallel

Figure 11.46

Notch filter

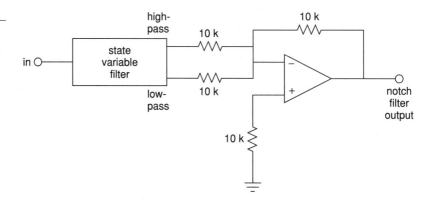

summing amplifier, as shown in Figure 11.46. For reasonable Q values, there will be tight correlation between the calculated band-pass center and −3 dB frequencies, and the notch center and −3 dB frequencies. The component calculations proceed as in the band-pass filter.

Example 11-7

A filter is needed to remove induced 60 Hz hum from a transducer's signal. The rejection bandwidth of the filter should be no more than 2 Hz.

From the specifications we know that the center frequency is 60 Hz and the Q is $\frac{60}{2}$, or 30. For simplicity, we will use the fixed-gain form. (Note that the gain of the filter on either side of the notch will be unity.)

$$R_{damping} = 3Q - 1$$
$$R_{damping} = 3 \times 30 - 1$$
$$R_{damping} = 89 \: \Omega$$

$$\omega_o = 2\pi 60$$
$$\omega_o = 377 \text{ rad/S}$$

Scaling C produces

$$C = \frac{1}{377}$$
$$C = 2.65 \text{ milliFarads}$$

A practical value scaling of 10^4 produces the circuit of Figure 11.47. Note that the damping resistors have only been scaled by 10^3, as an $R_{damping}$ value of 890 kΩ might be excessive. Remember, it is the ratio of these two resistors that is important, not their absolute values.

FIGURE 11.47

Completed notch filter for Example 11-7

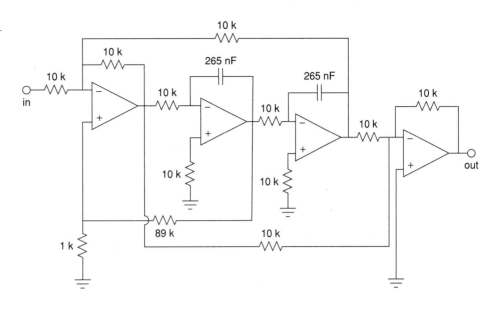

multiSIM COMPUTER SIMULATION

Figure 11.48 (pp. 468–470) shows a simulation of the circuit of Example 11-7 using MultiSIM. To keep the layout simple, an ideal op amp model was chosen. The AC analysis shows the very sharp notch centered at 60 Hz as expected. As filters of this type are designed to remove a single frequency without affecting surrounding material, high-precision tuning components are required. To see the effects of even modest component deviations in a production run, a Monte Carlo analysis proves invaluable. In Figure 11.48c a series of 10 runs is shown. Each resistor and capacitor in the filter has been given a 1% nominal tolerance. Further, the frequency plot range has been narrowed down to just 10 Hertz on either side of the target frequency. Even with these relatively tight tolerances, tuning deviations of more than 1 Hertz can be seen. Also, the response shape is not perfectly symmetrical in all cases.

A Note on Component Selection

Ideally, the circuit of Example 11-7 will produce −3 dB points at approximately 59 Hz and 61 Hz and will infinitely attenuate 60 Hz tones. In reality, component tolerances may alter the response and, therefore, high-quality parts are required for accurate, high-Q circuits such as this. Even simpler, less-demanding circuits such as a second-order Sallen and Key filter may not perform as expected if lower quality parts are used. As a general rule, component accuracy and stability becomes more important as filter Q and order increase. One percent tolerance metal film resistors are commonly used, with 5% carbon film types being satisfactory for the simpler circuits. For capacitors, film types such as polyethylene (mylar) are common for general-purpose work, with

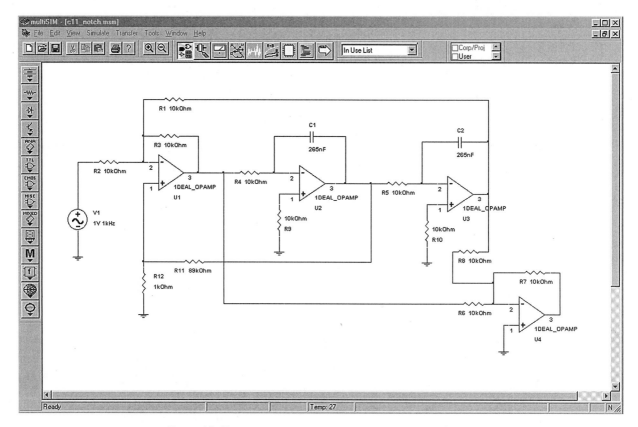

Figure 11.48a

State-variable notch filter in MultiSIM

polycarbonate, polystyrene, polypropylene, and teflon being used for the more stringent requirements. For small capacitance values (<100 pF), NPO ceramics may be used. Generally, large ceramic disc and aluminum electrolytic capacitors are avoided due to their wide tolerance and instability with temperature, applied voltage, and other factors.

Filter Design Tools

In order to further speed the process of active filter design, some manufacturers offer free filter design software. Examples include FilterCAD from Linear Technology and FilterPro from Burr-Brown. Some programs are rather generic and help you design Sallen and Key, multiple-feedback, and state-variable filters. Others are written expressly to support the manufacturer's specialized filter ICs. Typically, the programs will print out component values given desired filter types and break frequencies. Bode plots and pulse waveform simulations may also be available. Such programs can certainly reduce design tedium.

FIGURE 11.48b

Ideal notch response

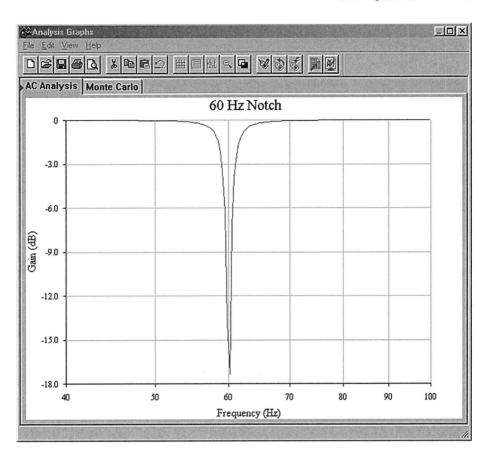

11.9 AUDIO EQUALIZERS

Another range of circuits that fall under the heading of filters are equalizers. Actually, equalizers are a class of adjustable filters that may produce gain as well as attenuation. Perhaps the most common uses for equalizers are in the audio, music, and communications areas. As the name suggests, equalizers are used to adjust or balance the input frequency spectrum. Equalizers range from complex *1/3 octave* and *parametric* types for use in recording studios and large public-address systems, to the simpler bass and treble controls found on virtually all home and car stereos. Many of the more complex equalizers are based on extensions of the state-variable filter. On the other hand, some equalizers are little more than modified amplifiers. We'll take a look at the very common bass and treble controls, which adjust low and high audio frequencies, respectively.

The purpose of bass and treble controls is to allow the listener some control of the balance of high and low tones. They may be used to help compensate for the acoustical shortcomings of a loudspeaker, or perhaps solely to compensate for personal taste. Unlike the filters previously examined, these controls will be manipulated by the user

Figure 11.48c

Typical response with 1% component variations

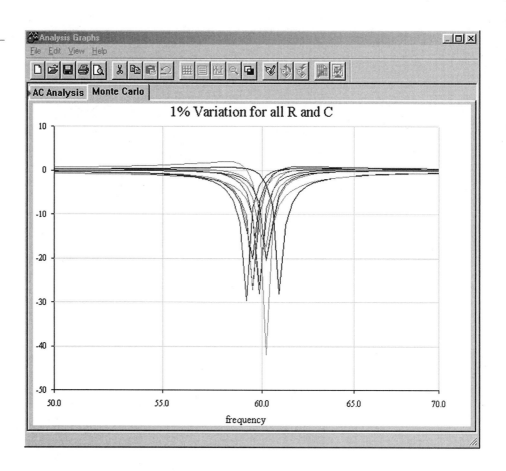

and must provide for signal boost as well as cut. A typical response curve is shown in Figure 11.49. Normally, bass and treble circuits are realized with parallel-parallel inverting amplifiers. In essence, the feedback network will change with frequency.

A simple bass control is shown in Figure 11.50. To understand how this circuit works, let's look at what happens at very low and at very high frequencies. First of all, at very high frequencies, capacitor C is ideally shorted. Thus, the setting of potentiometer R_b is inconsequential. In this case, the gain magnitude of the amplifier will be set at $\frac{R_c}{R_a}$. Normally, R_c is equal to R_a, so the gain is unity. At very low frequencies the exact opposite happens; the capacitor is seen as an open. Under this condition the gain of the amplifier depends on the setting of potentiometer R_b. If the wiper is set to the extreme right, the gain becomes $\frac{R_c}{R_a+R_b}$. If the wiper is moved to the extreme left, the gain becomes $\frac{R_c+R_b}{R_a}$. If R_b is set to nine times R_a, the total gain range will vary from .1 to 10 (−20 dB to +20 dB). A similar arrangement may be used to adjust the treble range.

When the bass and treble controls are combined, component loading makes the circuit somewhat more difficult to design. Basic models have already been derived by a number of sources, though, including the one shown in Figure 11.51. The following equations are used to design the desired equalizer. (Refer to Figure 11.52.)

AUDIO EQUALIZERS   471

FIGURE 11.49

Response of general bass/treble equalizer

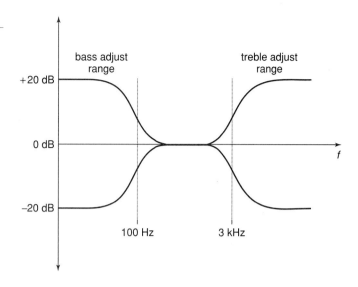

FIGURE 11.50

Simple bass section

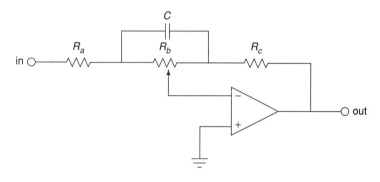

FIGURE 11.51

Bass/treble equalizer

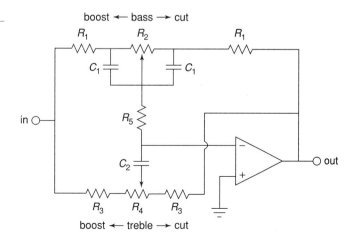

FIGURE 11.52

Response of bass/treble equalizer

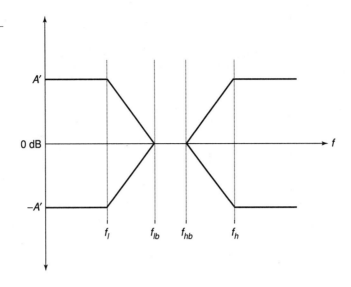

Bass section (assumes $R_2 \gg R_1$):

$$f_l = \frac{1}{2\pi R_2 C_1}$$

$$f_{lb} = \frac{1}{2\pi R_1 C_1}$$

$$A_{vb} = 1 + \frac{R_2}{R_1}$$

Treble section (assumes $R_4 \gg R_1 + R_3 + 2R_5$):

$$f_h = \frac{1}{2\pi R_3 C_3}$$

$$f_{hb} = \frac{1}{2\pi (R_1 + R_3 + 2R_5) C_3}$$

$$A_{vt} = 1 + \frac{R_1 + 2R_5}{R_3}$$

In actuality, it is very common to design these sorts of circuits empirically. In other words, the given equations are used as a starting point, and then component values are adjusted in the laboratory until the desired response range is obtained. Circuits like this may be altered further to include a midrange control. Generally, three adjustments is considered to be the maximum for this type of circuit. An example of a bass-midrange-treble equalizer is shown in the schematic for the Pocket Rocket amplifier in Chapter 6.

multiSIM COMPUTER SIMULATION

A simple bass equalizer is simulated in Figure 11.53 (pp. 473–475). The maximum cut and boost is set to a factor of approximately 10, or 20 dB. In order to plot the response curves, two AC analysis runs are performed: the first with the potentiometer at maximum, and the second at minimum. The two curves are plotted from 10 Hz up to 10 kHz. These curves are essentially mirror images. In both cases the response above 1 kHz is smooth and reasonably flat. The transition to the cut/boost region occurs in the 100 Hz to 1 kHz range, with nearly full action by 40 Hz. If the plot is rerun with the potentiometer in some other position (say, 25% of rotation), some smooth, scaled curve within these two extremes will be plotted. If the potentiometer is set at midpoint, the response will be flat across the entire range.

FIGURE 11.53a

Bass equalizer in MultiSIM

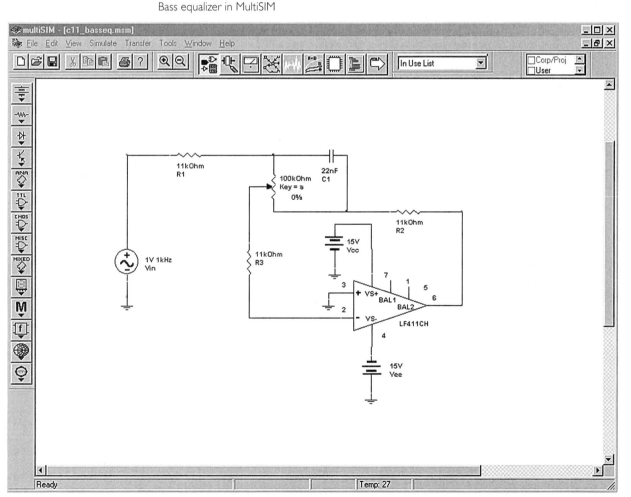

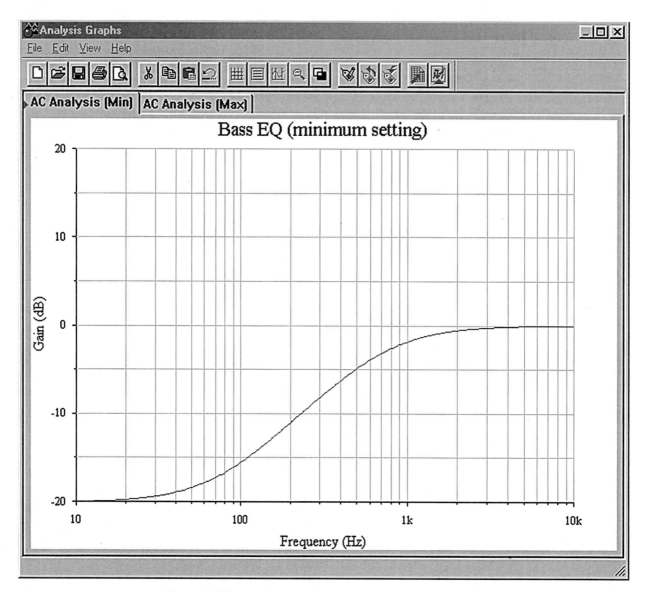

FIGURE 11.53b

Gain response at full cut

11.10 Switched-Capacitor Filters

Our final topic is the class of ICs known as *switched-capacitor filters*. These are just specific realizations of the types of filters that we have already examined. Generally, switched-capacitor filters come in two types: fixed order and alignment, and universal

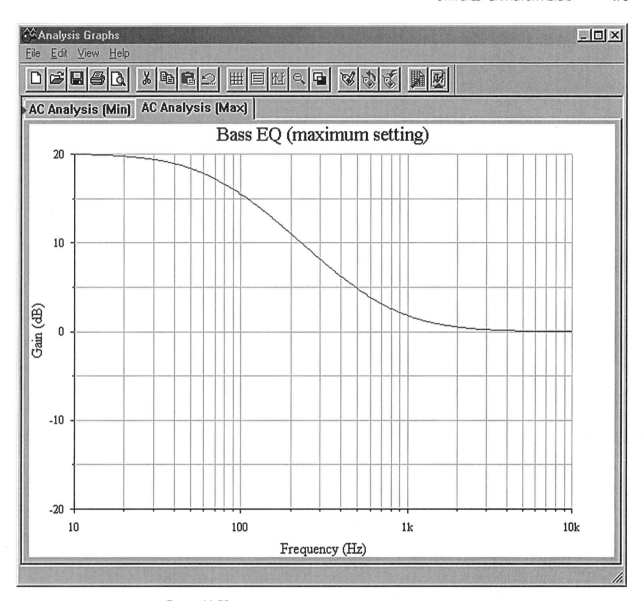

FIGURE 11.53c

Gain response at full boost

(state-variable based). A typical fixed IC might offer a sixth-order, low-pass Butterworth filter. The number of external components required is minimal. The universal types offer most of the flexibility of the state-variable designs discussed previously. Both types are tunable and are relatively easy to use. Tuning is accomplished by adjusting the frequency of an external clock signal. The higher the clock frequency, the higher

FIGURE 11.54

Switched capacitor circuit

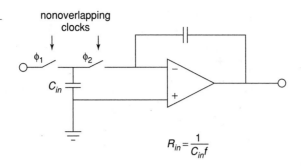

the resulting critical frequency. These ICs offer a convenient "black box" approach to general-purpose filter design. As is the case with most special-purpose devices, individual manufacturer's data sheets will give the specific application and design procedures for their parts.

The concept behind the switched-capacitor filter is quite interesting. The basic idea is to mimic a resistor through the use of a capacitor and a pair of alternating switches. As an example, a simple integrator is shown in Figure 11.54. As you have already seen, integrators are little more than first-order, low-pass filters. In this circuit, the input resistor has been replaced with a capacitor, C_{in}, and a pair of switches. These switches are controlled by a nonoverlapping biphase clock. This means that when one switch is closed, the other will be open, and that during switching, one switch will break contact before the other switch makes contact. (This is sometimes referred to as a "break before make" switch.)

During the first half of the clock cycle, C_{in} charges to the value of V_{in}. During the second half of the cycle, this charge is transferred to the integration capacitor. Therefore, the total charge transferred during one clock cycle is

$$Q = C_{in} V_{in} \tag{11.25}$$

The flow of charge versus time defines current, so the average input current is

$$I_{in} = \frac{Q}{T_{clock}} \tag{11.26}$$

Substituting Equation 11.25 into Equation 11.26 yields

$$I_{in} = C_{in} \frac{V_{in}}{T_{clock}} \tag{11.27}$$

Because the input resistance is defined as the ratio of V_{in} to I_{in} and recognizing that f_{clock} is the reciprocal of T_{clock}, Equation 11.27 is used to find R_{in}:

$$R_{in} = \frac{1}{C_{in} f_{clock}} \tag{11.28}$$

Equation 11.28 says two important things: first, because R_{in} sets the input impedance, it follows that input impedance is inversely proportional to the clock frequency. Second, because R_{in} is used to determine the corner frequency (in conjunction with the integration/feedback capacitor), it follows that the critical frequency of this circuit is directly proportional to the clock frequency. In other words, a doubling of clock frequency will halve the input impedance and double the critical frequency. Depending on the actual design of the IC, there will be a constant ratio between the clock frequency and the critical frequency. This is a very useful attribute. It means that you can make a tunable/sweepable filter by using one of these ICs and an adjustable square wave generator. For that matter, anything that can produce a square wave (such as a personal computer) can be used to control the filter response.

Typically, the ratio of clock frequency to critical frequency will be in the range of 50 to 100. The lower limit of clock frequency is controlled by internal leakage paths that create offset errors. The upper limit is controlled by switch settling time, propagation delays, and the like. A range of 100 Hz to 1 MHz is reasonable. This means that the entire audio frequency range is covered by these devices. For use at the highest frequencies, or with high impedance sources, an input buffer amplifier should be used.

An example of a switched capacitor filter IC is the MF4, shown in Figure 11.55. The MF4 is a fourth-order Butterworth low-pass filter. It comes in two variations: the MF4-50, which has a 50:1 clock to critical frequency ratio, and the MF4-100, which has a 100:1 ratio. Both variations produce unity gain in the pass band. The input capacitor for the MF4-50 is 2 pF, which, from Equation 11.28, produces an input impedance of 500 kΩ for a critical frequency of 20 kHz. The external component count is minimal (excluding the clock source). Input clock and level shift pins are provided for use with TTL or CMOS clocks. For less-demanding fixed-frequency work, it

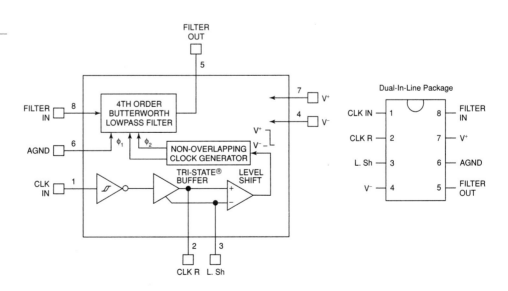

FIGURE 11.55

Block diagram of the MF4

Reprinted with permission of National Semiconductor Corporation

Figure 11.56

MF4 equivalent internal circuit for oscillator

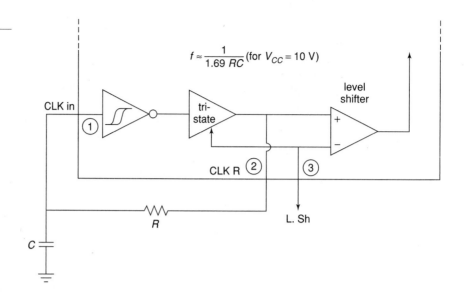

is possible to configure a simple *RC* oscillator with the internal Schmitt Trigger as shown in Figure 11.56. An example of a TTL clocked filter is shown in Figure 11.57. As you can see, the design is sparse, at best. Besides fourth-order Butterworth, filters of higher-order and other alignments are available. In short, these devices are good choices for general-purpose filter work, particularly when space and tuning considerations are important.

For designs requiring a bit more flexibility, universal filters such as the LTC1068 from Linear Technology are available. Because this filter is basically little more than a switched capacitor version of the state-variable, a wide range of response types are possible. These include the high-, low-, and band-pass outputs, as well as a variety of alignments including Bessel, Butterworth, and Chebyshev. The LTC1068 is modular, so its internal op amps may be used as gain blocks or for creating a notch output (as seen in Example 11-7). Normally, no more than four or five external components (resistors and capacitors) are needed to realize a given filter function. Component calculation

Figure 11.57

TTL clock input

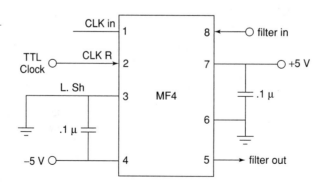

procedures are specified by the manufacturer. Note that this is a quad device, so a cascaded response of up to eighth-order is possible.

Although switched-capacitor filters offer relatively quick and physically small realizations, they are not perfect. First of all, clock feedthrough is typically in the range of 10 mV, meaning that 10 mV of clock signal "leaks" into the output. Fortunately, this signal is much higher than the critical frequency, but may cause some problems for low-noise applications. Another problem arises from the fact that switched capacitor filters are actually sampled data devices. As you will see in the next chapter, sampled data devices may suffer from a distortion producing phenomenon called *aliasing*. In order to avoid aliasing, the input signal must not contain any components that are greater than one-half of the clock frequency. For example, if the MF4-50 is used to create a 1 kHz filter, a 50 kHz clock is required. No component of the input signal may exceed 25 kHz (one-half of the clock) if aliasing is to be avoided. If this requirement cannot be guaranteed, some form of prefiltering is needed. Within these limits though, switched capacitor filters make light work out of many general-purpose applications.

// 11.11 Extended Topic: Voltage-Controlled Filters //

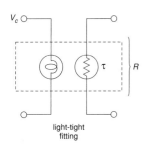

FIGURE 11.58

A photoresistor/lamp used as a variable resistance

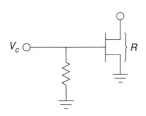

FIGURE 11.59

Using a JFET in the ohmic region

A voltage-controlled filter, or VCF, is nothing more than a standard filter whose tuning frequency is controlled by an external voltage. You might think of this concept as an extension of the clock control aspects of the switched capacitor filter. VCFs are used in a wide range of applications including instrumentation devices such as swept frequency analyzers and music synthesizers. Any application that requires precise or rapid control of tuning frequency calls for a VCF. Virtually any of the filters presented in this chapter may be turned into VCFs. All you need to do is substitute the tuning elements of the filter with a voltage-controlled version. Typically, this means replacing the tuning resistors with voltage-controlled resistances.

Two popular ways of creating a voltage-controlled resistance include the photoresistor/lamp combination (Figure 11.58), and the use of an FET in its ohmic region (Figure 11.59). To use these items, simply remove the tuning resistor(s) and replace them with a voltage-controlled resistance. As an example, a simple single-pole high-pass VCF is shown in Figure 11.60. As the control voltage (Vc) increases, the lamp brightness increases causing the photoresistor's value to drop. Because the photoresistor sets the tuning frequency, the net result is an increase in f_c. The FET version produces a resistance that is proportional to the magnitude of the gate voltage (V_c).

These two solutions are not without their problems. In the case of the lamp/photoresistor, response time is not very fast, and the lamp portion requires a fairly large drive current. The FET circuit eliminates these problems, but requires that the voltage across it remain fairly low (usually less than 100 mV). Larger signal swings will drive the FET out of the ohmic region, and distortion will increase dramatically. Also, the popular N channel variety requires a negative gate potential, which is generally not preferred.

FIGURE 11.60

Variable resistance connection for VCF

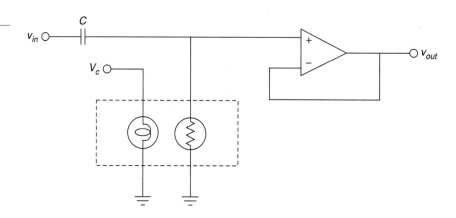

In both cases, one more problem remains: it is difficult to create a wide linear control range.

Another way to create the effect of a voltage-variable tuning element is through the use of an operational transconductance amplifier, or OTA (see Chapter 6). Remember, this device is essentially a voltage-to-current converter. Its output current is a function of its control current. (The control currrent is easily derived from a control voltage and resistor.) This device is ideally suited to "inverting" type inputs, where an input resistor is used as a voltage-to-current converter. One possible example is shown in Figure 11.61, a state-variable VCF. The boxed sections show where an OTA has replaced a standard single resistor. In this circuit, a large control voltage creates a large control current, thus increasing transconductance. This simulates a smaller tuning resistor value, and thus creates a higher tuning frequency. The OTA approach proves to be reliable, repeatable, and generally low in cost. It also offers a fairly wide linear tuning range.

FIGURE 11.61

Using OTAs as a controlled element in a VCF

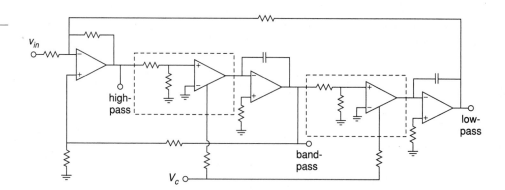

Summary

Filters are frequency-selective circuits. The basic forms are high-pass, low-pass, band-pass, and band-reject. Although filters may be constructed solely from resistors, capacitors, and inductors, active filters using op amps offer many advantages. These advantages include: modest component size, control over impedances and loading effects, elimination of inductors, and gain (if desired). The negative aspects include: frequency range limited by op amps used, power supply required, and the inability to handle large input/output powers. For many applications the advantages far outweigh the disadvantages, and therefore, active filters are used in a wide variety of modern products.

Filters are further defined by order and alignment. Order indicates the steepness of the attenuation slope. As a general rule, the eventual rolloff rate will equal 6 dB times the order, per octave. Order also indicates the minimum number of reactive elements needed to realize the filter. Alignment indicates the shape of the filter response in the frequency domain. Popular alignments include Bessel (constant time delay), Butterworth (maximally flat response in the pass band), and Chebyshev (ripples in the pass band, but with faster rolloff rates). There is generally a trade-off between fast attenuation rates and smooth phase response. Alignment is indicated by the damping or Q of the filter. Q is the reciprocal of damping. Filters with low damping factors (i.e., high Q) tend to be "peaky" in the frequency domain and produce ringing on pulse-type inputs. (Chebyshevs are in this category.) The filter's critical frequency and 3 dB down frequency are not the same for alignments other than the Butterworth. The actual amount of "skew" depends on the alignment and order of the filter.

Once filter performance is specified, there are a number of ways in which the circuit can be physically realized. Common high- and low-pass realizations use the Sallen and Key VCVS approach. There are two variations on this theme: the unity-gain form and the equal-component form. Both forms use a second-order building block section. For higher orders, several second-order sections (and optionally, a first-order section) are combined to produce the final filter. It is important to remember that higher-order filters are not simple combinations of identical lower-order filters. For example, a fourth-order 1 kHz Butterworth filter is not made by cascading a pair of identical second-order 1 kHz Butterworth filters. Rather, each section requires specific damping and frequency factors. A common design procedure utilizes lookup tables for these factors. The filters are designed by first scaling the general filter to the desired cutoff frequency, and then scaling the components for practical values.

For relatively low Qs (<1), band-pass filters are best realized as a cascade of high- and low-pass filters. For higher Qs, this technique is not satisfactory. Moderate Qs (up to 10) may be realized with the multiple-feedback filter. Very high Q applications (up to 100) may be realized with the state-variable filter. The state-variable is often known as the universal filter, as it produces high-, low-, and band-pass outputs. With the addition of a fourth amplifier, a band-reject filter may be formed. Fixed and adjustable gain versions of the state-variable may be utilized by the designer.

A somewhat more specialized group of filters are the equalizers commonly employed in audio recording and playback equipment. Unlike traditional filters, equalizers offer both boost and attenuation of frequencies. Generally, these circuits are based on parallel-parallel inverting amplifiers, utilizing an adjustable, frequency-selective feedback network.

Switched capacitor filter ICs offer the designer expedient solutions to general-purpose filter design. They are generally suited to the audio frequency range and require very few external components. The critical frequency is set by a clock input. The order and alignment may be either factory set or user adjustable (as in the universal state-variable types).

References

Lancaster, Don. *Active Filter Cookbook*, 2nd ed. Woburn, Mass.: Butterworth-Heinemann, 1996.

Bohn, Dennis. *Audio Handbook*, ed. Santa Clara, Calif.: National Semiconductor Corporation, 1976.

Audio Handbook. Norwood Mass.: Precision Monolithics Inc., 1990.

Review Questions

1. What are the four main types of filters?
2. What are the advantages of active filters versus passive filters?
3. What are the disadvantages of active filters versus passive filters?
4. What do the characteristics of order and poles indicate?
5. Name several popular filter alignments.
6. How do the popular filter alignments differ from one another in terms of phase and magnitude response?
7. Which alignment should be used if linear phase response is of particular importance?
8. Which alignment should be used if fastest rolloff rate is of particular importance?
9. When are the "3 dB down" and critical frequencies of a filter identical?
10. Outline the process for creating high-order filters and explain why cascades of similar lower-order filters do not give the appropriate results.
11. Explain where each of the following band-pass filters would be appropriate: high-pass/low-pass cascade, multiple-feedback, and state-variable.
12. Why is the state-variable often referred to as the universal filter?
13. Briefly explain the operation of an adjustable bass equalizer.
14. Briefly explain the concept behind the switched-capacitor filter.
15. What are the advantages and disadvantages of switched-capacitor ICs versus the more traditional op amp approach?

Problems

Analysis Problems

1. Using 11.24, determine the loss for a 1 kHz Butterworth second-order low-pass filter at 500 Hz, 1 kHz, 2 kHz, and 4 kHz.
2. Using Figure 11.29, determine the loss for a 2 kHz 3 dB ripple Chebyshev third-order low-pass filter at 500 Hz, 1 kHz, 2 kHz, 4 kHz, and 6 kHz.
3. Using Figure 11.24, determine the loss for a 500 Hz Bessel second-order high-pass filter at 200 Hz, 500 Hz, and 2 khz.
4. Using Figure 11.29, determine the loss one octave above the cutoff frequency for a fourth-order low-pass filter of the following alignments: Butterworth, Bessel, 1 dB-ripple Chebyshev.
5. Repeat Problem 4 for high-pass filters.
6. Using Figures 11.24 and 11.29, determine the loss at 8 kHz for 3 kHz low-pass Butterworth filters of orders 2 through 6.
7. Using Figures 11.24 and 11.29, determine the loss at 500 Hz for 1.5 kHz high-pass Bessel filters of orders 2 through 6.
8. Repeat Problem 7 using 3 dB-ripple Chebyshevs.
9. Using Figures 11.24 and 11.29, determine the loss at 500 Hz for 200 Hz high-pass 3 dB-ripple Chebyshev filters of orders 2 through 6.
10. Using Figure 11.29, determine the loss one octave below the cutoff frequency for a third-order high-pass filter of the following alignments: Butterworth, Bessel, 1 dB-ripple Chebyshev.
11. Repeat Problem 10 for low-pass filters.
12. An application requires that the stop-band attenuation of a low-pass filter be at least −15 dB at 1.5 times the critical frequency. Determine the minimum order required for the Butterworth and 1 dB and 3 dB-ripple Chebyshev alignments.
13. An application requires that the stop-band attenuation of a high-pass filter be at least −20 dB one octave below the critical frequency. Determine the minimum order required for the Butterworth and 1 dB-ripple and 3 dB-ripple Chebyshev alignments.
14. A band-pass filter has a center frequency of 1020 Hz and a bandwidth of 50 Hz. Determine the filter Q.
15. A band-pass filter has upper and lower break frequencies of 9.5 kHz and 8 kHz. Determine the center frequency and Q of the filter.
16. Design a second-order Butterworth low-pass filter with a critical frequency of 125 Hz. The pass band gain should be unity.
17. Repeat Problem 16 for a high-pass filter.
18. Repeat Problem 16 using a Bessel alignment.

Design Problems

19. A particular application requires that all frequencies below 400 Hz should be attenuated. The attenuation should be at least −22 dB at 100 Hz. Design a filter to meet this requirement.
20. Repeat Problem 19 for an attenuation of at least −35 dB at 100 Hz.
21. Audiophile quality stereo systems often use subwoofers to reproduce the lowest possible musical tones. These systems typically use an electronic crossover approach as explained in Example 11-4. Design an electronic crossover for this application using third-order Butterworth filters. The crossover frequency should be set at 65 Hz.
22. Explain how the design sequence of Problem 21 is altered if either a new crossover frequency is chosen, or a different alignment is specified.
23. Design a band-pass filter that will only allow frequencies between 150 Hz and 3 kHz. The attenuation slopes should be at least 40 dB per decade. (A filter such as this is useful for "cleaning up" recordings of human speech.)
24. Design a band-pass filter with a center frequency of 2040 Hz and a bandwidth of 400 Hz. The circuit should have unity gain. Also, determine the f_{unity} requirement of the op amp(s) used.
25. Repeat Problem 24 for a center frequency of 440 Hz and a bandwidth of 80 Hz.
26. Design a band-pass filter with upper and lower break frequencies of 700 Hz and 680 Hz.
27. Design a notch filter to remove 19 kHz tones. The Q of the filter should be 25. (This filter is useful in removing the stereo "pilot" signal from FM radio broadcasts.)
28. Design a second-order low-pass filter with a critical frequency of 30 kHz. Use a state-variable filter. The circuit should have a gain of +6 dB in the pass band.
29. Design a bass/treble equalizer to meet the following specification: maximum cut and boost = 25 dB below 50 Hz and above 10 kHz.
30. Using the MF4, design a fourth-order low-pass Butterworth filter with a critical frequency of 3.5 kHz. Do not use an external oscillator.
31. Using the MF4, design a low-pass filter that is adjustable from 200 Hz to 10 kHz. Do not ignore the oscillator design.

Challenge Problems

32. Design a low-pass second-order filter that may be adjusted by the user from 200 Hz to 2 kHz. Also, make the circuit switchable between Butterworth and Bessel alignments.
33. Design a subsonic filter that will be 3 dB down from the pass band response at 16 Hz. The attenuation at 10 Hz must be at least 40 dB. Although pass band ripple is permissible, the gain should be unity.
34. Design an adjustable band-pass filter with a Q range from 10 to 25, and a center frequency range from 1 kHz to 5 kHz.
35. Modify the design of the previous problem so that as the Q is varied, the pass-band gain remains constant at unity.

Computer Simulation Problems

36. Verify the magnitude response of the circuit designed in Problem 16 by using a simulator. Check both the critical frequency and the rolloff rate.
37. Verify the magnitude response of the electronic crossover designed in Problem 21 by using a simulator. Plot both outputs simultaneously on one graph.
38. Verify the magnitude and phase of the filter designed in Problem 24 by using a simulator.
39. Compare the simulations of the circuit designed in Problem 28 using the relatively slow LM741, versus the medium-speed LF411. Is there any noticeable change? What can you conclude from this? Would the results be similar if the break frequency was increased by a factor of 50?
40. It is very common to plot the adjustment range of equalizers on a single graph, as shown in Figure 11.49. Use a simulator to create a plot of the adjustment range of the equalizer designed in Problem 29.
41. Simulate and verify the design of challenge Problem 32.
42. Simulate and verify the design of challenge Problem 33.
43. Verify the design of challenge Problem 34 using a simulator. Include four separate plots, showing maximum and minimum Q with maximum and minimum center frequency.
44. Verify the design of challenge Problem 35 using a simulator. Include two simultaneous plots, one showing minimum Q with maximum and minimum center frequency, and the other showing maximum Q with maximum and minimum center frequency.

CHAPTER 12

ANALOG-TO-DIGITAL-TO-ANALOG CONVERSION

CHAPTER OBJECTIVES

After completing this chapter, you should be able to:

- Outline the concept of *pulse code modulation*.
- Detail the advantages and disadvantages of signal processing in the digital domain.
- Define the terms *resolution*, *quantization*, and *Nyquist frequency*.
- Define an *alias*, and detail how it is produced and subsequently avoided.
- Explain the operation of an *R/2R* digital-to-analog converter.
- Explain the need for antialias and reconstruction filters.
- Explain the operation of a successive approximation analog-to-digital converter.
- Detail the need for and operation of a track-and-hold amplifier.
- Explain the operation of a flash analog-to-digital converter.
- Compare the different analog-to-digital converters in terms of speed, size, and complexity, and detail typical applications for each.

12.1 INTRODUCTION

Up to now, all of the circuits you have studied in this book were analog circuits. That is, the input waveforms were time continuous and had infinite *resolution* along the time and amplitude axes. That is, you could discern increasingly smaller and finer changes as you examined a particular section. No matter whether the circuit was a simple amplifier, function synthesizer, integrator, filter, or what have you, the analog nature of the signal was always true. Fundamentally, the universe is analog in nature (at least as far as we can tell—until someone discovers a quantum time particle). Our only real deviation from the pure analog system was the use of the comparator. Although the input to the comparator was analog, the output was decidedly digital; its output was either a logic high ($+V_{sat}$) or a logic low ($-V_{sat}$). You can think of the comparator's output as having very low resolution, as only two states are possible. The comparator's output is still time continuous in that a logic transition can occur at any time. This is in contrast to a pure digital system where transitions are time discrete. This means that logic levels can only change at specific times, usually controlled by some form of master clock. A purely digital system, then, is the antithesis of a pure analog system. An analog system is time continuous and has infinite amplitude resolution. A digital system is time discrete and has finite amplitude resolution (two states in our example).

As you have no doubt noticed in your parallel work, digital systems have certain advantages and benefits relative to analog systems. These advantages include noise immunity, storage capability, and available numeric processing power. It makes sense, then, that a combination of analog and digital systems could offer the best of both worlds. This chapter examines the processes of converting analog signals into a digital format and turning digital words into an analog signal. A few representative examples of processing the signal in the digital domain are presented as well. Some examples with which you might already be familiar include the stereo compact disk (CD) and the digital storage oscilloscope. We will break down this topic into two broad sections: analog-to-digital conversion (AD) and digital-to-analog conversion (DA). As many AD systems require digital-to-analog converters, we will examine DA systems first.

The Advantages and Disadvantages of Working in the Digital Domain

Given enough time, an analog circuit may be designed and manufactured for virtually any application. Why, then, would anyone desire to work in the digital domain? Perhaps the major reason for working in the digital domain is the flexibility it offers. Once signals are represented in a digital form, they may be manipulated by various means, including software programs. You have probably discovered that replicating a computer program is far easier than replicating an analog circuit. What's more, a program is much easier to update and customize than a hardware circuit. Because of this, it is possible to manipulate a signal in many different ways with the same digital/computer hardware; all that needs to be altered is the manipulation instructions (i.e., the program). The analog circuit, in contrast, needs to be rewired and extra components need to be added or old portions removed. This can be far more costly and time intensive than just updating software. By working in the digital domain, processing circuits do not exist per se; rather, a generic IC such as a CPU is used to create a "virtual circuit." With a certain amount of intelligence in the system design, the virtual circuit may be able to alter its own performance in order to precisely adapt to various signals. This all boils down to the fact that a digital scheme may offer much greater flexibility for involved tasks and allows a streamlined, generic hardware solution for complex applications. Because of this attribute, the digital solution may wind up being significantly less expensive than its analog counterpart.

When an analog signal is transferred to the digital domain, it is represented as a series of numbers (usually, high/low binary logic levels). One nice property of this representation is that it is exactly repeatable. In other words, an infinite number of copies of the data may be generated, and no distortions or deviations from the original will appear. The last copy will be identical to the first. Compare this to a simple analog copy. For example, if you were to record a song with a cassette recorder and then make a copy of the tape, the second-generation copy would suffer from increased noise and distortion. A copy of the second copy would produce even worse results. Every time the signal is copied, some corruption occurs. It is for this reason that early long-distance telephone calls were of such low quality. Modern communications systems employ digital techniques that allow much higher quality, even if one person is in New York and the other is in Australia, halfway around the planet.

Besides being a desirable mathematical attribute, repeatability also lends itself to the problem of long-term storage. A storage medium for a binary signal only needs to resolve two levels, whereas the analog medium needs to resolve very fine changes in signal strength. As you might guess, deterioration of the analog medium is a serious problem and results in information loss. The digital medium can theoretically survive a much higher level of deterioration without information loss. In a computerized system, data may be stored in a variety of formats including RAM (Random Access Memory) and magnetic tape or disk. For playback only (i.e., read only), data may be stored in ROM (Read Only Memory) or laser disk formats (such as DVD or audio CD).

As always, the benefits of the digital scheme arrive with specific disadvantages. First, for simpler applications, the cost of the digital approach is very high and cannot be justified. Second, the process of converting a signal between the analog and digital domains is an inexact one. Some information about the signal will be lost during the conversion. This is because the digital representation has finite resolution. This means that only signal changes larger than a certain minimum size (the resolution step size) are discernable, and therefore, some form of round-off error is inevitable. This characteristic helps determine the range of allowable signals, from the smallest detectable signal to the maximum signal before overload occurs. Third, analog systems are inherently faster than digital systems. Analog solutions can process input signals at much higher frequencies than digital schemes. Also, analog systems work in real time, whereas digital systems might not. Digital systems can only perform in real-time if the input signal is not a very high frequency, if the processing task is not overly complex, or if specialized processing circuits are added. Not all applications require real-time performance, so this limitation is not always a problem. Also, because we can expect computing power to get less and less expensive in the coming years, cost-effective digital processing will undoubtedly expand into new areas.

12.2 THE SAMPLING THEOREM

There are many different ways in which an analog signal may be turned into a digital form. This process is referred to as *AD conversion* or, more simply, as *digitization*. We shall only examine the most popular method, called *pulse code modulation*, or PCM for short. The reverse of this process, or turning the digital information back into analog

FIGURE 12.1

Sampling input signals over time

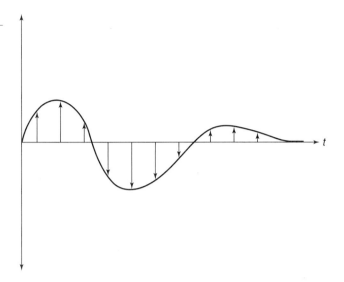

form, is called *DA conversion*. The two most important charracteristics in the conversion process are *sampling frequency* and *amplitude resolution*. Let's look at the basic idea behind PCM.

In essence, PCM measures and encodes the value of the input signal at specific points in time. Normally, the time spacing is constant, and several points are used over the length of one input cycle. In other words, the process involves taking representative samples of the input signal over time. This is shown graphically in Figure 12.1 and is referred to as *sampling*. The result of the sampling procedure is a list of times and corresponding amplitude values. A sequence may look something like this: at $T = 1$ mS, $V_{in} = 23$ mV; at $T = 2$ mS, $V_{in} = 45$ mV; at $T = 3$ mS, $V_{in} = -15$ mV; etc. If the time interval between samples is held constant (i.e., constant rate of sampling), then all we need to know is a starting time and the sampling rate in order to reconstruct the actual sample times.[1] This is much more efficient than recording each sample time. The resulting amplitude values may be manipulated in a variety of ways since they are now in a numeric form. The ultimate accuracy of this conversion will depend on two primary factors: how often we sample the signal and the *accuracy* and *resolution* of the sample measurement. Theoretically, the conversion will never be 100% accurate; that is, once converted, a finite amount of information will be lost forever. Another way of stating this is that when the digital representation is converted back to analog, the result will not be identical to the original waveform. In practical terms, it is possible to reduce the error to such small values that it may be ignored in many applications. It is important then, that we investigate the implications of sample rate and accuracy/resolution on the quality of conversion.

---

[1] The terms *sample rate* and *sample frequency* are often used interchangeably, and are usually denoted by f_s.

12.3 Resolution and Sampling Rate

Perhaps the most obvious source of error is the finite measurement accuracy of the individual signal levels. The major problem here is one of resolution. Resolution represents the finest discernible change in the signal and is often specified in terms of a *number of bits*, although a voltage specification is also possible. Because the signal level is represented with a binary number, it follows that a large number of bits are needed in order to achieve fine resolution. For example, if an 8-bit word is used, there will be 256 distinct values available. If the maximum peak-to-peak value of the input signal is 1 V, it works out that each *step* in the word represents about 3.9 mV (1 V/256). Under these conditions, it would be impossible to perfectly encode a value of 14 mV. The nearest values available would be binary 11, which yields 11.7 mV (3 × 3.9 mV), and binary 100, which yields 15.6 mV (4 × 3.9 mV). Obviously, the resulting round-off creates some error in the digital representation. This error can be reduced by increasing resolution so that finer steps may be detected. If 16 bits are used for the same 1 V range, a total of 65,536 values are available, with each step working out to 15.26 μV. Now, although we still may not be able to exactly represent the 14 mV level, we are guaranteed to be within ± 7.63 μV, instead of ± 1.95 mV as in the 8-bit case. Because the round-off errors tend to be random in magnitude and polarity, this effect may be viewed as a noise source. In other words, lower resolutions (i.e., fewer bits) produce noisier signals. The number of bits required for a particular application may vary from fewer than 6 for high speed video applications to more than 20 for high quality audio or measurement purposes.

It is important to note that once the size of the digital word is chosen and the peak amplitude fixed, the input signal must stay within specific bounds or gross distortion will occur. For example, if a peak input of 1 V produces the maximum numeric value, there is no way that the digital word can represent a level greater than 1 V. Likewise, if the step size is set at 1 mV, any signals less than 1 mV are lost. Also, low-level signals will suffer from reduced resolution. For best results, the signal peak should produce the maximum numeric value. If the peak is significantly less, the result is akin to using fewer bits in the representation. Finally, even if the resolution is adequate, the absolute accuracy of the conversion must be considered, as it is in any measurement device.

Example 12-1

A certain system uses a 12-bit word to represent the input signal. If the maximum peak-to-peak signal is set for 2 V, determine the resolution of the system and its dynamic range.

A *12-bit word* means that 2^{12}, or 4096, levels are possible. As these levels are equally spaced across the 2 V range, each step is

$$\text{Step Size} = \frac{2 \text{ V}}{4096}$$
$$\text{Step Size} = 488 \ \mu\text{V}$$

Therefore, the system can resolve changes as small as 488 μV.

Dynamic range represents the ratio of the largest value possible to the smallest.

$$\text{Dynamic Range} = \frac{2\text{ V}}{488\ \mu\text{V}}$$
$$\text{Dynamic Range} = 4096$$
$$\text{Dynamic Range} = 20\log_{10} 4096$$
$$\text{Dynamic Range} = 72\text{ dB}$$

Note that the voltage range affects the step size, but does not affect the bit resolution. The actual number of discrete steps that may be resolved is set by the number of bits available. You may notice that each additional bit adds approximately 6 dB of range (a doubling of voltage). Consequently, the dynamic range calculation may be streamlined to

$$\text{Dynamic Range} \approx 6\text{ dB} \times \text{Number of Bits} \qquad (12.1)$$

EXAMPLE 12-2

Audio compact disks use a 16-bit representation of the music signal. Determine the dynamic range. Also, if the maximum output level is .775 V peak, determine the step size.

$$\text{Dynamic Range} \approx 6\text{ dB} \times \text{Number of Bits}$$
$$\text{Dynamic Range} \approx 6\text{ dB} \times 16$$
$$\text{Dynamic Range} \approx 96\text{ dB}$$

For 16 bits, the total number of steps is 2^{16}, or 65536. Assuming that the signal is bipolar, the total signal range will be from $-.775$ V to $+.775$ V, or 1.55 V.

$$\text{Step Size} = \frac{1.55\text{ V}}{65536}$$
$$\text{Step Size} = 23.65\ \mu\text{V}$$

At this point, we must consider the effect of sampling rate on the quality of the signal. It should be intuitively obvious that higher sampling rates afford greater overall conversion accuracy. Of course, there is a trade-off associated with high sampling rates, and that is the accompanying high data rate. In other words, greater resources will be required to store and process the larger volume of digital information. The real question is, just how fast does the sampling rate need to be for optimum efficiency? The Nyquist sampling theorem states that at least two samples are needed per cycle for proper signal conversion. If the signal is not a sinusoid, then at least two samples are required per cycle of the highest frequency component. For example, if a range of signals up to 10 kHz needs to be digitized, then a sample rate of at least 20 kHz is required. Normally, a certain amount of "breathing room" is added to this figure. Another way of looking at this relationship is to state that the highest input-frequency component can be no more than one-half of the sampling rate. Because this is such

FIGURE 12.2

The aliasing effect (sampling rate too low)

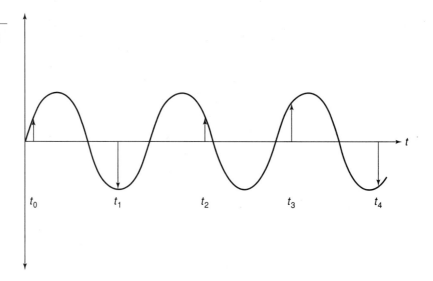

an important parameter, the value of one-half of the sampling rate is given the name *Nyquist frequency*.

$$\text{Nyquist frequency} = \frac{f_s}{2} \qquad (12.2)$$

If an input frequency component is greater than the Nyquist frequency, a unique form of distortion called *alias distortion*, is produced. The resulting distortion product, called an *alias*, is a new signal at a frequency that is equal to the difference between the input and Nyquist frequencies. Normally, this new signal is not harmonically related to the input signal, and thus, is easily detected. The aliasing effect is shown graphically in Figure 12.2. Here we see a sampling rate that is only about 1.5 times the input frequency, rather than the required factor of 2 times minimum. In Figure 12.3 the sample points are redrawn and connected as simply as possible. Note that the resulting outline is that of a lower-frequency wave. What we notice here is that the data points produced in Figure 12.2 are identical to the points produced by a lower-frequency input wave. When these data points are converted back to analog form, the DA converter will produce this lower-frequency wave. Oddly enough, the original waveform has completely disappeared; hence the term *alias*. Any signal component that is greater than the Nyquist frequency will produce aliases.

When considering the possibility of alias distortion, it is worth repeating that the components of the signal must be investigated, not just the base frequency. For example, a 1 kHz square wave has a 1 kHz fundamental and an infinite series of odd-numbered harmonics (3 kHz, 5 kHz, 7 kHz, 9 kHz, etc.). If a 12 kHz sampling rate is used to digitize this signal, a fair amount of alias distortion will be seen. In this example, the Nyquist rate is 6 kHz. All of the harmonics above 6 kHz will produce an alias. In order to prevent this, the input signal must be frequency-band limited. That is, a low-pass filter must be used to attenuate all components above the Nyquist frequency before AD conversion takes place. The amount of filtering required depends

FIGURE 12.3

Alias production

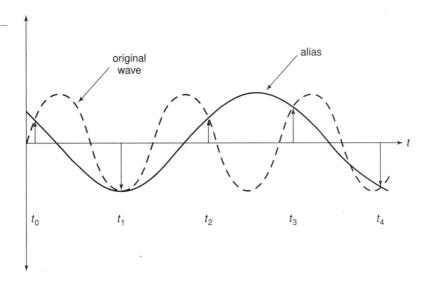

on the resolution of the conversion and the relative strength of the above-band signals. Because filters cannot roll off infinitely fast, as noted in Chapter 11, sampling rates are normally set more than twice as high as the highest needed input component. In this way, the Nyquist frequency will be somewhat greater than the maximum desired input frequency. The low-pass filter (often referred to as an *antialias filter*) will use this frequency range as its transition band. Even though the attenuation in the transition band is less than optimum, alias distortion will not be a problem. This is shown graphically in Figure 12.4. As you can see, high rolloff rate filters are desirable in order to attenuate the out-of-band signals as quickly as possible. Very fast filter rolloff rates mean that the sampling frequency need only be as little as 10% greater than the theoretical minimum in order to maintain sufficient alias rejection.

EXAMPLE 12-3

Suppose that you need to digitize telephone signals. Assuming that you would like to maintain a dynamic range of at least 50 dB, with an upper frequency limit of 3 kHz, determine the minimum acceptable sampling rate and number of bits of required.

The minimum Nyquist rate is equal to the highest desired input frequency. In this case, that's 3 kHz. Because the sampling rate is twice the Nyquist frequency, the sampling rate must be at least 6 kHz. In reality, if input components above 3 kHz exist, an antialias filter will be needed, and the sampling rate will have to be increased somewhat.

Because dynamic range is set by the number of bits used, we find that

$$\text{Bits Required} = \frac{\text{Dynamic Range}}{6 \text{ dB}}$$

$$\text{Bits Required} = \frac{50 \text{ dB}}{6 \text{ dB}}$$

$$\text{Bits Required} = 9$$

Figure 12.4

Filtering to eliminate alias effects

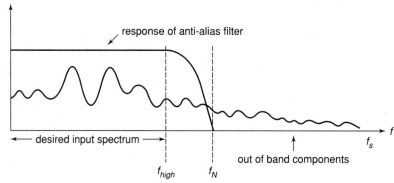

a. Spectrum before filtering

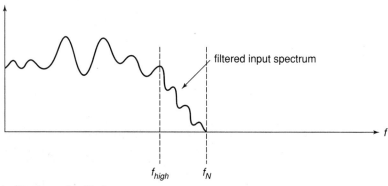

b. Spectrum after filtering

Because we cannot have a fractional bit, the value is rounded up. The final system specification is a minimum rate of 6 kHz with a 9-bit resolution. Note that this represents a data rate of 9 bits per sample × 6000 samples per second, or 54000 bits per second (6750 bytes per second).

12.4 Digital-to-Analog Conversion Techniques

The basic digital-to-analog converter is little more than a weighted summing amplifier. Each successive bit in the digital word represents a level that is twice as large as the preceding bit. If each bit is taken as a given current or voltage, the increasing levels may be produced by using different gains in the summing inputs. A simple four-bit converter is shown in Figure 12.5. This system can represent 2^4, or 16, different levels. Each input is driven by a simple high/low logic level that represents a 1 or 0 for that particular bit. Note that the input resistors vary by factors of 2. The gain for the upper most path is $\frac{R_f}{R_f}$, or unity. This input is used for the most significant bit of the input word (MSB). The next input shows a gain of $\frac{R_f}{2R_f}$, or .5. The third input shows a gain of .25, and the final input shows a gain of .125. The final input has the lowest gain and is

FIGURE 12.5

A simple 4-bit converter

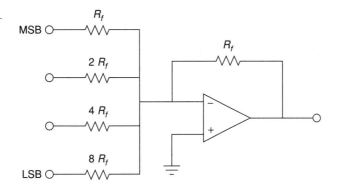

used for the least significant bit of the input word (LSB). If the input word had a higher resolution (i.e., more bits), extra channels would be added, each having half the gain of the preceding input. To better understand the conversion process, let's take a look at a few representative inputs and outputs.

The circuit of Figure 12.5 may be driven by simple 5 V TTL-type logic circuits. 5 V represents a logical high, whereas 0 V represents a logical low. What is the output level if the input word is 0100? Because a logical high represents 5 V, 5 V is being applied to the second input. All other inputs receive a logical low, or 0 V. The output is the summation of the input signals (remember, this is an inverting summer, so the final output should have its sign reversed).

$$V_{out} = -(V_{in1}A_1 + V_{in2}A_2 + V_{in3}A_3 + V_{in4}A_4)$$
$$V_{out} = -(0 \text{ V} \times 1 + 5 \text{ V} \times .5 + 0 \text{ V} \times .25 + 0 \text{ V} \times .125)$$
$$V_{out} = -2.5 \text{ V}$$

So, a value of 4 (binary 100) is equivalent to a potential of 2.5 V. If we increase the word value to 9 (binary 1001), we see

$$V_{out} = -(V_{in1}A_1 + V_{in2}A_2 + V_{in3}A_3 + V_{in4}A_4)$$
$$V_{out} = -(5 \text{ V} \times 1 + 0 \text{ V} \times .5 + 0 \text{ V} \times .25 + 5 \text{ V} \times .125)$$
$$V_{out} = -5.625 \text{ V}$$

The minimum output occurs at binary 0000 (0 V), and the maximum at binary 1111 (−9.375 V). The step size is equal to the logic level times the minimum gain; in this case that's .625 V. Notice that the output value may be found by simply multiplying the value of the input word by the minimum step size. Also, it is important to note that the output signal is unipolar (in this example, always negative).

A digital representation, of course, is made up of a sequence of words, not just one word. In reality, the logic circuits are constantly feeding the summing amplifier new words at a predetermined rate. Because of the changing inputs, the output of the converter is constantly changing as well. Using our previously calculated values, if the converter is fed the sequence 0000, 0100, 1001, 1111, the output will move from

FIGURE 12.6

Output with four digital words

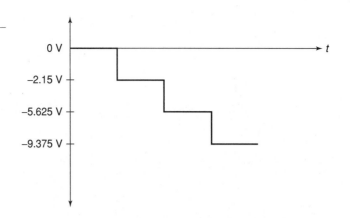

0 V to −2.5 V, to −5.625 V, to a final value of −9.375 V. This output is graphed in Figure 12.6. If this sequence is repeated over and over, the waveform of Figure 12.7 is the result. Note that a "stair-step" type wave is created. You might also think of this as a very rough form of a ramp function. A better ramp would be produced if we used all of the available values for the input sequence, as in 0000, 0001, 0010, 0011,..., 1111. In order to remove the negative DC offset and make the signal bipolar, all we need to do is pass the signal through a coupling capacitor. The frequency of this waveform is controlled by the rate at which the words are fed to the converter. Note that by increasing the resolution and the number of words fed to the converter per cycle, a very close approximation to the ideal ramp function may be achieved. For that matter, by changing the input words to other sequences, we can create a wide variety of output wave shapes. This is the concept behind the digital *arbitrary function generator*. An arbitrary function generator allows you to create wave shapes beyond the simple sine/square/triangle found on the typical laboratory function generator. We'll take a closer look at this particular piece of test equipment a little later.

In order to increase resolution, it appears that all you need to do to the summing amplifier is add extra channels with larger and larger resistors. Unfortunately, the resistor sizes soon become impractical and another approach is required. For example,

FIGURE 12.7

Cycled output

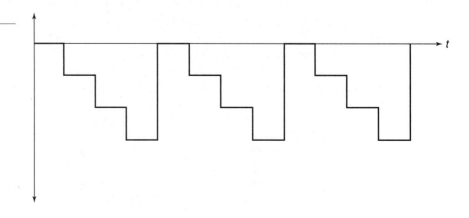

FIGURE 12.8

R/2R ladder network

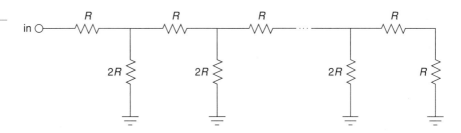

a 16-bit system would require that the LSB resistor be equal to 65,536 R_f. One problem is that the resulting small input current may be dwarfed by input bias and offset currents. Also, high component accuracy is needed for the more significant inputs in terms of the input resistors and the drive signals. The excessively large resistors may also contribute added noise. The standard solution to this problem involves of the use of an $R/2R$ resistive divider network.

An $R/2R$ network is shown in Figure 12.8. This circuit exhibits the unique attribute of constant division by 2 for each stage. You may think of this as either a division of voltage at each successive node or a division of current in each successive leg. An example of a four-stage (i.e., 4-bit) network is shown in Figure 12.9. In order to find the voltage at any given node, the loading effects of the following stages must be taken into account. This is much easier to do than it first appears. If we need to find the voltage at point A, we must first find the resistance in parallel with the initial 2 kΩ resistor. A quick inspection shows that each stage is loaded by the following stages, so it is easiest if we start at the last stage and work toward the input. The effective resistance to the right of node C is 1 kΩ in series with 1 kΩ, or 2 kΩ. This resistance is placed in parallel with the 2 kΩ resistor seen from node C to ground. The result is 1 kΩ. In other words, from C to ground, we see 1 kΩ. This creates a 2:1 voltage divider with the 1 kΩ resistor placed from B to C, so the voltage at C must be half the voltage at B. This also points up the fact that the current entering node C splits into two equal portions: one that travels toward point D, and the other that travels through the 2 kΩ resistor to ground. This is shown graphically in Figure 12.10. If you look at the equivalent circuit section of Figure 12.10, you will notice that this portion now looks exactly like the final portion of the original network. That is, every time a section is simplified and analyzed, the result will be a halving of voltage and current. It is already apparent that the voltage at D must be half the voltage at C, which in turn, must be half the voltage at B. As you can now prove, it follows that the voltage at B must be half the voltage at A. In a similar fashion, the current passing through each $2R$ leg is half the preceding current.

FIGURE 12.9

A 4-stage ladder

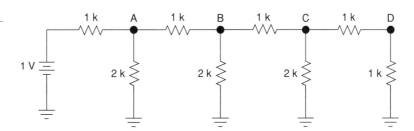

FIGURE 12.10

Ladder analysis

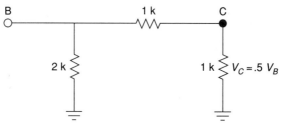

a. Equivalent circuit

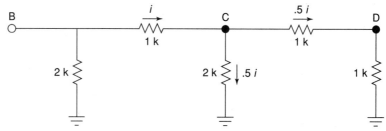

b. Current division

Adapting the $R/2R$ network to the DA converter is relatively easy. The network is fed from a stable current source, with each $2R$ element feeding into a summing amplifier. In series with each $2R$ element is a solid-state switch, which sets the appropriate logic level. This is shown in Figure 12.11, with the network effectively on its side. When a logical high is presented to a given bit, the switch is closed and current flows through the $2R$ element and into the op amp. Note that the right end of the resistor is effectively at ground, as the summing node of the op amp is a virtual ground. If a logical low is presented, the switch shunts the current to ground, bypassing the op amp. In this way, the appropriately weighted currents are summed and used to produce the output voltage.

This technique offers several advantages over the simpler weighted gain version. First, all branches are fed by one common current source. Because of this, there is no need for output level matching. Second, only two different values of resistors are required for any number of bits used, rather than the impractically wide range seen earlier. It is more economical to control the tolerance of just two different parts than 12 or 16. Note that small input currents are still generated for the least significant bits, so attention to input bias and offset currents remains important.

Practical Digital-to-Analog Converter Limits

Perhaps the most obvious limit associated with the DA converter is its speed. The op amp used in the DAC must be much faster than the final signals it is meant to produce. A given output waveform may contain several dozen individual sample points per cycle. The op amp must respond to each sample point. Consequently, wide bandwidth and high slew rates are required.

FIGURE 12.11

Converter with R/2R ladder

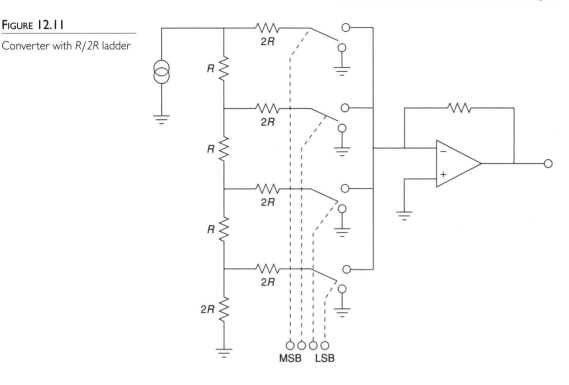

Integrated DAC spec sheets offer a few important parameters of which you should be aware. First of all, there is *conversion speed*. This figure tells how long it takes the DAC to turn the digital input word into a stable analog output voltage. This sets the maximum data rate. Next come the *accuracy* and *resolution*. Resolution indicates the number of discrete steps that may be produced at the output, and is set by the number of bits available. This is not the same as accuracy. Accuracy is actually comprised of several different factors including *offset error, gain error*, and *nonlinearity*. Offset error is normally measured by applying the all-zero input word and then measuring the output signal. Ideally, this signal will be zero volts. The deviation from zero is taken as the offset error. This has the effect of making all output levels inaccurate by a constant voltage. Offset error is relatively easy to compensate for in many applications by applying an equal offset of opposite polarity. Gain error is a deviation that affects each output level by a constant percentage. It is as if the signal were passed through a small amplifier or attenuator. This error may be compensated for by using an amplifier with a gain equal to the reciprocal of the error. The two gains will effectively cancel. The effect of offset and gain error are shown in Figures 12.12 and 12.13.

Nonlinearity errors may be broken into two forms: *integral nonlinearity* and *differential nonlinearity*. Integral nonlinearity details the maximum offset between the ideal outputs and the actual outputs for all possible inputs. Differential nonlinearity details the maximum output deviation relative to one LSB caused by two adjacent input words. If differential nonlinearity is beyond ±1 LSB, the system may be nonmonotonic. In other words, a higher digital input word may actually produce a lower analog output

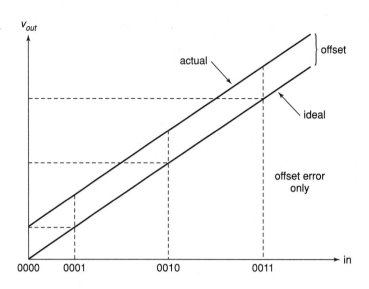

FIGURE 12.12

Offset error

voltage. These two forms of error are shown in Figure 12.14. Note that it is possible to have high integral nonlinearity and yet still have modest differential nonlinearity. This is the case in Figure 12.14b.

As you can see, accuracy is dependent on rather complex factors. In an effort to boil this down to a single number, some manufacturers give an *effective number of bits* specification. For example, a 16-bit DAC may be specified as having 14-bit accuracy. This means that the 14 most significant bits behave in the idealized fashion,

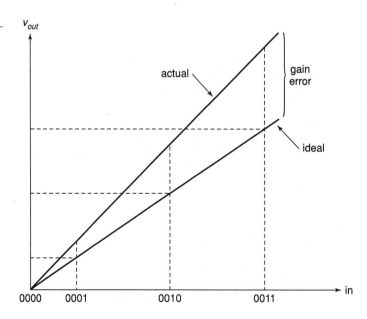

FIGURE 12.13

Gain error only

Figure 12.14

Linearity error

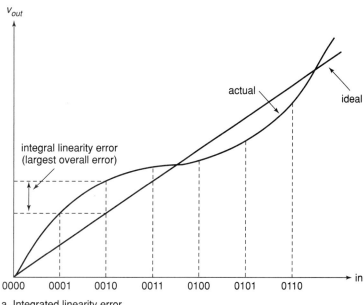

a. Integrated linearity error

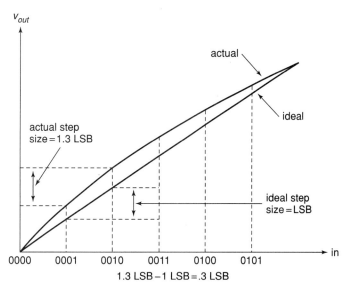

b. Differential linearity error (relative-adjacent error)

but the lowest 2 bits may be swamped out by linearity errors. Another spec that you will sometimes see is *no missing codes*. This means that for every increase in the input word, there will be an appropriate positive output level change.

In practice, the standard DA converter is used with an output filter. As you can see from the previous figures, the waveforms produced by the DAC contain a stair-step side effect. Generally, this is not desirable. The abrupt changes in output level indicate

FIGURE 12.15

Output reconstruction

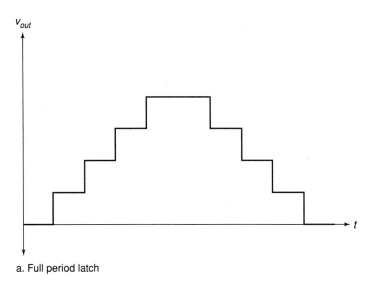

a. Full period latch

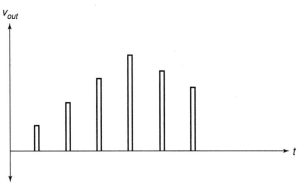

b. Partial period latch

that higher frequency components are present. All components above the Nyquist rate should be filtered out with an appropriate low-pass filter. This filter is sometimes referred to as a *reconstruction* or *smoothing* filter. In an improperly designed system, the reconstruction filter will remove some of the highest in-band frequency components (i.e., components immediately below the Nyquist frequency). To compensate for this, logic levels are often latched to the DAC for shortened periods, thus creating a more spiked appearance, rather than the stair-step form. This effect is shown in Figure 12.15. Although this spiked waveform appears to be less desirable than the stair-step form, it creates higher levels for the uppermost components, and after filtering, the result is a smoother overall frequency response.

To further increase the quality of the output waveform, a technique known as *oversampling* is sometimes employed. The basic idea is to create new sample points in between the existing ones. The result is a much denser data rate, which hopefully, will yield more exacting results after filtering. Also, the higher data rate may loosen the requirements of the reconstruction filter. A typical system might use four times

FIGURE 12.16

Oversampled output

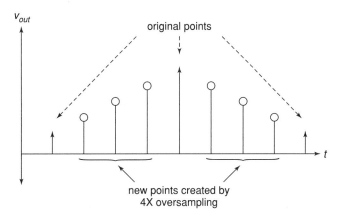

oversampling, meaning that the output data rate is four times the original. Therefore, for each input word, three new words have to be added. This effect is shown in Figure 12.16. There are a number of ways to create the new sample points. The most obvious way is via simple interpolation, but this does not achieve the best results. Another technique involves initializing the new values to zero and then passing the data stream through a digital low-pass filter, which effectively calculates the proper values. An extension of the oversampling principle is the *delta-sigma* technique. In delta-sigma, very high rates of oversampling are used in conjunction with specialized digital filter algorithms. The algorithms essentially trade the higher data rate for a slower rate with increased resolution. The design and analysis of delta-sigma systems is fairly advanced and is beyond the scope of this text. Suffice to say that these techniques can increase the quality of the output signal and are widely used in applications such as high-quality audio CD and DVD players.

Digital-to-Analog Converter Integrated Circuits

There are many possible applications for digital-to-analog converters, and a number of different chips have evolved to meet specific needs. Generally, you can group these into specific classes, such as high speed, high resolution, or low cost. We shall examine three representative types. The devices we will look at are the DAC0832, a basic 8-bit unit; the DAC7545, a microprocessor-compatible 12-bit unit; and the PCM1716, a 24-bit high-quality converter used in the audio industry.

DAC0832 This IC is a popular microprocessor-compatible 8-bit converter. The DAC0830 and DAC0831 are similar, but with somewhat reduced performance. It is a multiplying DAC. In other words, the output signal is a function of the digital input word and a reference input. In some applications, the reference input is not fixed, but rather, is a variable input signal. A feature list and pin-out are shown in Figure 12.17. Notable items are a settling time of only 1 μS, low power requirements, and high linearity. The DAC0832 may be used in either stand-alone mode or with a microprocessor. The switching waveforms are shown in Figure 12.18.

Figure 12.17
DAC-0832

Reprinted with permission of National Semiconductor Corporation

DAC0830/DAC0831/DAC0832 8-Bit μP Compatible, Double-Buffered D to A Converters

General Description

The DAC0830 is an advanced CMOS/Si-Cr 8-bit multiplying DAC designed to interface directly with the 8080, 8048, 8085, Z80®, and other popular microprocessors. A deposited silicon-chromium R-2R resistor ladder network divides the reference current and provides the circuit with excellent temperature tracking characteristics (0.05% of Full Scale Range maximum linearity error over temperature). The circuit uses CMOS current switches and control logic to achieve low power consumption and low output leakage current errors. Special circuitry provides TTL logic input voltage level compatibility.

Double buffering allows these DACs to output a voltage corresponding to one digital word while holding the next digital word. This permits the simultaneous updating of any number of DACs.

The DAC0830 series are the 8-bit members of a family of microprocessor-compatible DACs (MICRO-DAC™). For applications demanding higher resolution, the DAC1000 series (10-bits) and the DAC1208 and DAC1230 (12-bits) are available alternatives.

Features

- Double-buffered, single-buffered or flow-through digital data inputs
- Easy interchange and pin-compatible with 12-bit DAC1230 series
- Direct interface to all popular microprocessors
- Linearity specified with zero and full scale adjust only—NOT BEST STRAIGHT LINE FIT.
- Works with ±10 V reference-full 4-quadrant multiplication
- Can be used in the voltage switching mode
- Logic inputs which meet TTL voltage level specs (1.4 V logic threshold)
- Operates "STAND ALONE" (without μP) if desired
- Available in 20-pin small-outline or molded chip carrier package

Key Specifications

- Current settling time — 1 μs
- Resolution — 8 bits
- Linearity (guaranteed over temp.) — 8, 9, or 10 bits
- Gain Tempco — 0.0002% FS/°C
- Low power dissipation — 20 mW
- Single power supply — 5 to 15 V_{DC}

Typical Application

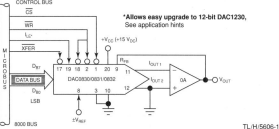

Connection Diagrams (Top Views)

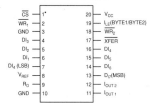

†This is necessary for the 12-bit DAC1230 series to permit interchanging from an 8-bit to a 12-bit DAC with no PC board changes and no software changes, See application section.

Switching Waveform

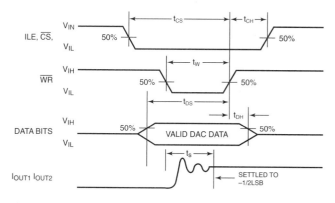

Definition of Package Pinouts

Control Signals (All control signals level actuated)

$\overline{CS}$: **Chip Select** (active low). The $\overline{CS}$ in combination with ILE will enable $\overline{WR}_1$.

LE: **Input Latch** Enables (active high). The ILE in combination with $\overline{CS}$ enables $\overline{WR}_1$.

$\overline{WR1}$: **Write 1.** The active low $\overline{WR}_1$ is used to load the digital input data bits (DI) into the input latch. The data in the input latch is latched when $\overline{WR}_1$ is high. To update the input latch CS and $\overline{WR}_1$ must be low while ILE is high.

$\overline{WR2}$: **Write 2** (active low). This signal, in combination with $\overline{XFER}$, causes the 8-bit data which is available in the input latch to transfer to the DAC register.

$\overline{1FER}$: **Transfer control signal** (active low). The $\overline{XFER}$ will enable $\overline{WR}_2$.

Other Pin Functions

DI_0-DI_7: **Digital Inputs.** DI_0 is the least significant bit (LSB) and DI_7 is the most significant bit (MSB).

I_{OUT1}: **DAC Current Output 1.** I_{OUT1} is a maximum for a digital code of all 1's in the DAC register, and is zero for all 0's in DAC register.

I_{OUT2}: **DAC Current Output 2.** I_{OUT2} is a constant minus I_{OUT}, or $I_{OUT1} + I_{OUT2}$ = constant (1 full scale for a fixed reference voltage).

R_{IO}: **Feedback Resistor.** The feedback resistor is provided on the IC chip for use as the shunt feedback resistor for the external op amp which is used to provide an output voltage for the DAC. This on-chip resistor should always be used (not an external resistor) since it matches the resistors which are used in the on-chip R-2R ladder and tracks these resistors over temperature.

V_{REF}: **Reference Voltage Input.** This input connects an external precision voltage source to the internal R-2R ladder. V_{REF} can be selected over the range of +10 to −10V. This is also the analog voltage input for a 4-quadrant multiplying DAC application.

V_{CC}: **Digital Supply Voltage.** This is the power supply pin for the part. V_{CC} can be from +5 to +15 V_{DC}. Operation is optimum for +15 V_{DC}.

GND: The pin 10 voltage must be at the same ground potential as I_{OUT1} and I_{OUT2} or current switching applications. Any difference of potential (V_{OS} pin 10) will result in a linearity change of

$$\frac{V_{OS} \text{pin } 10}{3 V_{REF}}$$

For example, If V_{REF} = 10 V and pin 10 is 9 mV offset from I_{OUT1} and I_{OUT2} the linearity change will be 0.03%.

Pin 3 can be offset −100 mV with no linearity change, but the logic input threshold will shift.

FIGURE 12.18

DAC-0832 switching waveforms
Reprinted with permission of National Semiconductor Corporation

FIGURE 12.19

DAC-0832 state variable filter application

Reprinted with permission of National Semiconductor Corporation

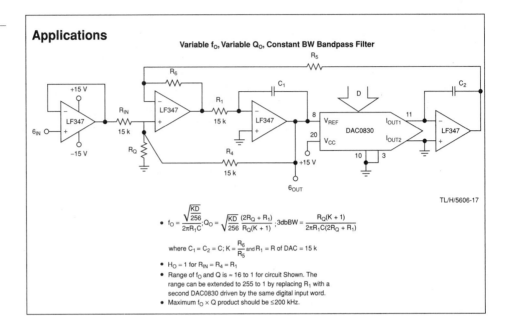

An interesting application of DAC0832 can be found in Figure 12.19. Basically, this is a digitally-controlled state variable filter. Note that the converter replaces the input resistor of the second integrator. Normally, that resistor would be used to convert the output voltage of the first integrator into an input current for the second integrator. This job is now handled by the DAC0832. The digital input word effectively sets the voltage-to-current conversion. Thus, a change in the input word alters the tuning frequency of the filter just as a potentiometer. Compare this circuit to the OTA-based voltage-controlled filter from Chapter 11. Conceptually, they are very similar.

DAC7545 The DAC7545 is a fairly standard 12-bit linear converter and is shown in Figure 12.20. Its interesting aspects are that it is a multiplying converter and that it is microprocessor-compatible. The multiplying effect comes from the fact that a reference is used to drive the $R/2R$ ladder network. If the reference is changed, the output is effectively rescaled. Consequently, you can think of the output signal as equal to the reference value times the digital input word. You may also think of this as a form of "digital volume control."

With the inclusion of a few extra logic lines, the IC has become microprocessor compatible. This means that the DAC7545 has chip select and read/write lines along with the 12 data input lines. This allows the converter to be connected directly to the microprocessor data bus. By using memory-mapped I/O, the microprocessor can write data to the converter just as it writes data to memory. A 16-bit microprocessor system can present the converter with all of the data it needs during one write cycle; however, an 8-bit microprocessor will need two write cycles and some form of latch. One address may be used for the lower 8 bits, and another address for the remaining 4 bits. A simplified system is shown in Figure 12.21 using a 16-bit microprocessor.

DIGITAL-TO-ANALOG CONVERSION TECHNIQUES 505

DAC7545

CMOS 12-Bit Multiplying
DIGITAL-TO-ANALOG CONVERTER
Microprocessor Compatible

FEATURES

- FOUR-QUADRANT MULTIPLICATION
- LOW GAIN TC: 2 PPM/°C typ
- MONOTONICITY GUARANTEED OVER TEMPERATURE
- SINGLE 5 V TO 15 V SUPPLY
- TTL/CMOS LOGIC COMPATIBLE
- LOW OUTPUT LEAKAGE: 10 nA max
- LOW OUTPUT CAPACITANCE: 70 pF max
- DIRECT REPLACEMENT FOR AD7545, PM-7545

DESCRIPTION

The DAC7545 is a low-cost CMOS, 12-bit four-quadrant multiplying, digital-to-analog converter with input data latches. The input data is loaded into the DAC as a 12-bit data word. The data flows through to the DAC when both the chip select ($\overline{CS}$) the write ($\overline{WR}$) pins are at a logic low.

Laser-trimmed thin-film resistors and excellent CMOS voltage switches provide true12-bit integral and differential linearity. The device operates on a single +5 V to +15 V supply and is available in 20-pin side-brazed DIP, 20-pin plastic DIP or a 20-lead plastic SOIC package. Devices are specified over the commercial, industrial, and military temperature ranges and are available with additional reliability screening.

The DAC7545 is well suited for battery of other low power applications because the power dissipation is less than 0.5 mW when used wih CMOS logic inputs and V_{DD} = +5 V.

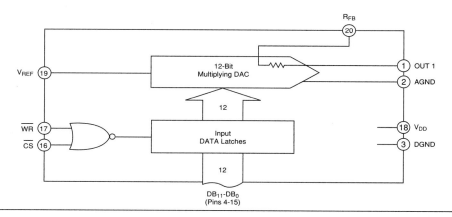

FIGURE 12.20

The DAC7545

Copyright 1989 Burr-Brown Corporation. Reprinted, in whole or in part, with the permission of Burr-Brown Corporation

Figure 12.21

Microprocessor to DAC7545 interface

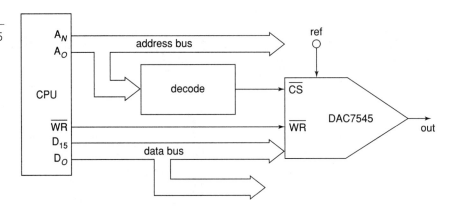

PCM1716 The PCM1716 is a stereo 24-bit converter designed specifically for high-quality digital audio applications. It comes in a 28-pin SSOP package. A block diagram and feature list are shown in Figure 12.22a. Unlike the other converters, the PCM1716 features serial input of data, not parallel. It includes its own on-board serial conversion circuitry and logic. This technique helps to reduce system cost. It is also surprisingly convenient as many specialized digital signal processing ICs that might be used with the PCM1716 utilize a serial-type output. This may be fed directly into the PCM1716 in 16-, 20-, or 24-bit format.

The PCM1716 specification sheet is shown in Figure 12.22b. Note that this device can be utilized with sampling rates from 16 kHz to 96 kHz. Total harmonic distortion plus noise is typically 97 dB below a full-scale output when used with a 44.1 kHz sample rate. The internal low-pass filters show less than ±0.002 dB ripple. This sort of performance was unheard of not too long ago. Due to its high resolution and 106 dB dynamic range, extra care must be taken during circuit layout to avoid hum pickup and RF interference.

Applications of Digital-to-Analog Converter Integrated Circuits

Example 12-4

Perhaps the first thing many people think of when they hear the terms *digital* or *digitized*, is the audio compact disk, or CD for short. Home CD players are excellent examples of the use of precision DA circuitry in our everyday lives. Music data is stored on the CD with 16-bit resolution and a sampling rate of 44.1 kHz. This produces a Nyquist frequency of 22.05 kHz, which is high enough to encompass the hearing range of most humans. Error correction and auxiliary data is also stored on the disk. The data is stored on disk in the form of very tiny pits, which are read by a laser. The signal is then converted into the common electronic logic form where it is checked for error and adjusted as need be. The data stream is then fed to the DA converter for audio reconstruction. A single converter may be multiplexed between the two stereo channels, or two dedicated converters may be used. Oversampling in the

FIGURE 12.22a

PCM1716
Courtesy of Burr-Brown Corporation

PCM1716

www.burr-brown.com/databook/PCM1716.html

Soundplus™ 24-Bit, 96 kHz Sampling CMOS Delta-Sigma Stereo Audio DIGITAL-TO-ANALOG CONVERTER

FEATURES

- **ENHANCED MULTI-LEVEL DELTA-SIGMA DAC**
- **SAMPLING FREQUENCY (fs): 16 kHz–96 kHz**
- **INPUT AUDIO DATA WORD:**
  16-, 20-, 24-Bit
- **HIGH PERFORMANCE:**
  THD + N: –96 dB
  Dynamic Range: 106 dB
  SNR: 106 dB
  Analog Output Range: $0.62 \times V_{cc}$ (Vp-p)
- **8x OVERSAMPLING DIGITAL FILTER:**
  Stop Band Attenuation: –82 dB
  Passband Ripple: ±0.002 dB
  Slow Roll Off
- **MULTI FUNCTIONS:**
  Digital De-emphasis
  L/R Independent Digital Attenuation
  Soft Mute
  Zero Detect Mute
  Zero Flag
  Chip Select
  Reversible Output Phase
- **+5 V SINGLE SUPPLY OPERATION**
- **SMALL 28-LEAD SSOP PACKAGE**

DESCRIPTION

The PCM1716 is designed for Mid to High grade Digital Audio applications which achieve 96 kHz sampling rates with 24-bit audio data. PCM1716 uses a newly developed, enhanced multi-level delta-sigma modulator architecture that improves audio dynamic performance and reduces jitter sensitivity in actual applications.

The internal digital filter operates at 8x over sampling at a 96 kHz sampling rate, with two kinds of roll-off performances that can be selected: sharp roll-off, or slow roll-off, as required for specific applications.

PCM1716 is suitable for Mid to High grade audio applications such as CD, DVD-Audio, and Music Instruments, since the device has superior audio dynamic performance, 24-bit resolution and 96 kHz sampling.

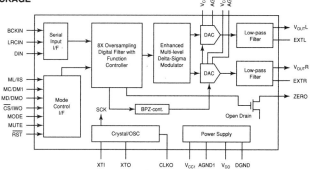

FIGURE 12.22b

PCM1716 specifications

SPECIFICATIONS

All specifications at +25°C, +V_{CC} = +V_{DD} = +5V, f_S = 44.1 kHz, and 24-bit input data, SYSCLK = 384 f_S, unless otherwise noted.

PARAMETER	CONDITIONS	PCM1716 MIN	PCM1716 TYP	PCM1716 MAX	UNITS
RESOLUTION			24		Bits
DATA FORMAT					
Audio Data Interface Format			Standard/I^2S		
Data Bit Length			16/20/24 Selectable		
Audio Data Format			MSB First, 2's Comp		
Sampling Frequency (f_S)		16		96	kHz
System Clock Frequency[1]			256/384/512/768 f_S		
DIGITAL INPUT/OUTPUT LOGIC LEVEL					
Input Logic Level V_{IH}		2.0			V
V_{IL}				0.8	V
Output Logic Level (CLKO) V_{OH}	I_{OH} = 2 mA	4.5			V
V_{OL}	I_{OL} = 4 mA			0.5	V
CLKO PERFORMANCE[2]					
Output Rise Time	20 ~ 80% V_{DD},10 pF		5.5		ns
Output Fall Time	80 ~ 20% V_{DD},10 pF		4		ns
Output Duty Cycle	10 pF Load		37		%
DYNAMIC PERFORMANCE[3] (24-Bit Data)					
THD+N V_O = 0 dB	f_S = 44.1 kHz		–97	–90	dB
	f_S = 96 kHz		–94		dB
V_O = –60 dB	f_S = 44.1 kHz		–42		dB
Dynamic Range	f_S = 44.1 kHz EIAJ A-weighted	98	106		dB
	f_S = 96 kHz A-weighted		103		dB
Signal-to-Noise Ratio[4]	f_S = 44.1 kHz EIAJ A-weighted	98	106		dB
	f_S = 96 kHz A-weighted		103		dB
Chanel Separation	f_S = 44.1 kHz	96	102		dB
	f_S = 96 kHz		101		dB
DYNAMIC PERFORMANCE[3] (16-Bit Data)					
THD+N V_O = 0 dB	f_S = 44.1 kHz		–94		dB
	f_S = 96 kHz		–92		dB
Dynamic Range	f_S = 44.1 kHz EIAJ A-weighted		98		dB
	f_S = 96 kHz A-weighted		97		dB
DC ACCURACY					
Gain Error			–1.0	–3.0	% of FSR
Gain Mismatch: Channel-to-Channel			–1.0	–3.0	% of FSR
Bipolar Zero Error	V_O = 0.5 V_{CC} at Bipolar Zero		–30	–60	mV
ANALOG OUTPUT					
Output Voltage	Full Scale (0 dB)		0.62 V_{CC}		V_{p-p}
Center Voltage			0.5 V_{CC}		V
Load Impedance	AC Load	5			kΩ
DIGITAL FILTER PERFORMANCE					
Filter Characteristics 1 (Sharp Roll-Off)					
Passband	–0.002 dB			0.454 f_S	
	–3 dB			0.490 f_S	
Stopband		0.546 f_S			
Passband Ripple				–0.002	dB
Stopband Attenuation	Stop Band = 0.546 f_S	–75			dB
	Stop Band = 0.567 f_S	–82			dB
Filter Characteristics 2 (Slow Roll-Off)					
Passband	–0.002 dB			0.274 f_S	
	–3 dB			0.454 f_S	
Stopband		0.732 f_S			
Passband Ripple				–0.002	dB
Stopband Attenuation	Stopband = 0.732 f_S	–82			dB
Delay Time			30/f_S		sec
De-emphasis Error				–0.1	dB
INTERNAL ANALOG FILTER					
–3 dB Bandwidth			100		kHz
Passband Response	f = 20 kHz		–0.16		dB
POWER SUPPLY REQUIREMENTS					
Voltage Range	V_{DD}, V_{CC}	4.5	5	5.5	VDC
Supply Current: I_{CC} + I_{DD}	f_S = 44.1 kHz		32	45	mA
	f_S = 96 kHz		45		mA
Power Dissipation	f_S = 44.1 kHz		160	225	mW
	f_S = 96 kHz		225		mW
TEMPERATURE RANGE					
Operation		–25		+85	°C
Storage		–55		+100	°C

NOTES: (1) Refer section of system clock. (2) External buffer is recommended. (3) Dynamic performance specs are tested with 20 kHz low pass filter and THD+N specs are tested with 30 kHz LPF, 400 Hz HPF, Average Mode. (4) SNR is tested at internally infinity zero detection off.

Burr-Brown IC Data Book

Figure 12.23

Audio compact disk playback system

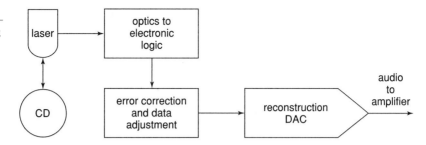

range of 2X to 8X is often used for improved signal quality. A block diagram of the system is shown in Figure 12.23. The actual DAC portion seems almost trivial when compared to some of the more sophisticated elements.

The storage density of the optical CD is quite remarkable. This small disk (less than 5 inches in diameter) can hold 70 minutes of music. Ignoring the auxiliary data, we can quickly calculate the total storage. We have two channels of 16-bit data, or 32 bits (4 bytes) per sample point. There are 44,100 samples per second for 70 minutes, yielding 185.22 megasamples. The total data storage is 5.927 gigabits, or 741 megabytes.

Example 12-5

As we have already mentioned, it is possible to connect DACs directly to microprocessor systems. Furthermore, the microprocessor may write to the DAC with no more effort than writing to a memory location. The microprocessor can write any series of data words we desire out to the DAC and can repeat a sequence virtually forever. Given this ability, we can make an arbitrary waveform generator. Instead of being locked into a set of predefined wave shapes as on ordinary function generators, this system allows for all manner of wave shapes. The accuracy and flexibility of the system will depend on its speed and the available DAC resolution.

The basic idea is one of table lookup. For example, let's say that we have a 16-bit system. We will create a table of data values for one cycle of the output waveform. For convenience sake, we might make the table size a handy power of 2, such as 256. In other words, a single output cycle will be chopped into 256 discrete time chunks. It is obvious, then, that the converter must be a few hundred times faster than the highest fundamental that we wish to produce. By increasing or decreasing the output data rate, we can change the frequency of the output fundamental. This is known as a *variable sample rate* technique. It is also possible to change the fundamental frequency with a *fixed rate* technique (this is somewhat more complex, but does offer certain advantages). An output flow chart is shown in Figure 12.24. Upon initialization, an address pointer is set to the starting address of the data table. The CPU reads the data from the table via the pointer. The pointer is incremented so that it now points to the next element in the table. (Some CPUs offer a postincrement addressing mode so that both steps may be performed in a single instruction.) Next, the CPU writes the data to the special DAC address. At this point, some form of software/hardware delay is invoked that sets the output data rate. After the delay, the CPU reads the next data element via the pointer and continues as in the first run. Once the 256th element is sent out, the

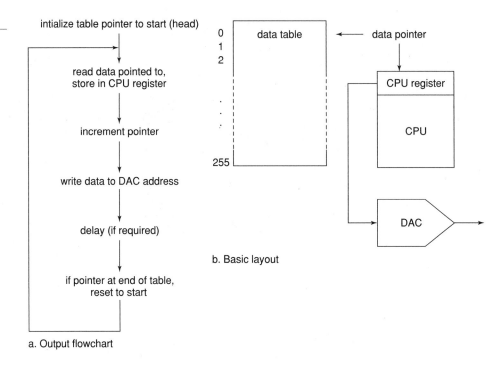

Figure 12.24

Arbitrary waveform generator

pointer is reset to the start of the table and the process continues on. In this way, the table can be thought of as circular, or neverending. If the system software is written in a higher level language, the pointer/data table may be implemented as a simple array where the array index is set by a counter. This will not be as efficient as a direct assembly level approach, though.

The real beauty of this system is that the data table may contain virtually any sequence of data. The data could represent a sine, pulse, triangle, or other standard function. More importantly, the data could represent a sine wave with an embedded noise transient, or a signal containing a hum component just as easily. This data could come from three basic sources. First, the data table can be filled through direct computation if the time-domain equation of the desired function is known. Second, the data could be manufactured by the user through some form of interaction with a computer, perhaps with a mouse or drawing pad. Finally, the data may be derived from a real-world signal. That is, an analog-to-digital converter may be used to record the signal in digital form. The data may then be loaded into the table and played back repeatedly. The arbitrary waveform generator allows its user to test circuits and systems with a range of wave shapes that would be impossible or impractical to generate otherwise.

Example 12-6

Under computer control, DA converters can be used as part of an automated test equipment system. In order to fully characterize an electronic product, a number of individual tests need to be run. Setting up each individual test can be somewhat time consuming and is subject to operator error. Automating this procedure may improve repeatability and decrease

FIGURE 12.25

Simple test set-up

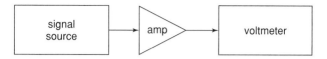

testing time. There are many ways in which this process may be automated. We'll look at one approach.

Let's assume that we would like to make frequency response measurements for an amplifier at 20 different frequencies. A circuit test system appropriate for this job is shown in Figure 12.25. To perform this test manually requires 20 distinct settings of the source signal, and 20 corresponding output readings. This can prove to be rather tedious if many units are to be tested. It would be very handy if there was some way in which the source frequency could be automatically changed to preset values. This is not particularly difficult. Most modern sources have control voltage inputs that may be used to set the frequency. The required control voltage can be created and accurately set through the use of a computer and DA converter. The computer can be programmed to send specific digital words to the DAC, which in turn feeds the signal source control input. In other words, the data word directly sets the frequency of the signal source. The computer can be programmed to send virtually any sequence of data words at almost any rate and do it all without operator intervention. All the operator needs to do is start the process. Test repeatability is very high with a system like this. A block diagram of this system is shown in Figure 12.26.

In order to record the data, the voltmeter may be connected to a strip chart recorder, or better yet, back to the computer. The data may be sent to the computer in digital form if the voltmeter is of fairly advanced design, or, with the inclusion of an analog-to-digital converter, the output signal may be directly sampled and manipulated by the computer. In either case, data files may be created for each unit tested and stored for later use. Also, convenient statistical analysis may be performed quickly at the end of a test batch. Note that because the DA converter only generates a control signal, very high resolution and low distortion is normally not required. If a high resolution converter is used, it is possible to create the test signals in the computer (as in the arbitrary function generator), and dispense with the signal source.

The automated test system is only one possible application of instrument control. Another interesting example is in the generation of laser "light shows." A block diagram of a simplified system is shown in Figure 12.27. In order to create the complex

FIGURE 12.26

Automated test set-up

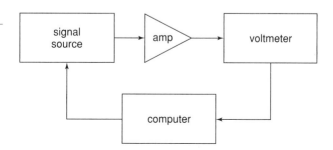

FIGURE 12.27

Computer-controlled lighting

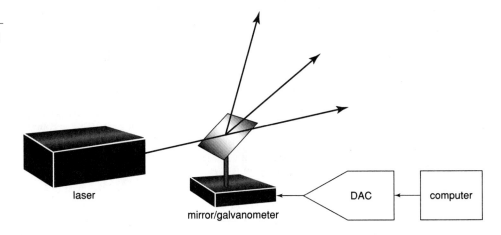

patterns seen by the audience, a laser beam is bounced off of tiny moving mirrors. The mirrors may be mounted on something as simple as a galvanometer. The galvanometer is fed by a DAC. The pattern that the laser beam makes is dependent on how the galvanometer moves the mirror, which is in turn controlled by the data words fed to the DAC. In practice, several mirrors may be used to deflect the beam along three axes.

DA converters can be used to adjust any device with a control-voltage type input. Also, they may be used to control electromechanical devices that respond to an applied voltage. Their real advantage is the repeatability and flexibility they offer.

12.5 ANALOG-TO-DIGITAL CONVERSION

Now that you know the basics behind digital-to-analog conversion, we may examine the converse system, analog-to-digital conversion. The analog-to-digital conversion process is sometimes referred to as *quantization*, implying the individual discrete steps that the output assumes. There are several techniques to produce the conversion. Some techniques are optimized for fastest possible conversion speed, and some for highest accuracy. We shall investigate the more popular types.

The concept of AD conversion is simple enough: you wish to measure the voltage of the incoming waveform at specific instances in time. This measurement will be translated into a digital word. One practical limitation is the fact that the conversion circuitry may require a small amount of computation or translation time. For ultimate accuracy, then, it is important that the measured waveform not change during the conversion interval. To do this, specialized subcircuits called *track-and-hold* (or *sample-and-hold*) amplifiers are used. They are usually abbreviated as T/H or S/H. Their job is to capture an input potential and produce a steady output to feed to the AD converter. The T/H schematic symbol is shown in Figure 12.28. When the T/H logic is in track mode, the circuit acts as a simple buffer, so its output voltage equals its input. When the logic goes to the hold state, the output voltage locks at its present potential and

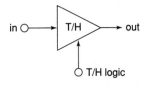

FIGURE 12.28

Symbol for the track-and-hold amplifier

FIGURE 12.29

Track-and-hold operation

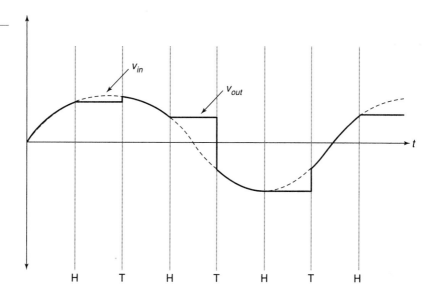

stays there until the circuit is switched back to track mode. A representation of this operation is shown in Figure 12.29.

In reality, the track-and-hold process is not perfect, and errors may arise. These errors are shown magnified in Figure 12.30. First of all, there is a small delay between the time the logic signal changes and the T/H starts to react. The exact amount of time is variable and is responsible for *aperture error*. This is equal to the voltage difference between the signals at the desired and actual times. Also, due to the dynamic nature of the process of switching from track to hold, some initial ringing may occur in the

FIGURE 12.30

Track-and-hold errors

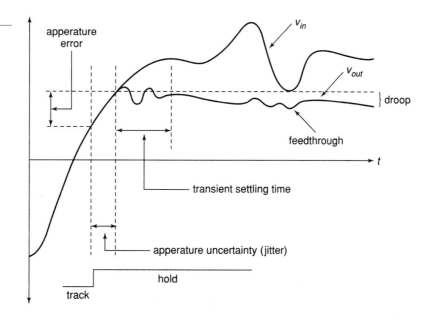

hold waveform. You can see that as time progresses, the hold waveform tends to decay toward zero. This is because the held voltage is usually formed across a capacitor. Although the discharge time constant can be very long, it cannot be infinite, and thus, the charge eventually bleeds off. This parameter is measured by the *droop rate* (fundamental units of volts per second). Finally, the possibility of *feedthrough error* exists. If a large change occurs on the input waveform during the hold period, it is possible that a portion of the signal may "leak" through to the T/H output. These errors are particularly troublesome when working with high-resolution converters. Lower-resolution systems may not be adversely affected by these relatively small aberrations.

In order to create a T/H, a pair of high-impedance buffers are normally used, along with some form of switch element and a capacitor to hold the charge. Examples are shown in Figure 12.31. Figure 12.31a shows the general voltage type, open-loop T/H.

FIGURE 12.31

Track-and-hold circuits

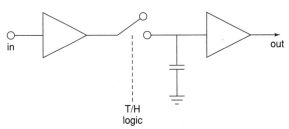

a. General open-loop voltage mode track-and-hold

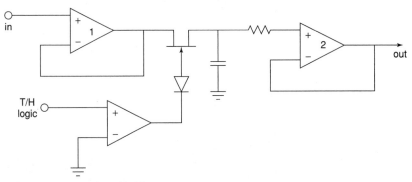

b. Op amp-based track-and-hold

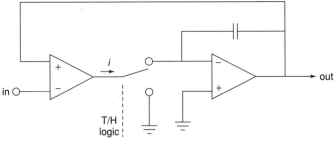

c. General closed-loop current-mode track-and-hold

The buffers exhibit high input impedance. When the switch is closed, op amp 1 directly feeds op amp 2, and therefore the output voltage equals the input voltage. Normally, the hold capacitor is relatively small and does not adversely affect the drive capability of op amp 1. A more detailed version of this circuit is shown in Figure 12.31b. Each op amp is a FET input type for minimum input current draw. The switch is a simple JFET, which is controlled by a comparator. When the T/H logic is high, the gate of the JFET is high, thus producing a low on resistance (i.e., a closed switch). When the comparator output goes low, the JFET is turned off creating a high impedance (i.e., open switch). In this state, op amp 2 is fed by the hold capacitor and buffers this potential to its output. For minimum droop, it is essential that the capacitor be a low leakage type and that op amp 2 have very low input bias current (e.g., FET input). The input resistor is used only to limit possible destructive discharge currents when the circuit is switched off. The diode positioned between the comparator and FET is used to prevent an excessively large positive comparator output potential from reaching the gate of the FET and possibly damaging it. (A comparator high will reverse-bias the series diode.) Figure 12.31c shows an alternate circuit using a closed-loop, current-mode approach. Note that the hold capacitor is now forming part of an integrator. While the open-loop form offers faster acquisition and settling times, the closed-loop system offers improved signal tracking. A closed-loop voltage-mode is also possible, but the current form generally offers fewer problems with leakage and switching transients. For general-purpose work, a variety of track-and-hold amplifiers are available in IC form from several manufacturers.

multiSIM COMPUTER SIMULATION

A simulation of a track-and-hold circuit similar to the one shown in Figure 12.31b is shown in Figure 12.32 (pp. 516–517). The 100 Hz input signal is being sampled at 2 kHz, or 20 times per cycle. Three waveforms of interest are shown in the Transient Analysis graph. The input waveform is seen in the center as the smoothly varying sine wave. Along the bottom of the graph, narrow spikes can be seen. This is the T/H logic signal. The high portion of the pulse (at 0 volts) is the track logic, and the wider low signal (dropping below the −1 volt limit of the graph) represents the hold logic. The stair-stepped sine wave is the output voltage. Note how the circuit acquires or tracks the input signal during the track pulse and then remains at that level when the hold logic is applied. It is during these hold times that the computations will be performed by the analog-to-digital converter.

Analog-to-Digital Conversion Techniques

Several different techniques have evolved for dealing with differing system requirements. These include *flash, staircase, successive approximation,* and *delta-sigma* forms.

Flash conversion is generally used for high-speed work, such as video applications. The circuits are usually low resolution. A flash converter is made up of a string of comparators as shown in Figure 12.33. The input signal is applied to all of the comparators simultaneously. Each comparator is also tied into a reference ladder. Effectively, there is one comparator for each quantization step. When a given signal is applied, a number of comparators toward the bottom of the string will produce a high level, as V_{in}

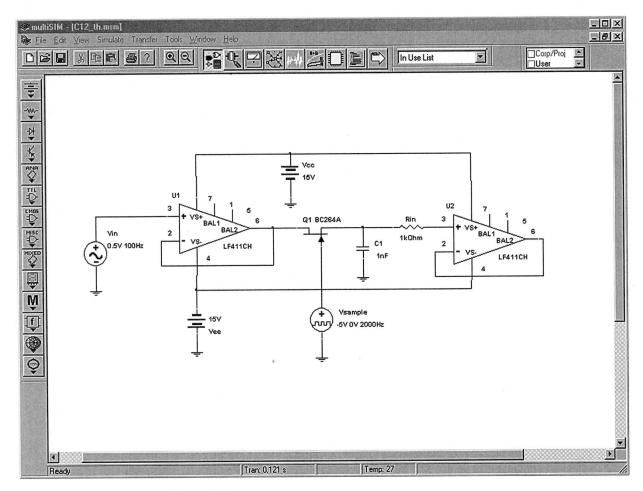

FIGURE 12.32a

Track and hold circuit in MultiSIM

will be greater than their references. Conversely, the comparators toward the top will indicate a low. The comparator at which the outputs shift from high to low indicates the step value closest to the input signal. The set of comparator outputs can be fed into a priority encoder that will turn this simple unweighted sequence into a normal binary word.

The only time delay involved in the conversion is that of logic propagation delay. Therefore, this conversion technique is quite useful for rapidly changing signals. Its downfall lies in the fact that one comparator is needed for each possible output step change. An 8-bit flash converter requires 256 comparators, whereas a 16-bit version requires 65,536. Obviously, this is rather excessive, and units in the 4- to 6-bit range are common. Although 6-bit resolution may appear at first to be too coarse for any application, it is actually quite useful for video displays.

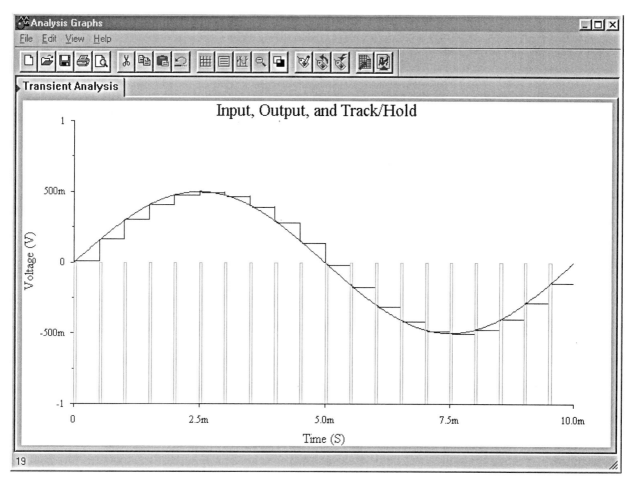

FIGURE 12.32b

MultiSIM waveforms for track-and-hold circuit

For high-resolution work, some other technique must be used. One possibility is shown in Figure 12.34. This is called a staircase converter. Its operation is fairly simple. When a conversion is first started, the output of the counter will be all zeros. This produces a DAC output of zero, and thus, the comparator output will be high. The next clock cycle will increment the up counter, causing the DAC output to increase by one step-level. This signal is compared against the input, and if the input is greater, the output will remain high. The clock will continue to increment the counter in this fashion until the DAC output just exceeds the input level. At this point the comparator output drops low, indicating that the conversion is complete. This signal can then be used to latch the output of the counter. This circuit gets its name from the fact that the waveform produced by the DAC looks like a staircase. The staircase technique can be used for very high resolution conversion, as long as an appropriate high resolution

518 ANALOG-TO-DIGITAL-TO-ANALOG CONVERSION

FIGURE 12.33

Flash converter

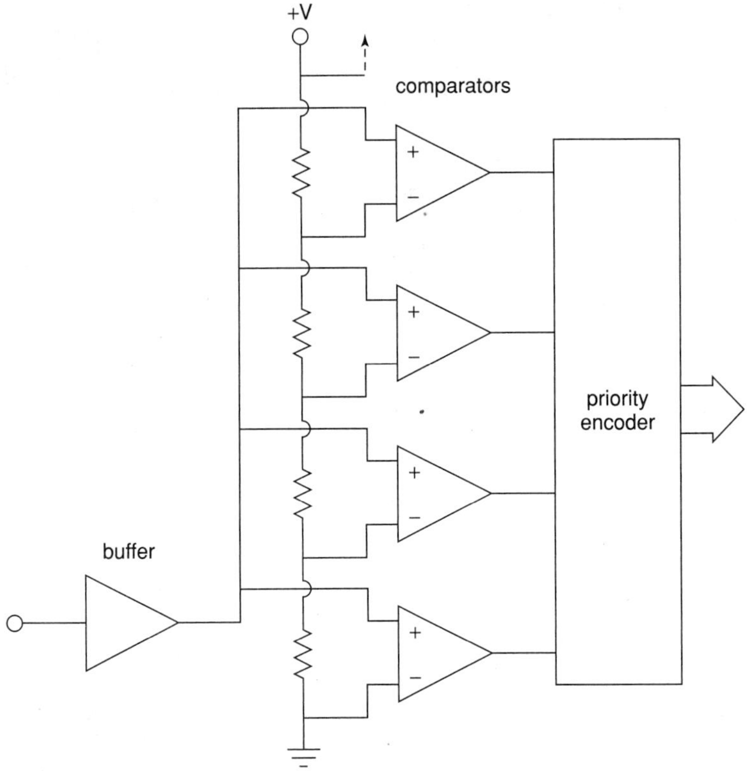

FIGURE 12.34

Staircase converter

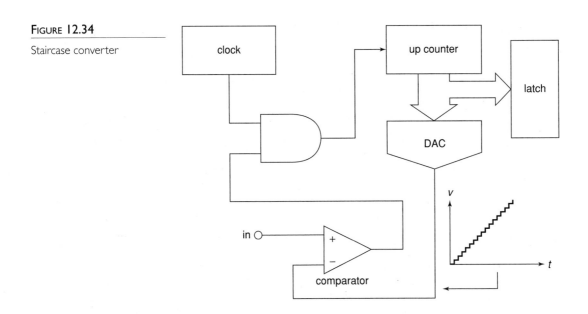

DAC is used. The major problem with this form is its very low conversion speed, due to the fact that there must be time to test every possible bit combination. Consequently, a 16-bit system requires 65,536 comparisons. Even if a fast 1 microsecond DAC is used, this would limit the sampling interval to nearly 66 milliseconds. This translates to a maximum input frequency before aliasing of only 7 Hz. Noting that the lowest frequency most humans can hear is about 20 Hz, this technique is hardly suitable for something like digital audio recording. In fact, for general-purpose work, the staircase system is avoided in favor of the successive approximation technique.

Successive approximation is a good general-purpose solution suitable for systems requiring resolutions in the 16-bit area. Instead of trying to convert the input signal at one instant like the flash converter, this technique creates the output bit by bit. Unlike the staircase converter, each individual level does not need to be tested. Instead, a binary search algorithm is used. You might think of it as making a series of guesses, each time getting a little closer to the result. Each guess results in a simple comparison. For an *n*-bit output, *n* comparisons need to be made.

A block diagram of a successive approximation converter is shown in Figure 12.35. Instead of a simple up counter, a circuit is used to implement the successive approximation algorithm. Here is how the circuit works. The first bit to be tested is the most significant bit. The DAC is fed a 1 with all of the remaining bits being set to 0 (i.e., 100000...). This word represents half of the maximum value capable by the system. The comparator output indicates whether V_{in} is greater or less than the resulting DAC output. If the comparator output is high, it means that the required digital output must be greater than the present word. If the comparator output is low, then the present digital word is too large, so the MSB is set to 0. At this point, the next most significant bit is tested by setting it to 1. The new word is fed to the DAC, and again, the comparator

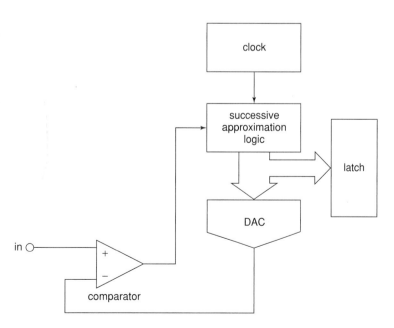

FIGURE 12.35

Successive-approximation converter

is used to determine whether or not the bit under test should remain at 1 or be reset. At this point the two most significant bits have been determined. The remaining bits are individually set to 1 and tested in a similar manner until the least significant bit is determined. In this way, a 16-bit system only requires 16 comparisons. With a 1 microsecond DAC, conversion takes only 16 microseconds. This translates to a sampling rate of 62.5 kHz; thus a maximum input frequency of 31.25 kHz is allowed. As you can see, this is far more efficient than the staircase technique.

For the highest resolutions combined with high sample rates, delta-sigma conversion techniques are popular. Basically, a low-resolution converter is run at a rate many times higher than the Nyquist frequency (perhaps 256 times higher). A special digital filter called a *decimator* converts the low-resolution high sample-rate data stream into a lower rate with higher resolution. This technique can achieve conversions in the 50 kHz range with 20-bit resolution. Design and analysis of delta-sigma modulators and digital filters is an advanced topic beyond the scope of this text. We will, however, look at a representative IC in the next section.

Analog-to-Digital Converter Integrated Circuits

In this section we shall examine a few specific ADC ICs, along with selected applications. The ICs include the ADC0844, an 8-bit microprocessor-compatible unit; the ADC12181, a very fast 12-bit converter utilizing pipelining techniques; and the CS5396, a 24-bit converter designed primarily for audio applications.

ADC0844 A block diagram of the ADC0844 is shown in Figure 12.36. Along with the built-in clock and 8-bit successive approximation register (SAR), the IC also includes an input multiplexer, tri-state latches and read, write, and chip select logic pins. This means that the ADC0844 is easily interfaced to a microprocessor data bus and may be used as a memory-mapped I/O device. This IC is a moderate-speed device, showing a typical conversion speed of 40 microseconds. The timing diagram for the ADC0844 is shown in Figure 12.37. A conversion is initiated by bringing both the chip-select and write-logic lines low. The falling edge of write resets the converter and its rising edge starts the actual conversion. After the conversion period, which is set internally, the digital data may be transferred to the output latches with the read-logic line. Note that the pulse repetition rate of the write-logic line sets the sampling rate. Therefore, a small program running on the host microprocessor that reads and writes to the ADC may be used to control the sampling rate and store the data for later use.

ADC12181 The ADC12181 is a 12-bit, 10 MHz converter with an internal sample-and-hold. Its data sheet is shown in Figure 12.38. In order to achieve its combination of high sample rate with relatively high resolution, the ADC12181 relies on a technique known as *pipelining*. In this particular chip, the pipeline consists of 15 stages. Each stage produces a one-bit digital signal and an error signal known as a *residual*. The residual is passed to the following stage where it is multiplied by a gain of 2, thus bringing it up to the former bit-weight. Depending on the size of an internally generated reference value, this reference may be subtracted from the residue. The newly resulting residue is passed to the next stage where the process is repeated. In essence, the

FIGURE 12.36

The ADC0841

Reprinted with permission of National Semiconductor Corporation

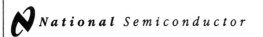

June 1999

ADC0844/ADC0848
8-Bit μP Compatible A/D Converters with Multiplexer Options

General Description

The ADC0844 and ADC0848 are CMOS 8-bit successive approximation A/D converters with versatile analog input multiplexers. The 4-channel or 8-channel multiplexers can be software configured for single-ended, differential or pseudo-differential modes of operation.

The differential mode provides low frequency input common mode rejection and allows offsetting the analog range of the converter. In addition, the A/D's reference can be adjusted enabling the conversion of reduced analog ranges with 8-bit resolution.

The A/Ds are designed to operate from the control bus of a wide variety of microprocessors. TRI-STATE® output latches that directly drive the data bus permit the A/Ds to be configured as memory locations or I/O devices to the microprocessor with no interface logic necessary.

Features

- Easy interface to all microprocessors
- Operates ratiometrically or with 5 V_{DC} voltage reference
- No zero or full-scale adjust required
- 4-channel or 8-channel multiplexer with address logic
- Internal clock
- 0V to 5V input range with single 5V power supply
- 0.3" standard width 20-pin or 24-pin DIP
- 28 Pin Molded Chip Carrier Package

Key Specifications

■ Resolution	8 Bits
■ Total Unadjusted Error	±½ LSB and ± 1 LSB
■ Single Supply	5 V_{DC}
■ Low Power	15 mW
■ Conversion Time	40 μs

Block and Connection Diagrams

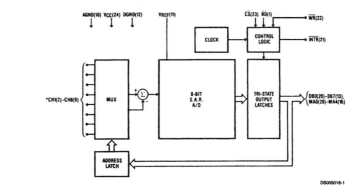

*ADC0848 shown in DIP Package CH5-CH8 not included on the ADC0844

TRI-STATE® is a registered trademark of National Semiconductor Corp.

© 1999 National Semiconductor Corporation   DS005016   www.national.com

FIGURE 12.37

Timing diagram for the ADC0841

Reprinted with permission of National Semiconductor Corporation

Timing Diagrams

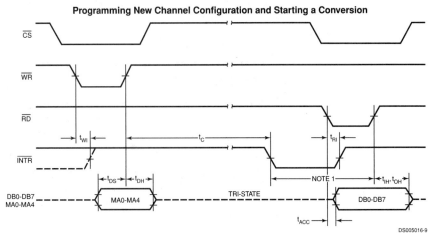

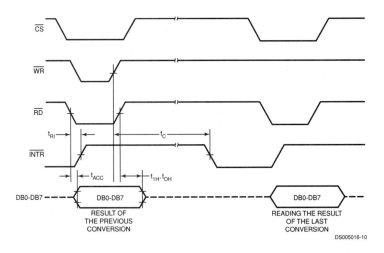

signal propagates down the pipeline in a fashion conceptually similar to the successive approximation technique.

The ADC12181 is used for applications requiring both high speed and high resolution. It also offers self-calibration, single +5 V power supply operation, and low power consumption.

CS5396 The CS5396 offers stereo 24-bit resolution with a maximum sampling rate of 96 kHz. This is ideal for a wide range of high-quality digital audio systems. The CS5396 data sheets are shown in Figure 12.39. This IC uses the delta-sigma conversion technique and achieves a dynamic range of 120 dB. Note that the noise plus distortion rating (THD+N) is greater than 105 dB. The CS5396 can operate in stand-alone mode

FIGURE 12.38a

ADC12181

Reprinted with permission of National Semiconductor Corporation

 National Semiconductor

January 1999

ADC12181
12-Bit, 10 MHz Self-Calibrating, Pipelined A/D Converter with Internal Sample & Hold

General Description

The ADC12181 is a monolithic CMOS analog-to-digital converter capable of converting analog input signals into 12-bit digital words at 10 megasamples per second (MSPS). The ADC12181 utilizes an innovative pipeline architecture to minimize die size and power consumption. Self-calibration and error correction maintain accuracy and performance over temperature.

The ADC12181 converter operates on a 5 V power supply and can digitize analog input signals in the range of 0 to 2 V. A signle convert clock controls the conversion operation. All digital I/O is TTL compatible.

The ADC12181 is designed to minimize external components necessary for the analog input interface. An internal sample-and-hold circuit samples the analog input and an internal amplifier buffers the reference voltage input.

The ADC12181 is available in the 32-lead TQFP package and is designed to operate over the extended commercial temperature range of −40°C to +85°C.

Features

- Single 5 V power supply
- Simple analog input interface
- Internal Sample-and-hold
- Internal Reference buffer amplifier
- Low power consumption

Key Specifications

- Resolution — 12 Bits
- Conversion Rate — 10 Msps (min)
- DNL — ±0.4 LSB (typ)
- SNR — 65 dB (typ)
- ENOB — 10.4 Bits (typ)
- Analog Input Range — 2 Vpp (min)
- Supply Voltage — +5V −5%
- Power Consumption, 10 MHz — 235 mW (typ)

Applications

- Image processing front end
- PC-based data acquisition
- Scanners
- Fax machines
- Waveform digitizer

Connection Diagram

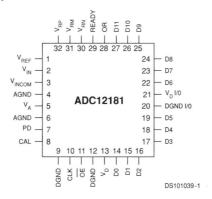

DS101039-1

FIGURE 12.38b

ADC12181 block diagram

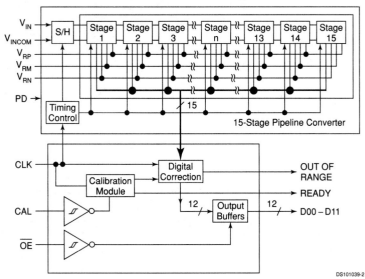

FIGURE 12.39a

CS5396

Copyright © Cirrus Logic, Inc.
1997 (All Rights Reserved)

**CS5396
CS5397**

120 dB, 96 kHz Audio A/D Converter

Features

- 24-Bit Conversion
- 120 dB Dynamic Range (A-Weighted)
- Low Noise and Distortion
  >105 dB THD + N
- Complete CMOS Stereo A/D System
  Delta-Sigma A/D Converters
  Digital Anti-Alias Filtering
  S/H Circuitry and Voltage Reference
- CS5396 - digital filter optimized for audio
- CS5397 - non-aliasing digital filter
- Adjustable System Sampling Rates
  including 32, 44.1, 48 & 96 kHz
- Differential Analog Architecture
- Linear Phase Digital Anti-Alias Filtering
- 10 Tap Programmable Psychoacoustic
  Noise Shaping Filter
- Single +5 V Power Supply

General Description

The CS5396 and CS5397 are complete analog-to-digital converters for stereo digital audio systems. They perform sampling, analog-to-digital conversion and anti-alias filtering, generating 24-bit values for both left and right inputs in serial form at sample rates up to 100 kHz per channel.

The CS5396/97 use a patented 7th-order, tri-level delta-sigma modulator followed by digital filtering and decimation, which removes the need for an external anti-alias filter. The ADCs use a differential architecture which provides excellent noise rejection.

The CS5396 has a linear phase filter optimized for audio applications with ± 0.005 dB passband ripple and >117 dB stopband rejection. The CS5397 has a non-aliasing filter response with ± 0.005 passband ripple and >117 dB stopband attenuation. Other features available in both the CS5396 and CS5397 are an optional low group delay filter and a unique psychoacoustic noise shaping filter which subjectively truncates the output to 16, 18 or 20 bits while 24-bit sound quality is preserved.

The CS5396/97 are targeted for the highest performance professional audio systems requiring wide dynamic range, negligible distortion and low noise.

ORDERING INFORMATION:
CS5396-KS   28-pin SOIC   −10 to 50°C
CS5397-KS   28-pin SOIC   −10 to 50°C
CDB5396/97   Evaluation Board

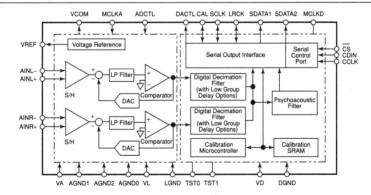

Preliminary Product Information | This document contains information for a new product.
Cirrus Logic reserves the right to modify this product without notice.

or with a micro controller. In the latter case, several additional features are available including data reduction to 16, 18, or 20 bits, mute, high-pass filter defeat, and a peak input signal level monitor for driving bar graph displays or pseudo VU meters.

As with any high-resolution converter, you must be particularly careful with the circuit layout or excessive noise may result. This IC has separate digital and analog common pins that must be connected together as close as possible to the package,

Figure 12.39b
CS5396 connection diagram

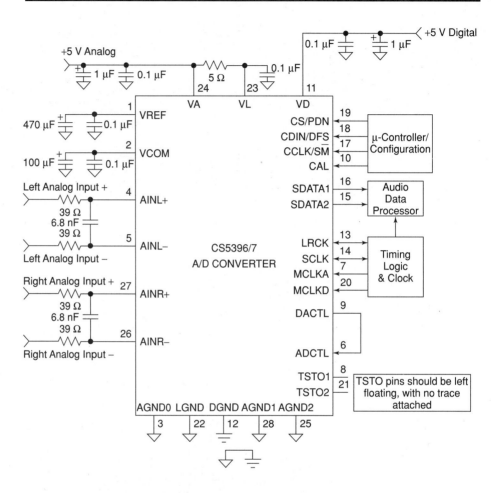

preferably via large ground planes passing under the IC. These ground planes should only be connected at this one spot. Also, attention must be paid to proper power supply bypassing. Many ICs in this category require separate analog and digital power supplies. Although the IC only requires 5 V, note that there are separate connections for analog and digital power supplies.

Applications of Analog-to-Digital Converter Integrated Circuits

Example 12-7

Perhaps the most straightforward application of analog-to-digital conversion is the acquisition of a signal. Once in the digital domain, the signal may be processed in a variety of ways. A

FIGURE 12.40

Basic digital storage oscilloscope

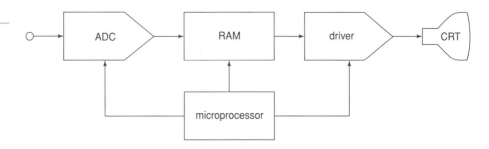

good example of this is the **D**igital **S**torage **O**scilloscope (DSO). A simplified block diagram of a DSO is shown in Figure 12.40. Because the final output of the system is a simple graph, it generally does not make sense to resolve the input beyond 8 bits (256 steps), and often even fewer bits may be used. Typically, high sampling rates are more important than fine resolution in this application. Consequently, 12- and 16-bit converters are not found here. Instead, lower-resolution converters capable of sampling at tens or hundreds of MHz are used. A commercial DSO is shown in Figure 12.41.

DSOs can usually be run in one of two modes: continuous, or single shot. In single-shot mode, the ADC acquires the signal and stores it in memory. This data can then be routed to the display circuits continuously in order to create a trace on the CRT. The signal is captured and stored in computer memory, thus it can be replayed virtually forever without a loss of clarity. This is very useful for catching quick, nonrepetitive transients. Once the signal is captured, it may be examined at leisure. For that matter, if some form of level-sensing logic is included, sampling can be initiated by specific transient events. This is known as *baby-sitting*. For example, you might suspect that a circuit you have designed occasionally emits an undesirable voltage spike. It is not practical for you to hook up the circuit and stare at the face of an oscilloscope, perhaps for hours, waiting for the system to misbehave. Instead, the DSO can be programmed to wait for the transient event before recording. In this way, you can leave the system, and when you return some time later, the spike will have been recorded and will be waiting your inspection. An extension of this concept is pretrigger recording. In this variation,

FIGURE 12.41

Commercial DSO
Copyright 2000, Tektronix, Inc. Reprinted with permission. All rights reserved

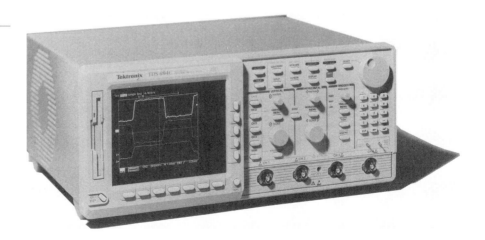

the DSO is constantly recording and "throwing away" data. When the transient spike finally occurs, the DSO has a snapshot of the events leading up to the spike, as well as the spike itself. This can provide very useful information in some applications. In either case, since the data is in digital form, it may be off-loaded to a computer for further analysis (if the DSO has facility for this). Indeed, many DSOs offer some interesting on-board analysis functions, including signal smoothing (noise reduction), and trace cursors which allow for easy delta time/delta voltage computation.

In continuous mode, a DSO appears to operate in much the same fashion as an ordinary analog oscilloscope. In this mode, the DSO samples the input signal and passes the data out to memory. From here the data is relayed to the CRT circuitry where a trace is produced. The trace is being constantly updated, so any change in the input will be quickly displayed. Trace updates may not actually be in real time, and thus, may not be as well suited to rapidly changing waveforms as are analog oscilloscopes.

Example 12-8

Another interesting application of the analog-to-digital converter is in the digital sampling music keyboard or drum computer (generally referred to as a *sampler*). The usage of a sampler is quite straightforward, actually. The idea is to mimic the sound of a given instrument (such as a trumpet or flute) from the keyboard. Before the advent of the sampler, this was done by properly setting the filters, amplifiers, and oscillators of a keyboard synthesizer. Although the resulting sounds were reasonably close to the desired instrument, they usually weren't close enough to fool the average listener. The sampler bypasses the problems of synthesizers by directly recording an instrument with an analog-to-digital converter. For example, a trumpet player might sound an *A* into a microphone that is connected to the sampler. This note is digitized and stored in RAM. Now, when the keyboard player hits the *A* key, the data is retrieved from RAM and fed to a DAC where it is reconstructed. The result is the exact same note that the trumpet player originally produced. By recording several different pitches from many different instruments, the keyboard player literally has the power of an entire orchestra at his fingertips. Of course, there is no limit to the sounds that might be sampled, and the keyboardist could just as easily use the sampler to "play" a collection of dog barks, door slams, and bird calls. A sampling drum computer is similar to a sampling keyboard, but replaces the standard musical keyboard with a series of buttons that allow the musician to create a programmed sequence of notes. This sequence can be played back at any time, and at virtually any tempo. In this manner, an entire percussion section may be simulated using digital recordings of real drums. A block diagram of a typical musical sampling system is shown in Figure 12.42.

Further enhancements to the digital sampling musical instrument include direct connections to personal computers. In this way, the computer can be used to analyze and augment the sound samples in new ways. For example, the personal computer may be used to create a graphic display of the waveform, or compute and display the spectral components of the sound using a popular technique called the *Fast Fourier Transform*. An example is shown in Figure 12.43. Finally, the computer can be programmed to "play" the sampler. In this way, even nonkeyboardists can take advantage of the inherent musical flexibility this system offers.

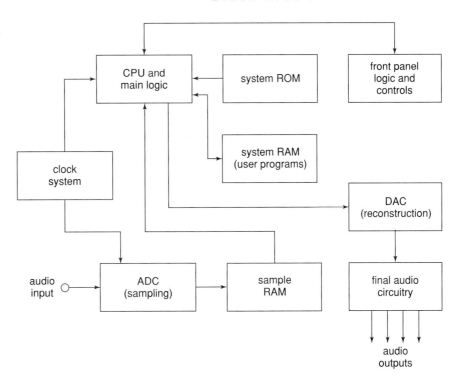

FIGURE 12.42

Block diagram of musical instrument sampling and playback device

12.6 EXTENDED TOPIC: DIGITAL SIGNAL PROCESSING

Much has been said in this chapter concerning the ability of personal computers to store and manipulate data in the digital domain. This process is called *digital signal processing*, or DSP. The sampled data is generally stored in RAM as a large array of values. Normally, the data is in integer form. This is because microprocessors are generally faster at performing integer calculations than floating point calculations. Also, AD converters produce integer values, so this is doubly convenient. As an example, if 8-bit resolution is used, each sample point will require one byte (8 bits). The byte is the fundamental unit of computer memory storage, and thus storage as a simple byte multiple (8 bits, 16 bits, 32 bits) is straightforward. The data may be stored in either unipolar (unsigned) or bipolar (signed) formats. For unsigned systems, the data value "0" means "most negative peak." In signed systems, a two's compliment form is normally used. Here, a data value of "0" means "0 V". All negative values have their most significant bit set (for an 8-bit representation, $-1 = 11111111$, $-2 = 11111110$, etc.). Input signals are often AC, thus signed representations are quite common.

In BASIC, the data array might be called data(), and individual elements are accessed by properly setting the array index. The first element is data(1), the second is data(2), and the Nth element is data(N). Although BASIC is useful, it is not as powerful as assembly, or languages such as C, which offer direct manipulation of memory addresses via pointers. In C, the starting, or base, address of the data might be given the name data_ptr. To access different elements in the array, an offset is added to the base. The address of the second element is data_ptr + 1, the third is

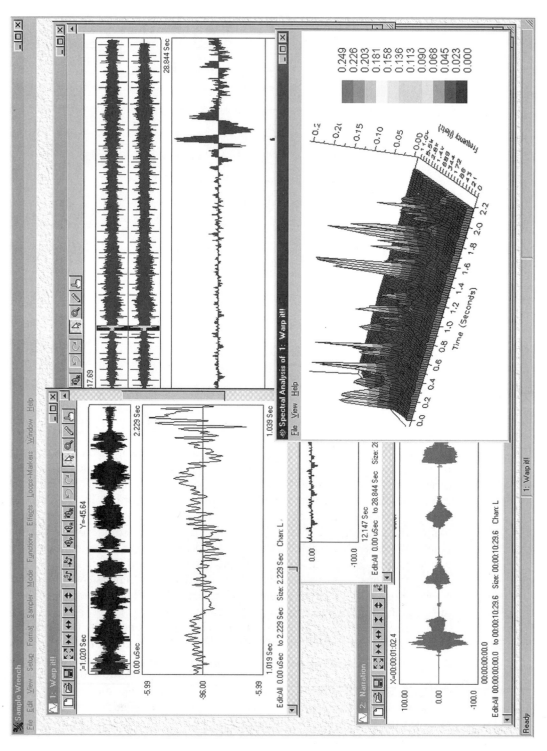

FIGURE 12.43
Commercial waveform display and editing software
Reprinted courtesy of dissidents

data_ptr + 2, and so on. Direct manipulation of pointers is generally faster for the microprocessor than array indexing. (To find the value stored at a given address, C uses the indirection operator, *. To assign the second data value to the variable x, you would say x = *(data_ptr + 1). The BASIC equivalent is x = data(2).)

No matter how the data is accessed, any mathematical function may be applied to the data elements. In the examples below, BASIC is used to illustrate the concepts. Only the processing portion of the code is shown. The array is called data() and is assumed to have TOTAL number of elements. Also, questions concerning integer versus floating point representation are bypassed in order to keep the examples as straightforward as possible. In the real world, problems such as execution speed, round-off error, and overflow cannot be ignored.

Let's start with something simple. Suppose we would like to simulate the operation of an inverting buffer. An inverting buffer simply multiplies the input signal by −1. Here is how we do this in the digital domain: each point is individually multiplied by −1.

```
10 For element = 1 to TOTAL
20 data(element) = -1 * data(element)
30 Next element
```

We can extend this concept a bit further by replacing "−1" with a variable called "gain." In this way, we can alter the signal amplitude digitally.

Another useful calculation is average value. For signed data, the average value is equal to the DC offset component.

```
10 sum = 0
20 For element = 1 to TOTAL
30 sum = sum + data(element)
40 Next element
50 average = sum/TOTAL
```

In order to add a DC offset, a constant is added to each data element in turn.

```
10 For element = 1 to TOTAL
20 data(element) = offset + data(element)
30 Next element
```

We might wish to combine two different waveforms into a third waveform. Simple addition is all that is required. (The example assumes that both arrays are the same size.)

```
10 For element = 1 to TOTAL
20 data3(element) = data1(element) + data2(element)
30 Next element
```

If the two source arrays are multiplied instead of added, signal modulation may be produced.

These examples are relatively simple. With proper programming, a wide variety of functions may be simulated including spectral analysis, translation to new sampling rates, and the realization of complex filter functions that are impractical to implement in the analog domain. The major drawback of DSP techniques is the inherent computing power required. Thus, real-time application of complex DSP principles is not possible without using very powerful computers. Specialized ICs have been designed to implement DSP subfunctions very quickly. A classic example of the breed is the Motorola 56000 DSP IC. These chips make real-time DSP practical. For example, an all-DSP graphic equalizer for home or professional audio use can be designed using the 56000. The spectral shaping is performed directly in the digital domain instead of using dedicated analog circuitry. The DSP ICs are programmable, thus they offer the distinct advantage of being able to produce different functions at different times by loading new programs. In this way, they are far more flexible than dedicated analog circuits.

Summary

In this chapter we have covered the basic concepts of analog-to-digital and digital-to-analog conversion. By placing a signal in the digital domain, a variety of new analysis, storage, and transmission techniques becomes available.

The process of transforming an analog signal into a digital representation is referred to as quantization, or more simply, as digitizing or sampling. One of the most popular methods used is PCM, pulse code modulation. In this scheme, the input signal is measured, or sampled, at a constant rate. Each sample point is represented as a digital word. The sequence of words describes the input signal and is used to recreate it, if necessary. The ultimate accuracy of the conversion is dependent on the resolution and sampling rate of the system. Resolution refers to the number of bits present in the digital word. An 8-bit word can represent 256 steps, whereas a 16-bit word is much finer, offering 65,536 discrete steps. With so many steps, round-off error is much less of a problem in 16-bit systems than in, say, 8- or 12-bit systems.

The minimum allowable sampling rate is twice the highest input frequency. In other words, at least two samples per cycle are required for the highest harmonic in the input signal. Another way of stating this is that no input frequency component can exceed the Nyquist frequency, which is defined as one-half of the sampling frequency. If the input exceeds this limit, alias distortion may occur. Aliases are nonharmonically related frequencies that are effectively created by improper sampling. In order to remove all possibility of alias distortion, special high-order low-pass filters, called antialias filters, are normally placed before the sampling circuitry.

Two popular methods of analog-to-digital conversion are the flash and successive approximation techniques. Flash converters are very fast, but require one comparator per output step, so they are not normally used where high resolution is required. Successive approximation takes longer than flash conversion, but can produce resolutions in excess of 16 bits. In either case, the conversion process is not instantaneous, and any fluctuation of the input signal during the conversion can produce errors. In order to alleviate this difficulty, special track-and-hold amplifiers are used between the antialias filters and the AD converter to create a nonvarying signal.

Once the signal has been digitized, it may be stored in RAM or some other media for future use, or directly analyzed. Often, a personal computer can prove to be very useful for waveform analysis.

To reconstruct the waveform, a digital-to-analog converter is used. In essence, this is usually little more than a weighted summing amplifier. Due to accuracy and construction constraints, an $R/2R$ ladder technique is often employed. In order to smooth out the resulting waveform and remove any remaining digital "glitches," the signal is passed through a reconstruction filter. This is a low-pass filter and is often called a smoothing filter. Generally, the process of digital-to-analog conversion is much faster than analog-to-digital conversion. Indeed, the successive approximation analog-to-digital scheme requires an internal digital-to-analog converter.

Review Questions

1. What is *PCM*?
2. Define the term *resolution*.
3. What is *quantization*?
4. What is an *alias*, and how is it produced? How is an alias avoided?
5. Define *Nyquist frequency*, and discuss the importance of this parameter.
6. Explain how a summing amplifier may be used to create a digital-to-analog converter.
7. Explain the difference between integral nonlinearity and differential nonlinearity.
8. What is a smoothing (reconstruction) filter?
9. What is the purpose of a track-and-hold amplifier?
10. Detail the differences between flash conversion and the successive approximation technique. Where would each type be used? What are their limitations?
11. Give several examples of possible DSP functions.

Problems

Analysis

1. Determine the number of quantization steps for a 10-bit system.
2. A 14-bit converter produces a maximum peak-to-peak output of 2.5 V. What is the step size?
3. Determine the dynamic range of the converters in Problems 1 and 2.
4. We wish to resolve a 1 V peak-to-peak signal to at least 1 mV. What is the minimum allowable number of bits in the converted data?
5. We wish to create analog signals using an arbitrary waveform generator. If we send out digital words at the rate of 50 kHz, what is the maximum allowable conversion speed for the DAC?
6. Assume that we are trying to digitize ultrasonic signals lying between 25 kHz and 45 kHz.
    a. What is the Nyquist frequency?
    b. What is the minimum acceptable sampling rate?
7. Determine the maximum allowable conversion time for the ADC of Problem 6.
8. Given a 14-bit ADC,
    a. Determine the number of comparators required for the flash technique.
    b. Determine the number of comparisons required if the successive approximation technique is used.
9. Assume that a single 16-bit ADC is connected to a personal computer. The sampling rate is 10 kHz. Determine the data rate in bytes per second.
10. Referring to Problem 9, if the personal computer has 350 k bytes of RAM available for data storage, how much time does this represent?
11. DAT (digital audio tape) recorders normally use a 16 bit representation with a sampling rate of 48 kHz. If the unit is used to record a performance of Stravinsky's "Rite of Spring" (35 minutes total), what is the required storage capacity in bytes?
12. If the data is transferred serially from the DAT of Problem 11 to a digital signal processing IC in real time, what is the width of each individual pulse?
13. A 12-bit 2-microsecond DAC is used as part of a discrete successive approximation analog-to-digital converter. Assuming that logic delays and signal settling times are negligible, determine:
    a. The minimum time allowable between sample points.
    b. The maximum input signal frequency without aliasing.
14. A 6-bit video DA converter produces a maximum output swing of approximately 1.25 V (unipolar). Determine the output voltage for the following digital input words.
    a. 000001
    b. 100000
    c. 111111
    d. 011101
15. A 10-bit instrumentation DAC produces an output of 16 mV with an input of 0000000100. Determine:
    a. The step size.
    b. The maximum output signal.

16. An 8-bit ADC produces a full scale output of 11111111 with a 2 V input signal. Determine the output word given the following inputs. (Assume that this converter rounds to the nearest output value and is unipolar.)
    a. 100 mV
    b. 10 μV
    c. 0 V
    d. 1.259 V

17. Assume that comparator/logic delays, amplifier settling times, and other factors require .4 μS total in a particular IC fabrication technique. If this technology is used to create AD converters, determine the maximum conversion time for the following 8-bit converters:
    a. Flash
    b. Successive approximation
    c. Staircase/Ramp type

Design

18. We wish to digitize human voice signals. The maximum input frequency is to be limited to 3 kHz and resolution to better than .5% of the maximum input value is required.
    a. Draw a block diagram of the complete system.
    b. Determine the minimum bit requirement.
    c. Determine the minimum sampling rate if the Nyquist rate is set to 25% greater than the theoretical minimum.
    d. Determine the antialias filter tuning frequency.
    e. Determine the preferred conversion technique.
    f. Determine which of the ICs presented in the chapter is best suited for this system.

19. We wish to design a system capable of digitizing complex signals with a spectrum ranging from DC to 400 kHz. Accuracy must be at least .2% of full scale.
    a. Draw a block diagram of the complete system.
    b. Determine the minimum bit requirement.
    c. Determine the minimum sampling rate if the Nyquist rate is set to 20% greater than the theoretical minimum.
    d. Determine the antialias filter tuning frequency.
    e. Determine the preferred conversion technique.
    f. Determine which of the ICs presented in the chapter is best suited for this system.

20. Write a computer algorithm that can be used to "flip" digital data back to front (i.e., play it backwards).

21. Write a computer algorithm that will determine the maximum peak value of the digital data.

22. Write a computer algorithm that will determine the RMS value of the digital data.

APPENDIX A

DATA SHEETS

A selection of data sheets follows for your convenience. For safety and reliability, be sure to obtain the most current release of product specifications before you use these (or any) products in an actual application. New products, and updated versions of existing ones, are released often.

Manufacturer's manuals and technical guides are usually inexpensive. Some are even free. Most companies offer product information on CD-ROM and many have searchable databases on their Web sites. Some Web sites that you may be interested in are:

Analog Devices	www.analog.com
Burr-Brown	www.burr-brown.com
Crystal Semiconductor	www.crystal.com
Harris Semiconductor	www.semi.harris.com
Linear Technology	www.linear-tech.com
National Semiconductor	www.national.com
Maxim Integrated Circuits	www.maxim-ic.com
Motorola Semiconductor Products Sector	www.mot-sps.com
Philips Semiconductors	www.philips.semiconductors.com
Texas Instruments	www.ti.com
THAT	www.thatcorp.com

The following data sheets for the LF411, LM386, and LM13700 are reprinted with permission of National Semiconductor Corporation. The data sheets for the μA741 and NE5533/5534 are reprinted with permission of Philips Semiconductors.

National Semiconductor

April 1998

LF411
Low Offset, Low Drift JFET Input Operational Amplifier

General Description

These devices are low cost, high speed, JFET input operational amplifiers with very low input offset voltage and guaranteed input offset voltage drift. They require low supply current yet maintain a large gain bandwidth product and fast slew rate. In addition, well matched high voltage JFET input devices provide very low input bias and offset currents. The LF411 is pin compatible with the standard LM741 allowing designers to immediately upgrade the overall performance of existing designs.

These amplifiers may be used in applications such as high speed integrators, fast D/A converters, sample and hold circuits and many other circuits requiring low input offset voltage and drift, low input bias current, high input impedance, high slew rate and wide bandwidth.

Features

- Internally trimmed offset voltage: 0.5 mV(max)
- Input offset voltage drift: 10 $\mu V/°C$(max)
- Low input bias current: 50 pA
- Low input noise current: 0.01 $pA/\sqrt{Hz}$
- Wide gain bandwidth: 3 MHz(min)
- High slew rate: 10V/μs(min)
- Low supply current: 1.8 mA
- High input impedance: $10^{12}\Omega$
- Low total harmonic distortion A_V=10,
  R_L=10k, V_O=20 Vp-p, BW=20 Hz–20 kHz: <0.02%
- Low 1/f noise corner: 50 Hz
- Fast settling time to 0.01%: 2 μs

Typical Connection

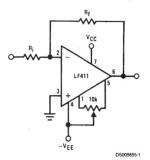

Connection Diagrams

Metal Can Package

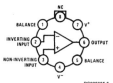

Note: Pin 4 connected to case.

Top View
Order Number LF411ACH
or LF411MH/883 (Note 1)
See NS Package Number H08A

Ordering Information

LF411XYZ

X indicates electrical grade
Y indicates temperature range
  "M" for military
  "C" for commercial
Z indicates package type
  "H" or "N"

Dual-In-Line Package

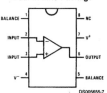

Top View
Order Number LF411ACN,
LF411CN or LF411MJ/883 (Note 1)
See NS Package Number N08E or J08A

BI-FET II™ is a trademark of National Semiconductor Corporation.

© 1999 National Semiconductor Corporation   DS005655   www.national.com

Simplified Schematic

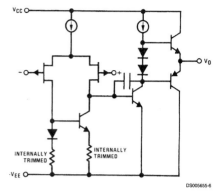

Note 1: Available per JM38510/11904

Absolute Maximum Ratings (Note 2)

If Military/Aerospace specified devices are required, please contact the National Semiconductor Sales Office/Distributors for availability and specifications.

	LF411A	LF411
Supply Voltage	±22V	±18V
Differential Input Voltage	±38V	±30V
Input Voltage Range (Note 3)	±19V	±15V
Output Short Circuit Duration	Continuous	Continuous
	H Package	N Package
Power Dissipation (Notes 4, 11)	670 mW	670 mW

	H Package	N Package
$T_j$max	150°C	115°C
θ_jA	162°C/W (Still Air)	120°C/W
	65°C/W (400 LF/min Air Flow)	
θ_jC	20°C/W	
Operating Temp. Range	(Note 5)	(Note 5)
Storage Temp. Range	$-65°C \leq T_A \leq 150°C$	$-65°C \leq T_A \leq 150°C$
Lead Temp. (Soldering, 10 sec.)	260°C	260°C
ESD Tolerance	Rating to be determined.	

DC Electrical Characteristics (Note 6)

Symbol	Parameter	Conditions		LF411A			LF411			Units
				Min	Typ	Max	Min	Typ	Max	
V_{OS}	Input Offset Voltage	R_S=10 kΩ, T_A=25°C			0.3	0.5		0.8	2.0	mV
$\Delta V_{OS}/\Delta T$	Average TC of Input Offset Voltage	R_S=10 kΩ (Note 7)			7	10		7	20 (Note 7)	μV/°C
I_{OS}	Input Offset Current	V_S=±15V (Notes 6, 8)	T_j=25°C		25	100		25	100	pA
			T_j=70°C			2			2	nA
			T_j=125°C			25			25	nA
I_B	Input Bias Current	V_S=±15V (Notes 6, 8)	T_j=25°C		50	200		50	200	pA
			T_j=70°C			4			4	nA
			T_j=125°C			50			50	nA
R_{IN}	Input Resistance	T_j=25°C			10^{12}			10^{12}		Ω
A_{VOL}	Large Signal Voltage Gain	V_S=±15V, V_O=±10V, R_L=2k, T_A=25°C		50	200		25	200		V/mV
		Over Temperature		25	200		15	200		V/mV
V_O	Output Voltage Swing	V_S=±15V, R_L=10k		±12	±13.5		±12	±13.5		V
V_{CM}	Input Common-Mode Voltage Range			±16	+19.5		±11	+14.5		V
					−16.5			−11.5		V
CMRR	Common-Mode Rejection Ratio	$R_S \leq$10k		80	100		70	100		dB
PSRR	Supply Voltage Rejection Ratio	(Note 9)		80	100		70	100		dB
I_S	Supply Current				1.8	2.8		1.8	3.4	mA

AC Electrical Characteristic (Note 6)

Symbol	Parameter	Conditions	LF411A			LF411			Units
			Min	Typ	Max	Min	Typ	Max	
SR	Slew Rate	V_S=±15V, T_A=25°C	10	15		8	15		V/μs
GBW	Gain-Bandwidth Product	V_S=±15V, T_A=25°C	3	4		2.7	4		MHz
e_n	Equivalent Input Noise Voltage	T_A=25°C, R_S=100Ω, f=1 kHz		25			25		nV/$\sqrt{Hz}$
i_n	Equivalent Input Noise Current	T_A=25°C, f=1 kHz		0.01			0.01		pA/$\sqrt{Hz}$

Note 2: "Absolute Maximum Ratings" indicate limits beyond which damage to the device may occur. Operating Ratings indicate conditions for which the device is functional, but do not guarantee specific performance limits.

Note 3: Unless otherwise specified the absolute maximum negative input voltage is equal to the negative power supply voltage.

AC Electrical Characteristic (Note 6) (Continued)

Note 4: For operating at elevated temperature, these devices must be derated based on a thermal resistance of θ_{jA}.

Note 5: These devices are available in both the commercial temperature range $0°C \leq T_A \leq 70°C$ and the military temperature range $-55°C \leq T_A \leq 125°C$. The temperature range is designated by the position just before the package type in the device number. A "C" indicates the commercial temperature range and an "M" indicates the military temperature range. The military temperature range is available in "H" package only.

Note 6: Unless otherwise specified, the specifications apply over the full temperature range and for $V_S = \pm 20V$ for the LF411A and for $V_S = \pm 15V$ for the LF411. V_{OS}, I_B, and I_{OS} are measured at $V_{CM} = 0$.

Note 7: The LF411A is 100% tested to this specification. The LF411 is sample tested to insure at least 90% of the units meet this specification.

Note 8: The input bias currents are junction leakage currents which approximately double for every 10°C increase in the junction temperature, T_j. Due to limited production test time, the input bias currents measured are correlated to junction temperature. In normal operation the junction temperature rises above the ambient temperature as a result of internal power dissipation, P_D. $T_j = T_A + \theta_{jA} P_D$ where θ_{jA} is the thermal resistance from junction to ambient. Use of a heat sink is recommended if input bias current is to be kept to a minimum.

Note 9: Supply voltage rejection ratio is measured for both supply magnitudes increasing or decreasing simultaneously in accordance with common practice, from $\pm 15V$ to $\pm 5V$ for the LF411 and from $\pm 20V$ to $\pm 5V$ for the LF411A.

Note 10: RETS 411X for LF411MH and LF411MJ military specifications.

Note 11: Max. Power Dissipation is defined by the package characteristics. Operating the part near the Max. Power Dissipation may cause the part to operate outside guaranteed limits.

Typical Performance Characteristics

Input Bias Current

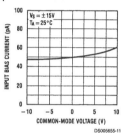

Input Bias Current

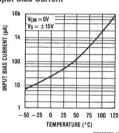

Supply Current

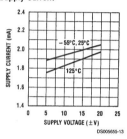

Positive Common-Mode Input Voltage Limit

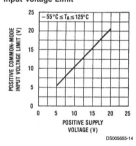

Negative Common-Mode Input Voltage Limit

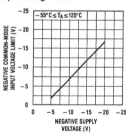

Positive Current Limit

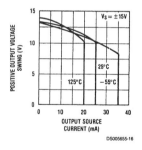

Typical Performance Characteristics (Continued)

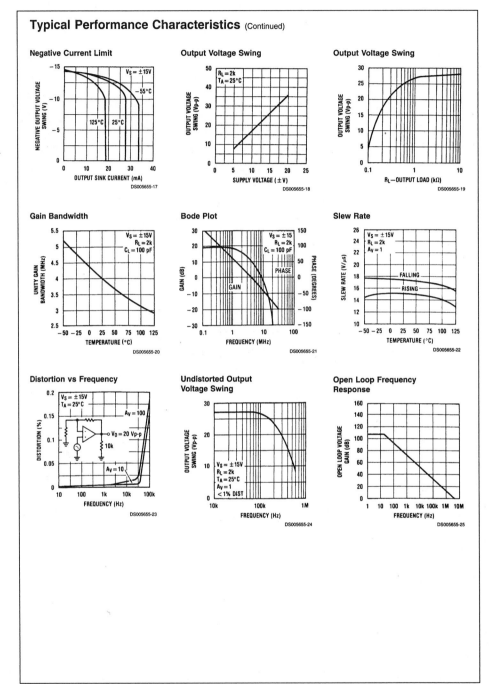

Typical Performance Characteristics (Continued)

Common-Mode Rejection Ratio

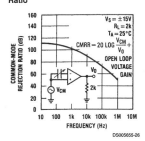

Power Supply Rejection Ratio

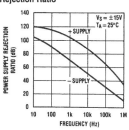

Equivalent Input Noise Voltage

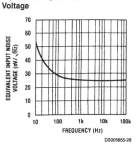

Open Loop Voltage Gain

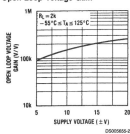

Output Impedance

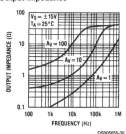

Inverter Settling Time

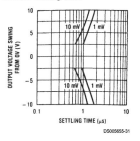

Pulse Response $R_L = 2\ k\Omega$, $C_L\ 10\ pF$

Small Signal Inverting

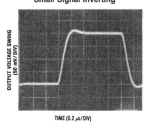

Small Signal Non-Inverting

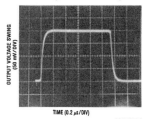

National Semiconductor

September 1997

LM386
Low Voltage Audio Power Amplifier

General Description

The LM386 is a power amplifier designed for use in low voltage consumer applications. The gain is internally set to 20 to keep external part count low, but the addition of an external resistor and capacitor between pins 1 and 8 will increase the gain to any value up to 200.

The inputs are ground referenced while the output is automatically biased to one half the supply voltage. The quiescent power drain is only 24 milliwatts when operating from a 6 volt supply, making the LM386 ideal for battery operation.

Features

- Battery operation
- Minimum external parts
- Wide supply voltage range: 4V–12V or 5V–18V
- Low quiescent current drain: 4 mA
- Voltage gains from 20 to 200
- Ground referenced input
- Self-centering output quiescent voltage
- Low distortion
- Available in 8 pin MSOP package

Applications

- AM-FM radio amplifiers
- Portable tape player amplifiers
- Intercoms
- TV sound systems
- Line drivers
- Ultrasonic drivers
- Small servo drivers
- Power converters

Equivalent Schematic and Connection Diagrams

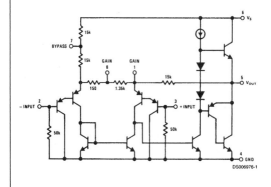

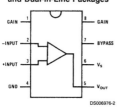

Small Outline,
Molded Mini Small Outline,
and Dual-In-Line Packages

Top View
Order Number LM386M-1,
LM386MM-1, LM386N-1,
LM386N-3 or LM386N-4
See NS Package Number
M08A, MUA08A or N08E

© 1999 National Semiconductor Corporation   DS006976   www.national.com

Absolute Maximum Ratings (Note 2)

If Military/Aerospace specified devices are required, please contact the National Semiconductor Sales Office/Distributors for availability and specifications.

Supply Voltage	
(LM386N-1, -3, LM386M-1)	15V
Supply Voltage (LM386N-4)	22V
Package Dissipation (Note 3)	
(LM386N)	1.25W
(LM386M)	0.73W
(LM386MM-1)	0.595W
Input Voltage	±0.4V
Storage Temperature	−65°C to +150°C
Operating Temperature	0°C to +70°C
Junction Temperature	+150°C
Soldering Information	
Dual-In-Line Package	
Soldering (10 sec)	+260°C
Small Outline Package	
(SOIC and MSOP)	
Vapor Phase (60 sec)	+215°C
Infrared (15 sec)	+220°C

See AN-450 "Surface Mounting Methods and Their Effect on Product Reliability" for other methods of soldering surface mount devices.

Thermal Resistance	
θ_{JC} (DIP)	37°C/W
θ_{JA} (DIP)	107°C/W
θ_{JC} (SO Package)	35°C/W
θ_{JA} (SO Package)	172°C/W
θ_{JA} (MSOP)	210°C/W
θ_{JC} (MSOP)	56°C/W

Electrical Characteristics (Notes 1, 2)

$T_A = 25°C$

Parameter	Conditions	Min	Typ	Max	Units
Operating Supply Voltage (V_S)					
LM386N-1, -3, LM386M-1, LM386MM-1		4		12	V
LM386N-4		5		18	V
Quiescent Current (I_Q)	$V_S = 6V$, $V_{IN} = 0$		4	8	mA
Output Power (P_{OUT})					
LM386N-1, LM386M-1, LM386MM-1	$V_S = 6V$, $R_L = 8\Omega$, THD = 10%	250	325		mW
LM386N-3	$V_S = 9V$, $R_L = 8\Omega$, THD = 10%	500	700		mW
LM386N-4	$V_S = 16V$, $R_L = 32\Omega$, THD = 10%	700	1000		mW
Voltage Gain (A_V)	$V_S = 6V$, $f = 1$ kHz		26		dB
	10 µF from Pin 1 to 8		46		dB
Bandwidth (BW)	$V_S = 6V$, Pins 1 and 8 Open		300		kHz
Total Harmonic Distortion (THD)	$V_S = 6V$, $R_L = 8\Omega$, $P_{OUT} = 125$ mW		0.2		%
	$f = 1$ kHz, Pins 1 and 8 Open				
Power Supply Rejection Ratio (PSRR)	$V_S = 6V$, $f = 1$ kHz, $C_{BYPASS} = 10$ µF		50		dB
	Pins 1 and 8 Open, Referred to Output				
Input Resistance (R_{IN})			50		kΩ
Input Bias Current (I_{BIAS})	$V_S = 6V$, Pins 2 and 3 Open		250		nA

Note 1: All voltages are measured with respect to the ground pin, unless otherwise specified.

Note 2: Absolute Maximum Ratings indicate limits beyond which damage to the device may occur. Operating Ratings indicate conditions for which the device is functional, but do not guarantee specific performance limits. Electrical Characteristics state DC and AC electrical specifications under particular test conditions which guarantee specific performance limits. This assumes that the device is within the Operating Ratings. Specifications are not guaranteed for parameters where no limit is given, however, the typical value is a good indication of device performance.

Note 3: For operation in ambient temperatures above 25°C, the device must be derated based on a 150°C maximum junction temperature and 1) a thermal resistance of 80°C/W junction to ambient for the dual-in-line package and 2) a thermal resistance of 170°C/W for the small outline package.

Application Hints

GAIN CONTROL

To make the LM386 a more versatile amplifier, two pins (1 and 8) are provided for gain control. With pins 1 and 8 open the 1.35 kΩ resistor sets the gain at 20 (26 dB). If a capacitor is put from pin 1 to 8, bypassing the 1.35 kΩ resistor, the gain will go up to 200 (46 dB). If a resistor is placed in series with the capacitor, the gain can be set to any value from 20 to 200. Gain control can also be done by capacitively coupling a resistor (or FET) from pin 1 to ground.

Additional external components can be placed in parallel with the internal feedback resistors to tailor the gain and frequency response for individual applications. For example, we can compensate poor speaker bass response by frequency shaping the feedback path. This is done with a series RC from pin 1 to 5 (paralleling the internal 15 kΩ resistor). For 6 dB effective bass boost: R ≅ 15 kΩ, the lowest value for good stable operation is R = 10 kΩ if pin 8 is open. If pins 1 and 8 are bypassed then R as low as 2 kΩ can be used. This restriction is because the amplifier is only compensated for closed-loop gains greater than 9.

INPUT BIASING

The schematic shows that both inputs are biased to ground with a 50 kΩ resistor. The base current of the input transistors is about 250 nA, so the inputs are at about 12.5 mV when left open. If the dc source resistance driving the LM386 is higher than 250 kΩ it will contribute very little additional offset (about 2.5 mV at the input, 50 mV at the output). If the dc source resistance is less than 10 kΩ, then shorting the unused input to ground will keep the offset low (about 2.5 mV at the input, 50 mV at the output). For dc source resistances between these values we can eliminate excess offset by putting a resistor from the unused input to ground, equal in value to the dc source resistance. Of course all offset problems are eliminated if the input is capacitively coupled.

When using the LM386 with higher gains (bypassing the 1.35 kΩ resistor between pins 1 and 8) it is necessary to bypass the unused input, preventing degradation of gain and possible instabilities. This is done with a 0.1 μF capacitor or a short to ground depending on the dc source resistance on the driven input.

Typical Performance Characteristics

Quiescent Supply Current vs Supply Voltage

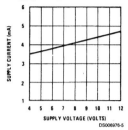

Power Supply Rejection Ratio (Referred to the Output) vs Frequency

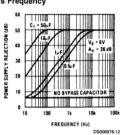

Peak-to-Peak Output Voltage Swing vs Supply Voltage

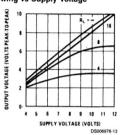

Voltage Gain vs Frequency

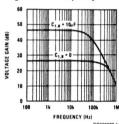

Distortion vs Frequency

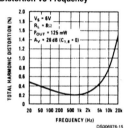

Distortion vs Output Power

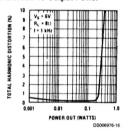

Device Dissipation vs Output Power — 4Ω Load

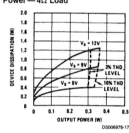

Device Dissipation vs Output Power — 8Ω Load

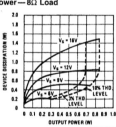

Device Dissipation vs Output Power — 16Ω Load

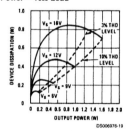

Typical Applications

Amplifier with Gain = 20 Minimum Parts

Amplifier with Gain = 200

Amplifier with Gain = 50

Low Distortion Power Wienbridge Oscillator

Amplifier with Bass Boost

Square Wave Oscillator

May 1998

LM13700/LM13700A
Dual Operational Transconductance Amplifiers with Linearizing Diodes and Buffers

General Description

The LM13700 series consists of two current controlled transconductance amplifiers, each with differential inputs and a push-pull output. The two amplifiers share common supplies but otherwise operate independently. Linearizing diodes are provided at the inputs to reduce distortion and allow higher input levels. The result is a 10 dB signal-to-noise improvement referenced to 0.5 percent THD. High impedance buffers are provided which are especially designed to complement the dynamic range of the amplifiers. The output buffers of the LM13700 differ from those of the LM13600 in that their input bias currents (and hence their output DC levels) are independent of I_{ABC}. This may result in performance superior to that of the LM13600 in audio applications.

Features

- g_m adjustable over 6 decades
- Excellent g_m linearity
- Excellent matching between amplifiers
- Linearizing diodes
- High impedance buffers
- High output signal-to-noise ratio

Applications

- Current-controlled amplifiers
- Current-controlled impedances
- Current-controlled filters
- Current-controlled oscillators
- Multiplexers
- Timers
- Sample-and-hold circuits

Connection Diagram

Dual-In-Line and Small Outline Packages

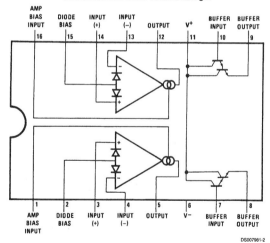

Top View
Order Number LM13700M, LM13700N or LM13700AN
See NS Package Number M16A or N16A

© 1999 National Semiconductor Corporation   DS007981

Absolute Maximum Ratings (Note 1)

If Military/Aerospace specified devices are required, please contact the National Semiconductor Sales Office/Distributors for availability and specifications.

Supply Voltage (Note 2)	
LM13700	36 V_{DC} or ±18V
LM13700A	44 V_{DC} or ±22V
Power Dissipation (Note 3) T_A = 25°C	
LM13700N, LM13700AN	570 mW
Differential Input Voltage	±5V
Diode Bias Current (I_D)	2 mA
Amplifier Bias Current (I_{ABC})	2 mA
Output Short Circuit Duration	Continuous
Buffer Output Current (Note 4)	20 mA
Operating Temperature Range	
LM13700N, LM13700AN	0°C to +70°C
DC Input Voltage	+V_S to −V_S
Storage Temperature Range	−65°C to +150°C
Soldering Information	
Dual-In-Line Package	
Soldering (10 sec.)	260°C
Small Outline Package	
Vapor Phase (60 sec.)	215°C
Infrared (15 sec.)	220°C

See AN-450 "Surface Mounting Methods and Their Effect on Product Reliability" for other methods of soldering surface mount devices.

Electrical Characteristics (Note 5)

Parameter	Conditions	LM13700 Min	LM13700 Typ	LM13700 Max	LM13700A Min	LM13700A Typ	LM13700A Max	Units
Input Offset Voltage (V_{OS})			0.4	4		0.4	1	
	Over Specified Temperature Range			2				mV
	I_{ABC} = 5 µA		0.3	4		0.3	1	
V_{OS} Including Diodes	Diode Bias Current (I_D) = 500 µA		0.5	5		0.5	2	mV
Input Offset Change	5 µA ≤ I_{ABC} ≤ 500 µA		0.1	3		0.1	1	mV
Input Offset Current			0.1	0.6		0.1	0.6	µA
Input Bias Current			0.4	5		0.4	5	µA
	Over Specified Temperature Range		1	8		1	7	
Forward Transconductance (g_m)		6700	9600	13000	7700	9600	12000	µmho
	Over Specified Temperature Range	5400			4000			
g_m Tracking			0.3			0.3		dB
Peak Output Current	R_L = 0, I_{ABC} = 5 µA		5		3	5	7	
	R_L = 0, I_{ABC} = 500 µA	350	500	650	350	500	650	µA
	R_L = 0, Over Specified Temp Range	300			300			
Peak Output Voltage								
Positive	R_L = ∞, 5 µA ≤ I_{ABC} ≤ 500 µA	+12	+14.2		+12	+14.2		V
Negative	R_L = ∞, 5 µA ≤ I_{ABC} ≤ 500 µA	−12	−14.4		−12	−14.4		V
Supply Current	I_{ABC} = 500 µA, Both Channels		2.6			2.6		mA
V_{OS} Sensitivity								
Positive	$\Delta V_{OS}/\Delta V^+$		20	150		20	150	µV/V
Negative	$\Delta V_{OS}/\Delta V^-$		20	150		20	150	µV/V
CMRR		80	110		80	110		dB
Common Mode Range		±12	±13.5		±12	±13.5		V
Crosstalk	Referred to Input (Note 6) 20 Hz < f < 20 kHz		100			100		dB
Differential Input Current	I_{ABC} = 0, Input = ±4V		0.02	100		0.02	10	nA
Leakage Current	I_{ABC} = 0 (Refer to Test Circuit)		0.2	100		0.2	5	nA
Input Resistance		10	26		10	26		kΩ
Open Loop Bandwidth			2			2		MHz
Slew Rate	Unity Gain Compensated		50			50		V/µs
Buffer Input Current	(Note 6)		0.5	2		0.5	2	µA
Peak Buffer Output Voltage	(Note 6)		10			10		V

Note 1: "Absolute Maximum Ratings" indicate limits beyond which damage to the device may occur. Operating Ratings indicate conditions for which the device is functional, but do not guarantee specific performance limits.

Note 2: For selections to a supply voltage above ±22V, contact factory.

Electrical Characteristics (Note 5) (Continued)

Note 3: For operation at ambient temperatures above 25°C, the device must be derated based on a 150°C maximum junction temperature and a thermal resistance, junction to ambient, as follows: LM13700N, 90°C/W; LM13700M, 110°C/W.

Note 4: Buffer output current should be limited so as to not exceed package dissipation.

Note 5: These specifications apply for $V_S = \pm 15V$, $T_A = 25°C$, amplifier bias current (I_{ABC}) = 500 µA, pins 2 and 15 open unless otherwise specified. The inputs to the buffers are grounded and outputs are open.

Note 6: These specifications apply for $V_S = \pm 15V$, I_{ABC} = 500 µA, R_{OUT} = 5 kΩ connected from the buffer output to $-V_S$ and the input of the buffer is connected to the transconductance amplifier output.

Schematic Diagram

One Operational Transconductance Amplifier

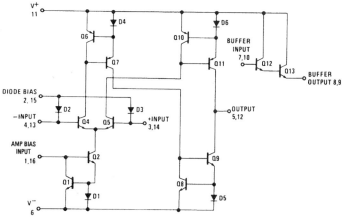

Typical Performance Characteristics

Input Offset Voltage

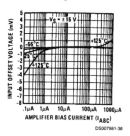

Input Offset Current

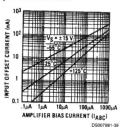

Input Bias Current

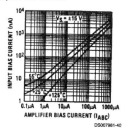

Typical Performance Characteristics (Continued)

Peak Output Current

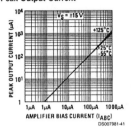

Peak Output Voltage and Common Mode Range

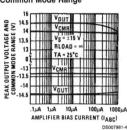

Leakage Current

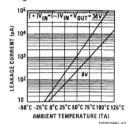

Input Leakage

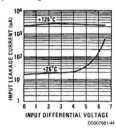

Transconductance

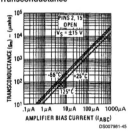

Input Resistance

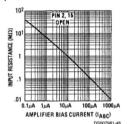

Amplifier Bias Voltage vs Amplifier Bias Current

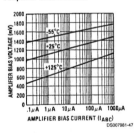

Input and Output Capacitance

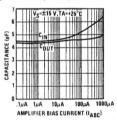

Output Resistance

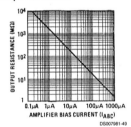

Typical Performance Characteristics (Continued)

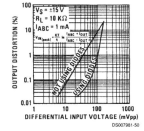

Distortion vs Differential Input Voltage

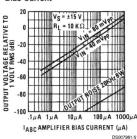

Voltage vs Amplifier Bias Current

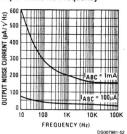

Output Noise vs Frequency

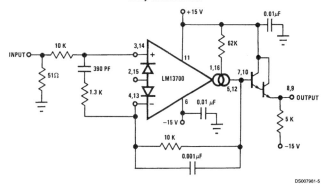

Unity Gain Follower

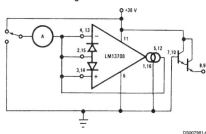

Leakage Current Test Circuit

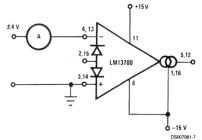

Differential Input Current Test Circuit

Circuit Description

The differential transistor pair Q_4 and Q_5 form a transconductance stage in that the ratio of their collector currents is defined by the differential input voltage according to the transfer function:

$$V_{IN} = \frac{kT}{q} \ln \frac{I_5}{I_4} \quad (1)$$

where V_{IN} is the differential input voltage, kT/q is approximately 26 mV at 25°C and I_5 and I_4 are the collector currents of transistors Q_5 and Q_4 respectively. With the exception of

Philips Semiconductors

Product specification

General purpose operational amplifier

µA741/µA741C/SA741C

DESCRIPTION
The µA741 is a high performance operational amplifier with high open-loop gain, internal compensation, high common mode range and exceptional temperature stability. The µA741 is short-circuit-protected and allows for nulling of offset voltage.

FEATURES
- Internal frequency compensation
- Short circuit protection
- Excellent temperature stability
- High input voltage range

PIN CONFIGURATION

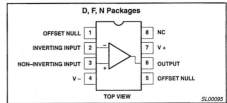

Figure 1. Pin Configuration

ORDERING INFORMATION

DESCRIPTION	TEMPERATURE RANGE	ORDER CODE	DWG #
8-Pin Plastic Dual In-Line Package (DIP)	-55°C to +125°C	µA741N	SOT97-1
8-Pin Plastic Dual In-Line Package (DIP)	0 to +70°C	µA741CN	SOT97-1
8-Pin Plastic Dual In-Line Package (DIP)	-40°C to +85°C	SA741CN	SOT97-1
8-Pin Ceramic Dual In-Line Package (CERDIP)	-55°C to +125°C	µA741F	0580A
8-Pin Ceramic Dual In-Line Package (CERDIP)	0 to +70°C	µA741CF	0580A
8-Pin Small Outline (SO) Package	0 to +70°C	µA741CD	SOT96-1

ABSOLUTE MAXIMUM RATINGS

SYMBOL	PARAMETER	RATING	UNIT
V_S	Supply voltage		
	µA741C	±18	V
	µA741	±22	V
P_D	Internal power dissipation		
	D package	780	mW
	N package	1170	mW
	F package	800	mW
V_{IN}	Differential input voltage	±30	V
V_{IN}	Input voltage[1]	±15	V
I_{SC}	Output short-circuit duration	Continuous	
T_A	Operating temperature range		
	µA741C	0 to +70	°C
	SA741C	-40 to +85	°C
	µA741	-55 to +125	°C
T_{STG}	Storage temperature range	-65 to +150	°C
T_{SOLD}	Lead soldering temperature (10sec max)	300	°C

NOTES:
1. For supply voltages less than ±15V, the absolute maximum input voltage is equal to the supply voltage.

General purpose operational amplifier

µA741/µA741C/SA741C

DC ELECTRICAL CHARACTERISTICS
$T_A = 25°C$, $V_S = \pm15V$, unless otherwise specified.

SYMBOL	PARAMETER	TEST CONDITIONS	µA741 Min	µA741 Typ	µA741 Max	µA741C Min	µA741C Typ	µA741C Max	UNIT
V_{OS}	Offset voltage	$R_S=10k\Omega$		1.0	5.0		2.0	6.0	mV
		$R_S=10k\Omega$, over temp.		1.0	6.0			7.5	mV
$\Delta V_{OS}/\Delta T$				10			10		µV/°C
I_{OS}	Offset current			20	200		20	200	nA
		Over temp.						300	nA
		$T_A=+125°C$		7.0	200				nA
		$T_A=-55°C$		20	500				nA
$\Delta I_{OS}/\Delta T$				200			200		pA/°C
I_{BIAS}	Input bias current			80	500		80	500	nA
		Over temp.						800	nA
		$T_A=+125°C$		30	500				nA
		$T_A=-55°C$		300	1500				nA
$\Delta I_B/\Delta T$				1			1		nA/°C
V_{OUT}	Output voltage swing	$R_L=10k\Omega$	±12	±14		±12	±14		V
		$R_L=2k\Omega$, over temp.	±10	±13		±10	±13		V
A_{VOL}	Large-signal voltage gain	$R_L=2k\Omega$, $V_O=\pm10V$	50	200		20	200		V/mV
		$R_L=2k\Omega$, $V_O=\pm10V$, over temp.	25			15			V/mV
	Offset voltage adjustment range			±30			±30		mV
PSRR	Supply voltage rejection ratio	$R_S \leq 10k\Omega$					10	150	µV/V
		$R_S \leq 10k\Omega$, over temp.		10	150				µV/V
CMRR	Common-mode rejection ratio					70	90		dB
		Over temp.	70	90					dB
I_{CC}	Supply current			1.4	2.8		1.4	2.8	mA
		$T_A=+125°C$		1.5	2.5				mA
		$T_A=-55°C$		2.0	3.3				mA
V_{IN}	Input voltage range	(µA741, over temp.)	±12	±13		±12	±13		V
R_{IN}	Input resistance		0.3	2.0		0.3	2.0		MΩ
P_D	Power consumption			50	85		50	85	mW
		$T_A=+125°C$		45	75				mW
		$T_A=-55°C$		45	100				mW
R_{OUT}	Output resistance			75			75		Ω
I_{SC}	Output short-circuit current		10	25	60	10	25	60	mA

Philips Semiconductors　　　　　　　　　　　　　　　　　　　　　　　　　　Product specification

General purpose operational amplifier　　　　μA741/μA741C/SA741C

DC ELECTRICAL CHARACTERISTICS
$T_A = 25°C$, $V_S = ±15V$, unless otherwise specified.

SYMBOL	PARAMETER	TEST CONDITIONS	SA741C Min	SA741C Typ	SA741C Max	UNIT
V_{OS}	Offset voltage	$R_S=10kΩ$		2.0	6.0	mV
		$R_S=10kΩ$, over temp.			7.5	mV
$ΔV_{OS}/ΔT$				10		μV/°C
I_{OS}	Offset current			20	200	nA
		Over temp.			500	nA
$ΔI_{OS}/ΔT$				200		pA/°C
I_{BIAS}	Input bias current			80	500	nA
		Over temp.			1500	nA
$ΔI_B/ΔT$				1		nA/°C
V_{OUT}	Output voltage swing	$R_L=10kΩ$	±12	±14		V
		$R_L=2kΩ$, over temp.	±10	±13		V
A_{VOL}	Large-signal voltage gain	$R_L=2kΩ$, $V_O=±10V$	20	200		V/mV
		$R_L=2kΩ$, $V_O=±10V$, over temp.	15			V/mV
	Offset voltage adjustment range			±30		mV
PSRR	Supply voltage rejection ratio	$R_S≤10kΩ$		10	150	μV/V
CMRR	Common mode rejection ration		70	90		dB
V_{IN}	Input voltage range	Over temp.	±12	±13		V
R_{IN}	Input resistance		0.3	2.0		MΩ
P_d	Power consumption			50	85	mW
R_{OUT}	Output resistance			75		Ω
I_{SC}	Output short-circuit current			25		mA

AC ELECTRICAL CHARACTERISTICS
$T_A=25°C$, $V_S = ±15V$, unless otherwise specified.

SYMBOL	PARAMETER	TEST CONDITIONS	μA741, μA741C Min	μA741, μA741C Typ	μA741, μA741C Max	UNIT
R_{IN}	Parallel input resistance	Open-loop, f=20Hz	0.3			MΩ
C_{IN}	Parallel input capacitance	Open-loop, f=20Hz		1.4		pF
	Unity gain crossover frequency	Open-loop		1.0		MHz
t_R	Transient response unity gain Rise time	$V_{IN}=20mV$, $R_L=2kΩ$, $C_L≤100pF$		0.3		μs
	Overshoot			5.0		%
SR	Slew rate	$C≤100pF$, $R_L≥2kΩ$, $V_{IN}=±10V$		0.5		V/μs

Philips Semiconductors

Product specification

General purpose operational amplifier

µA741/µA741C/SA741C

EQUIVALENT SCHEMATIC

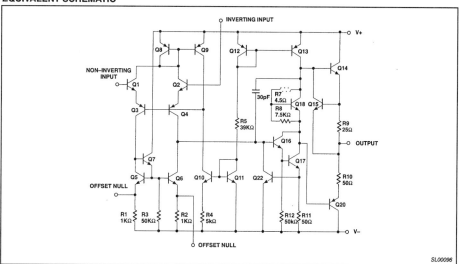

Figure 2. Equivalent Schematic

Philips Semiconductors

General purpose operational amplifier

µA741/µA741C/SA741C

TYPICAL PERFORMANCE CHARACTERISTICS

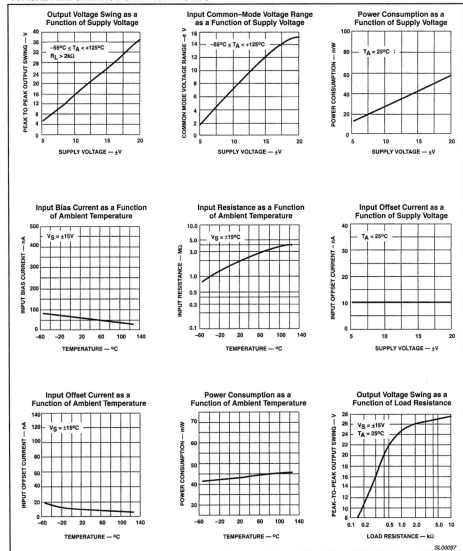

Figure 3. Typical Performance Characteristics

General purpose operational amplifier — µA741/µA741C/SA741C

TYPICAL PERFORMANCE CHARACTERISTICS (Continued)

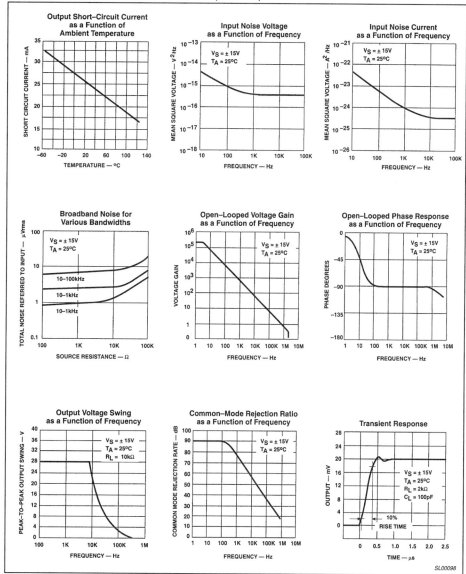

Figure 4. Typical Performance Characteristics (cont.)

Philips Semiconductors — Product specification

General purpose operational amplifier — µA741/µA741C/SA741C

TYPICAL PERFORMANCE CHARACTERISTICS (Continued)

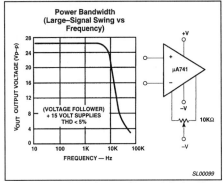

Figure 5. Typical Performance Characteristics (cont.)

Philips Semiconductors Linear Products

Product specification

Dual and single low noise op amp

NE5533/5533A/ NE/SA/SE5534/5534A

DESCRIPTION
The 5533/5534 are dual and single high-performance low noise operational amplifiers. Compared to other operational amplifiers, such as TL083, they show better noise performance, improved output drive capability and considerably higher small-signal and power bandwidths.

This makes the devices especially suitable for application in high quality and professional audio equipment, in instrumentation and control circuits and telephone channel amplifiers. The op amps are internally compensated for gain equal to, or higher than, three. The frequency response can be optimized with an external compensation capacitor for various applications (unity gain amplifier, capacitive load, slew rate, low overshoot, etc.) If very low noise is of prime importance, it is recommended that the 5533A/5534A version be used which has guaranteed noise specifications.

FEATURES
- Small-signal bandwidth: 10MHz
- Output drive capability: 600Ω, 10V$_{RMS}$ at VS=±18V
- Input noise voltage: $4nV/\sqrt{Hz}$
- DC voltage gain: 100000
- AC voltage gain: 6000 at 10kHz
- Power bandwidth: 200kHz
- Slew rate: 13V/μs
- Large supply voltage range: ±3 to ±20V

PIN CONFIGURATIONS

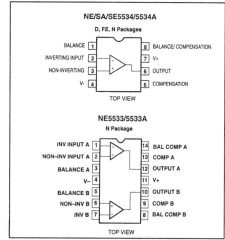

APPLICATIONS
- Audio equipment
- Instrumentation and control circuits
- Telephone channel amplifiers
- Medical equipment

ORDERING INFORMATION

DESCRIPTION	TEMPERATURE RANGE	ORDER CODE	DWG #
14-Pin Plastic Dual In-Line Package (DIP)	0 to +70°C	NE5533N	0405B
14-Pin Plastic Dual In-Line Package (DIP)	0 to +70°C	NE5533AN	0405B
8-Pin Plastic Small Outline (SO) package	0 to +70°C	NE5534D	0174C
8-Pin Hermetic Ceramic Dual In-Line Package (CERDIP)	0 to +70°C	NE5534FE	
8-Pin Plastic Dual In-Line Package (DIP)	0 to +70°C	NE5534N	0404B
8-Pin Plastic Small Outline (SO) package	0 to +70°C	NE5534AD	0174C
8-Pin Hermetic Ceramic Dual In-Line Package (CERDIP)	0 to +70°C	NE5534AF	
8-Pin Plastic Dual In-Line Package (DIP)	0 to +70°C	NE5534AN	0404B
8-Pin Plastic Dual In-Line Package (DIP)	-40°C to +85°C	SA5534N	0404B
8-Pin Plastic Small Outline (SO) package	-40°C to +85°C	SA5534AD	0174C
8-Pin Plastic Dual In-Line Package (DIP)	-55°C to +125°C	SE5534N	0404B
8-Pin Hermetic Ceramic Dual In-Line Package (CERDIP)	-55°C to +125°C	SE5534AF	
8-Pin Plastic Dual In-Line Package (DIP)	-55°C to +125°C	SE5534AN	0404B
8-Pin Plastic Dual In-Line Package (DIP)	-40°C to +85°C	SA5534AN	0404B

Philips Semiconductors Linear Products

Product specification

Dual and single low noise op amp

NE5533/5533A/ NE/SA/SE5534/5534A

ABSOLUTE MAXIMUM RATINGS

SYMBOL	PARAMETER	RATING	UNIT
V_S	Supply voltage	±22	V
V_{IN}	Input voltage	±V supply	V
V_{DIFF}	Differential input voltage[1]	±0.5	V
T_A	Operating temperature range		
	SE	-55 to +125	°C
	SA	-40 to +85	°C
	NE	0 to +70	°C
T_{STG}	Storage temperature range	-65 to +150	°C
T_J	Junction temperature	150	°C
P_D	Power dissipation at 25°C[2]		
	16D Pkg	1350	mW
	16N Pkg	1500	mW
	8D Pkg	750	mW
	8FE Pkg	800	mW
	8N Pkg	1150	mW
	Output short-circuit duration[3]	Indefinite	
T_{SOLD}	Lead soldering temperature (10sec max)	300	°C

NOTES:
1. Diodes protect the inputs against over voltage. Therefore, unless current-limiting resistors are used, large currents will flow if the differential input voltage exceeds 0.6V. Maximum current should be limited to ±10mA.
2. For operation at elevated temperature, derate packages based on the following junction-to-ambient thermal resistance:
    8-pin ceramic DIP 150°C/W
    8-pin plastic DIP 105°C/W
    8-pin plastic SO 160°C/W
    16-pin plastic DIP 80°C/W
    16-pin plastic SO 90°C/W
3. Output may be shorted to ground at V_S=±15V, T_A=25°C. Temperature and/or supply voltages must be limited to ensure dissipation rating is not exceeded.

Philips Semiconductors Linear Products

Product specification

Dual and single low noise op amp

NE5533/5533A/ NE/SA/SE5534/5534A

DC ELECTRICAL CHARACTERISTICS
$T_A=25°C$, $V_S=\pm15V$, unless otherwise specified. [1, 2, 3]

SYMBOL	PARAMETER	TEST CONDITIONS	SE5534/5534A Min	SE5534/5534A Typ	SE5534/5534A Max	NE5533/5533A NE/SA5534/5534A Min	NE5533/5533A NE/SA5534/5534A Typ	NE5533/5533A NE/SA5534/5534A Max	UNIT
V_{OS}	Offset voltage	Over temperature		0.5	2		0.5	4	mV
					3			5	mV
$\Delta V_{OS}/\Delta T$				5			5		µV/°C
I_{OS}	Offset current	Over temperature		10	200		20	300	nA
					500			400	nA
$\Delta I_{OS}/\Delta T$				200			200		pA/°C
I_B	Input current	Over temperature		400	800		500	1500	nA
					1500			2000	nA
$\Delta I_B/\Delta T$				5			5		nA/°C
I_{CC}	Supply current per op amp	Over temperature		4	6.5		4	8	mA
					9			10	mA
V_{CM}	Common mode input range		±12	±13		±12	±13		V
CMRR	Common mode rejection ratio		80	100		70	100		dB
PSRR	Power supply rejection ratio			10	50		10	100	µV/V
A_{VOL}	Large-signal voltage gain	$R_L \geq 600\Omega$, $V_O=\pm10V$	50	100		25	100		V/mV
		Over temperature	25			15			V/mV
V_{OUT}	Output swing	$R_L \geq 600\Omega$	±12	±13		±12	±13		V
		Over temperature	±10	±12		±10	±12		V
		$R_L \geq 600\Omega$, $V_S=\pm18V$	±15	±16		±15	±16		V
		$R_L \geq 2k\Omega$	±13	±13.5		±13	±13.5		V
		Over temperature	±12	±12.5		±12	±12.5		V
R_{IN}	Input resistance		50	100		30	100		kΩ
I_{SC}	Output short circuit current			38			38		mA

NOTES:
1. For NE5533/5533A/5534/5534A, $T_{MIN} = 0°C$, $T_{MAX} = 70°C$
2. For SE5534/5534A, $T_{MIN} = -55°C$, $T_{MAX} = +125°C$
3. For SA5534/5534A, $T_{MIN} = -40°C$, $T_{MAX} = +125°C$

Philips Semiconductors Linear Products

Product specification

Dual and single low noise op amp

**NE5533/5533A/
NE/SA/SE5534/5534A**

AC ELECTRICAL CHARACTERISTICS
$T_A=25°C$, $V_S=±15V$, unless otherwise specified.

SYMBOL	PARAMETER	TEST CONDITIONS	SE5534/5534A			NE5533/5533A NE/SA5534/5534A			UNIT
			Min	Typ	Max	Min	Typ	Max	
R_{OUT}	Output resistance	A_V=30dB closed-loop f=10kHz, R_L=600Ω, C_C=22pF		0.3			0.3		Ω
	Transient response	Voltage-follower, V_{IN}=50mV R_L=600Ω, C_C=22pF, C_L=100pF							
t_R	Rise time			20			20		ns
	Overshoot			20			20		%
	Transient response	V_{IN}=50mV, R_L=600Ω C_C=47pF, C_L=500pF							
t_R	Rise time			50			50		ns
	Overshoot			35			35		%
A_V	Gain	f=10kHz, C_C=0		6			6		V/mV
		f=10kHz, C_C=22pF		2.2			2.2		V/mV
GBW	Gain bandwidth product	C_C=22pF, C_L=100pF		10			10		MHz
SR	Slew rate	C_C=0		13			13		V/μs
		C_C=22pF		6			6		V/μs
	Power bandwidth	V_{OUT}=±10V, C_C=0		200			200		kHz
		V_{OUT}=±10V, C_C=22pF		95			95		kHz
		V_{OUT}=±14V, R_L=600Ω C_C=22pF, V_{CC}=±18V		70			70		kHz

ELECTRICAL CHARACTERISTICS
$T_A=25°C$, $V_S = 15V$, unless otherwise specified.

SYMBOL	PARAMETER	TEST CONDITIONS	5533/5534			5533A/5534A			UNIT
			Min	Typ	Max	Min	Typ	Max	
V_{NOISE}	Input noise voltage	f_O=30Hz		7			5.5	7	nV/√Hz
		f_O=1kHz		4			3.5	4.5	nV/√Hz
I_{NOISE}	Input noise current	f_O=30Hz		2.5			1.5		pA/√Hz
		f_O=1kHz		0.6			0.4		pA/√Hz
	Broadband noise figure	f=10Hz-20kHz, R_S=5kΩ					0.9		dB
	Channel separation	f=1kHz, R_S=5kΩ		110			110		dB

Philips Semiconductors Linear Products	Product specification

Dual and single low noise op amp

**NE5533/5533A/
NE/SA/SE5534/5534A**

EQUIVALENT SCHEMATIC

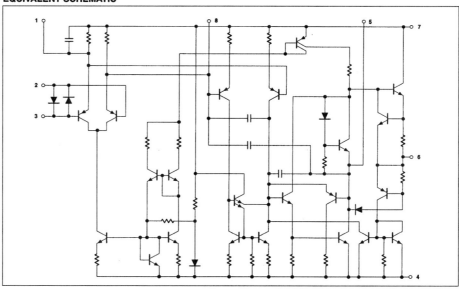

Philips Semiconductors Linear Products

Product specification

Dual and single low noise op amp

NE5533/5533A/
NE/SA/SE5534/5534A

TYPICAL PERFORMANCE CHARACTERISTICS

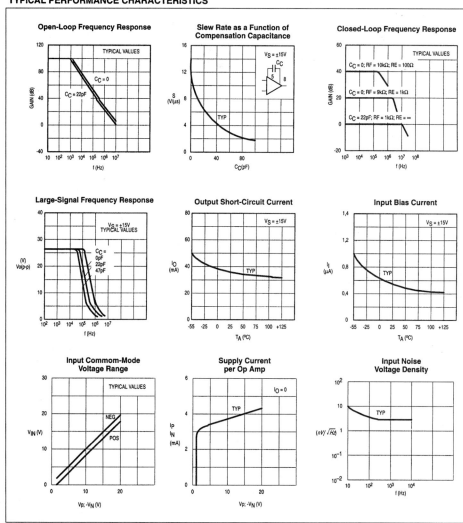

Philips Semiconductors Linear Products

Dual and single low noise op amp

NE5533/5533A/
NE/SA/SE5534/5534A

TYPICAL PERFORMANCE CHARACTERISTICS (Continued)

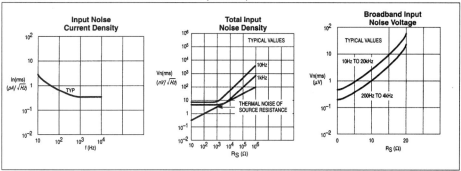

TEST LOAD CIRCUITS

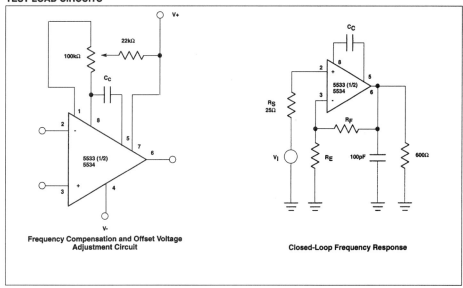

Frequency Compensation and Offset Voltage Adjustment Circuit

Closed-Loop Frequency Response

APPENDIX B

Answers to Odd-Numbered Review Questions

Chapter 1

1. The log-based decibel represents ratios of power. As such, decibel-based gains and signals add instead of multiply, simplifying calculations. Further, the decibel scheme tends to focus on the magnitude of change and relates well to the human hearing mechanism.
3. The third letter represents the reference level. dBV is referenced to 1 volt, whereas dBm is referenced to 1 milliwatt.
5. At a minimum, lead networks are comprised of a single capacitor (or inductor) and a resistor. An example is a coupling capacitor. Lead networks block DC and low frequencies, allowing only higher frequencies to pass through. The phase response at low frequencies approaches +90 degrees and approaches 0 degrees at high frequencies (hence its name).
7. f_1 refers to the lower half-power, or −3 dB frequency. f_2 refers to the upper half-power, or −3 dB frequency. $f_2 - f_1$ defines the system's bandwidth. f_1 and f_2 are alternately referred to as *corner* or *break* frequencies as well.
9. A single lead network varies from +90 degrees at low frequencies to 0 degrees at high frequencies (+45 degrees at the break frequency). A single lag network varies from 0 degrees at low frequencies to −90 degrees at high frequencies (−45 degrees at the break frequency).
11. The response of each network is graphically added to the midband response of the system in order to achieve the overall system response. Rolloff slopes will be additive as will the ultimate phase changes.
13. Common-mode rejection is the ability of a differential amplifier to ignore an identical signal (i.e., a common-mode signal) placed on both inputs. The ratio of the differential gain to the common-mode gain yields the common-mode rejection ratio (CMRR). The ideal value for CMRR is infinite.
15. An active load maximizes voltage gain by forcing all signal current to the following stage.

Chapter 2

1. An op amp is a multistage amplifier. The circuit is usually DC coupled and is packaged as a single unit. It can be used as is or as part of a much larger system.
3. An op amp is a multistage amplifier typically consisting of a differential input stage, voltage gain stages, and a class B output stage.
5. The output stage is usually a class B or class AB voltage follower.
7. A comparator is in essence a very high gain differential amplifier used as an interface between analog signals and digital logic. Even very small differences between the input signals is enough to force the output into either positive or negative saturation. Thus, a comparator is used to determine if one signal is more positive than the other, with the output saturation states corresponding to a simple true/false answer.
9. A mask is used to expose only certain areas of a wafer to the dopant. In this way, other areas are masked or shielded and are not affected. A series of masks will be used to create transistors. For example, one mask may be used to prepare transistor emitter areas.
11. Because an entire circuit is made at once, monolithic ICs tend to be inexpensive and very reliable.

Chapter 3

1. Maintaining a constant speed while driving, walking in a straight line, etc.
3. Sacrifice factor indicates the amount of gain given up. It is equal to the difference between the open-loop and closed-loop gains in decibels.
5. Series-parallel (VCVS or ideal voltage amplifier), parallel-series (CCCS or ideal current amplifier), parallel-parallel (CCVS or ideal current-to-voltage transducer), and series-series (VCCS or ideal voltage-to-current transducer).
7. Noise and device-dependent parameters such as output clipping level.

Chapter 4

1. The inverting form uses a modified parallel-parallel scheme, whereas the noninverting form uses series-parallel negative feedback.
3. Parallel-series feedback is used in the inverting current amplifier.
5. A virtual ground is a point that is very close to 0 volts. It is normally found at the inverting input of parallel-input feedback forms and behaves as an ideal current summing node.
7. Output current can be increased by adding a discrete class B follower stage after the op amp and wrapping it inside of the feedback loop. The op amp delivers current to the follower, which is then responsible for delivering final load current.
9. All parameters stay unchanged with the exception of the system input impedance (set by the input divider network), low frequency response (set by coupling and feedback capacitors), and the current drawn (additional draw through the bias network).
11. A floating load is one that is not tied to ground at one end.

Chapter 5

1. Gain-Bandwidth product is also known as f_{unity} and is defined as the frequency at which the open-loop response falls to unity (0 dB). It is useful because the product of closed-loop noise gain and f_2 must equal f_{unity}. It can be seen, therefore, that gain can be traded for bandwidth and vice versa.
3. For lower breaks, the combination will produce an f_1 higher than the individual break frequencies. For upper breaks, the combination will produce an f_2 lower than the individual break frequencies.
5. Power bandwidth (f_{max}) is proportional to slew rate and inversely proportional to peak sine amplitude. f_{max} = Slew Rate/($2\pi V_{peak}$).
7. On the downside, noncompensated amplifiers require external compensation circuits (a capacitor at minimum). The advantage is that the frequency response and slew rate can be optimized for a given gain.
9. Output offset voltage is caused by less-than-perfect matching of the internal components of the op amp, as well as in the surrounding feedback network.
11. CMRR stands for Common-Mode Rejection Ratio. It is a measure of how well the op amp suppresses identical signals on the inputs relative to differential input signals.
13. In general, noise is proportional to temperature, bandwidth, and resistance. The key parameters to look for on an op amp's data sheet are the input noise voltage density and input noise current density.

Chapter 6

1. Instrumentation amplifiers generally offer very high gain, isolated high-input impedances for both inputs, and superior common-mode rejection when compared to a single op amp-based differential amplifier.
3. The device is able to go into a "power down" or "sleep" mode and thereby drastically reduce power consumption. Further, the performance can be fine-tuned against power consumption.
5. Direct motor drive and audio power amplifiers.
7. OTA stands for Operational Transconductance Amplifier. The transconductance (and hence the gain) of the device is set by an external control current, I_{ABC}.
9. Any application requiring a controlled-gain amplifier would suffice, such as an audio compressor.
11. Input differencing is achieved through the use of a current mirror. The input circuit generates the difference between the input currents by subtracting the current presented to one input (via the current mirror) from the other input current.
13. The construction of the current feedback amplifier differs considerably from that of an ordinary op amp. The current feedback amplifier does not rely on negative feedback to maintain its inverting and noninverting inputs at the same level; rather a unity gain buffer is used. Also, the output is modeled as a current-controlled source rather than as a voltage-controlled source. This combination leads to a situation where the feedback factor β does not enter into the gain equation and thus to a first approximation, the feedback factor (via sacrifice factor) does not influence the closed-loop gain.

Chapter 7

1. Active rectifiers have the primary advantage of accurately rectifying small signal levels. They also exhibit a high-input impedance and a low-output impedance.
3. A peak detector produces an output that is proportional to the peak value of the input waveform. It can be used as an envelope detector in AM systems.
5. A clamper adds or subtracts a DC potential to an input signal in order to shift to a new level, for example, making the entire waveform positive by shifting it up.
7. A transfer function generator has an arbitrary input/output function curve. In comparison, the function for an amplifier is a straight line (the slope of which indicates the gain). Function generation circuits can be used to create some desired transfer characteristic or to predistort or compensate for nonlinearities that exist elsewhere, creating an overall linear system response.
9. A Schmitt trigger is a comparator with hysteresis. In effect there are different threshold levels depending on the current output state. The circuit winds up being relatively immune to generating false outputs due to input noise.
11. Dedicated comparators are usually designed to produce very fast switching times. Further, external connections are usually available for logic interfacing, strobe, and so forth.
13. A log amplifier produces an output signal that is proportional to the log of the input. This has the effect of compressing the signal. In contrast, the antilog amplifier produces a signal that is proportional to the antilog of the input, producing an expansion of the input signal.
15. A multiplier can be used in a variety of ways. In the simplest case, the multiplier can be used as a controlled gain amplifier. It can also be used for mathematically squaring a signal (e.g., for RMS calculations) or for other mathematical operations such as division.

Chapter 8

1. A voltage regulator is used to turn the fluctuating output of a rectified and filtered AC power signal into a smooth and stable DC level.
3. Regulators need something to measure the output voltage against. Without a reference, there is no way of determining the present output level.
5. Linear regulators are relatively simple to design and use. They also exhibit very fast transient response and do not radiate interference signals as switching regulators do.
7. The primary function of a pass transistor is that of a control element. In a linear regulator it is used to effectively absorb the difference between the unregulated input and the desired output. In a switching regulator it is used as a controlled switch that allows current pulses through which in turn, charge the reactive elements and produce the desired output.
9. The normal "ground" pin can be tied to the middle of a voltage divider connected across the output. In this way, the regulator only "sees" a portion of the output voltage.
11. The average area of the PWM current pulses is equal to the DC current demand. If the current demand goes up, the pulses will widen to keep the equality. If the current demand drops, the opposite happens and the pulses will narrow.
13. Thermal resistance is a measure of how easily an item can transfer heat to its surrounding environment. A low thermal resistance means that the item conducts heat easily.
15. Use a heat sink size appropriate for the target power dissipation. Do not block air flow around heat sinks. Use a modest amount of heat sink grease (more is not better). Excessive torque is not required on the mounting hardware. Use insulators to prevent the heat sink from becoming electrically "live." In high-power situations, consider using forced air cooling (fans).

Chapter 9

1. Positive feedback reinforces change. It causes instability and, hence, oscillation. In contrast, negative feedback tends to create a stabilizing effect on amplifiers.
3. The oscillator exploits the Wien bridge, consisting of a series RC leg, a parallel RC leg, and a pair of resistors making up the other half of the bridge. At only one frequency will the bridge be in balance; that is, the phase shift produced by the RC legs will be 0 degrees. As long as sufficient gain is available to overcome the loss of the RC portion, the circuit will oscillate at this frequency.

5. Simply pass the output of the sine or triangle wave generator into a comparator.
7. Primarily, the accuracy of the Wien bridge output frequency is set by the accuracy of the timing capacitors and resistors. It is also influenced by any extra phase shift produced by the op amp.
9. (See text for block diagram.) The PLL consists of three major components: A VCO, a phase comparator, and a loop filter. Negative feedback is used to get the VCO to track the input signal. The phase comparator has two inputs: one for the input signal and the other for the VCO output. It produces an error signal that is proportional to the phase difference between these two signals. A control voltage is derived from this signal and is used to drive the VCO. The loop will be stable as long as the error signal remains small.
11. Fixed frequency: A local or carrier frequency generator in an AM or FM system, audio calibration oscillator, etc. VCO: Music synthesizer, tracking generator in a spectrum analyzer, etc.
13. Monostable operation is one-shot or single event. Something has to occur in order to trigger the pulse. Astable operation is continuous. In other words, a nonstop waveform is being generated.

Chapter 10

1. Integrators find the "area under the curve" of the input waveform.
3. The capacitor has a differential characteristic between voltage and current, $i = C\, dv/dt$. When placed in the feedback loop, this gives rise to the differential (C as R_i) or the integral (C as R_f) of the input.
5. The basic integrator has very high DC gain and will eventually saturate due to offsets. For this reason, a lower gain limit is imposed. This has the side effect of producing a lower frequency limit of accurate integration.
7. Primarily the additional components reduce the range of frequencies that can be accurately processed. In the case of the integrator, a lower limit is created, and in the case of the differentiator, an upper limit is established.
9. One advantage of the analog computer is its immediacy; a technician or engineer can "tweak the knobs" in real-time if desired. An advantage of using a digital computer is its accuracy and repeatability.

Chapter 11

1. High-pass, low-pass, band-pass, and band-reject (or notch).
3. Active filters require DC power, their frequency response is limited by the active circuitry, and they do not handle large amounts of output power (usually).
5. Bessel, Butterworth, Chebyshev, Elliptic (Cauer).
7. The Bessel alignment is linear phase.
9. f_c and f_{3dB} are identical when using a Butterworth alignment.
11. The cascade is appropriate for wide bandwidth applications. Multiple feedback is appropriate for narrow bandwidth applications with modest Qs of 10 or less. For higher Q response, the state-variable is preferred.
13. In essence, an inverting voltage amplifier is used. A potentiometer is situated between R_i and R_f so that the gain is a function of the potentiometer's wiper arm. The feedback network is frequency sensitive due to the inclusion of a capacitor (or capacitors). At high frequencies, the capacitor shorts out the potentiometer leaving the original R_i and R_f, and a gain of unity.
15. A primary advantage is the simplicity of design. Usually, very few additional parts are required to create complex filters. On the downside, switched-capacitor filter ICs exhibit limited frequency range compared to the more general op amp approach and there is also the possibility of clock noise leaking into the output signal.

Chapter 12

1. Pulse Code Modulation. This is the technique of measuring an input signal at regular intervals and storing the results as a sequence of digital words.
3. Quantization is the process of measuring and thus defining the signal via a series of discrete steps.
5. The Nyquist frequency is the highest frequency that can be sampled and recovered without concern of aliasing. The sample rate must be at least twice the Nyquist frequency, although it is normally somewhat more than twice. This extra factor is due to the fact that low-pass

antialiasing filters do not rolloff infinitely fast, and this transition band must be taken into consideration. In essence, the Nyquist frequency defines the upper limit frequency of the system.
7. Integral nonlinearity is the worst-case offset between the actual and ideal conversions for all possible outputs. Differential nonlinearity is also known as relative-adjacent error. It defines the maximum output deviation between two adjacent input words relative to the ideal 1 LSB. For the ideal case, both the integral and differential nonlinearity would be 0 LSB. If the differential nonlinearity is less than 1 LSB, the system response will be monotonic (i.e., increasing digital words always produce a higher analog output).
9. The track-and-hold is used to keep the signal to be measured constant during the measurement interval.
11. Signal compression, filtering, waveform shaping, Fourier Analysis, noise reduction, etc.

APPENDIX C

Answers to Odd-Numbered Problem sets

Chapter 1

1. a) 10 dB b) 19 dB c) 26.99 dB d) 0 dB e) −6.99 dB f) −15.23 dB
3. 33 dB
5. $G = 501$, $P_{out} = 12.53$ W
7. a) 1.06 b) 1 c) 199.5 d) 3.43 e) .398 f) .188
9. $A = 8.57$ $A' = 18.66$ dB
11. a) 0 dBW b) 13.6 dBW c) 8.13 dBW d) −7 dBW e) −26.4 dBW f) 30.8 dBW
    g) −43.5 dBW h) −65.2 dBW i) −172.5 dBW
13. a) 150 dBf b) 163.6 dBf c) 158.1 dBf d) 143 dBf e) 123.6 dBf f) 180.8 dBf
    g) 106.5 dBf h) 84.8 dBf i) −22.5 dBf
15. $G'_{total} = 28$ dB, $G = 631$
17. For $P'_{in} = 4$ dBm: output stage 1 = 6 dBm, stage 2 = 0 dBm, stage 3 = 15 dBm
    For $P'_{in} = -34$ dBm: output stage 1 = −24 dBW, stage 2 = −30 dBW,
    stage 3 = −15 dBW
19. a. 200 mW
21. $V'_{out} = 21$ dBV, 21 dBV = 11.2 V (final output)
    For stage 1: 4 dBV = 1.58 V
    For stage 2: 9 dBV = 2.82 V
23. a. 15 V
25. At 50 kHz: −.022 dB, −4.09 degrees
    At 700 kHz: −3 dB, −45 degrees
    At 10 MHz: −23.1 dB, −86 degrees
    $T_r = .5$ μsec
27. The amplitude portion does not change. Phases are: At 30 kHz −188.5 degrees,
    at 200 kHz −225 degrees, at 1 MHz −258.7 degrees
29.

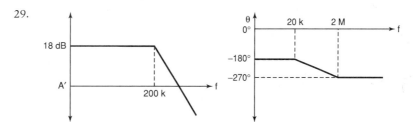

31. At 4 kHz: 78.7 degrees, at 20 Hz: 45 degrees, at 100 Hz: 11.3 degrees

33.

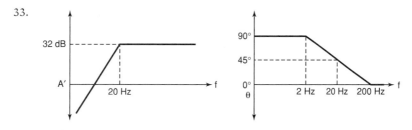

35. Net gain at 20 kHz = 35.87 dB
    Net phase at 20 kHz = 51.7 degrees
    At 100 kHz: phase = −5.1 degrees, $A'_v = 40$ dB
    At 800 kHz: phase = −70.5 degrees, $A'_v = 30.5$ dB

37.

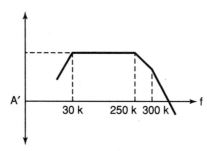

39.

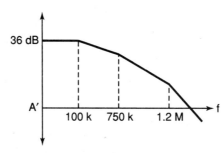

41. Each lag network rolls off at 20 dB/Decade for a 60 dB/Decade total (i.e., above 1.2 MHz).
43. $r'_e = 42.3$ ohms, $A'_v = 65$
45. $.48 \sin(2\pi 1000t)$
47. $1.656 \sin(2\pi 2000t)$
49. .775 V
51. 360 W
53. 71.5 dBV.
55. Greater than 30 Hz.

Chapter 2

1. a) V_{out} is unknown b) $-V_{sat}$ c) $+V_{sat}$
3. a) $-V_{sat}$ b) $+V_{sat}$ c) $-V_{sat}$ d) $-V_{sat}$
5.

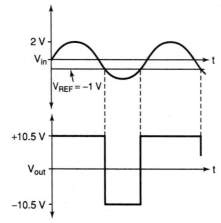

7. 55 degrees C
9. 4.967V
11. 125 ohms

Chapter 3

1. $S = 100$ at low frequencies, $S = 10$ at 1 kHz
3. $A_{cl} = 19.96$ Approximate $A_{cl} = 20$ At 1 kHz, the approximate $A_{cl} = 20$, the exact value $= 19.6$
5. $S = 1000$ $F_{2sp} = 25$ kHz
7. The margins are 60 degrees and 12 dB to spare.
9. $Z_{incl} = 20$ M ohms, $Z_{ocl} = .25$ ohms
11. $S = 2000$
13. $5\sin(2\pi 500t)$
15. Typical values might be $R_i = 1$ k ohm, $R_f = 9$ k ohms.
17. a) Decreased high frequency gain. b) Gain increase for high frequencies.
    c) The output signal will appear to be clipped, in the same manner as an overdriven amplifier.

Chapter 4

1. $A_v = 7.67$ $Z_{in} = 470$ k
3. $A_v = 10$ $Z_{in} =$ infinite
5. -9.7 dBV
7. -660 mV
9. R_i is 20 k, $R_f = 50$ k
11. $I_{out} = 100\,\mu A$ (1/3 full scale)
13. $R = 333.3$ ohms
15. $I_{outmax} = 2.29$ mA $I_{inmax} = 208\,\mu A$
17. $R'_i = 3.33$ k, $R'_f = 16.67$ k
19. 0 dB is a gain of one.

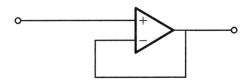

21. $R_f = 50$ k ohms

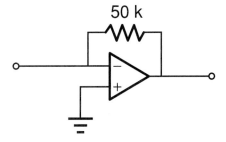

23. $A_v = 20$, $R_i = 19R_f$ One possibility is:

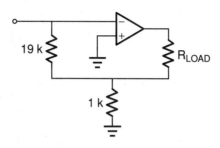

25. $R = 1\text{ k ohm}$ $I_{out} = 200\,\mu A$

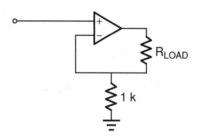

27. The figure below is one possible answer.

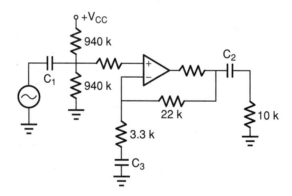

29.

```
         1.5 k        15 k
1 o────/\/\──┬────/\/\──┐
         1 k │           │
2 o────/\/\──┤   |\      │
         3 k │   | \─────┴──o
3 o────/\/\──┴───|+/
                 |/
                  │
                 ═══
```

31. For C_1 dominant (C_2 and C_3 increased 10×) $C_1 = 3.18\,\mu F$
    $C_2 = 7.23\,\mu F$
    $C_3 = 1.45\,\mu F$

33. $A_{max} = 9.15$ $A_{min} = 6.45$
35. The figure below is one possible answer.

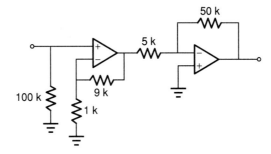

37. 1.88 V to 2.12V
39. Set $A_v = 20$

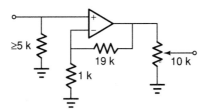

41. The loss is approximately .0119 (ignoring loading and phase angle).
43. $A_v = 11$, Z_{in} approaches infinity.

Chapter 5

1. $A_n = 6$, $f_2 = 667$ kHz
3. $A_n = 6$, $f_{unity} = 1.5$ MHz
5. $f_{max} = 239$ kHz
7. $SR = 1.26 V/\mu S$
9. $SR = 6.79 V/\mu S$
11. $SR = 1.25 V/\mu S$
13. $V_{outoffset} = +/-4$ mV (approximately)
15. $V_{outripple} = .5 \mu V$
17. $V_{outnoise} = 16.8 \mu V$, Input referred noise $= .84 \mu V$
19. $V_{outnoise} = 290 \mu V$, S/N $= 6901$ (76.8 dB)
21. The result is a 1.25 V peak 100 kHz triangle wave.

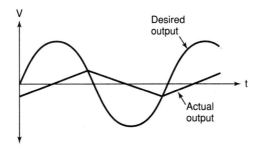

23. $R_{off} = 952$ ohms, $V_{drift} = +/-7.49$ mV
25. $C = 2.4\,\mu F$
27. For a single stage, a 411 will do. For gain, set $R_f = 49R_i$.
29. Try 2 stages with equal gain. $A = 12.25$ each. $f_{unity} = 7.35$ MHz 318s will do. $R_f = 11.25R_i$ (A three-stage design is also possible.)
31. $E_n = 5.68\,\mu V$ RMS
33. Total offset $= +/-28.5$ mV
35. 741 stage 1, $f_2 = 1$ MHz
    411 stage 2, $f_2 = 667$ kHz
    318 stage 3, $f_2 = 750$ kHz
    System f_2 is approximately equal to 667 kHz.

Chapter 6

1. Desired output $= 200$ mV, Hum $= 4.5\,\mu V$
3. $V_{out} = 1.69V$
5. $I_{set} = 14.5\,\mu A$
7. $I_{set} = 14.75\,\mu A$, $A_v = 10.4$, $SR = .3V/\mu S$, $f_{unity} = 250$ kHz
   $f_2 = 24$ kHz, $f_{max} = 4.77$ kHz
9. $C >= 1\,\mu F$
11. $P_d = .8W$
13. $I_{abc} = 218\,\mu A$
15. $C = 212$ nF
17. Set R_i to 100 k ohms. $R_f = 400$ k ohms, $R_b = 800$ k ohms, $C = 63.7$ nF

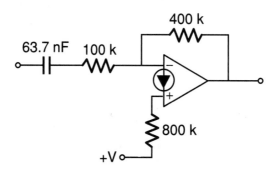

19. $R_f = 1.5$ k, $R = 38.5$ ohms

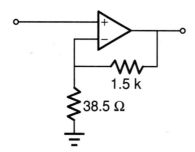

21.

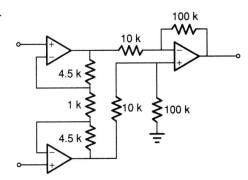

23. As outlined in textbook Figure 6.15, but add a capacitor and 1080 ohm in series from pin 1 to 8.
25. $I_{abc} = .15\,\text{mA}$, $gm = 3\,\text{mS}$, net gain $= .924$

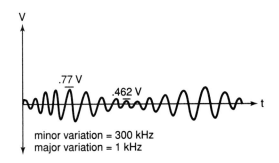

Chapter 7

1.

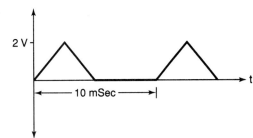

3.

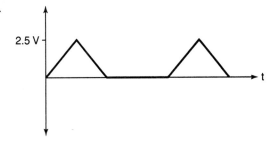

5. 1V DC

7.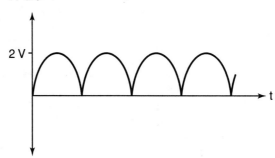

9.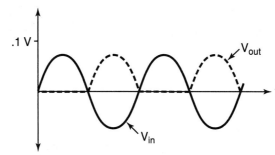

11.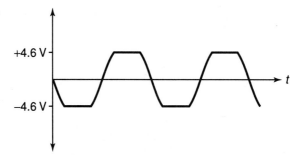

13.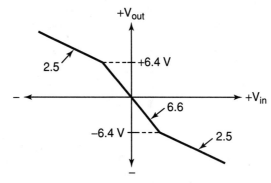

15. 8.97V peak square wave

17.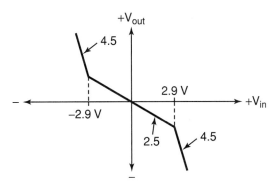

19.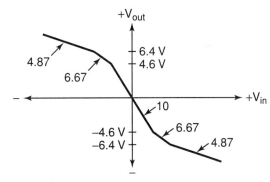

21. $+/-10.4V$
23. $V_{out} = 5V$
25. $V_{out} = -72.9$ mV
27. $V_{out} = .2 \sin(2\pi\, 1000t), \quad .4 \sin(2\pi\, 1000t), \quad 1 \sin(2\pi\, 1000\, t)$
29. $V_{out} = -12.5 \sin(2\pi\, 2000t)$

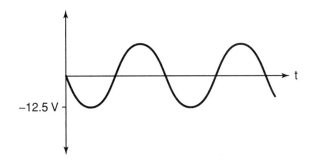

31. $R_1 = 500$ k ohms
33. $R_f = 60$ k, $\quad R_i = 6$ k, $\quad R_a = 6$ k
35. $R_1 = 20$ k, $\quad R_2 = R_6 = 5$ k, $\quad R_4 = R_8 = 2.87$ k
    $R_3 = R_7 = 34.1$ k, $\quad R_5 = R_9 = 13.7$ k
37. $R_i = 22.1$ k
39. $R_1 = R_2 = 1.35$ k

43. Use Figures 7.41 and 7.42 as a guide. $R_f = R_i = 10$ k, $R_6 = R_2 = 90$ k, $R_8 = R_4 = 72$ k, $R_3 = R_7 = 785$ k, $R_5 = R_9 = 279$ k
45. Threshold $= +/-2.88$ V

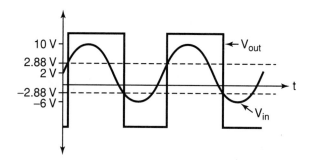

Chapter 8

1. $P_d = 9$ W

3.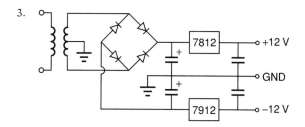

5. $P_d = 3.86$ W
7. $R_3 = 10$ k, $R_2 = 35.5$ k, $R_1 = 5.85$ k, beta $>= 25$
9. $R_2 = 2064$ ohms
11. Follow Figure 8.11b for this. For 5 V $R_2 = 720$ ohms, for 12 V $R_2 = 2064$ ohms, for 15 V $R_2 = 2640$ ohms
13. $R_{sc} = 6.5$ ohms The general form is shown in text Figure 8.14. Choosing $R_2 = 10$ k, $R_1 = 2.59$ k, $R_3 = 2.06$ k
15. $R_{pos} = 9.86$ ohms $R_{neg} = 8.43$ ohms. Refer to text Figure 8.19.
17. Choose $R_2 = 10$ k, $R_1 = 40$ k, $R_3 = .15$ ohms, $C_1 = 1$ nF, $C_3 = 20$ pF, $L_1 = .278$ mH, $C_2 >= 26.6\,\mu F$

19.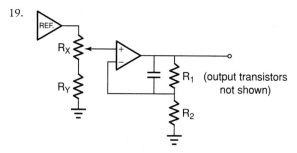

21. $P_d = 8.33$ W

Chapter 9

1. $f_o = 2.4$ kHz
3. $f_{o\text{-max}} = 4081$ Hz, $f_{o\text{-min}} = 371$ Hz
5. $f_o = 229$ Hz
7. $f_o = 9917$ Hz
9. 13.25 kHz,   6583 Hz
11. $f_o = 27.9$ kHz
13. 66.9 kHz and 14.8 kHz

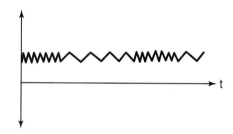

15. $V_{thres} = 9$ V, $f_o = 10.3$ kHz
17. 13.6 Hz to 25 Hz,   136 Hz to 250 Hz,   1360 Hz to 2500 Hz
19. $R_3 = R_4 = 8.47$ k ohms
21. $R_3 = R_4 = 81.2$ ohms,   P1 = P2 = 730.8 ohms
23. $R_2 = 1.5$ k, $R_{b1} = 2.2$ k, $R_{b2} = 1.8$ k (All else unchanged)
25. $C_1 = 4.59$ nF, $C_2 = .459$ nF, $C_3 = 45.9$ pF
27. $R = 3938$ ohms
29. $C = 5.56$ nF
31. $R_2/R_3 = .37 : 1$
33. $R = 10$ k, $C_1 = 523$ pF
35. $C_1 = 1.15$ nF, $F_1 = +/-86.7$ kHz, $C_2 = 2.04$ nF
37. Increase all tuning capacitors by 71.7.
39. Feed the sine wave into a comparator that has a variable reference.
41.

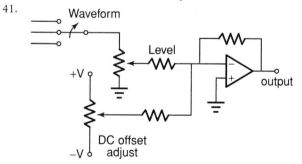

Chapter 10

1. A 4.17 Vpp triangle at 1 kHz
3. $V_{out} = .663 \sin(2\pi\, 10000t)$
5. $A = -20$
7. A 6.06 mVpp triangle at 50 kHz

9. V_{out} eventually reaches -1 V
11. An 80 mV peak-to-peak square wave at 100 Hz
13. $V_{out} = 2.26\sin(2\pi 60 t)$
15. $V_{out} = -.04t$
17.

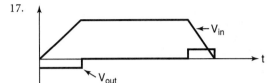

19. A 150 mV peak square wave at 2.5 kHz
21. $R = 1$ k ohm
23. $C_f = 31.8$ nF
25. $R_i = 50$ ohms, $C = 31.8$ nF, $R_f = 3774$ ohms, $C_f = 422$ pF
27. $V_{out}(t) = 6.37 \sum_{n=1}^{\infty} 1/((2n-1)^2)\cos(2\pi 500(2n-1)t)$
29. $V_{out}(t) = -R_f/L \int V_{in}(t)dt$
31. $V_{out} - .314 \sum_{n=1}^{\infty} 1/(2n-1)\sin(2\pi 500(2n-1)t)$

Chapter 11

1. 500 Hz: 0 dB   1 kHz: 3 dB   2 kHz: 13 dB   4 kHz: 24 dB
3. 200 Hz: 10 dB   500 Hz: 3 dB   2 kHz: .2 dB
5. Butterworth: 0 dB   Bessel: .75 dB   1 dB Chebyshev: 1 dB
7. Order 2: 15 dB   3: 20 dB   4: 24 dB   5: 27 dB   6: 30 dB
9. Order 2: 1.75 dB   3: 2.5 dB   4: 0 dB   5: 2.5 dB   6: 2 dB
11. Butterworth: 0 dB   Bessel: 1 dB   1 dB Chebyshev: 3 dB
13. Butterworth: 4th (23 dB)   1 dB Chebyshev: 3rd (26 dB)   3 dB Chebyshev: 3rd (28 dB)
15. $BW = 1.5$ kHz, $f_o = 8.72$ kHz, $Q = 5.81$
17.

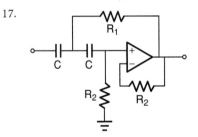

19.

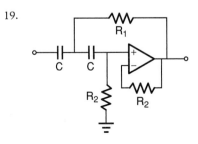

ANSWERS TO PROBLEMS   585

21.
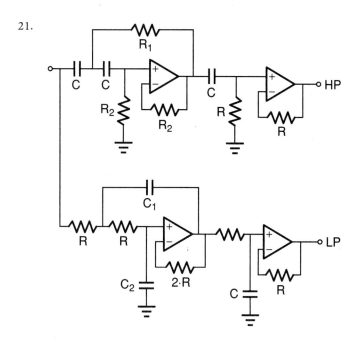

23.
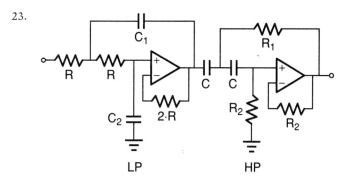

25.
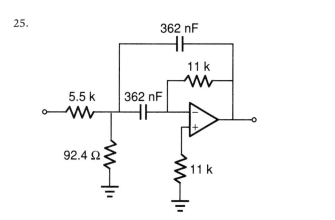

27.

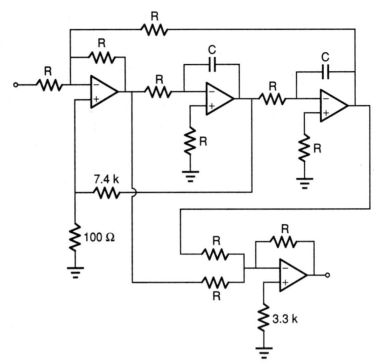

29.

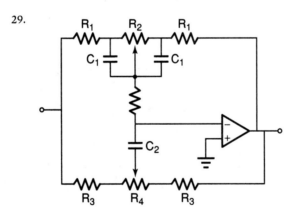

31. Using the internal clock, $C = 11.8$ nF, $R = 100$ ohms in series with a 5 k pot.
33. Cascade three second-order HP stages
    Stage 1: $C = .345\,\mu F$, $R_1 = 6.57$ k, $R_2 = 15.22$ k
    Stage 2: $C = .729\,\mu F$, $R_1 = 2.28$ k, $R_2 = 44$ k
    Stage 3: $C = .972\,\mu F$, $R_1 = 625$ ohms, $R_2 = 160$ k

Chapter 12

1. 1024
3. 60.2 dB, 84.3 dB (for 16384)

5. $20\,\mu S$
7. $11.1\,\mu S$
9. 20 k bytes per second
11. 403.2 M bytes for stereo
13. $f_n = 20.8$ kHz
15. 4.09 V
17. a) $.4\,\mu S$ b) $3.2\,\mu S$ c) $102.4\,\mu S$
19. a)

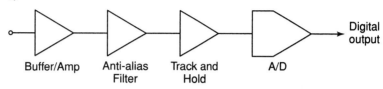

    Buffer/Amp   Anti-alias Filter   Track and Hold   A/D

   b) 9 bits
   c) $f_s = 960$ kHz
   d) $f_c = 400$ kHz
   e) Subranging/pipelined
   f) ADC12181

21. In BASIC, if $A(\ )$ is the data array and N is its size,
```
Max = 0
FOR X = 1 to N
   Temp = ABS(A(X))
   IF Temp > Max THEN Max = Temp
NEXT X
```

INDEX

Page numbers followed by *f* or *t* indicate a corresponding figure or table number.

AC
  differential amplifier and, 30–36, 31*f*1.22, 35*f*1.25
  operational amplifier, 55
  ripple, 282
accelerometer, 391–393, 392*f*10.11, 393*f*10.12
adder/substractor, 125–126, 126*f*4.28
aliasing, 479
aliasing effect, 490, 490*f*12.2, 491*f*12.3, 492*f*12.4
alignment(s), 422*f*11.9a, 422*f*11.9b, 422–426
amplifier, differential, 1, 26, 27*f*1.17, 41, 123–125, 124*f*4.25, 124*f*4.26, 125*f*4.27
  AC analysis of, 30*f*1.20, 30–36, 31*f*1.21, 31*f*1.22, 32*f*1.23, 33*f*1.24, 35*f*1.25
  AC equivalent circuit and, 30, 31*f*1.22, 35*f*1.25
  buffers with, 184, 185*f*6.1
  common mode rejection for, 36–38, 37*f*1.26, 37*f*1.27, 38*f*1.28
  current mirror and, 38*f*1.29, 38–40, 39*f*1.30, 39*f*1.31, 40*f*1.32, 40*f*1.33
  DC analysis of, 27*f*1.18, 27–29, 28*f*1.19
  electrical operation of, 2
  input offset current and voltage for, 29
  negative feedback, 67, 68*f*3.1
  operational amplifier and, 46, 46*f*2.2
  variations of, 30, 31*f*1.21
amplifier, operational, 1–2, 26, 44, 63
  AC characteristics of, 55
  audio uses and, 197
  block diagram of, 46*f*2.2, 46–50, 47*f*2.3, 48*f*2.4, 50*f*2.5
  chip manufacturing process of, 61*f*2.16, 61–62
  comparator and, 55–60, 56*f*2.10, 57*f*2.11a, 58*f*2.11b, 58*f*2.12, 59*f*2.13, 60*f*2.14
  current feedback, 214–217, 215*f*6.31, 218*f*6.32
  data sheet and interpretation for, 52–55, 53*f*2.8
  DC characteristics of, 54–55
  five connections of, 45
  four-part theme of, 49
  general, 45, 45*f*, 2.1
  high current devices for, 197–201, 198*f*6.15, 199*f*6.16, 200*f*6.17, 201*f*6.18
  high speed, 201–202, 202*f*6.19
  high voltage devices for, 201
  hybrid construction of, 63
  inverted and noninverted signals of, 45
  LF411, 46–47, 47*f*2.3, 52–54, 153, 153*t*5.1, 157, 159, 160*f*5.15c, 161, 162*f*5.19, 165, 165*t*5.2, 354*f*9.25b, 536–541
  LM6364, 197
  LM386, 542–546
  manufacture of, 60–63, 60*f*2.15, 60–64, 61*f*2.16, 62*f*2.17
  monolithic construction of, 61–62, 62*f*2.17
  noise, 559–565
  Norton (LM3900), 207*f*6.23, 207–211, 208*f*6.24, 209*f*6.25, 209*f*6.26, 211*f*6.27, 212*f*6.28a, 213*f*6.28b, 213*f*6.29, 214*f*6.30
  numbering system for, 52
  output of, 224–226, 226*f*7.5
  package styles of, 60, 60*f*2.15
  programmable, 193–196, 194*f*6.10, 194*f*6.11, 195*f*6.12, 197*f*6.14, 741, 15b, 47–49, 48*f*2.4, 52, 153, 153*t*5.1, 157, 158, 159*f*5, 161*f*5.18, 165*t*5.2, 355*f*9.25c, 552–558
  simplified, 49, 50*f*2.5
  simulation model for simple, 50–52, 51*f*2.6, 51*f*2.7
  specialized, 183–184
  specialize devices and, 217–220, 218*f*6.32, 219*f*6.33
  three stages of, 46, 46*f*2.2
  transconductance (LM13700), 204–207, 205*f*6.21, 206*f*6.22, 480, 480*f*11.61, 547–551
  ultra-fast, 197, 202*f*6.19
amplifier circuits, operational, 101, 134–135
  adder/substractor and, 125–126, 126*f*4.28
  adjustable gain amplifier and, 126–128, 127*f*4.29
  adjustable inverter/noninverter and, 126–128, 127*f*4.29, 127*f*4.30

589

current boosting and, 131–133, 132*f4.34*, 133*f4.35a*, 134*f4.35b*
differential amplifier and, 123–125, 124*f4.25*, 124*f4.26*, 125*f4.27*
frequency response of, 139
inverting current amplifier and, 114–117, 115*f4.16*, 116*f4.17*, 117*f4.18*, 125
inverting current-to-voltage transducer and, 110, 111*f4.12*
inverting voltage amplifier and, 105*f4.6*, 105–110, 106*f4.7*, 106*f4.8*, 106*f4.9*, 108*f4.10a*, 109*f4.10b*, 109*f4.11*, 130, 130*f4.30*
non and compensated devices for, 161–164, 162*f5.19*, 163*f5.20*
noninverting current-to-voltage transducer, 111*f4.13*, 111–113, 113*f4.14a*, 114*f4.14b*, 114*f4.15*
noninverting summing amplifier and, 120–122, 121*f4.22*, 121*f4.23*, 123*f4.24*, 125
noninverting voltage amplifier and, 101*f4.1*, 101–105, 103*f4.2*, 103*f4.3*, 104*f4.4*, 105*f4.5*
single-supply biasing and, 128–130, 129*f4.31*, 130*f4.32*, 130*f4.33*
slew rate and, 152–157, 153*t5.1*, 158*f5.15a*, 158–161, 159*f15b*, 160*f15c*, 161*f5.18a*, 161*t5.17*
summing amplifier, 117–120, 118*f4.19*, 118*f4.20*, 120*f4.21*
amplifier(s), 380
antilog, 266, 268, 268*f7.58*
Bode Plot and, 12
buffer, 202–204
communications, 173
complete gain plot and, 22–24, 23*f1.15*
current, 92, 92*f3.17*
dBW and, 9–11
decibels and, 5, 6–8
high-speed, 217
instrumentation, 125
inverting, 141, 208–210, 209*f6.25*
log, 266–269, 267*f7.56*, 268*f7.57*
log (precision), 273–276, 274*f7.65*, 275*f7.66*, 276*f7.67*
lower break frequency and, 15
microphone, 77, 79*f3.7*
noninverting, 212, 214, 214*f6.30*
parallel-series (PS) and, 92, 92*f3.17*
summing, 117–120, 265, 465
track-and-hold, 512*f12.28*, 512–515, 513*f12.29*, 513*f12.30*, 514*f12.31*, 516*f12.32a*, 517*f12.32b*
voltage-controlled, 205, 206*f6.22*
amplifiers, instrumentation, 125, 184
guard drive and, 192*f6.6*, 192–193
improved, 185–188, 186*f6.3*, 188*f6.4*, 189–190, 193*f6.9*
MultiSIM and, 190*f6.5a*, 190–193, 191*f6.5b*, 191*f6.5c*
three-amplifier design of, 184–185, 185*f6.2*
analog
dis/advantages of, 485–486
resolution of, 484–485
analog-to-digital (AD) conversion, 485–486, 533–534
converter (ADC) integrated circuits and, 520–529, 521*f12.36*, 522*f12.37*, 523*f12.38a*, 524*f12.38b*, 525*f12.39a*, 526*f12.39b*
delta-sigma converter and, 520
digital signal processing (DSP) and, 529–532
flash converter, 515–516, 518*f12.33*
MultiSIM and, 515, 516*f12.32a*, 517*f12.32b*
pulse code modulation (PCM) for, 486–487
staircase converter and, 517, 518*f12.34*, 519
successive-approximation converter and, 519*f12.35*, 519–520
track-and-hold and, 512*f12.28*, 512–515, 513*f12.29*, 513*f12.30*, 514*f12.31*, 516*f12.32a*, 517*f12.32b*
analysis, 25
asymptotes, 13, 13*f1.5*
attenuator(s)
decibels and, 5–6

bandwidth
noise, 176
power, 152–164
risetime and, 20, 204
Barkhausen Criterion, 327, 329
BASIC, 529, 531
Bel gain, 3–4, 7
bias
optimized, 212, 213*f6.29*
single-supply, 128–130, 129*f4.31*, 130*f4.32*, 130*f4.33*
Bode Plot, 24, 40
active filters and, 436
definition of, 12
gain and, 162*f5.19*, 163*f5.20*
gain-bandwidth (GBW) and, 143, 143*f5.4a*, 144*f5.4b*, 145*f5.4c*, 145*f5.4d*
lag network response and, 17*f1.8*, 17*f1.9*, 17–20, 18*f1.1*, 18*f1.10*, 19*f1.12*, 20*f1.20*
lead network gain response and, 12*f1.3*, 12–15, 13*f1.4*, 13*f1.5*
lead network phase response and, 15–17, 16*f1.6*, 16*f1.7*
margins and, 70, 71*f3.3*
risetime vs. bandwidth for, 20–22, 21*f1.14*
typical, 12*f1.2*
boosting
current, 131–133, 132*f4.34*, 133*f4.35a*, 134*f4.35b*
break
dominant, 149
break point, 246, 248, 251, 252–254
brownout, 280
buffer, 123*f4.24*, 184, 185*f6.1*, 202–204, 228
output, 376
Burr-Brown 35, 84, 201

cable effects, 192–193
capacitance
   stray, 202
capacitor
   compensation, 161
   coupling, 151
   differentiators and, 394
   input coupling, 211
   as integration element, 382*f10.2*, 382–383
   Miller compensation, 139–140, 140*f5.1*
   nomenclature of, 2
   stability, 291
CD, 506–509, 509*f12.23*
chip
   manufacturing process of, 61*f2.16*, 61–62
Chopper Zero amplifier (CAZ), 217, 220
circuit gain(s)
   dB form for, 10–11
circuit simulations, 24–25, 25*f1.16a*, 26*f1.16b*
circuit values
   symmetrical, 33
circuit(s). *see* amplifier circuits, operational
   high-gain, 161
   log, 266–268, 267*f7.56*, 275
   MOS logic, 263
   motor drive, 200, 201*f6.18*
   nulling, 180
   realization of, 408*f10.27*, 408*f10.28*, 408–409
   safety of, 70
   sigmoid, 266
   "square-up," 59, 59*f2.13*, 60*f2.14*
   virtual, 485
circuits, nonlinear, 276–277
   comparators and, 254–266
   definition of, 223
   divider, 270–273, 272*f7.63*
   function generation, 244–254
   log/antilog amplifiers with, 266–276
   multiplication, 270–273, 271*f7.61*
   power amplifier overload detector and, 234–235, 235*f7.18*

precision rectifiers and, 15, 224*f7.1*, 224*f7.2*, 224–235, 232*f7.13*, 233*f7*, 233*f7.14*, 233*f7.15*
   squaring, 270–273, 271*f7.62*, 273*f7.64*
   wave shaping in, 223–224, 235–243
clamper(s), 235–240, 236*f7.19*, 236*f7.20*
   active positive, 236–240, 237*f7.21*, 238*f7.23*, 238*f7.24*, 239*f7.25*
   MultiSIM and, 240, 241*f7.26a*, 241*f7.26b*
   negative, 236
   offset with, 237, 237*f7.22*, 238*f7.23*
clipping, 157–158, 160*f5.16*, 161*f5.17*, 198
closed-loop gain, 69
CMRR (Common Mode Rejection Ratio), 173–174, 180, 184, 185, 188, 189, 192
Commutating Auto Zero. *See* Chopper Zero amplifier (CAZ)
compact disk. *See* CD
comparator, 55–60, 258, 259*f7.49*
   analog/digital system and, 485
   dual, 263
   function generations and, 264–266, 265*f7.55*
   general purpose (LM311), 259–263, 261*f7.52*, 261*f7.52a*, 262*f7.52b*, 264*f7.56*
   high-speed (LM360), 259–260
   hysteresis and, 255–259, 257*f7.46*, 257*f7.47*, 258*f7.48*
   low-power (LM393), 259–260
   MultiSIM and, 56–57, 57*f2.11a*, 58*f2.11b*, 259, 260*f7.51a*, 260*f7.51b*
   operational amplifier and, 55–60, 56*f2.10*, 57*f2.11a*, 58*f2.11b*, 58*f2.12*, 254
   phase, 359
   simple, 255–257, 256*f7.45*
   Threshold, 364, 367
   Trigger, 365, 367
   waveforms for, 258, 259*f7.50*
   window, 263, 264*f7.54*

compensation
   decompensated devices and, 161–162, 162*f5.19*
   feedforward, 162–163
   noncompensated devices and, 161–162, 162*f5.19*
compensation components, 275, 275*f7.66*
computer simulation. *See* MultiSIM
computer(s). *See also* microprocessor
   analog, 407*f10.26*, 407–409, 408*f10.27*, 408*f10.28*
   circuit simulations by, 24–25, 25*f1.16a*, 26*f1.16b*
   control of lighting by, 511–512, 512*f12.27*
   digital, 409–410
   drum, 528, 529*f12.41*, 530*f12.43*
conversion. *See* analog-to-digital conversion; digital-to-analog conversion
conversion speed, 497
CPU, 509–510, 510*f12.24*
crossover system, 450*f11.32*, 450–451, 451*f11.33*, 453*f11.35*
crosstalk, 122, 123*f4.24*
current(s)
   bleed off, 292
   boasting of, 292, 292*f8.8*
   collector, 28–29, 32
   emitter, 28–29, 32
   high, 197–201, 198*f6.15*, 199*f6.16*, 200*f6.17*, 201*f6.18*
   improved, 38*f1.28*
   input bias, 164, 166, 195
   input noise, 196
   input offset, 164
   nomenclature of, 2–3
   tail, 28, 32, 37
   Widlar, 48
current differencing amplifier. *see* Norton (LM3900) amplifier
current limiting, 305*f8.19*, 305–306
current mirror, 38, 38*f1.29* 39*f1.31*, 41
   bias and active load for, 40, 40*f1.32*, 40*f1.33*
   transfer curve mismatch and, 39, 39*f1.30*

curve
  area under, 381
  transfer, 249–250, 251f7.40

damping, 434f11.15, 433–435, 436f11.18, 442, 442f11.25, 443, 446, 458, 458f11.39
data sheet, 52–55, 53f2.8
dB form, 8–9, 11
dBf, 9
dBm, 9, 11
dBV, 8–11, 11f1.1
dBW, 8, 10
DC
  active filters and, 420, 433
  clampers and, 235–236
  CMRR (Common Mode Rejection Ratio) and, 173–174
  differential amplifier and, 27f1.18, 27–29, 28f1.19
  instrumentation amplifier, 191f6.5b
  linear variable differential transformer (LVDT) and, 403
  MultiSIM voltage output and, 80f3.8b
  offsets and, 164, 383–384, 531
  operational amplifier, 54–55
  potentials and, 36
  return resistor, 104
  undesirable levels of, 164
  voltmeter, 112–113, 114f4.15
DC restorer. See clampers
decibel(s), 40
  computation of, 4–5
  dBW and, 8
  dBW and dBV signal representation and, 8–11, 11f1.1
  power or voltage gain measured in, 3–8
  ratios of change in, 4
delta-sigma technique, 500–501, 501f12.16
design hint, 157–158
detector. See peak detector
  power amplifier overload, 234–235, 235f7.18
differentiator(s), 380–381, 412
  accuracy/usefulness of, 394–395

alternatives to, 409–410
analog computers and, 407f10.26, 407–409, 408f10.27, 408f10.28
basic, 393, 393f10.13, 395f10.15
frequency and, 395
linear variable differential transformer (LVDT) and, 403, 404f10.22c, 405f10.24, 405–406, 406f10.25
MultiSIM and, 398, 399f10.20a, 400f10.20b, 401, 402f10.22a, 403f10.22b, 404f10.22c
optimizing, 395–396, 396f10.16, 396f10.17, 397f10.18
practical, 395–396, 396f10.17, 397f10.18
simple op amp, 394, 394f10.1, 395f10.15
square wave and, 401–403, 404f10.23
summing, 410f10.29, 410–411
time-continuous method and, 397f10.19, 397–398, 399f10.20a, 400f10.20b
time-discrete method and, 398–401, 401f10.21, 402f10.22a, 403f10.22b, 404f10.22c
diffusion, 61–62, 62f2.17
digital domain
  dis/advantages of, 485–486
digital-to-analog (DA) conversion, 485, 487, 492–496, 493f12.5, 494f12.6, 494f12.7
  converter (DAC) integrated circuits and, 501–512, 502f12.17, 503f12.18, 504f12.19, 505f12.20, 506f12.21, 507f12.22a, 508f12.22b
  converter (DAC) limits for, 496–501, 500f12.15
digitization. See analog-to-digital (AD) conversion
diode, 224f7.1, 224–225, 228, 236, 240, 245, 246, 309, 335
  biased, 244, 252, 252f7.42, 252–254, 253f7.43, 255f7.44a
  blocking, 252

germanium, 254
shunting, 369
Zener, 240, 252, 283f8.3
distortion
  slewing induced, 155–156
divider
  decoupling, 212
  simple voltage, 12–13, 14, 76, 76f3.5
  voltage, 109–110, 124, 199, 255, 283
divider network
  resistive, 495
domain technique, 428
drift, 170–173
droop rate, 514
duplication, 486

emitter bias technique, 28
envelope detector. See peak detector
equalizer(s)
  audio, 469–473, 471f11.49, 471f11.50, 471f11.51, 472f11.52, 473f11.53a, 474f11.53b, 475f11.53c
equations, 2
error
  feedthrough, 514
  gain, 497, 498f12.13
  nonlinearity, 497–498, 499f12.14
  offset, 497, 498f12.12
error band, 257
error signal, 68

false turn-off spike, 254–255, 256f7.45
feedback. See negative feedback; positive feedback
feedback signal, 68
femtowatt. See dBf
filter(s)
  analog, 417
  antialias, 491, 492f12.4
  digital, 417
  high rolloff rate, 491
  LC, 308–309, 324
  loop, 359–360
  types of, 417–419
  voltage-controlled, 207

filters, active, 416–417, 481. *See also* Sallen filters
 advantages of, 419–420
 alignment or class, 422–426
 audio equalizers and, 469–473, 471*f11.49*, 471*f11.50*, 471*f11.51*, 472*f11.52*, 473*f11.53a*, 474*f11.53b*, 475*f11.53c*
 band-pass, 417, 418*f11.3*, 419, 419*f11.7*, 429, 452–465, 454*f11.36*
 band-reject (notch), 417, 418*f11.4*, 465–469, 466*f11.46*, 467*f11.47*, 468*f11.48a*, 469*f11.48b*, 470*f11.48c*
 Bessel, 423, 441, 478
 Butterworth, 423, 433, 434, 442, 451, 474–475, 478
 Chebyshev, 424*f11.10a*, 424–425, 425*f11.10b*, 478
 elliptic (Cauer), 425, 425*f11.11*
 filter design tools and, 443, 445*f11.29a*, 446*f11.29b*, 447*f11.29c*, 448*f11.29d*, 449*f11.29e*
 higher-order, 442–452, 450*f11.31*, 450*f11.32*
 high-pass, 417, 417*f11.1*, 444, 449*f11.30*
 high-pass equal-component, 439–440, 441*f11.23*
 high-pass unity-gain, 439–441, 440*f11.22*, 441*f11.23*
 loudspeakers and basic, 451–452, 452*f11.34*, 453*f11.35*
 low-pass, 417, 418*f11.2*, 419, 419*f11.6*, 421*f11.8*, 426*f11.12*, 428–431, 491
 low-pass equal-component, 431–432, 432*f11.13*, 434*f11.15*, 435*f11.16*, 435
 low-pass unity-gain, 432–436, 433*f11.14*, 436*f11.18*, 436*f11.19*, 437*f11.20*, 437*f11.21a*, 438*f11.21b*, 439*f11.21c*,
 multiple feedback band-pass, 454*f11.37*, 454–458, 455*f11.38*, 459, 460*f11.41a*, 461*f11.41b*, 462*f11.41c*
 multiple feedback band-pass unity-gain, 455, 455*f11.38*
 MultiSIM and, 436, 437*f11.21a*, 438*f11.21b*, 439*f11.21c*, 459, 460*f11.41a*, 461*f11.41b*, 462*f11.41c*, 468, 468*f11.48a*, 469*f11.48b*, 470*f11.48c*
 order and poles of, 420–422, 421*f11.8*
 practical, 426–452
 second-order, 441, 441*f11.24*, 443
 single pole, 443, 444*f11.28*
 state-variable, 459–461, 463*f11.42*, 463*f11.43*, 504, 504*f12.19*
 state-variable (fixed-gain), 463*f11.43*, 463–465, 464*f11.44*, 465*f11.45*
 switched-capacitor (MF4), 474–479, 476*f11.54*, 477*f11.55*, 478*f11.56*
 VCVS (Key Voltage-Controlled Voltage Source), 426*f11.12*, 426–428, 432*f11.13*, 437*f11.21a*, 438*f11.21b*, 439*f11.21c*, 440*f11.22*, 441*f11.23*
 voltage-controlled (VCF), 479*f11.58*, 479*f11.59*, 480*f11.60*, 480*f11.61*, 479–480
555 timer, 364–365, 365*f9.38*
followers
 voltage, 202–204, 203*f6.20*, 225
frequency
 break, 149, 417, 422, 442–444
 center, 360, 362, 364
 critical, 151*f5.8*, 151–152
 free-running, 360, 362, 364
 high, 120, 139, 192, 226, 227*f7.6*
 integrator and, 385–386
 low, 150–152, 151*f5.7*, 151*f5.8*, 348, 420
 lower break, 15, 211
 moderate, 420
 noise corner, 176
 normalized, 146–147, 429
 Nyquist, 490–491, 506, 520
 open-loop, 139–140, 140*f5.1*
 oscillation, 371
 response, 139
 sampling, 487, 487*f12.1*
 upper break, 19
frequency domain, 2
frequency factor, 440
frequency scaling, 435*f11.16*, 434–435, 436*f11.19*, 442, 443*f11.26*, 458, 458*f11.40*
function generator
 biased diode, 155*f7.44a*, 252*f7.42*, 252–254, 253*f7.43*
 comparator and, 264–266, 265*f7.55*
 increasing gain-function circuit and, 252–254
 multiple-section, 244–245, 247*f7.34*, 248*f7.37*, 250*f7.38*
 MultiSIM and, 254, 255*f7.44a*, 256*f7.44b*
 simple, 244*f7.32*, 244–245, 245*f7.32*, 246*f7.33*, 247*f7.35*, 248*f7.36*
 temperature transducer and, 249–251, 250*f7.39*, 251*f7.40*, 251*f7.41*

gain
 closed-loop, 102, 141, 141*f5.2*, 217
 high-frequency, 19, 41
 loop, 69, 74, 75
 low-frequency, 12, 15, 41
 mismatched, 124, 124*f4.26*
 noise, 141, 172, 179
 open-loop, 69, 75, 76, 78*f3.6b*, 90, 94, 140, 141, 141*f5.2*, 196, 384
 signal, 173
 voltage, 3–4, 7, 33–34, 36, 141, 214
gain plot
 complete, 22–24, 23*f1.15*
gain sweep, 76, 77*f3.6a*
gain-bandwidth (GBW), 139–143, 140*f5.1*, 141*f5.2*, 142*f5.3*, 146*f5.5*, 196
 MultiSIM and, 139–146, 143*f5.4a*, 144*f5.4b*, 145*f5.4c*, 145*f5.4d*
 multistage considerations of, 146–150, 148*f5.6*

594 INDEX

gain/phase output
  MultiSIM schematic for, 26*f1.16b*
generator(s), 376
  arbitrary waveform, 509–510, 510*f12.24*
  digital arbitrary function, 494
  555 astable operation and, 367–370, 368*f9.41*, 368*f9.42*, 369*f9.43*
  monostable operation and, 365*f9.39*, 365–367, 366*f9.40*
  MultiSIM and, 348, 349*f9.22a*, 350*f9.22b*, 353, 353*f9.25a*, 354*f9.25b*, 355*f9.25c*
  ramp function and, 343–345, 345*f9.17*, 346*f9.18*, 358*f9.30*, 358–359
  signal, 346–348, 347*f9.20*, 347*f9.21*, 349*f9.22a*, 350*f9.22b*
  square wave, 349–353, 350*f9.23*, 351*f9.24*, 353*f9.25a*, 354*f9.25b*, 355*f9.25c*
  triangle/square function, 343–354, 346*f9.19*, 347*f9.20*, 349*f9.22a*, 350*f9.22b*, 371
  waveform, 370–375, 371*f9.44*, 372*f9.45*, 373*f9.46*, 375*f9.47*
guard drive, 192*f6.6*, 192–193

heat sink, 317
  thermal resistance for, 317–322, 318*f8.31*, 319*f8.32*, 320*f8.33*, 321*f8.34*
hysteresis, 255–259, 257*f7.46*, 257*f7.47*, 258*f7.48*

IC sockets, 202
idealizations, 101, 138, 216–217
impedance
  high, 184, 225
  high internal, 40
  input, 106–107, 110, 118
  low, 308
  parallel-series and, 93*f3.18*, 93–95, 94*f3.19*
  series-parallel and, 79, 81*f3.9*, 81–84, 82*f3.10*, 82*f3.11*

impedance scaling, 435, 435*f11.17*, 437*f11.20*, 442, 443*f11.27*, 458, 458*f11.40*
inductor, 394
inputs
  inverting/noninverting, 208
  noninverting/inverting, 255
integrator(s), 344, 380–381, 386*f10.7*, 412
  accelerometer with, 391–393, 392*f10.11*, 393*f10.12*
  accuracy/usefulness of, 383, 386
  alternatives to, 409–410
  augmenting, 411, 411*f4.31*
  basic, 381*f10.1*, 381–382
  double, 411*f4.32*, 411–412
  frequency and, 385–386, 391
  MultiSIM and, 388, 389*f10.8a*, 390*f10.8b*
  noise and, 386
  optimizing, 383–387, 384*f10.4*
  practical, 384–385, 385*f10.5*, 385*f10.6*
  simple op amp, 382*f10.3*, 382–384, 384*f10.4*
  summing, 410, 410*f4.29*
  time-continuous method and, 387–388, 389*f10.8a*, 390*f10.8b*
  time-discrete method and, 388–393, 390*f10.9*, 391*f10.10*
inverter/noninverter
  adjustable, 126–128, 127*f4.29*, 127*f4.30*
inverting amplifiers, 141, 208–210, 209*f 6.35*
inverting current amplifier, 114–117, 115*f4.16*, 116*f4.17*, 117*f4.18*, 125
inverting voltage amplifier, 105*f4.6*, 105–110, 106*f4.7*, 106*f4.8*, 106*f4.9*, 108*f4.10a*, 109*f4.10b*, 109*f4.11*, 130, 130*f4.30*, 232

keyboard, digital sampling, 528, 529*f12.41*, 530*f12.43*

ladder (R/2R) network, 336*f9.11*, 341*f9.15*, 494–496, 495*f12.8*, 495*f12.9*, 496*f12.10*, 497*f12.11*, 504

lag network, 17, 17*f1.8*, 40–41
  approximate gain response and, 17–18, 18*f1.10*
  approximate phase response and, 19, 19*f1.12*
  exact gain response and, 17*f1.9*, 17–18
  150kHz, 19, 20*f1.13*
  low-pass filter and, 419, 419*f11.6*, 428
  MultiSIM schematic for, 25*f1.16a*
  multistage effects and, 20–22
  operational amplifier and, 46, 46*f2.2*
  oscillators and, 329–330, 330*f9.4*
  risetime and, 21*f1.15*, 22
  unbalanced, 192–193, 193*f6.8*
Laplace transform, 428
lead network, 12, 12*f1.3*, 20, 40–41
  approximate gain response, 13*f1.5*, 13–15
  approximate phase response and, 16*f1.7*, 16–17
  capacitors and, 139
  critical frequency, 151*f5.8*, 151–152
  exact gain response, 13, 13*f1.4*
  exact phase response and, 15–16, 16*f1.6*
  high-pass filters, 438
  oscillators and, 329–330, 330*f9.4*, 335
LED, 230–232, 235
level shifter, 120
light alarm, 12, *57*, 58*f23*
light transmission, 276, 276*f7.67*
limiter
  active, 240–243, 241*f7.27*, 241*f7.29*, 243*f7.28*, 243*f7.30*
linear regulators, 283–286, 284*f8.4*
  adjustable regulators (LM317), 293–296, 294*f8.10*, 295*f8.11a*, 296*f8.11b*
  current boosting and, 292, 292*f8.8*
  dual tracking (LM325) regulators and, 305–307, 306*f8.20*

fixed (LM7805) regulators and, 296–297, 297f8.13
LM723 regulators and, 296–305, 297f8.13, 297f8.14, 299–300f8.15, 301f8.16, 302f8.17, 303f8.18
low dropout regulators and, 292–293
MultiSIM and, 286, 287f8.5a, 288f8.5b, 295, 295f8.11a, 296f8.11b
programmable and tracking regulators and, 293–307
three pin regulators (LM78XX), 287–291, 289f8.6a, 290f8.6b, 291f8.7, 315
load regulation, 282
logarithm, 3–4, 8
loss
insertion, 420
loudspeakers, 420, 450–452

margins
gain, 70, 71f3.3
phase, 70, 71f3.3
measurement
light transmission, 276f7.67, 726
meter, digital
underflow of, 11
microprocessor, 504, 506f12.21, 509–510, 510f12.24
midband
response, 22
milliwatt. *see* dBm
modulator
pulse-width, 266
multiplier, 270f7.60, 270–273, 271f7.61, 271f7.62, 272f7.63, 273f7.64
MultiSIM, Electronics Workbench, 24, 25f1.16a, 26f16b, 41
active filters and, 436, 437f11.21a, 438f11.21b, 439f11.21c, 467, 468f11.48a, 469f11.48b, 470f11.48c
adjustable regulators (LM317) and, 295, 295f8.11a, 296f8.11b
analog-to-digital (AD) conversion and, 515, 516f12.32a, 517f12.32b

audio equalizers and, 473, 473f11.53a, 474f11.53b, 475f11.53c
clamper(s) and, 240, 241f7.26a, 241f7.26b
comparator and, 56–57, 57f2.11a, 58f2.11b, 259, 260f7.51a, 260f7.51b
current boosting and, 131–133, 133f4.35a, 134f4.35b
differentiators and, 398, 399f10.20a, 400f10.20b, 401, 402f10.22a, 403f10.22b, 404f10.22c
distortion negative feedback a, 87f3.14a, 87–88, 88f3.14b, 88f3.14c, 89f3.14d
filters, active, 459, 460f11.41a, 461f11.41b, 462f11.41c
gain-bandwidth and, 139–146, 143f5.4a, 144f5.4b, 145f5.4c, 145f5.4d
impedance negative feedback and, 76–80, 77f3.6a, 78f3.6b, 80f3.8
integrators and, 388, 389f10.8a, 390f10.8b
inverting voltage amplifier and, 108, 108f4.10a, 109f4.10b
linear regulators and, 286, 287f8.5a, 288f8.5b
negative feedback and, 76–80, 77f3.6a, 78f3.6b, 80f3.8
noninverting current-to-voltage transducer and, 112, 113f4.14a, 114f4.14b
Norton amplifier and, 207f6.23, 208f6.24, 209f6.25, 209f6.26, 211–214
offset(s) and, 170, 171f5.26a, 172f5.26b, 172f5.26c
oscillators and, 342, 342f9.16a, 343f9.16b, 344f9.16c, 345f9.16d
precision rectifiers and, 228f7.8a, 229, 229f7.8b
signal generators and, 348, 349f9.22a, 350f9.22b
slew rate and, 157, 158f5.15a, 159f15b, 160f5.15c

square wave generators and, 353, 353f9.25a, 354f9.25b, 355f9.25c
triangle/square generators and, 348, 349f9.22a, 350f9.22b
multistage effects
elements combination and, 22–24, 23f1.15
music. *See* preamplifier
amplifiers and, 198–199
digital sampling, 528, 529f12.41, 530f12.43
music synthesizer, 355, 356f9.26

negative feedback, 97, 215
basic concepts of, 67–68, 68f3.1
definition of, 66–67
effects of, 68–70, 69f3.2, 71f3.3
four variants of, 69f3.2, 70–72, 71f3.3
limitations of, 96–97
MultiSIM computer simulation and, 76–80, 77f3.6a, 78f3.6b, 80f3.8
parallel-parallel (PP) and, 95, 95f3.20, 105, 111
parallel-series (PS) and, 89–92, 90f3.15, 90f3.16, 92f3.17, 114
parallel-series impedance and, 93f3.18, 93–95, 94f3.19
precision rectifiers and, 224–225
series-parallel (SP) connection and, 72f3.4, 72–79, 76f3.5, 77f3.6a, 78f3.6b, 79f3.7, 80f3.8, 101, 426
series-parallel distortion and, 84–88, 85f3.12, 85f3.13, 87f3.14a, 88f3.14b, 88f3.14c, 89f3.14d
series-parallel impedance and, 79–84, 81f3.9, 82f3.10, 82f3.11
series-series (SS) and, 95, 96f3.21, 111
negative saturation, 226–227
noise, 176f5.27, 176–181, 177f5.28, 178f5.29
circuit, 174
integrator and, 386
Johnson, 175

shot, 175
signal-to-noise ratio (S/N) and, 175, 179–180
thermal, 175
nomenclature, 2
noninverting amplifiers, 141, 334
noninverting summing amplifier, 120–122, 121f4.22, 121f4.23, 123f4.24, 125
noninverting voltage amplifier, 101f4.1, 101–105, 103f4.2, 103f4.3, 104f4.4, 105f4.5
nonlinearity, 497–498

octave, 13
octave slope, 20
offset(s)
clamper and, 237, 237f7.22, 238f7.23
compensation and sources of, 164–170, 165f5.21, 165t5.2, 167f5.22, 168f5.23, 169f5.24, 169f5.25, 180
definition of, 164
drift and, 170
nulling circuits and, 180
resistor and, 170, 171f5.26a, 172f5.26b, 172f5.26c
Ohm's Law, 30, 116
operation
continuous, 309–310
step down/inverter, 309
oscillation, 67, 70
oscillator(s), 376, 477
555 timer and, 364–365, 365f9.38
adjustable, 333f9.7, 334–335
Barkhausen Criterion and, 327, 329
basic, 328–329, 329f9.3
definition of, 326–327
MultiSIM and, 342, 342f9.16a, 343f9.16b, 344f9.16c, 345f9.16d
phase shift, 335–342, 336f9.10, 336f9.11, 337f9.12, 338f9.13, 338f9.14, 341f9.15
phase-locked loop (PLL) and, 354, 355f9.25c, 359f9.32, 359–364, 360f9.33

positive feedback and, 327f9.1, 327–328
practical, 327f9.1, 327–328
square wave, 365
voltage-controlled (VCO) (NE566), 327, 354–364, 356f9.26, 356f9.27, 357f9.28, 357f9.29, 358f9.30, 359f9.31
Wien bridge, 328–334, 330f9.4, 332f9.5, 332f9.6, 333f9.7, 333f9.8
oscilloscope
digital storage (DSO), 526–528, 527f12.40, 527f12.41
overloads, 49
oversampling, 500–501, 501f12.16, 506, 508

parallel-parallel (PP), 95, 95f3.20, 105, 111
parallel-series (PS), 89–92, 90f3.15, 90f3.16, 92f3.17
pass transistor. See linear regulator
peak detector, 227–232, 229f7.9, 230f7.10, 231f7.11, 232f7.9
phase shift, 338, 338f9.14
photoresist, 62
pixel, 203
planar process, 62
Pocket Rockit, 198, 199f6.16
polarity, 344–345, 389, 391, 392
inversion of, 110
negative, 270
positive feedback, 67, 70, 110, 259, 327f9.1, 327–328, 334
potentials
collector, 29, 35
DC, 36
potentiometer, 199, 302–303, 372, 409
power, 40
consumption of, 193–194
decibels and, 3–4, 7
dissipation of, 307, 319–320, 324
low, 197
voltage and, 7
watts measurement of, 8
power down mode, 194
Power Law, 11

power supply, 280, 281f8.1
bipolar, 291, 291f8.7
preamplifier
musical instrument, 108, 109f4.11
precision rectifiers
full-wave, 232f7.13, 232–235, 233f7.14, 234f7.17
half-wave, 224f7.2, 224–232, 225f7.3, 225f7.4, 226f7.5, 227f7.6, 227f7.7, 228f7.8a, 229f7.8b
inverting half-wave, 232–235, 233f7.13, 234f7.16
MultiSIM and, 228f7.8a, 229, 229f7.8b
passive, 224, 224f7.1
peak detector and, 227–232, 229f7.9, 230f7.10, 231f7.11, 232f7.9
PSpice, 24, 41
PSRR (Power Supply Rejection Ratio), 174–175, 180, 199
pulse
duty cycle of, 308, 372, 372f 9.45
pulse shape, 230f7.10
pulse signal(s)
slew rate, 153
push-pull configuration, 322–323

quantization. See analog-to-digital (AD) conversion

RAM (Random Access Memory), 486, 528, 529
range
capture, 360, 361
tracking, 360
rectifiers. see precision rectifiers
refresh rate, 203
regulator(s). See also linear regulators; switching regulators
series mode, 282, 283f8.2
shunt mode, 282, 283f8.2, 283f8.3
resistance
dynamic base-emitter, 275
input noise, 176–177
resistor(s)
bias, 252
compensation, 165, 177

DC-return, 151f5.8, 151–152
gain, 252
nomenclature of, 2
offsets and, 170, 171f5.26a, 172f5.26b, 172f5.26c
programmable amplifiers, 194, 194f6.11
resolution, 487–492
amplitude, 487
risetime
bandwidth and, 20, 204
rollover
open-loop, 140
ROM (Read Only Memory), 486

sacrifice factor, 69, 73, 75, 91
Sallen filters
high-pass and, 438–442
low-pass and, 426–439
sample-and-hold amplifier. see amplifiers, track-and-hold
sampling rate, 487–492, 490f12.2
Schmitt Trigger, 257, 263
secondary switcher. see switching regulators
series-parallel (SP)
connection of, 72f3.4, 72–84, 76f3.5, 77f3.6a, 78f3.6b, 79f3.7, 80f3.8
distortion and, 84–88, 85f3.12, 85f3.13, 87f3.14a, 88f3.14b, 88f3.14c, 89f3.14d
impedance effects of, 79, 81f3.9, 81–84, 82f3.10, 82f3.11
series-series (SS), 95, 96f3.21, 111
signal(s)
adjustable DC, 205
common-mode, 36–38, 37f1.26, 37f1.27, 38f1.28
dB form for, 10
differential input, 189
error, 359–360
fast, 393
interference, 464
output, 225, 225f7.3
pulse, 153
sinusoidal, 153–157, 154f5.9, 154f5.10, 154f5.11, 154f5.12, 155f5.13, 156f5.14
start-up, 327f9.1, 328

simulation model
MultiSIM, 50, 51f2.7, 52
simple SPICE, 50, 51f2.6
sine shaper, 371
signal
error, 359–360
slew rate, 203, 226, 229–230
definition of, 152–153, 153t5.1
multiple stages and, 158–161, 161f5.18a, 161t5.17
power bandwidth and, 152, 156–157
pulse signals and, 153
sinusoidal signal(s), 153–157, 154f5.9, 154f5.10, 154f5.11, 154f5.12
Slewing Induced Distortion (SID), 155–156
SPICE, 24, 41
strobe, 263, 264f7.53
summers. See summing amplifiers
summing amplifier, 117–120, 118f4.19, 118f4.20, 120f4.21, 265, 465
switching regulators, 307–309, 308f8.21, 308f8.22, 309f8.23, 323–324
full bridge switcher and, 323, 323f8.36
inverting switcher and, 314f8.28, 314–315
LM3578, 310–313, 311f8.25, 312f8.26, 313f8.27
primary switcher and, 322f8.35, 322–323, 323f8.36
step down, 309f8.24, 309–313, 313f8.27, 315, 316f8.30
step up, 314–315, 315f8.29

test equipment, 510–511, 511f12.25, 511f12.26
thermal resistance, 317–322, 318f8.31, 319f8.32, 320f8.33, 321f8.34
three pin regulators (LM78XX), 287–291, 289f8.6a, 290f8.6b, 291f8.7, 315
threshold(s), 263
TIMD (Transient Inter-Modulation Distortion), 97

time-continuous method, 387–388, 389f10.8a, 390f10.8b, 397f10.19, 397–398, 399f10.20a, 400f10.20b
time-discrete method, 388–393, 390f10.9, 391f10.10
transducer(s), 95
high-frequency, 451
inverting current-to-voltage, 110
lag network and, 19, 20f1.13
noninverting current-to-voltage, 111f4.13, 111–113, 113f4.14a, 114f4.15, 114f14b
temperature, 249–250, 250f7.39, 251f7.40, 251f7.41
transimpedance, 214
transistor(s)
overloads and, 49
pass, 292, 292f8.8, 292f8.9, 307
transresistance, 110
trimmers, volume, 451
TTL clock, 477, 478f11.58

upper-breaks, 74

virtual ground, 105, 114, 121, 166, 382
voltage, 40
collector, 29, 32, 33
common-mode gain of, 36–37, 37f1.27
constant, 282
dBV and, 8, 10–11, 11f1.1
decibels and, 7
differential input, 188
drift and, 170–173
error, 186
gain of, 3–4, 7, 33–34, 36, 141
high, 201
input offset, 195
measurements of, 10–11
nomenclature of, 2–3
noise, 175, 176f5.27, 177–178
over, 281
source of, 49
voltage follower, 104, 104f4.4
Voltage Law (Kirchhoff), 28
voltage regulation, 280–281, 323–324, 380
heat sink usage and, 317–322

line regulation and, 282
linear regulators and, 283–307
load regulation and, 282
need for, 280–283
primary switcher and, 322f8.35, 322–323
switching regulators and, 307–316
voltage source
  voltage-controlled, 49

voltmeter, 511, 511f12.25, 511f12.26
  DC, 112–113, 114f4.15

watts
  dBW and, 9–10
waveform(s)
  common mode and input-output, 35f1.25, 35–36
  single input and, 33f1.24, 33–34

wave(s)
  shaping of, 223–224
  square, 346, 359, 359f9.31, 372, 372f9.45, 401–403, 404f10.23
  triangle, 345–346, 371–372, 389, 400
Websites, manufacturer, 535
Widlar current source, 48

Zener potential, 240–241, 244–245, 269, 283–285